AF183640

272
Current Topics in Microbiology and Immunology

Springer-Verlag Berlin Heidelberg GmbH

W. Doerfler P. Böhm (Eds.)

Adenoviruses: Model and Vectors in Virus-Host Interactions

Virion-Structure, Viral Replication and Host-Cell Interactions

With 60 Figures and 8 Tables

Springer

Prof. Dr. WALTER DOERFLER
Institut für Klinische und Molekulare Virologie
Universität Erlangen-Nürnberg
91054 Erlangen, Germany
e-mail: doerfler@viro.med.uni-erlangen.de

PETRA BÖHM
Institut für Genetik
Universität zu Köln
Weyertal 121
50931 Köln, Germany
e-mail: pboehm@scan.genetik.uni-koeln.de

Cover Illustration by M. Hösel (this volume):

Overexpression of the E4-ORF3 gene of Ad12 (C) or Ad2 (D) leads to the reorganization of the PML protein domains in BHK21 hamster cells. In untransfected cells (A) or in cells transfected with the pIRES2-EGFP vector (B), PML protein domains retain their spherical structures. All sucessfully transfected cells are green due to the expression of EGFP (enhanced expression of the green fluorescent protein). The PML protein is detectable by staining with primary anti-PML protein antibodies and secondary Cy3-anti mouse IgG. Bars designate magnifications. This figure was reproduced from Hösel et al. (2001a) see Chapter 14.

ISSN 0070-217X
ISBN 978-3-642-05517-1 ISBN 978-3-662-05597-7 (eBook)
DOI 10.1007/978-3-662-05597-7

Library of Congress Catalog Card Number 72-152360

http://www.springer.de

Originally published by Springer-Verlag Berlin Heidelberg New York in 2003
Softcover reprint of the hardcover 1st edition 2003

Library of Congress Catalog Card Number 15-12910

Cover Design: Design & Production GmbH, Heidelberg
Typesetting: Fotosatz-Service Köhler GmbH, Würzburg
Production Editor: Christiane Messerschmidt, Rheinau
Printed on acid-free paper 27/3020 5 4 3 2 1 0

Preface

After volumes 109, 110 ("The Molecular Biology of Adenoviruses"), and 199 I–III ("The Molecular Repertoire of Adenoviruses"), volumes 272 and 273 of the *Current Topics in Microbiology and Immunology* in 2003 are the third series devoted to the analyses of molecular mechanisms elicited by adenovirus infections. Adenovirus research continues at a rigorous pace.

It is true, for many researchers in the biomedical sciences, adenoviruses are nowadays best known as possible vectors for human somatic gene therapy. However, our goal in editing the current volumes has been to present an overview on basic research dealing with the molecular biology of this group of viruses. Of course, much of the essential information on the adenoviral genomes, their replication and gene expression as well as information on other biologically relevant problems, such as adenoviral oncogenesis, virus–host interactions, and the immune response against adenoviruses, had already reached a highly sophisticated level when we edited the previous adenovirus volume in 1995. Nevertheless, many of the most exciting problems in adenovirology and on the host responses to the infection with these viruses both at the cellular and organismic levels are still unresolved but presently actively worked on. Possible applications of adenoviruses such as gene transfer vectors in human somatic gene therapy will vitally depend on advances in basic virology. Perhaps we can expect progress in foreign DNA transfer into mammalian cells and organisms only when we have derived all the necessary information from molecular adenovirus research. In this way, we might learn how the virus transports its DNA into an environment for which the virus has become specialized in the course of perhaps millions of years. Adenoviruses are experts when it comes to invading the human organism. This fact should also caution us when developing these viruses into vectors for human gene therapy.

Although predictions on future trends and realities in biomedical research can be risky at best, we venture to speculate that it will most likely not be the infectious virus itself that will eventually be directly applied in human gene therapy. The virus as a pathogen with its high degree of expertise in maneuvering inside the human organism cannot be tamed sufficiently, even by the most ambitious gene technologists, to become an innocuous transfer vector.

Individual contributions to the two current volumes on adenoviruses encompass the entire gamut of molecular virology and of problems related to virus–host interactions, viral oncology and host defenses against viral infections. The chapters in this volume have been assigned to sections on virion structure, viral replication, and host–cell interactions; problems discussed in the next volume will relate to oncogenesis, immune responses, and gene therapy.

We hope that these books will again serve as a resource of information on current adenovirus research and help motivate novel and productive research on adenoviruses as efficient tools in mammalian molecular biology.

We wish to express our gratitude to our colleagues, the authors of these chapters, for taking the time and effort to share the results of their exciting research and write these chapters. We are also indebted to Ms. Clauss at Springer Verlag in Heidelberg for her help in editing the manuscripts.

Köln, November 2002

Walter Doerfler
Petra Böhm

List of Contents

List of Contributors

(Their addresses can be found at the beginning of their respective chapters.)

Akusjärvi, G. 253
Anderson, C.W. 331
Benkö, M. 3
Brenkman, A.B. 187
Brockmann, D. 97
Burnett, R.M. 57
Chroboczek, J. 37
Doerfler, W. 415
Esche, H. 97
Favier, A.-L. 37
Flint, S.J. 287
Freimuth, P. 331
Galinier, R. 37
Gonzalez, R.A. 287
Gout, E. 37
Harrach, B. 3
Hay, R.T. 131
Hearing, P. 165
Hösel, M. 415
Howitt, J. 331
de Jong, R.N. 187
Liu, H. 131
Matthews, D.A. 399
Naismith, J.H. 131
Ostapchuk, P. 165
Padmanabhan, R. 365
Russell, W.C. 399
San Martín, C. 57
Sasaguri, Y. 365
Schröer, J. 415
Stévenin, J. 253
Tanimoto, A. 365
van der Vliet, P.C. 187
Webb, D. 415
Young, C.S.H. 213

1
Virion and Structure

Chapter 1

Molecular Evolution of Adenoviruses

M. Benkö, B. Harrach

Abstract. New advances in the field of genetic characterization of adenoviruses originating from different animal species are summarized. Variations seen in the host range and specificity, pathogenicity, genomic arrangement or gene complement are much wider than expected based on previous studies of human adenoviruses. Several exceptional adenoviruses from the two traditional conventional genera are now removed, and proposed to form at least two new genera. The eventual host origin of the new genera, however, is not clarified. Novel results from the genomic and phylogenetic analyses of adenoviruses originating from lower vertebrate species (including reptiles, amphibians and fish) seem to imply that

M. Benkö, B. Harrach
Veterinary Medical Research Institute of the Hungarian Academy of Sciences,
P.O. Box 18, 1581 Budapest, Hungary
e-mail: benko@novell.vmri.hu

probably five major clusters of adenoviruses exist corresponding to the five major classes of Vertebrata. Adenoviruses, which are now suspected to have common origin with enterobacterium phages from the family *Tectiviridae*, are perhaps very ancient indeed, and may have undergone a co-evolution with vertebrate hosts.

1 Introduction

Although adenoviruses have been isolated from every class of vertebrates including fishes, amphibians and reptiles (Russell and Benkõ 1999), according to the International Committee on Taxonomy of Viruses-approved official taxonomy (Benkö et al. 2000) the family *Adenoviridae* contains only two genera (*Mastadenovirus* and *Aviadenovirus*). For decades, the two genera seemed to be sufficient for the proper classification of the majority of the known mammalian and avian adenovirus types, nevertheless the system was compromised by several puzzling exceptions. Namely, half of the ten recognized bovine types did not fit properly into the genus *Mastadenovirus*. These viruses, compared to human adenoviruses (HAdVs) and other mammalian isolates showed unusual biological properties such as restricted growth, peculiar inclusion body morphology, elevated heat resistance and lack of the common genus specific antigen. Similarly, certain highly pathogenic, non-conventional avian adenovirus isolates were also recognized, which did not possess the genus specific antigen. Therefore, the genus *Aviadenovirus* had to be divided into three groups, and close relatedness was seen within the individual groups only.

To date, the full genomic sequence of one or more representatives of each conventional and atypical mammalian and avian adenovirus group is known. Analysis of the genomic organizations combined with results of phylogenetic calculations implies that at least two (probably three) new genera should be created within the family *Adenoviridae*. We have presumed that DNA sequence data of adenoviruses of lower vertebrate host species (frog, snake, fish) might shed light on possible reasons for the occurrence of the exceptional adenoviruses in mammals and birds. Analysis of the frog adenovirus genome revealed that one atypical avian adenovirus is indeed a close relative of the frog isolate (Davison et al. 2000). For a better understanding of adenoviral phylogeny, we have initiated molecular biological examinations of adenoviruses of lower vertebrate species. Presently, two adenovirus isolates, one from a corn snake and another from a white sturgeon, are subjects of genome mapping and DNA sequencing studies in our laboratory. The accumulating DNA sequence and phylogenetic data provide evidence for the existence of five distinct groups, which should be classified into five different genera within the family *Adenoviridae*. The transition of these new concepts into scientific public opinion and recognition of the necessity of radical reform in adenoviral taxonomy, however, appear to be very slow and hampered likely by old stereotypes. This reluctance to accept the new taxonomic proposals is especially inconceivable because according to the original concept of adenovirus classification – in which the genera were determined and named by the classes of the host species (mammals and birds) – the viruses of lower vertebrate classes should also have been assigned to separate genera if differences in the biological and serological properties had warranted it.

The vast majority of our present knowledge about the adenoviral virion structure, genomic organization and replication strategies comes from studies performed on different human serotypes, most frequently on type 2 or 5 (HAdV-2, HAdV-5) that belong to the former subgenus C recently established as species *Human adenovirus C* (Benkö et al. 2000). Members of the other HAdV species (A–F) might show significantly different properties in their replication ability, cell, tissue or organ tropism, receptor usage, pathogenicity, etc. Several slight differences in the genomic organization have also been recognized, nevertheless the six human adenovirus species containing more than 50 serotypes constitute a relatively uniform cluster. Primate, non-human adenoviruses comprising at least 26 different types are obviously closely related to HAdVs. However, adenoviruses originating from non-primate mammalian host species exhibit much greater heterogeneity, as illustrated by the excellent example of bovine adenoviruses (BAdVs).

In the next few pages summarizing our and others' recent results, we try to present lines of evidence that justify the establishment of the proposed new genera and re-classification of several adenovirus types. Based on the emerging picture, a hypothesis on the possible molecular evolution of adenoviruses will be described with special emphasis given to different evolutionary events that might have led to the somewhat unclear present-day situation, which is not easy to interpret. After a short overview on the exceptional adenoviruses, the four different types of genomic organization observed in adenoviruses studied so far will be discussed according to the main transcription units. Then the results of former and more recent phylogenetic studies will be presented and contrasted with each other. The occurrence and distribution of different adenoviruses from the different host species will be shown with the possible explanations. Finally, an even less firmly based – but certainly fascinating – theory on the origin of adenoviruses will be presented taking us back in time hundreds of millions of years.

2
Unusual Adenoviruses

In the genus *Mastadenovirus*, five bovine isolates (serotypes 4, 5, 6, 7, and 8) distinguished as subgroup 2 BAdVs (Bartha 1969) were recorded as exceptions. By restriction enzyme analysis (Benkö et al. 1988), the genome of these viruses was found to be smaller than the genome of the typical (so-called subgroup 1) BAdVs (types 1, 2, 3, and 9), and no cross-reaction could be detected between the members of the two subgroups in Southern blot hybridizations (Benkö et al. 1990). BAdV-4 has been chosen for further study, and in its very first DNA sequences determined, an interesting codon usage heavily biased towards AT rich triplets was recognized (Benkö and Harrach 1994). Additionally, a strange isolate referred to as strain OAV287 originating from diseased lambs in Australia has also been described (Boyle et al. 1994). Because of the partial cross-neutralization with BAdV-7, no official ovine adenovirus type number has been assigned to it. Nevertheless, OAV287 became the subject of detailed studies as a potential gene delivery vector (Vrati et al. 1995; Khatri et al. 1997; Xu et al. 1997), and was the very first fully sequenced representative of these 'atypical mastadenoviruses' recently proposed to form a third genus (Benkö and Harrach 1998). Most of the characteristic features in the genome organization of this putative new genus were originally derived from the unique genome arrangement found in OAV287 (Both 2002).

In the genus *Aviadenovirus*, only the members of group I (comprising the largest number of virus types isolated from different poultry or wild bird species) were similar to each other and shared common antigens, whereas groups II and III each represented by only one serotype seemed to be very different (Hess 2000; McFerran and Smyth 2000). Group II was assigned to allocate different virus isolates related to three distinguished disease entities, namely the hemorrhagic enteritis of turkey (THE), the marble spleen disease (MSD) of pheasant, and the splenomegaly of chicken. The isolates originating from the different diseases, however, were indistinguishable by serum neutralization. The official taxonomic name of these viruses is turkey adenovirus type 3, while two earlier turkey isolates (types 1 and 2) are conventional and belong to group I. For ease of reference, the abbreviation THEV will be used in the following. In the first published DNA sequence of THEV including the penton base and core protein genes, a high genomic AT content was also noted (Jucker et al. 1996). Group III contained one virus, which caused severe disease in layer hen flocks all over the world commencing in 1976; its colloquial name is 'egg drop syndrome '76' or, in short, EDS virus. With retrospective studies, the presence of antibodies to EDS virus was detected in the sera collected before 1976 from a number of bird species including duck, which was supposed to be the reservoir of the virus. Therefore, the official name of the EDS virus is duck adenovirus type 1, but it will be referred to here as EDS virus.

To resolve the problems of taxonomy, subgroup 2 BAdVs and OAV287, as well as the group II and III avian adenoviruses were recently proposed to be re-classified (Benkö and Harrach 1998). As a first step, however, only their removal from the respective genera could be achieved (Benkö et al. 2000). Two pending proposals concern the establishment of two new genera. The traditional (*Mastadenovirus* and *Aviadenovirus*) genera contain viruses isolated from host species of the respective class of vertebrates (mammals and birds). The phylogenetic distance and differences between their genomic organization are accepted to reflect the evolutionary distance between the hosts, i. e., the two vertebrate classes. In addition, there are two new genera proposed to contain viruses that are phylogenetically closely related but of miscellaneous (mammalian, avian, amphibian, etc.) origin. The genus *Atadenovirus* was proposed for the allocation of the atypical BAdVs (serotypes 4–8) together with the former avian group III EDS virus. The other new proposed genus *Siadenovirus* contains only two putative members (Davison et al. 2000), but both of them, the former group II aviadenovirus THEV and frog adenovirus, have been fully sequenced. Because of the unclear host origin, the proposed names of the two new genera refer to other common properties of the candidate members. The genomes of atadenoviruses studied so far have very high (>60%) AT content, while siadenoviruses contain a seemingly unique gene, a sialidase gene homolog at the left-hand side of their genomes (Davison and Harrach 2002). In the following comparison of the different genomes, these putative genus names will be used.

3
Four Major Types of Genome Organization

Initially, based on partial DNA sequences from animal (bovine, canine, murine, etc.) mastadenoviruses, the adenoviral genome in general was thought to be very conservative and similar to that of HAdV-2. Variations were seen especially in the E1 and E3 regions, but such differences were frequently encountered also among the different HAdV species. Due to the increase of interest in non-human adenoviral expression vectors and gene delivery systems (Löser et al. 1999), the study of animal adenoviruses intensified considerably resulting in many new partial or complete DNA sequences during the past decade. By now, as many as 16 full genomic sequences are available from animals compared to the previously overwhelming number (five) of HAdVs.

Analysis of the animal adenovirus sequences revealed several unexpected features including unusual genome organizations. The genome of the CELO (chicken embryo lethal orphan) virus, the prototype of conventional group I avian adenoviruses officially called fowl adenovirus type 1 (FAdV-1) was found to be considerably longer and organized somewhat differently to that of the mastadenoviruses (Chiocca et al. 1996). In contrary, the genome size of OAV287 (Vrati et al. 1996b) is much shorter than that of HAdVs and is similar to that estimated for subgroup 2 BAdVs (Benkö et al. 1988). Moreover, OAV287 also has a special genome arrangement, which was analyzed and characterized in detail by Khatri and Both (1998). The full genome sequence of the EDS virus (Hess et al. 1997) and THEV (Pitcovsky et al. 1998) was also determined, and they both were very different from that demonstrated for FAdV-1. Additional full genomic sequences from typical mammalian and avian adenoviruses including mouse adenovirus type 1 (Meissner et al. 2000), BAdV-3 (Reddy et al. 1998b), porcine adenovirus type 3 (Reddy et al. 1998a) and porcine adenovirus type 5 (Nagy et al. 2001), FAdV-9 (Ojkic and Nagy 2000), as well as the most recently completed genome sequence of BAdV-4 (Benkö et al. 2001) all strengthened the observation that basically four major genome organization patterns exist (Davison et al. 2000).

In addition to the genome arrangements of the mastadenoviruses and the aviadenoviruses (typical members of the two traditional genera), two novel genomic organizations were recognized. In accordance with the results of the phylogenetic analyses discussed later, the genome organization of EDS virus, the atypical bovine and some other ruminant adeno-

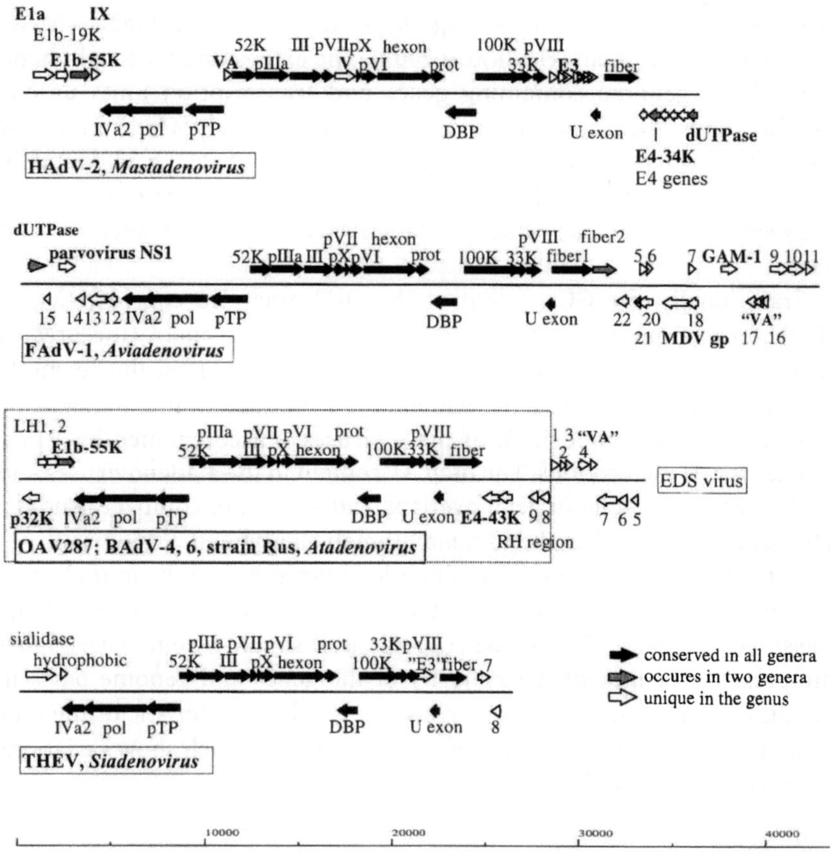

Fig. 1. Schematic presentation of the four different genome organization patterns

viruses was found to be very similar to that seen in OAV287, while the genome arrangement of an amphibian adenovirus isolate originating from leopard frog (*Rana pipiens*) turned out to be almost identical to that of the former group II 'avian' adenovirus THEV.

A schematic presentation of the four types of genome maps is shown in Fig. 1. Apparently, the highest degree of conservation in the adenovirus genome is in the middle part, which encodes the structural proteins and viral enzymes, whereas the two ends – containing mainly regulatory early regions – might exhibit larger variations, and these variations account for the remarkable differences seen in the overall genome sizes. Aviadenoviruses possess the largest genome, in spite of the fact that a couple of structural proteins and early regions described in mastadenoviruses are missing from them (Harrach 2002). The extension at the right hand part of the genome containing genes and transcription units unique for the genus *Aviadenovirus* add up to approximately 44–45 kilo base pairs (kb), whereas the smallest genome of about 26 kb is found in the two siadenoviruses. The genome size in mastadenoviruses ranges between 31 and 36 kb, while in atadenoviruses it is between 29 and 33 kb.

Traditionally, the GC content of the DNA was also considered as a characteristic marker of the genera and HAdV subgenera (the present species). With the analysis of the atypical adenoviruses, the formerly known range of GC content has widened and the lowest value is now 34%. The biased base composition of these viruses is accompanied by a preference for AT rich codons. The high AT content in the atadenoviruses and siadenoviruses results in gene overlaps and in a more compact genome. The actual cause of this phenomenon remains unknown.

In the following sections, the genomic organization of the four clusters will be summarized according to the transcription units conventionally classified into early (E) and late (L) regions. It should be emphasized that the presence or lack of the majority of the genes and genome parts in the different clusters seem to be consistent and characteristic of the four genera. Nevertheless, there also exist certain units, which show variations even within the individual genera.

3.1 Early Regions

The left-hand end of the genome contains different genes in the different clusters. In mastadenoviruses, after the left inverted terminal repeat (ITR), the E1A and E1B regions are found without exception. Candidate members of the proposed genera, as well as aviadenoviruses lack the E1A transcription unit. In aviadenoviruses, several open reading frames (ORFs) on both strands have been described at this location, but the function of most of them has not yet been elucidated (Chiocca et al. 1996). In atadenoviruses, at the position of E1A, the complementary strand encodes a gene of a novel structural protein p32K, which is transcribed leftward (Vrati et al. 1996b). A homolog of the p32K gene at the same position has been identified in every candidate atadenovirus examined so far including the corn snake isolate (Farkas et al. 2002) and BAdV-4 through 8

(P. Élő, unpublished results), therefore it can be considered as a marker of atadenoviruses. In the two sequenced siadenoviruses, a putative sialidase gene was found on the *r* strand in the place of the E1 region. Beside mastadenoviruses, only the atadenoviruses seem to have a homolog of one of the E1B genes, namely the 55K protein, but this homology can be identified with special search options only, e.g., using the Conserved Domain database within GenBank.

The delayed early gene of protein IX, the only structural protein that is not transcribed from the major late promoter (MLP) was found only in mastadenoviruses. Recent studies revealed, that besides cementing the hexons and thus stabilizing the capsid, protein IX also has transcriptional activity (Rosa-Calatrava et al. 2001). The lack of protein IX (and V, contained in the late region) in the avi-, at- and siadenoviruses suggests that the virions of these animal adenoviruses are slightly different structurally. Perhaps other yet uncharacterized proteins (e.g., p32K in atadenoviruses) substitute protein IX (and/or V). Nevertheless, an interesting contradiction is that protein IX negative HAdV mutants show thermolabile virion phenotype (Jones and Shenk 1979), while atadenoviruses also lacking this protein were demonstrated to have increased heat resistance (Bartha 1969). The other delayed early gene, IVa2, is present in every studied member of all four genera.

The conservation level of the E2 region on the *l* strand encoding proteins [DNA-binding-protein (DBP) precursor terminal protein (pTP) and DNA polymerase] essential for replication and maturation seems to be exceptional. Among the few genes identified thus far from the white sturgeon adenovirus genome, two E2 genes (pTP and DNA polymerase) were found, and their estimated relative map position corresponds to that in other genera (M. Benkő et al. 2002).

The E3 region apparently exists only in the members of the genus *Mastadenovirus* and no homologous genes were detected in any members of any other genera. The size and content of the E3 region is highly variable within the genus and differs even among HAdVs of different species. As this region is non-essential for in vitro virus replication, it can be deleted and replaced by foreign gene expression constructs. E3 is one of the most commonly used insertion sites in adenovirus gene delivery systems (Russell 2000) and has been extensively studied in numerous animal adenoviruses too (e.g., Raviprakash et al. 1987; Esford and Haj-Ahmad 1994; Evans et al. 1998; Idakamanti et al. 1999). In avi- and atadenoviruses, no ORFs are found on the *r* strand between the genes of pVIII and the fiber. The putative E3 region (one ORF at the conventional location) of siadenoviruses has no homology with mastadenoviral E3 genes. However, a novel short ORF termed the U exon (Davison et al. 1993) was identified on the *l* strand between the genes of pVIII and the fiber in all four genera. This small coding region extends from an initiation codon to a splice donor site, and seems to code for the amino-terminal part of a protein; however, downstream exons have not been identified. U exon is present in each examined type of all four genera with the exception of mouse adenovirus type 1, porcine adenovirus type 5 and BAdV-10. The deduced amino acid sequences are conserved within, but not between the genera. A homolog of atadenoviral U exon was also identified in the snake adenovirus genome (Farkas et al. 2002).

From the E4 region, a single gene of the 34K protein of mastadenoviruses has its homolog in the atadenoviruses only. However, this short E4 region in the atadenoviruses is not at the right-hand end of the genome, but is followed by a unique 'E3' region (encoded on the *l* strand) that has no homology with the E3 regions of the mastadenoviruses or siadenoviruses, and was named E3 only because of its deletable nature (Xu et al. 1997). The corresponding right-hand end of the EDS virus, aviadenoviruses and siadenoviruses are partially characterized and numerous putative ORFs need confirmation and further studies (Chiocca et al. 1997; Payet et al. 1998; Davison et al. 2000).

3.2 Late Regions

As mentioned above and illustrated in Fig. 1, the late regions – occupying the middle part of the genomic DNA – seem to be much better conserved throughout the family. The single exception is protein V, which occurs only in the mastadenoviruses. Protein V is a structural core protein to which important role is attributed during the viral replication cycle. It remains associated with the viral DNA during its transport to the nucleus of infected cells (Matthews and Russell 1998a, b). Protein V is missing from the avi-, at- and siadenoviruses, and no other gene is found in its location as a replacement.

The presence and order of 12 genes (of the following proteins: 52K, pIIIa, III, pVII, pX, pVI, hexon, protease, 100K, 33K, pVIII, fiber) are fully conserved throughout the family, although the number of late mRNA classes in which clusters of genes are transcribed might be different, and has not been elucidated in every virus genus or type. Nevertheless, all late genes are coupled with the MLP, which also seems to be conserved (Song et al. 1996; Vrati et al. 1996a; Sheppard et al. 1998a).

Although the larger late genes are appropriate for phylogenetic calculations, the fiber gene is an exception, because of the repetitive structure of the shaft, and it can be used effectively in structural alignments only. In most adenoviruses, there is only one fiber protein inserted in the penton base; however, in aviadenoviruses, two fiber projections per vertex are found (Gelderblom and Maichle-Lauppe 1982). In FAdV-1, the two fibers are of different sizes and are encoded by two tandem fiber genes (Chiocca et al. 1996). In other aviadenoviruses however, the duplicate fiber projections are of identical length and duplication of the fiber gene has not been confirmed (Sheppard et al. 1998a; Ojkic and Nagy 2000). EDS virus and THEV have only one fiber gene and one fiber protein per vertex. This is true for every examined member of the genera *Atadenovirus* and *Siadenovirus*. Interestingly, in the species *Human adenovirus F*, HAdV-40 and 41 have two fiber genes, but only one protein per vertex occurs, although of two alternating sizes (Kidd et al. 1993). This is an exceptional feature among mastadenoviruses.

Comparative analysis of the protease cleavage sites of the precursor viral proteins revealed novel recognition and cleavage sequences in OAV287 (Vrati et al. 1996b). In other members of the proposed genus *Atadenovirus* including the corn snake isolate, identical motifs were recognized, which could probably be regarded as markers of this genus (Farkas et al. 2002).

3.3
VA RNA and ITR

VA RNAs are non-protein coding sequences, which can antagonize the antiviral activity of interferons (Mathews and Shenk 1991). VA RNA genes homologous to those of HAdV-2 were found only in human, chimpanzee and some (but not all) monkey adenoviruses (Kidd et al. 1995). In some human and chimpanzee adenoviruses, two VA RNA genes occur. Non-primate mastadenoviruses studied so far lack such a gene. On the other hand in one avian adenovirus (FAdV-1), there exists a non-homologous VA RNA gene, but on the *l* strand close to the right-hand end of the genome. In the other aviadenoviruses examined, no VA RNA could be identified. Interestingly, EDS virus from among the atadenoviruses has a

VA RNA gene homologous to that of FAdV-1, in a similar genomic location but in the opposite orientation (Hess et al. 1997). The genes surrounding the VA RNA gene, however are non-homologous in the EDS virus and FAdV-1. Apparently, the presence or absence, or the number (one or two) of VA RNA genes is not a marker of genera. On the other hand it might be useful data in the demarcation of species.

The length and complexity (concerning the number and variety of binding site motifs) of ITRs show dramatic differences in the different genera (Dán et al. 2001). Among mastadenoviruses, the length of ITRs ranges from 93 to 369 nucleotides (nt), and the ITRs are especially complex in HAdVs. The avi- and atadenoviruses have significantly shorter ITRs (from 53 to 71 nt, and from 46 to 75 nt, respectively) and very few binding sites are recognizable. Siadenoviruses have the shortest and simplest ITRs of 36 (frog) and 39 nt (THEV).

4 Phylogenetic Calculations

The very first attempts to estimate phylogenetic relationships among adenoviruses of different host species were made almost 15 years ago. The most easily accessible ITR sequences were determined from a number of human, mammalian and avian adenovirus types, and with the maximum likelihood calculation (Felsenstein 1981) phylogenetic trees were produced (Shinagawa et al. 1987). The authors were aware of the inadequacy of ITR sequences for such purposes (Dán et al. 2001), but no polypeptide gene sequences were known from a reasonable number of adenoviruses at that time. Furthermore, the two avian adenovirus representatives (FAdV-1 and EDS virus) have exceptionally short ITRs, therefore they could not be included in these early phylogenetic calculations. Thus the evolutionary distance between FAdV-1 and EDS virus was not revealed. Nevertheless, the first evidence for the substantial difference existing between the members of the two genera (*Mastadenovirus* and *Aviadenovirus*) were provided. Similarly, the close genetic relationship between canine adenovirus type1 and 2, and the distance between BAdV-1 and 3 were estimated correctly. However, some other conclusions, such as the polyphyletic origin of HAdVs were later rebutted.

In a more comprehensive publication based on phylogenetic analyses of a number of different genomic portions including E1A, E1B and E3 regions, the fiber, hexon and protease gene sequences (Bailey and

MAUTNER 1994), the correct demonstration of the monophyletic origin of HAdVs was achieved; in addition the puzzling distinctness of the atypical BAdVs was graphically presented, but it was not addressed at all in the discussion. Among the 12 adenoviruses from which the protease gene sequence was available at that time, two BAdVs (serotypes 3 and 7) and FAdV-1 were also included. Both the unrooted parsimony and the distance matrix trees clearly illustrated that BAdV-7 is at a much greater distance from BAdV-3 than is FAdV-1 (member of the genus *Aviadenovirus*). A straightforward inference could have been that BAdV-7 probably represents a third genus. Yet, this conclusion was to be drawn later. Phylogenetic comparisons based on the VA RNA sequences aimed at clarifying the relationships among HAdVs and between adenoviruses of human and simian origin (KIDD et al. 1995; MA and MATHEWS 1996) partially confirmed the former results, and established a basis for the more recently introduced species category of HAdVs and simian adenoviruses (BENKÖ et al. 2000).

Phylogenetic results implying the existence of a third genus were published first in connection with the taxonomic place of the EDS virus (HARRACH et al. 1997), and soon the full genomic sequence of this virus was described (HESS et al. 1997). In this latter paper, however, the relatedness of EDS virus to BAdVs in general was emphasized, but the genus-level diversity of BAdVs was ignored, thus the conclusions were somewhat confusing. Thanks to the enormously increased number of available adenoviral gene sequences, phylogenetic analyses are now able to provide highly reliable results (HARRACH and BENKÖ 1998). Different phylogenetic calculations performed on numerous different gene sequences, consistently result in four well-distinguishable clusters (Fig. 2) that neatly correspond to the four characteristic genome organization patterns.

Those genes and regions, which were studied intensively in order to determine their exact function and structural properties responsible for their specific activity became favorable subjects for phylogenetic calculations. The protease and hexon genes were included into such studies especially frequently. In every analysis performed either on protein or nucleotide sequences, the non-conventional BAdVs, OAV287 and the EDS virus always clustered together (DÁN et al. 1998; MATIZ et al. 1998; BARBEZANGE et al. 2000). In the past couple of years, further candidate atadenoviruses have been described. An epizootic hemorrhagic disease among mule deer (*Odocoileus hemionus*) in California killed thousands of animals (WOODS et al. 1996). The disease was reproduced with the isolated adenovirus, which was recently characterized by partial sequenc-

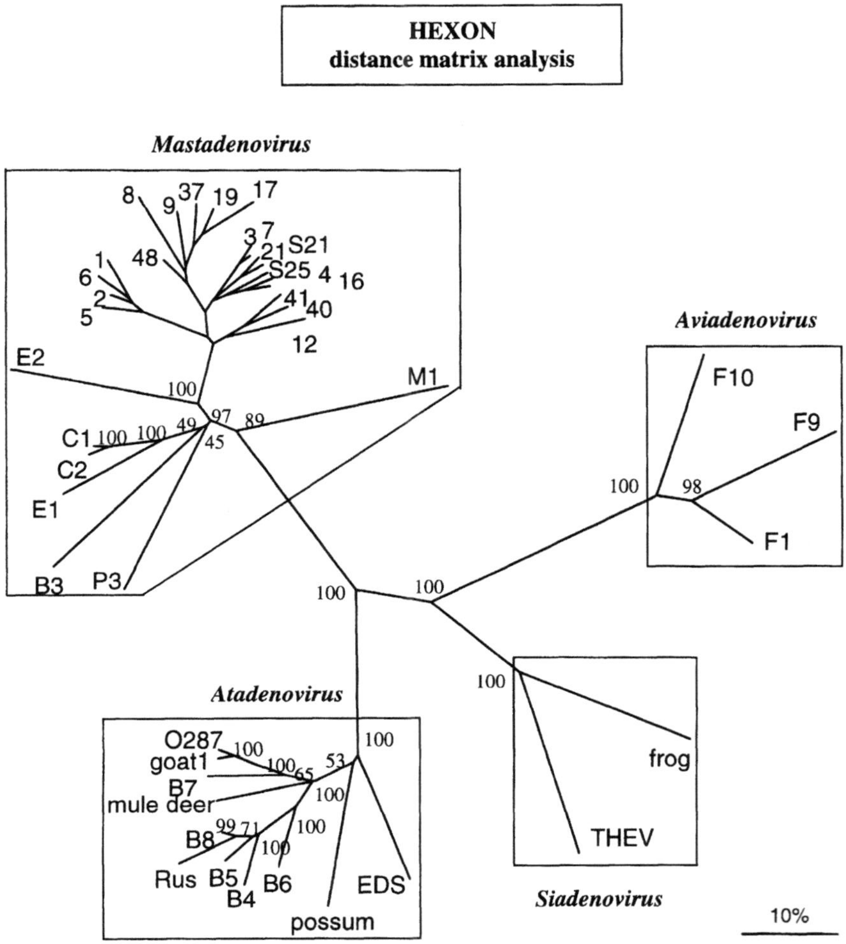

Fig. 2. Phylogenetic tree (on the amino acid sequence of hexon proteins) presenting the clear separation of the four clusters. ProtDist, Fitch-Margoliash (with global rearrangements); bootstrap values shown for 100 data sets (Felsenstein 1989)

ing and described as a new atadenovirus (Lehmkuhl et al. 2001). A new serotype of goat adenovirus was confirmed as belonging to atadenoviruses (Lehmkuhl and Cutlip 1999). Adenovirus-like particles were detected in the intestinal contents of brushtail possums (*Trichosurus vulpecula*) during a general survey in New Zealand. This marsupial animal is protected in Australia but became an unwanted pest shortly after having been introduced to New Zealand. The putative adenovirus could not be

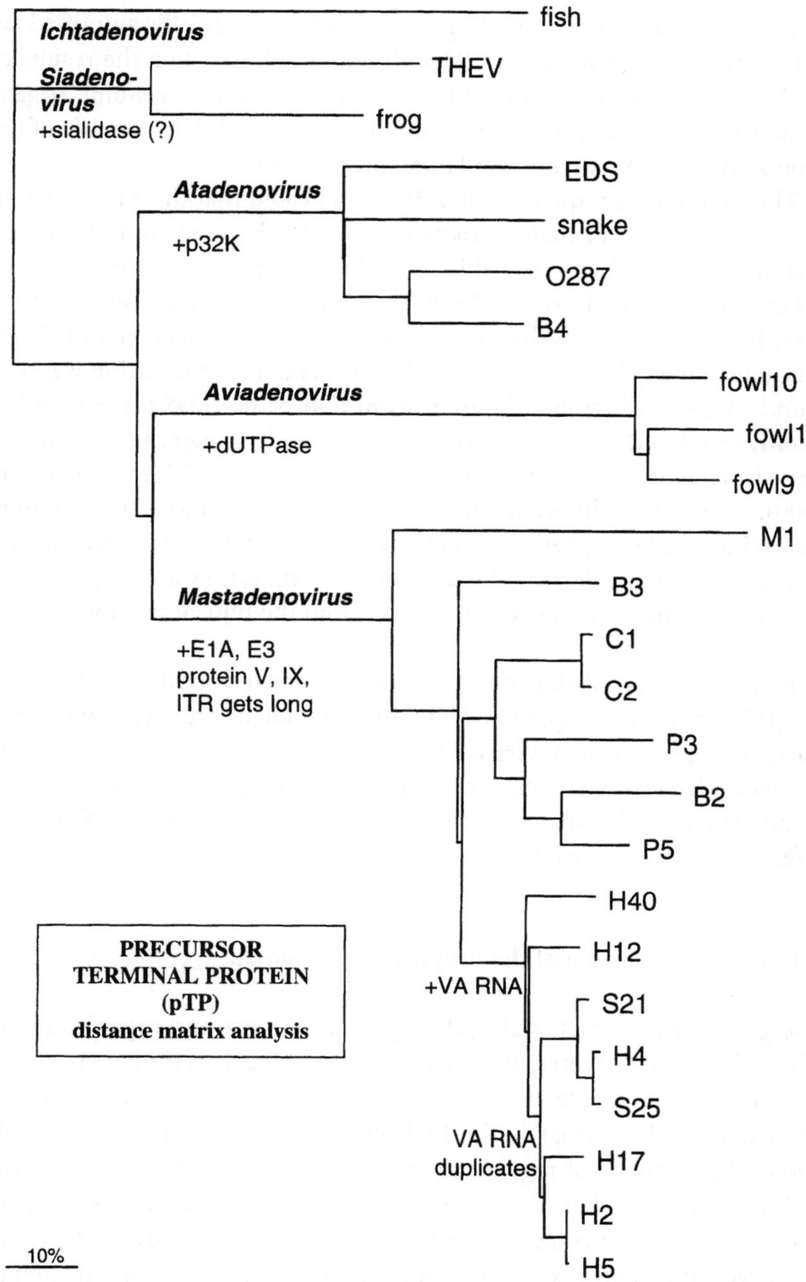

Fig. 3. Phylogenetic tree including the fish and snake adenovirus isolates. The appearance of newer genes are marked

isolated, but large parts of its genome were successfully amplified by polymerase chain reaction (PCR) and sequenced. Based on the results of phylogenetic analysis and on the partially determined genome organization, the possum adenovirus is undoubtedly a candidate member of the proposed genus *Atadenovirus* (Thomson et al. 2002).

The first partial sequences of THEV implicated that the AT content of this virus is also very high (Jucker et al. 1996), therefore, initially THEV was also proposed as a possible candidate of the third (*Atadenovirus*) genus (Benkö and Harrach 1998). The phylogenetic calculations, however, did not confirm this hypothesis, and when the full sequence of THEV was determined (Pitcovski et al. 1998) its genome organization was also found to be considerably different from that of the EDS virus or other putative atadenoviruses. Interestingly, the recently sequenced frog adenovirus isolate turned out to be the closest relative of THEV (Davison et al. 2000). Beside the almost identical genomic organization, the common clustering in phylogenetic calculations convincingly shows that the two viruses are genetically very close. At the same time, they are very far from the conventional bird adenoviruses, i.e., from the typical members of the genus *Aviadenovirus*.

Figure 3 shows a distance matrix tree calculated from the alignment of the pTP amino acid sequences from different adenovirus types including the white sturgeon adenovirus. The separation of five clusters implies that the fish adenovirus represents an independent lineage, possibly a new genus. The tentative name for this putative new genus is *Ichtadenovirus* referring to the host origin.

5
Presumed Host Origin of the Proposed New Genera

The great variability in coding strategy, genome size and organization of adenoviruses raised intriguing questions about the direction of evolution. A characteristic feature of viruses in two new proposed genera (*Atadenovirus* and *Siadenovirus*) is the high genomic AT content, which easily allows the insertion of regulatory motifs, such as TATA boxes, within the coding sequences. Thus gene overlaps or very short intergenic sequences occur resulting in a compact genome. Could such a strategy be the result of an evolutionary process or is it rather an atavistic feature? As an attempt to clarify the question, we have compared two phylogenetic trees: one constructed on adenovirus protease sequences and the other on the mitochondrial small subunit ribosomal RNA gene sequences from their hosts. A relatively good overlap of the viruses with the respective hosts was seen, i.e., on one end of the tree the human and murine adenoviruses occupied similar position as primates and mice on the host animals' tree, while the fowl adenoviruses overlapped the branch of different birds. However, the cluster of atadenoviruses having branches for viruses isolated from ruminants, birds or the possum was at a large distance corresponding to the position of lower vertebrate species. We have hypothesized that atadenoviruses might originate from reptiles (Harrach 2000).

Although the occurrence of adenovirus-like particles confirmed by electron microscopy has been described in a number of different reptilian species such as crocodile (Jacobson et al. 1984), chameleon (Jacobson and Gardiner 1990), bearded dragon (Jacobson et al. 1996), and boa (Ramis et al. 2000), there are very few isolates available. Winfried Ahne provided us with two snake strains propagated in large amounts. One isolate originated from a corn snake (Juhasz and Ahne 1992), while the other was from a royal python (Ogawa et al. 1992). By random cloning and sequencing, a putative genetic map of the corn snake isolate was compiled, which is in accordance with the genome arrangement of atadenoviruses. The strongest evidence for a common origin are the presence of a p32K gene homolog, the lack of proteins V and IX, and the characteristic protease cleavage sequences shared with other atadenoviruses. Moreover, the results of phylogenetic calculations clearly indicate (Fig. 3), that this snake adenovirus belongs to the genus *Atadenovirus* (Farkas et al. 2002). Perhaps it would be premature to declare that atadenoviruses represent the adenoviruses of reptiles, yet the hypothesis seems to be supported. For the confirmation of the theory, we would need more DNA sequences from other reptilian isolates. Reptiles have become popular pets during the past few years and consequently recognition of different viruses, including adenovirus-like particles, has increased (Ahne and Essbauer 2000; Essbauer and Ahne 2001; Perkins et al. 2001); thus we plan to analyze further samples in the future.

In the case of siadenoviruses, the host origin is less certain. Because of the very low number of (no more than two) candidate members, we are confined to mere speculations. As mentioned above, THEV was initially also presumed to belong to the genus *Atadenovirus*, although the former classification and serological results contradicted this. The striking similarity between the genomes of THEV and the frog adenovirus is firm proof for the common evolutionary past of these viruses. For the time being, however, we might only suppose that siadenoviruses represent the typical adenoviruses of amphibians. The fact that the frog adenovirus could be propagated in turtle heart cells only and not in any cells of amphibian origin leaves a dash of uncertainty. We cannot exclude the possibility that both of the presently seen hosts are new, and the origin of siadenoviruses must be searched for in other lower vertebrates such as turtles. This theory could be challenged by the examination of other adenovirus strains of amphibian origin. Fortunately, there is at least one more report on the isolation of adenoviruses from frogs (Cunningham et al. 1996), but it would seem to be prudent to search for additional amphibian samples in further species, such as salamanders and newts. Based on the full genomic sequence, powerful PCR methods can be elaborated, thus the question will certainly be solved in the near future.

The data at hand on the white sturgeon adenovirus isolate provide an even less firm basis for assumptions. Accepting the above described straightforward and tempting theory that adenovirus genera correspond to the vertebrate host classes, the isolate from the white sturgeon should prove to possess a fifth type of genomic organization. Our preliminary sequence data were obtained on highly preserved genes, thus the conclusion on the possible existence of the proposed genus *Ichtadenovirus* could be based on the results of phylogenetic calculations (Fig. 3).

6
Evolutionary Events Explaining the Present Heterogeneity

The rise of adenoviruses has been presumed to happen before or shortly after the host species separated from each other (Wadell et al. 1980). The possibility that lower vertebrate classes possess their own adenoviruses would mean that members of this virus family existed already more than 400 million years ago, when the divergence of vertebrate classes started (Kumar and Hedges 1998). Because in virology no fossils are available, we have no tools to study directly the ancestral viruses, and are confined to be contented with the examination of the present-day existing viruses. The overwhelming majority of adenoviruses known today are from mammalian and avian hosts, and represent the highest level of evolution. To better understand the relationships, many more adenovirus types from lower vertebrate animals would be needed. Nevertheless, we have already several data and tools at hands and may try to draw several conclusions from the picture as is it seen today.

The ancestral adenovirus genome was supposedly very short, containing a minimum of essential information for replication and survival. The gradual enlargement and evolution of the genome might have taken place by a variety of mechanisms similar to those hypothesized in the evolution of other large double stranded (ds) DNA virus families (McGeoch and Davison 1995). Such hypothetic events include recombination, gene capture, gene duplication, gene generation de novo, etc.

Indeed, we deem to see the traces of different ways of gene acquisition. There are numerous examples for the occurrence of gene capture. From the evolutionary standpoint useful genes might have been acquired from the chromosomes of the host animal or from concurrent infective agents. The DNA polymerase, sialidase, dUTPase, or the gene resembling the NS1 of parvoviruses (Fig. 1) are among the presumed candidates. Another example could be protein IX of mastadenoviruses, which besides being a structural protein is also a transcriptional activator and shares homology with leucine zippers (Lutz et al. 1997). EDS virus also gives an example for gene capture. Its VA RNA gene seems to have a common origin with that of FAdV-1. But in the two viruses, these genes are in different orientation and were probably acquired independently by the two viruses.

There are many examples also for recombination especially between closely related serotypes that are now classified into a common species (a category corresponding to the former subgenus or subgroup of HAdVs). So-called inter-typic recombinants have been recognized repeatedly. In scarce cases, however, recombination between more distantly related types, or even between members of different genera can be observed. Typically, fiber genes seem to have been prone to recombination events as exemplified by porcine adenovirus type 5, the fiber gene of which was recently recognized to be a recombinant (Nagy et al. 2002). The tail and shaft of the porcine adenovirus type 5 fiber gene show typical mastadenoviral structures, whereas the sequence of the fiber knob can be aligned with the corresponding part of atadenoviruses. EDS virus, a member of the genus *Atadenovirus*, has a fiber knob similar to that of typical aviadenoviruses.

Enlargement of the genome could also occur through gene duplication. The VA RNA gene is duplicated in some but not all human and monkey adenoviruses. The fiber genes in FAdV-1 or in the members of species *Human adenovirus F* seem to be the results of duplication. BAdV-10, from which serologically identical isolates are available with smaller or larger variations in the size and sequence of their fiber genes (Ursu et al. 1999) provides a good example for present-day evolution. Since BAdV-10 represents a separate and remote lineage of BAdVs, we hypothesize that the insertions and reiterations in its fiber gene sequence might indicate a process of actual adaptation to a new host, the cattle.

The mastadenoviral genome should be considered as being the most highly organized and most efficient protein synthesis system, which also has acquired sophisticated regulatory mechanisms for optimal interaction with the host cells and with the immune system of the animal. Our opinion opposes the hypothesis of Rosa-Calatrava et al. (2001) who speculate, that „viruses, as obligatory cell parasites, usually evolved toward the highest possible degree of simplification of their structure and components…" and "adenoviruses comply with this rule".

We propose that each of the present day existing four (possibly five) major lineages of adenoviruses co-evolved with their respective host classes, and the occurrence of unusual adenovirus types in a number of animals is the consequence of several inter-class host switches.

If we accept that atadenoviruses originate from reptiles, several independent host switches should be presumed. Nowadays, atadenoviruses can be seen in birds, in the brushtail possum and in a number of different ruminant species. We assume that these animals acquired atadenoviruses directly from reptilian hosts, and thus at least two, probably three independent host switches might have happened. The number and variety of serotypes in ruminant species would suggest that ruminants were infected early. Beside cattle and sheep, recently numerous other ruminant hosts were recognized to harbor atadenoviruses. The visible divergence among the atadenoviruses of ruminants implies that these viruses had a rather long history in the present hosts. EDS virus, although able to infect several different bird species, has no serovariants. This would suggest a shorter history in birds. Finally, the possum adenovirus needs further study in order to decide its direct origin. Phylogenetic trees on a limited number of gene sequences imply that the possum adenovirus is a separate lineage, and so is EDS virus, which, in spite of the long genome, seems to be the closest relative of the corn snake virus (Fig. 3).

The difference between THEV and the frog adenovirus is significant, but we do not yet have data concerning differences between different amphibian adenoviruses. For the time being, it is perhaps not too unfounded to suppose that the genus *Siadenovirus* represents the typical adenoviruses of amphibians. Theoretically, an original avian adenovirus could have infected a new, amphibian host – the leopard frog. In favor of this theory is the relative larger abundance of avian isolates (THEV, MSD

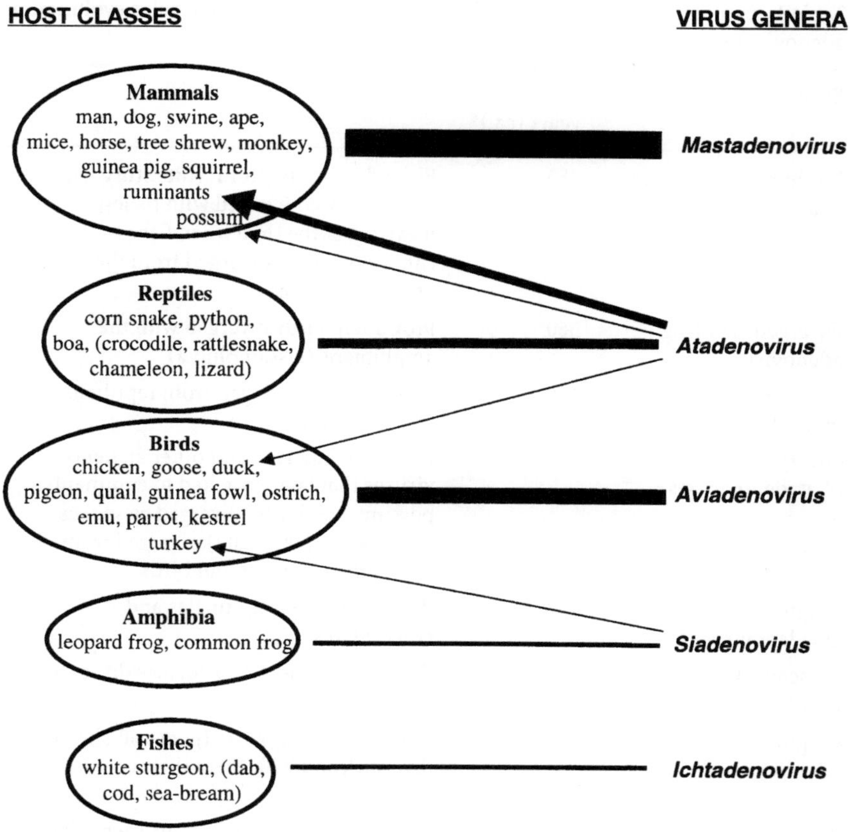

Fig. 4. Presentation of the recognized occurrence of adenoviruses from five proposed genera in host animals of five different classes. The *width of the lines* is intended to be proportional to the number of known virus serotypes. *Arrows* indicate presumed host switches

virus, splenomegaly); yet, these isolates are serologically indistinguishable, thus probably they do not have a too long history in the present avian hosts.

The present understanding of the incidence of infection by members of the five putative adenovirus genera in animals of the five vertebrate classes is summarized in Fig. 4. Perhaps this theory will be rebutted in the future; at present, however, there is no evidence against it and agrees with the existing virus genus concept. Based on such a hypothesis, the time point of certain assumed evolutionary events can be estimated as shown in the Table 1. The appearance of certain genes are shown in Fig. 3.

Table 1. Estimated possible time-scale of certain steps in the molecular evolution of adenoviruses

Event	Million years (ca.)[a]	Explanation
Minimal provable age of adenoviruses	450	Fish adenovirus (from white sturgeon) supposedly co-evolved with fishes, it existed at the time when fishes (Actinopterygii) diverged from the other vertebrates
Divergence of frog adenovirus	360	Frog adenovirus diverged with the amphibians (Lissamphibia)
Origin of mastadenoviruses	310	Mammalians diverged from reptilians (including birds)
Divergence of atadenoviruses	245	Snake adenovirus (and other atadenoviruses that later switched to ruminants, possum, and birds) diverged as snakes and lizards (Squamata) diverged from crocodilians, turtles and birds
Origin of the aviadenoviruses	228	Birds diverged from turtles and crocodilians
Muscovy duck AdV (DAdV-2)	112	Galliformes and Anseriformes diverged
Origin of ovine adenoviruses	20	Bovinae and Caprinae (goat and sheep) diverged, and with them their co-evolving adenoviruses
Acquisition of VA RNA	5.5–20	The chimpanzee and human adenoviruses have VA RNA (man diverged 5.5 million years ago), but not all old-world monkey adenoviruses (23 million years) and none of other animals

[a] Molecular time estimates on the phylogeny of vertebrates are from KUMAR and HEDGES (1998), and HEDGES and POLING (1999). In case of conflict, data of the later publication were considered.

7
Pathogenicity and Host Specificity

Adenoviruses in general are considered as facultative pathogens, i.e., usually the presence of some predisposing factors are required for the development of sever or fatal diseases. This also means that adenoviral disease can seldom be reproduced experimentally in healthy individuals.

The predisposing conditions include compromised immune status (AIDS, anti-tumor radiation or cytostatic therapy, organ transplantation), individual susceptibility (e.g., for conjunctivitis or urogenital infections), concurrent infections with other agents, old age, etc. In farm animals (such as calves or piglets), problems in management, housing or feeding might lead to overt adenoviral diseases. Most often, the mixing of adenovirus carrier populations with immunologically naive individuals provokes mild acute disease (nursery schools, military recruits, feed lots, fattening units).

In veterinary practice, however, there are several striking exceptions, predominantly among the atypical adenoviruses, which can cause experimentally reproducible specific diseases. EDS virus has caused world-wide epizooty starting in 1976, when the transition from ducks to chicken probably happened. The diseases attributed to turkey hemorrhagic enteritis virus can be reproduced experimentally in different bird species (chicken, pheasant, turkey) if they are not immune. The most recent example for high pathogenicity is the hemorrhagic epizooty among wild-living mule deer in California (Woods et al. 1996). Partial molecular characterization of the isolated virus strain, including DNA sequencing and phylogenetic analysis, clearly confirmed that this virus belongs to the genus *Atadenovirus* (Lehmkuhl et al. 2001).

Considering that those adenoviruses that are capable of infecting multiple host species seem to be more pathogenic, it is tempting to speculate, that in these cases the virus has not had the time for successful adaptation to the present host. It could also mean that the time of co-existence with the actual host is short: a recent host switch from a very different animal species might have taken place. The E3 region, which did not exist before the development of mastadenoviruses, is certainly an indicator of a high level adaptation to the host. A further characteristic of the atypical adenoviruses that we propose to have undergone host switch is their generally poor in vitro replication ability. Atadenovirus BAdVs were recognized first as differing from other mastadenoviruses because of the difficulties of their propagation. They require primary or low passage number fibroblast type cells, and cannot be grown in tissue cultures of epithelial type cells. THEV could be propagated in embryos or young birds, but not in conventional cell lines. The poor replication ability might be also the result of less sophisticated (developed?) genomes having primitive ITR sequences, lacking E1A region, protein IX and V genes.

These facts are not widely known among molecular virologists or they are considered as exceptions. Realizing that closely related viruses can be found in hosts of very distant origin, the theory of inter-class host switches would become more easily acceptable. Unfortunately, the stereotype approach – that an adenovirus isolated from, for example, birds, is necessarily an avian adenovirus – seems hard to be overcome.

8
Origin of Adenoviruses: The Fifth Cluster – The Missing Link?

For many years, the family *Adenoviridae* stood alone in the virus world. The almost artistically structured capsid with the protruding fibers, as well as the dsDNA genome in its complexity resembling the chromosomes, were unparalleled among viruses. In the early 1970s, a dsDNA bacteriophage PRD1, the prototype of a novel virus family was discovered. Members of the new virus family *Tectiviridae* share similar architecture with adenoviruses: an icosahedral capsid with fiber-like projections. The capsid of the phages is slightly smaller (63 nm) than the adenoviral particles and has a unique, lipid containing inner membrane, which is derived from the bacterial plasma membrane (Bamford et al. 1995). The dsDNA genome of phage PRD1 is almost 15 kb. It has ITRs, and contains the genes of two important proteins (terminal protein and DNA polymerase) in the same order as they are in adenoviruses (Bamford and Ackermann 2000). The terminal protein of phage PRD1 also acts as protein primer in viral DNA replication, a rare feature that occurs, besides in adenoviruses, only in *Φ*29 phage (Salas 1983).

A recent study on the resolution of the hexon (P3) protein structure of phage PRD1 revealed an arrangement very similar to that found in the hexon of HAdV-2 (Burnett 1997) suggesting an evolutionary link between adenoviruses and tectiviruses (Benson et al. 1999). However, no homology in the nucleotide or amino acid sequences of the two virus genes could be shown (Belnap and Steven 2000).

In an attempt to find some similarity with adenoviral hexon sequences, we have searched with the P3 sequence of PRD1 our dedicated adenovirus database, which contains all of the available adenoviral sequences (www.vmri.hu/~harrach). A short stretch of amino acid sequence of the hexon gene of THEV was found with a rather low probability value. We have tried to align the PRD1 P3 sequence around the found homology with a complete adenovirus hexon alignment. In this region, almost 25% of the amino acids of P3 seemed to be identical with the amino acids in the same position of at least one adenovirus type. In Fig. 5 from the approximately 200 amino acid long region, only two-thirds (~ 150 amino acids) are presented, where the homology is more convincing. Interestingly, scattered identical amino acids continue to occur also further downstream, where the first loop of the hexon starts. But this region is more variable among the adenovirus types, and the original alignment started to contain longer insertions and increased padding. Nevertheless, resembling motifs, which are also presented in Fig. 5, were still seen especially with members of the at- and siadenovirus genera.

Surprisingly, however, the deemed homology was found between the very conservative amino-terminal part of the hexon and the carboxyl-terminal portion of P3, therefore the authenticity needs thorough confirmation with special regard to the structural role of the protein domains examined. Also, the conservation level in the gaps between the identical amino acids is very poor. On the other hand, this very weak homology was demonstrated with the THEV sequence, which, according to our theory, is an amphibian adenovirus, and thus represents the most ancient type of adenovirus studied to date. We would expect to see an even more obvious similarity with a fish adenovirus.

As mentioned above, adenovirus-like particles have been observed in different fishes (Bloch et al. 1986; Miyazaki et al. 2000), but only one isolate is known from white sturgeon (*Acipenser tansmontatus*) (LaPatra et al. 1998). By electron microscopic examination, the virions of the putative white sturgeon adenovirus also appear to have double-layered shell, although the nature of the inner shell is completely unknown. We have started to analyze the genome of this white sturgeon adenovirus. Because of the low virus yields, a random cloning and sequencing approach was chosen. So far, partial sequences of the DNA polymerase, 100K protein genes and the full gene of the pTP have been identified. To date, no other genes have yet been found, but work is in progress. The pTP sequence was included in a distance matrix calculation, and as a result, the fish adenovirus was placed alone on an individual branch indicating its large distance from the other four clusters (Fig. 3). We have obtained similar tree topology with partial sequences of the DNA polymerase (Benkö et al. 2002) and the 100K genes. Following the earlier established hypothesis, the adenoviruses of fishes should represent a fifth genus.

The presence of pTP gene in the putative fish adenovirus implies that protein priming exists also in fish adenoviruses. If we accept the above-described evolutionary pathway of adenoviruses, and believe that adeno-

```
     8                                                                             85
H2   PQWSYMHISGQDASEYLSPGLVQFARATETYFSLNNKFRNPTVAPTHDVTTDRSQRLTLRFIPVDREDTAYSYKARFT
H3   PQWAYMHIAGQDASEYLSPGLVQFARATDTYFSMGNKFRNPTVAPTHDVTTDRSQRLMLRFVPVDREDNTYSYKVRYT
H4   PQWAYMHIAGQDASEYLSPGLVQFARATDTYFSLGNKFRNPTVAPTHDVTTDRSQRLSLRFVPVDREDNTYSYKVRYT
H12  PQWSYMHIAGQDASEYLSPGLVQFARATDTYFTLGNKFRNPTVAPTHDVTTDRSQRLTLRFVPVDREDTTYSYKARFT
H40  PQWSYMHIAGQDASEYLSPGLVQFARATDTYFSLGNKFRNPTVAPTHDVTTDRSQRLTLRFVPVDREETAYSYKVRFT
H48  PQWAYMHIAGQDASEYLSPGLVQFARATDTYFSLGNKFRNPTVAPTHDVTTDRSQRLTLRFVPVDREDTTYSYKARFT
S21  PQWAYMHIAGQDASEYLSPGLVQFARATDTYFNLGNKFRNPTVAPTHDVTTDRSQRLMLRFVPVDREDNTYSYKVRYT
S25  PQWAYMHIAGQDASEYLSPGLVQFARATDTYFSLGNKFRNPTVAPTHDVTTDRSQRLTLRFVPVDREDNTYSYKVRYT
P3   PQWSYMHISGQDASEYLSPGLVQFSQATETYFNLNNKFRNPTVAPTHDVTTERSQRLQLRFVPVDKEDTQYTYKTRFQ
C1   PQWSYMHIAGQDAAEYLSPALVQFAQATSSYFKLDNKFRNPTVAPTHDVTTERSQRLQLRFVPVMQEDGQYTYKTRFQ
E1   PQWSYMHIAGQDAAEYLSPALVQFAQATNSYFRLDNKFRNPTVAPTHDVTTERSQRLQLRFVPVMQEDGQYTFKTRFQ
B3   PQWSYMHIAGQDASEYLSPGLVQFAQATESYFNIGNKFRNPTVAPTHDVTTERSQRLQLRFVPVDREDTQYSYKTRFQ
M1   PQWAFMHIAGQDASQYLSPGVVQFAAATDTYFSLGNKFRNPTVAPTHDVTTDRSQRLTLRIVPVDREDSQYTYKTRFQ
F10  PRLQYFHIAGPGTREYLSEDLQQFISATGSYFDLKNKFRQTVVAPTRNVTTEKAQRLQIAFTPSKPTTRRRATRA-YN
F1   PRLQYFHIAGPGTREYLSEDLQQFISATGSYFDLKNKFRQTVVAPTRNVTTEKAQRLQIRFYPIQTDDTPNSYRVRYS
F9   PRLQYFHIAGPSTREYLSEDLQQFIAATGSYFELKNKFRQTVVAPTRNVTTEKAQRLQIRFYPTQTDDTPNSYRVRYS
O287 PQREFFHIAGRNAREYLSENLVQFISATQSFFNLGEKFRDPFVAPSTGVTTDRSQKLQLRIVPIQTEDNENFYKTRFT
B7   PQREFFHIAGRNAREYLSENLVQFISATQSFFNLGEKFRDPFVAPSSGVTTDRSQKLQLRIVPIQTEDNDNFYKTRFT
B8   PQREFFHIAGRNAREYLSENLVQFITATQSFFNLGEKFRDPFVAPTSGVTTDRSQKLQLRTVPIQVEDNENFYKARFT
B4   PQREFFHIAGRNAREYLSENLVQFITATQSFFNLGEKFRDPFVAPTSGVTTDRSQKLQLRIVPIQVEDNENFFKARFT
Rus  PQREFFHIAGRNAREYLSENLVQFITATQSFFNLGEKFRDPFVAPTSGVTTDRSQKLQLRIVPIQVEDNENFYKARFT
B5   PQREFFHIAGRNAREYLSENLVQFITATQSFFNLGEKFRDPFVAPTSGVTTDRSQKLQLRIVPIQVEDNENFYKARFI
B6   PQREFFHIAGRNAREYLSENLVQFITATQSFFNLGEKFRDPFVAPSSGVTTDRSQKLQLRIVPIQIDDNENFYKARFT
goat PQRDFFHIAGRNAREYLSENLVQFISATQSFFNLGEKFRDPFVAPSTGVTTDRSQKLQLRIVPIQTEDNENFYKTRFT
deer PQREFFHIAGRNAREYLSENLVQFITATQSFFNLGEKFRDPFVAPSSGVTTDRSQKLQLRIVPIQTEDNENFYKARFT
poss PQREFFHIAGRNAREYLSENLVQFISATQTFFNLGDKFRDPFVAPSSGVTTDRSQKLQLRIVPIQTEDNESFYKARFT
EDS  PQREFFHIAGRNAKEYLSENLVQFITATQNYFNLGDKFRDPYVAPTNGVTTDRSQKLHLRIVPIQTEDNDSFFKARFT
frog PKLDIFHIAGRDASEYLSENLIDFMSSTESYFSINKKFRETIVAPTKGVTSEQSQRLQVRIIPVQTQDAEQTYTARFT
THEV PKLDIFHIAGPDASEYLSENLVNFISSTESYFPINKKFRETIVAPTKGVTTEQSQKLQVKIVPTLTQDLENSFTARFT
PRD1 LKLEFANNNTAFAAVGANPLEAIYQGAGAADCEFEEISYTVYQSYLDQLPVGQNGYILPLIDLSTLYNLENSAQAGLT
   192                                                                             268

     86                                           131
H2   LAVGDNRVLDMASTYFDIRGVLDRGPTFKPYSGTAYNALAPKGAPN
H3   LAVGDNRVLDMASTFFDIRGVLDRGPSFKPYSGTAYNSLAPKGAPN
H4   LAVGDNRVLDMASTYFDIRGVLDRGPSFKPYSGTAYNSLAPKGAPN
H12  LAVGDNRVLDMASSYFDIRGVLDRGPSFKPYSGTAYNSLAPKGAPN
H40  LAVGDNRVLDMASTYFDIRGVLDRGPSFKPYSGTAYNSLAPKGAPN
H48  LAVGDNRVLDMASTYFDIRGVLDRGPSFKPYSGTAYNSLAPKGAPN
S21  LAVGDNRVLDMASTFFDIRGVLDRGPSFKPYSGSAYNSLAPKGAPN
S25  LAVGDNRVLDMASTYFDIRGVLDRGPSFKPYSGTAYNSLAPKGAPN
P3   LAVGDNRVLDMASTFFDIRGTLDRGPSFKPYSGTAYNIMAPKSAPN
C1   LAVGDNRVLDMASTYFDIRGTLDRGPSFKPYSGTAYNALAPRAGAN
E1   LSVGDNRVLDMGSTFFDIRGTLDRGPSFKPYSGTVYNSLAPKSAPN
B3   LAVGDNRVLDMASTYFDIRGTLDRGASFKPYSGTAYNSFAPKSAPN
M1   LSVGDNRVLDMASTYFDIRGRLDRGPSFKPYSGTAYNSLAPRAAPN
F10  INVGDGWVLDMGSTYFDIKGILDRGPSFKPYCGTAYNPLAPKESMF
F1   VNVGDSWVLDMGATYFDIKGVLDRGPSFKPYGGTAYNPLAPREAIF
F9   LNVGDSWVLDMGATYFDIKGVLDRGPSFKPYGGTAYNPLAPREAFF
O287 LNVGDNRVADLGSAYFDIEGVIDRGPTFKPYGGTAYNPLAPKSAFPNAAFMDTDEA-TTIYIAQLPNAYNAQN--KGV
B7   LNVGDNRVADLGSAYFDIEGVIDRGPSFKPYGGTAYNPLAPKSALPNTAFKDNDE--T-IYVAQLPQIYTAND--KGV
B8   LNVGDNRIADLGSAFFDIEGFLDRGQSFKPYGGTAYNPLAPKSAMPNMAFKSQEN--TT-YIAQLPQVYPAEN--KGI
B4   LNVGDNRIADLGSAFFDIEGFIDRGPSFKPYGGTAYNPLAPKSALPNMAFKNNEE--TT-YIAQLPQIYAAED--KGL
Rus  LNVGDNRIADLGSAFFDIEGFLDRGQSFKPYGGTAYNPLAPKSAMPNMAFKSQEN--TT-YIAQLPQVYPAEN--KGI
B5   LNVGDNRIADLGSAFFDIEGFVDRGSSFKPYGGTAYNPLAPKSAMPNMAFKSNEN--TT-YIAQLPQVYAAAD--KGV
B6   LNVGDNRVADLGSAYFDIEGFLDRGFSFKPYGGTAYNPLAPKSAMPNMAFQNNED--T-VYLAQLPQIYSANG--KGV
goat LNVGDNRVADLGSAYFDIEGVIDRGPTFKPYGGTAYNPLAPKSAFPNAAFMDNDEA-TTVYIAQLPNAYKAQN--KGV
deer LNVGDNRVADLGSAYFDIEGTLDRGPSFKPYGGTAYNPLAPKSAMPNITVTKGGD--T-IYIAQLPQTYAAND--KGV
poss LNIGDNRVADLGSAYFDIEGILDRGPSFKPYGGTAYNPLAPKSALPNTAVTEGT---TLKYYAQLPQLYSITD--NGV
EDS  LNVGDNRIVDMGSSYFDIQGILDRGPSFKPYRGTAYNPLAPTSGMPNMAYTHSNN--TT-YIGQLPQIYSVTD--KGV
frog IAVGEGRVLDMGSTYFDIRGIVDRGPSFKPYSGTAYNSLAPKSAVFNNVKTVSIGN-TSVLTAQAPHLYSINTGDNGC
THEV IAVGDGRVLDMGSTYFDIRGNIDRGPSFKPYGGTAYNPLAPRSAQFNNIKTVGG--KTY-LTAQATKFFSTS-G-NGC
PRD1 PNVDFVVQYANLYRYLSTIAVFDNGGSFNA--GTDINYLSQRTANFSDTRKLDP--KTWA--AQ-TRRRIATDFPKGV
   269                                                                             339
```

Fig. 5. Part of the alignment of adenovirus hexon amino acid sequences with the P3 sequence of phage PRD1. The conserved amino acids are printed in *bold*. *Underlined* residues are conserved at least in several types of all the four genera

viruses and tectiviruses have common origin, the fish adenovirus could be expected to be the closest known relative of the phages. We have tried to demonstrate sequence homology between the pTP of fish adenovirus and the terminal protein of PRD1, but without success. At the moment, no hexon gene sequence is available from the fish adenovirus for comparison. Based on the electron microscopic picture of the virions, and on the fact that hexon-specific PCR primers designed for the general or selective detection of mast-, avi-, at- or siadenoviruses, failed to amplify any product from the genome of the sturgeon adenovirus, the existence of a significantly different hexon protein in fish adenoviruses cannot be excluded. White sturgeon belongs to Chondrostei, which is an ancient group of fishes, thus it is supposed to have even more ancient type of adenoviruses than bony fishes (Teleostei).

9 Conclusions

Perhaps the title was too promising and the reader might feel a dash of disappointment, which is quite right, as the evolutionary pathway of adenoviruses is far from being fully explored. In contrary, we are just at the beginning of the process and there are many open questions concerning the confirmation and direction of host switches, adaptation strategies, and rearrangement possibilities. Nevertheless, we have perhaps succeeded in presenting the often ignored variability within this virus family, and also in showing how many and how efficient tools are already available for the investigations focused on adenoviral evolution.

The actual picture, however, seems to be attractive, perhaps too beautiful. It appears that the *Adenoviridae* is the first virus family in which molecular phylogeny is able to reveal a progressive development seemingly compatible with the evolutionary development of the full range of vertebrate hosts. Comprehensive phylogenetic analysis of the family *Herpesviridae* resulted in much more cautious and prudent conclusions, which concern mainly the present relationships between the members of this huge virus family (McGeoch et al. 2000), and no clear hypothesis inferring the direction of development could be elaborated for every subfamily.

There are many things yet to be examined and confirmed. If the common origin with tectiviruses is true, are we going to discover an intermediate type of virus in the invertebrates? Have adenoviruses, because of the complexity of their genome organization, an evolution pace which is in better harmony with that of the hosts? We hope that the presentation of the above data will increase awareness about the incredible heterogeneity of adenoviruses. This might stimulate further work on other subjects than HAdV-2 or 5, which would ultimately contribute to the recognition of new mechanisms used by the less sophisticated adenoviruses in their replication strategies. With the comparative analyses of distantly related viruses, the functional and structural studies will also be better fuelled. For confirmation of the provisional adenovirus species classification, many different strains are being examined, and these studies will contribute further to the clarification of the unanswered questions.

Acknowledgements. Part of this work was supported by grants (No T030073 and T034461) from the Hungarian Scientific Research Fund OTKA. The enthusiastic work of our young team was indispensable in the compilation and presentation of the data. The help of K. Ursu and Á. Dán in cloning and sequencing of ruminant adenoviruses, as well as in preparation of the line drawings is highly appreciated. We also wish to thank S.L. Farkas and P. Élö for their work on the snake and fish adenoviruses. Special thanks are due to W. Ahne, S. LaPatra and D. Thomson for generously approving the presentation of unpublished sequence results from our collaborative work on the corn snake, white sturgeon and brushtail possum adenovirus isolates, respectively.

Note added at the proof stage:
The ICTV approved the two new genera in Paris, June 2002, thus *Atadenovirus* and *Siadenovirus* are now official taxons.

References

Ahne W, Essbauer S (2000) Viruses of lower vertebrates. In: Brocchi E, Lavazza A (eds) Veterinary Virology in the New Millennium. Proceedings of the 5th International Congress of the European Society for Veterinary Virology. August 27–30th, Brescia, Italy, pp 38–61

Bailey A, Mautner V (1994) Phylogenetic relationships among adenovirus serotypes. Virology 205:438–452

Bamford DH, Ackermann H-W (2000) Family *Tectiviridae*. In: van Regenmortel MHV, Fauquet CM, Bishop DHL, Carstens EB, Estes MK, Lemon SM, Maniloff J, Mayo MA, McGeoch DJ, Pringle CR, Wickner RB (eds) Virus Taxonomy. Classification and Nomenclature of Viruses. Seventh Report of the International Committee on Taxonomy of Viruses. Academic Press, San Diego, pp 111–116

Bamford DH, Caldentey J, Bamford JKH (1995) Bacteriophage PRD1: a broad host range dsDNA tectivirus with an internal membrane. Adv Virus Res 45: 281–319

Barbezange C, Benkö M, Dán Á, Harrach B (2000): DNA sequencing and phylogenetic analysis of the protease gene of ovine adenovirus 3 suggest that adenoviruses of sheep belong to two different genera. Virus Res 66:79–85

Bartha A (1969) Proposal for subgrouping of bovine adenoviruses. Acta Vet Hung 19: 319–321

Belnap DM, Steven AC (2000) Déjá vu all over again: the similar structures of bacteriophage PRD1 and adenovirus. Trends Microbiol 8:91–93

Benkö M, Bartha A, Wadell G (1988) DNA restriction enzyme analysis of bovine adenoviruses. Intervirology 29:346–350

Benkö M, Dán Á, Bánrévi A, Harrach B (2001) Analysis of the complete (DNA) genome sequence of bovine adenovirus type 4 confirms the (genome organization) characteristic features of the members of the (proposed) new genus *Atadenovirus.* GenBank Accession number AF036092

Benkö M, Élö P, Ursu K, Ahne W, LaPatra SE, Thomson D, Harrach B (2002) First molecular evidence for the existence of distinct fish and snake adenoviruses. J Virol 76:10056–10059

Benkö M, Harrach B (1994) Identification of the proteinase gene of bovine adenovirus type 4. Acta Microbiol Hung 41:323

Benkö M, Harrach B (1998) A proposal for a new (third) genus within the *Adenoviridae* family. Arch Virol 143:829–837

Benkö M, Harrach B, D'Halluin JC (1990) Molecular cloning and physical mapping of the DNA of bovine adenovirus serotype 4: study of the DNA homology among bovine, human and porcine adenoviruses. J Gen Virol 71:465–469

Benkö M, Harrach B, Russell WC (2000) Family *Adenoviridae.* In: van Regenmortel MHV, Fauquet CM, Bishop DHL, Carstens EB, Estes MK, Lemon SM, Maniloff J, Mayo MA, McGeoch DJ, Pringle CR, Wickner RB (eds) Virus Taxonomy. Classification and Nomenclature of Viruses. Seventh Report of the International Committee on Taxonomy of Viruses. Academic Press, San Diego, pp 227–238

Benson SD, Bamford JK, Bamford DH, Burnett RM (1999) Viral evolution revealed by bacteriophage PRD1 and human adenovirus coat protein structures. Cell 98: 825–833

Bloch B, Mellergaard S, Nielsen E (1986) Adenovirus-like particles associated with epithelial hyperplasias in dab, *Limanda limanda* (L.). J Fish Dis 9:281–285

Boyle DB, Pye AD, Kocherhans R, Adair BM, Vrati S, Both GW (1994) Characterisation of Australian ovine adenovirus isolates. Vet Microbiol 41:281–291

Both GW (2002) Atadenovirus. Adenoviridae. In: Tidona CA, Darai G (eds) The Springer index of viruses. Springer-Verlag, Berlin, Heidelberg, New York, pp 2–8

Burnett RM (1997) The structure of adenovirus. In: Chiu W, Burnett RM, Garcea RL (eds) Structural biology of viruses. Oxford University Press, Oxford, pp 209–238

Chiocca S, Baker A, Cotten M (1997) Identification of a novel apoptotic protein, GAM-1, encoded by the CELO adenovirus. J Virol 71:3168–3177

Chiocca S, Kurzbauer R, Schaffner G, Baker A, Mautner V, Cotten M (1996) The complete DNA sequence and genomic organization of the avian adenovirus CELO. J Virol 70:2939–2949

Cunningham AA, Langton TE, Bennett PM, Lewin JF, Drury SE, Gough RE, Macgregor SK (1996) Pathological and microbiological findings from incidents of unusual mortality of the common frog (*Rana temporaria*). Philos Trans R Soc Lond B Biol Sci 351:1539–1557

Dán Á, Élő P, Harrach B, Zádori Z, Benkö M (2001) Four new inverted terminal repeat sequences from bovine adenoviruses reveal striking differences in the length and content of the ITRs. Virus Genes 22:175–179

Dán Á, Ruzsics Zs, Russell WC, Benkö M, Harrach B (1998) Analysis of the hexon gene sequence of bovine adenovirus type 4 provides further support for a new adenovirus genus (*Atadenovirus*). J Gen Virol 79:1453–1460

Davison AJ, Harrach B (2002) *Siadenovirus*. Adenoviridae. In: Tidona CA, Darai G (eds) The Springer index of viruses. Springer-Verlag, Berlin, Heidelberg, New York, pp 29–33

Davison AJ, Telford EAR, Watson MS, McBride K, Mautner V (1993) The DNA sequence of adenovirus type 40. J Mol Biol 234:1308–1316

Davison AJ, Wright KM, Harrach B (2000) DNA sequence of frog adenovirus. J Gen Virol 81:2431–2439

Esford LE, Haj-Ahmad Y (1994) Sequence analysis of the putative E3 region of bovine adenovirus type 2. Intervirology 37:277–286

Essbauer S, Ahne W (2001) Viruses of lower vertebrates. J Vet Med B Infect Dis Vet Public Health 48:403–475

Evans PS, Benkö M, Harrach B, Letchworth GJ (1998) Sequence, transcriptional analysis and deletion of the bovine adenovirus type 1 E3 region. Virology 244: 173–185

Farkas SL, Benkö M, Élö P, Ursu K, Dán Á, Ahne W, Harrach B (2002) Genomic and phylogenetic analyses of an adenovirus isolated from a corn snake (*Elaphe guttata*) imply common origin with the members of the proposed new genus *Atad Adenovirus*. J Gen Virol 83:2403–2410

Felsenstein J (1981) Evolutionary trees from DNA sequences: a maximum likelihood approach. J Mol Evol 17:368–376

Felsenstein J (1989) PHYLIP – Phylogeny inference package (version 3.2). Cladistics 5:164–166

Gelderblom H, Maichle-Lauppe I (1982) The fibers of fowl adenoviruses. Arch Virol 72:289–298

Harrach B (2000) Reptile adenoviruses in cattle? Acta Vet Hung 48:485–490

Harrach B (2002): *Aviadenovirus*. Adenoviridae. In: Tidona CA, Darai G (eds) The Springer index of viruses. Springer-Verlag, Berlin, Heidelberg, New York, pp 9–18

Harrach B, Benkö M (1998) Phylogenetic analysis of adenovirus sequences; proof of necessity of establishing a third genus in the *Adenoviridae* family. In: Wold WSM (ed) Adenovirus methods and protocols (Methods in molecular medicine, Vol. 21) Humana Press, Totowa, NJ, pp 309–339

Harrach B, Meehan BM, Benkö M, Adair BM, Todd D (1997) Close phylogenetic relationship between egg drop syndrome virus, bovine adenovirus serotype 7, and ovine adenovirus strain 287. Virology 229:302–306

Hedges SB, Poling LL (1999) A molecular phylogeny of reptiles. Science 283:998–1001

Hess M (2000) Detection and differentiation of avian adenoviruses: a review. Avian Pathol 29:195–206

Hess M, Blöcker H, Brandt P (1997) The complete nucleotide sequence of the egg drop syndrome virus: an intermediate between mastadenoviruses and aviadenoviruses. Virology 238:145–156

IDAMAKANTI N, REDDY PS, BABIUK LA, TIKOO SK (1999) Transcription mapping and characterization of 284R and 121R proteins produced from early region 3 of bovine adenovirus type 3. Virology 256:351–359
JACOBSON ER, GARDINER CH (1990) Adeno-like virus in esophageal and tracheal mucosa of a Jackson's chameleon (*Chamaeleo jacksoni*). Vet Pathol 27:210–212
JACOBSON ER, GARDINER CH, FOGGIN CM (1984) Adenovirus-like infection in two Nile crocodiles. JAVMA 185:1421–1422
JACOBSON ER, KOPIT W, KENNEDY FA, FUNK RS (1996) Coinfection of a bearded dragon (*Pogona vitticeps*) with adenovirus- and dependovirus-like viruses. Vet Pathol 33:343–346
JONES NC, SHENK T (1979) Isolation of adenovirus type 5 host range deletion mutants defective for transformation of rat embryo cells. Cell 17:683–689
JUCKER MT, MCQUISTON JR, VAN DEN HURK JV, BOYLE SM, PIERSON FW (1996) Characterization of the haemorrhagic enteritis virus genome and the sequence of the putative penton base and core protein genes. J Gen Virol 77:469–479
JUHASZ A, AHNE W (1992) Physicochemical properties and cytopathogenicity of an adenovirus-like agent isolated from corn snake (*Elaphe guttata*). Arch Virol 130:429–439
KHATRI A, BOTH GW (1998) Identification of transcripts and promoter regions of ovine adenovirus OAV287. Virology 245:128–141
KHATRI A, XU ZZ, BOTH GW (1997) Gene expression by atypical ovine adenovirus vectors during abortive infection of human and animal cells in vitro. Virology 239:226–237
KIDD AH, CHROBOCZEK J, CUSACK S, RUIGROK RW (1993) Adenovirus type 40 virions contain two distinct fibers. Virology 192:73–84
KIDD AH, GARWICZ D, ÖBERG M (1995) Human and simian adenoviruses: phylogenetic inferences from analysis of VA RNA genes. Virology 207:32–45
KUMAR S, HEDGES SB (1998) A molecular timescale for vertebrate evolution. Nature 392:917–920
LAPATRA SE, PARKER BL, GROFF JM, ENGELKING HM, KAUFMAN J, MUNN RJ (1998) Epidemiology of viral infections in white sturgeon from the Pacific Northwest. In: Proceedings of the 49th Annual Northwest Fish Culture Conference. Boise, Idaho, pp 27–32
LEHMKUHL HD, CUTLIP RC (1999) A new goat adenovirus isolate proposed as the prototype strain for goat adenovirus serotype 1. Arch Virol 144:1611–1618
LEHMKUHL HD, HOBBS LA, WOODS LW (2001) Characterization of a new adenovirus isolated from black-tailed deer in California. Arch Virol 146:1687–1196
LÖSER P, CICHON G, JENNINGS GS, BOTH GW, HOFMANN C (1999) Ovine adenovirus vectors promote efficient gene delivery in vivo. Gene Ther Mol Biol 4:33–43
LUTZ P, ROSA-CALATRAVA M, KEDINGER C (1997) The product of the adenovirus intermediate gene IX is a transcriptional activator. J Virol 71:5102–5109
MA Y, MATHEWS MB (1996) Structure, function, and evolution of adenovirus-associated RNA: a phylogenetic approach. J Virol 70:5083–5099
MATIZ K, URSU K, HARRACH B, ZÁDORI Z, BENKÖ M (1998) Sequencing and phylogenetic analysis of the protease gene, and genetic mapping of bovine adenovirus type 10 define its relatedness to other bovine adenoviruses. Virus Res 55:29–35
MATHEWS MB, SHENK T (1991) Adenovirus virus-associated RNA and translation control. J Virol 65:5657–5662

Matthews DA, Russell WC (1998a) Adenovirus core protein V is delivered by the invading virus to the nucleus of infected cell and later in infection is associated with nucleoli. J Gen Virol 79:1671–1675
Matthews DA, Russell WC (1998b) Adenovirus core protein V interacts with p32 – a protein which is associated with both the mitochondria and the nucleus. J Gen Virol 79:1671–1675
McFerran JB, Smyth JA (2000) Avian adenoviruses. Rev Sci Tech 19:589–601
McGeoch DJ, Davison AJ (1995) Origins of DNA viruses. In: Gibbs AJ, Calisher CH, Garcia-Arenal F (eds) Molecular basis of virus evolution. University Press, Cambridge, pp 67–75
McGeoch DJ, Dolan A, Ralph AC (2000) Toward a comprehensive phylogeny for mammalian and avian herpesviruses. J Virol 74:10401–10406
Meissner JD, Hirsch GN, LaRue EA, Fulcher RA, Spindler KR (2000) Completion of the DNA sequence of mouse adenovirus type 1: sequence of E2B, L1, and L2 (18–51 map units). GenBank Accession number NC_000942
Miyazaki T, Asai Y, Kobayashi T, Miyata M (2000) Lympholeukemia in madai *Pagrus major* in Japan. Dis Aquat Org 40:147–155
Nagy M, Nagy É, Tuboly T (2001) The complete nucleotide sequence of porcine adenovirus serotype 5. J Gen Virol 82:525–529
Nagy M, Nagy É, Tuboly T (2002) Sequence analysis of porcine adenovirus serotype 5 fiber gene: evidence for recombination. Virus Genes 24:177–181
Ogawa M, Ahne W, Essbauer S (1992) Reptilian viruses: adenovirus-like agent isolated from royal python (*Python regius*). J Vet Med B 39:732–736
Ojkic D, Nagy É (2000) The complete nucleotide sequence of fowl adenovirus type 8. J Gen Virol 81:1833–1837
Payet V, Arnauld C, Picault JP, Jestin A, Langlois P (1998) Transcriptional organization of the avian adenovirus CELO. J Virol. 72:9278–9285
Perkins LE, Campagnoli RP, Harmon BG, Gregory CR, Steffens WL, Latimer K, Clubb S, Crane M (2001) Detection and confirmation of reptilian adenovirus infection by *in situ* hybridization. J Vet Diagn Invest 13:365–368
Pitcovski J, Mualem M, Rei-Koren Z, Krispel S, Shmueli E, Peretz Y, Gutter B, Gallili GE, Michael A, Goldberg D (1998) The complete DNA sequence and genome organization of the avian adenovirus, hemorrhagic enteritis virus. Virology 249:307–315
Ramis A, Fernandez-Bellon H, Majo N, Martinez-Silvestre A, Latimer K, Campagnoli R, Harmon BG, Gregory CR, Steffens WL, Clubb S, Crane M (2000) Adenovirus hepatitis in a boa constrictor (*Boa constrictor*). J Vet Diagn Invest 12:573–576
Raviprakash KS, Grunhaus A, el Kholy MA, Horwitz MS (1989) The mouse adenovirus type 1 contains an unusual E3 region. J Virol 63:5455–5458
Reddy PS, Idamakanti N, Song JY, Lee JB, Hyun BH, Park JH, Cha SH, Bae YT, Tikoo SH, Babiuk LA (1998a) Nucleotide sequence and transcription map of porcine adenovirus type 3. Virology 251:414–426
Reddy PS, Idamakanti N, Zakhartchouk AN, Baxi MK, Lee JB, Pyne C, Babiuk LA, Tikoo SH (1998b) Nucleotide sequence, genome organization and transcription map of bovine adenovirus type 3. J Gen Virol 72:1394–1402

Rosa-Calatrava M, Grave L, Puvion-Dutilleul F, Chatton B, Kedinger C (2001) Functional analysis of adenovirus protein IX identifies domains involved in capsid stability, transcriptional activity, and nuclear reorganization. J Virol 75:7131–7141

Russell WC (2000) Update on adenovirus and its vectors. J Gen Virol 81:2573–2604

Russell WC, Benkö M (1999) Adenoviruses (*Adenoviridae*): Animal viruses. In: Webster RG, Granoff A (eds): Encyclopedia of virology. Academic Press, London, pp 14–21

Salas M (1983) A new mechanism for the initiation of replication of phi29 and adenovirus DNA: priming by the terminal protein. In: Doerfler W (ed) The molecular biology of adenoviruses. Current Topics in Microbiology and Immunology 109. Springer-Verlag, Berlin pp 89–106

Sheppard M, Werner W, Johnson MA (1998a) DNA sequence of the fowl adenovirus serotype 10 short fiber gene. DNA Seq 86:391–396

Sheppard M, Werner W, McCoy RJ, Johnson MA (1998b) The major late promoter and bipartite leader sequence of fowl adenovirus. Arch Virol 143:537–548

Shinagawa M, Iida Y, Matsuda A, Tsukiyama T, Sato G (1987) Phylogenetic relationships between adenoviruses as inferred from nucleotide sequences of inverted terminal repeats. Gene 55:85–93

Song B, Hu S-L, Darai G, Spindler KR, Young CSH (1996) Conservation of DNA sequence in the predicted major late promoter regions of selected Mastadenoviruses. Virology 220:390–401

Thomson D, Meers J, Harrach B (2002) Molecular confirmation of an adenovirus in brushtail possums (*Trichosurus vulpecula*). Virus Res 83:189–195

Ursu K, Zádori Z, Benkö M, Matiz K, Dán Á, Harrach B (1997) Mapping of naturally occurring deletions in bovine adenovirus isolates. Acta Microbiol Immunol Hung 44:33

Vrati S, Boyle D, Kocherhans R, Both GW (1995) Sequence of ovine adenovirus: homologs for 100K hexon assembly, 33K, pVIII, and fiber genes: early region E3 is not in the expected location. Virology 209:400–408

Vrati S, Brookes DE, Boyle DB, Both GW (1996a) Nucleotide sequence of ovine adenovirus tripartite leader sequence and homologues of the IVa2, DNA polymerase and terminal proteins. Gene 177:35–41

Vrati S, Brookes DE, Strike P, Khatri A, Boyle DB, Both GW (1996b) Unique genome arrangement of an ovine adenovirus: identification of new proteins and proteinase cleavage sites. Virology 220:186–199

Wadell G, Hammarskjöld M-L, Winberg G, Varsányi TM, Sundell G (1980) Genetic variability of adenoviruses. Ann N Y Acad Sci 354:16–42

Woods LW, Swift PK, Barr BC, Horzinek RC, Nordhausen RW, Stillian MH, Patton JF, Oliver MN, Jones KR, MacLachlan NJ (1996) Systemic adenovirus infection associated with high mortality in mule deer (*Odocoileus hemionus*) in California. Vet Pathol 33:125–132

Xu ZZ, Hyatt A, Boyle DB, Both GW (1997) Construction of ovine adenovirus recombinants by gene insertion or deletion of related terminal region sequences. Virology 230:62–71

Chapter 2

Novel Partner Proteins of Adenovirus Penton

J. Chroboczek, E. Gout, A.-L. Favier, R. Galinier

Abstract. Each of the 12 vertices of the adenovirus virion is made of penton, the complex of two oligomeric proteins: a pentameric penton base anchored in the capsid and an antenna-like trimeric fiber extending outwards. Adenovirus penton plays an essential role in the infection of host cells because it is indispensable for virus attachment and internalization. The initial interactions of penton with the primary and secondary receptors are well described. In contrast with that, the role of the penton components downstream of the initial cell contact is not known. This work shows for the first time that two adenovirus structural proteins, fiber and base, are able to interact intimately with different classes of cellular targets. In the case of penton base, a protein responsible for virus internalization, the partners include three ubiquitin-protein ligases that are involved in protein turnover, cell cycle control and endocytosis. Another base protein partner, BAG3, is involved in controlling Hsc70 chaperone

J. Chroboczek, E. Gout, A.-L. Favier, R. Galinier
Institut de Biologie Structurale, 41 rue Jules Horowitz,
38027 Grenoble, France
e-mail: wisia@ibsprive.ibs.fr

activity. Virus attachment protein, fiber, interacts with many different partners, some of them involved in signal transduction and cell growth. Further work will illustrate the implications of these interactions for both the viral and cellular life cycles.

1 Introduction

Each of the 12 vertices of the adenovirus (Ad) virion is made of penton, the complex of two oligomeric proteins: a pentameric penton base anchored in the capsid and an antenna-like trimeric fiber extending outwards (Fig. 1). Penton plays an essential role in the infection of host cells. Ad infection begins with the attachment of a viral particle to the cell surface, by interaction of the distal globular region of the fiber (C-terminal head domain) with a primary cell receptor (Philipson et al. 1968; Devaux et al. 1987; Louis et al. 1994; Henry et al. 1994). The penton base protein is involved in virus internalization, through direct interaction of an ArgGlyAsp (RGD) sequence, probably localized in the flexible loop region of the penton base, with cell integrins (Wickham et al. 1993; Schoehn et al. 1966; Stewart et al. 1997).

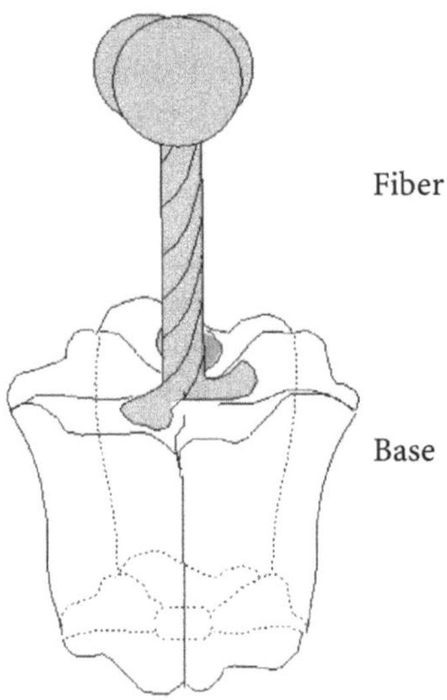

Fig. 1. Schematic view of adenovirus penton

2
Primary and Secondary Receptors

A closer look at the physical interactions that are necessary for virus attachment to the cell and its internalization, suggests that they are probably not general for all human adenoviruses and most probably not sufficient for the virus infectious cycle. As we know it now, the majority of human adenoviruses use the coxsackie and adenovirus receptor (CAR) protein as a primary receptor (Stevenson et al. 1995; Bergelson et al. 1997; Roelvink et al. 1998). However, Ad3 fiber (subgroup B), Ad40 and Ad41 short fibers (subgroup F) as well as Ad37 fiber (subgroup D) do not attach to CAR (Roelvink et al. 1998; Arnberg et al. 2000, 2002). Moreover, the short fibers of Ad40 and Ad41 are nearly as unrelated to their long fibers as they are to fibers of other subgroups (Chroboczek et al. 1995), which substantiates the data on the lack of interaction of these short fibers with CAR. Differences in primary receptors is reflected in virus tropism - CAR binding serotypes are predominantly respiratory viruses, whereas Ad3 is respiratory and ocular, Ad40 and Ad41 are enteric and Ad37 is ocular. Primary receptors for Ad3, Ad40 and Ad41 are unknown. Ad37 can use sialic acid as a cellular receptor (Arnberg et al. 2000), but it has been recently shown that an unidentified 50-kDa membrane protein of permissive Chang cells mediates sialic acid-independent binding of Ad37 (Wu et al. 2001). It has been suggested also that MHC-I alpha 2 chain might be a primary receptor for Ad5 (Hong et al. 1997); however, more recent data do not confirm it (McDonald et al. 1999; Davison et al. 1999).

It has been shown that the integrins $\alpha v\beta 3$ and $\alpha v\beta 5$ are the secondary receptors for serotypes 2, 3, 4 and 12 (Wickham et al. 1993; Bai et al. 1993; Mathias et al. 1994). However, Ad40 and Ad41 lack the $\alpha v\beta 3$ and $\alpha v\beta 5$ ligand (RGD motif) in their bases; instead the Ad40 base carries RGAD, and Ad41 carries IGDD. Furthermore, the uptake of Ad41 in cells which are efficiently producing commonly used serotypes was shown to be delayed, suggesting that Ad41 entry might be independent of αv integrins (Albinsson and Kidd 1999). More importantly, it seems unlikely that a process as complex as virus internalization would depend solely on the physical interaction of three contiguous residues of one viral protein with cell endocytotic apparatus.

Apart from interaction with primary and secondary receptors, penton complex is probably involved in many other physical interactions within the host cell. The Ad penton base protein is implicated in virus escape from the endosomes. It was observed that upon Ad infection, foreign molecules co-endocytosed with virus were released from the endosomes (Fitzgerald et al. 1993; Yoshimura et al. 1993). It has been suggested that the penton base undergoes an acid-induced conformational change in endosomes, resulting in the exposure of hydrophobic domains and leading to disruption of the endosomal membrane (Wohlfart et al. 1988). In addition, anti-base antibody inhibited Ad-dependent release of choline from the plasma vesicles (Seth, 1994). Furthermore, in the late phase of a viral cycle both penton base and fiber proteins are synthesized and oligomerized in the cytoplasm. Most probably, penton complex is formed in the cytoplasm once the two proteins are oligomerized. This assumption is based on the observation that during co-expression of fiber and base proteins in the baculovirus system, no free fiber could be observed – all newly synthesized fiber was immediately taken up in the complex with the penton base (Fender et al. 1997). Moreover, Ad pentons are easily isolated from the cytoplasmic extracts of infected cells. At the late phase of infectious cycle Ad proteins travel to the nucleus, where virion assembly takes place. Subsequently, a mature virus is released from the host cells. Collectively, these data suggest that during their life cycle in the host cell penton proteins are involved in plethora of interactions that permit virus attachment, cell entry, release from endosomes as well as penton components synthesis, folding, oligomerization and nuclear transport and possibly also virus release.

Analysis of the primary sequences of penton components permits the prediction of certain interactions. For example, the N-terminal tail of the fiber protein contains the conserved NPXY sequence, found in some coated pit localization signals (Kibbey et al. 1998; Brown and Breton 2000). Importance of this ^{11}NPVY^{14} tetra-peptide was confirmed by studies on a transfecting peptide composed of 20 N-terminal amino acids of Ad3 fiber and prolonged by 20 lysine residues (Zhang et al. 1999). The peptide was able to travel from the culture medium toward the cell nucleus, due to an extremely efficient fiber protein nuclear localization signal ^{1}AKRARLSTSF^{10} (Hong and Engler, 1991). Removal of the ^{11}NPVY^{14} tetra-peptide from the transfecting peptide resulted in 30-fold diminished transfecting activity. The NPXY sequence is known to interact with the tyrosine-binding domains in partner proteins (interestingly, cytoplasmic tails of both β3 and β5 integrin chains contain the NPXY sequence). This suggests that fiber protein could interact directly with coated pit components as well as with the nuclear transport system.

Furthermore, sequence analysis of penton base protein shows that it contains six YXXΦ motifs conserved among human serotypes. YXXΦ (where Φ is a large hydrophobic amino acid) is another motif involved in endocytosis (LI et al. 2000) which is quite often used by viruses (DESCHAMBEAULT et al. 1999; TIRABASSI and ENQUIST 1999). It is also implicated in intracellular signaling (SUN et al. 1996; DELAHAYE et al. 1998). In addition, the N-terminal part of the base protein contains two of the conserved sequence -PPXY which is known to interact with WW domains found in a wide range of signaling proteins (ROTIN 1998). These sequence considerations suggest that the Ad penton base protein should be involved in multiple intracellular interactions.

Several lines of experimental data substantiate above suggestions. By expression of Ad3 base protein or base and fiber proteins we have generated subviral particles – dodecahedra. The dodecahedra are composed either of 12 bases or of 12 pentons (complex of penton base and fiber) packed with icosahedral symmetry (SCHOEHN et al. 1996; FENDER et al. 1997). Like the native virus, these particles enter cells efficiently via the endocytotic pathway and accumulate at the nuclear membrane (FENDER et al. 1997; Fig. 2) which suggests that during their passage toward the nucleus they interact with adenovirus cellular partners.

Both types of dodecahedra enter cells, with base dodecahedra entering less efficiently. This suggests that bases alone are able to attach to the plasma membrane, to interact with coated pits, to escape to cytoplasm and to attach to the nuclear membrane. Similarly, it has been shown in BOULANGER's laboratory that the recombinant Ad2 base is internalized by HeLa cells and beginning 20 min after application the amount of this pentameric base protein (~350 kDa) increased with time in the nucleus, with a concomitant decrease in the membrane fraction (HONG et al. 1999). In addition, the Ad7 penton base was shown to travel toward the cell nucleus, transducing plasmid DNA tethered to the base through a bivalent peptide that was able to attach simultaneously the base and DNA (BAL et al. 2000). It has been shown recently that a recombinant bacteriophage bearing Ad penton base in its capsid was able to internalize and transduce mammalian cells (DI GOVINE et al. 2001). Interestingly, incorporation into the phage capsid of a central base protein fragment of 107 amino acids (residues 286–393; the total polypeptide chain has 571 residues) was sufficient for transducing activity. Furthermore, Ad5 penton base fusion protein containing a poly-lysine tail able to bind DNA was shown to deliver plasmid DNA to cultured cells (MEDINA-KAUWE et al.

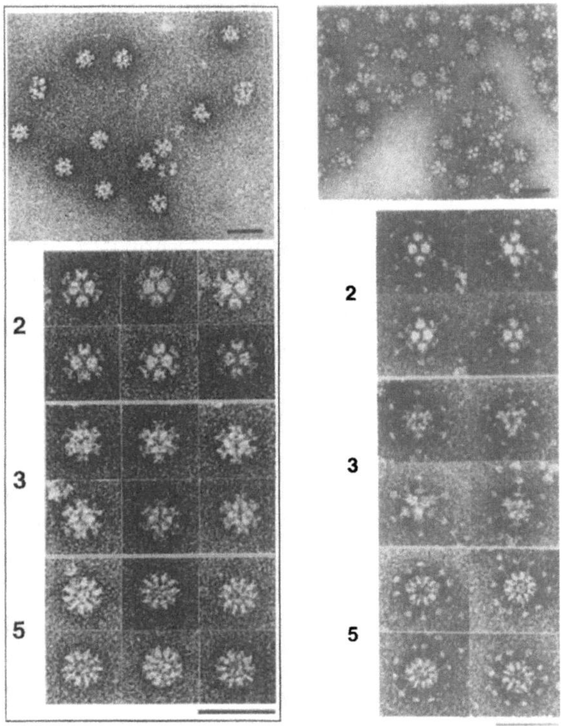

Fig. 2. Electron microscopy of dodecahedra (negative stain), without fiber (*left*) and with fiber (*right*). Axes of symmetry are given as 2, 3, 5

2001) All of these studies describe also entry inhibition with RGD peptides, showing the dependence of base protein entry on integrins. Nevertheless, they also point out possible base protein interactions with other cell components, downstream from the internalization event.

The identification of physical partners of penton components was attempted by panning a phage-display peptide library (HONG and BOULANGER 1995). Some of the results of this study suggested that both fiber and base could interact with phage bearing the hexapeptide with a sequence similar to that of the conserved motif found in human fibronectin. It is known that cells attach to fibronectin through $\alpha5\beta1$ integrin and data exist to suggest that integrin $\alpha5\beta1$ may act as an alternative secondary adenovirus receptor (DAVISON et al. 1997). However, results of panning suggest a direct interaction of penton with fibronectin, without $\alpha5\beta1$ as an intermediate. These results await a confirmation.

3
Novel Partner Proteins of Penton Base

In searching for the cell proteins that interact with penton base we used dodecahedra – adenovirus subviral particles composed of 12 bases (FENDER et al. 1997; Fig. 2) – to screen a human lung expression library for protein–protein interactions. This screen yielded five proteins (Fig. 3): the first three are ubiquitin-protein ligases, WWP3 belongs to MAGUK protein family (DIMITRATOS et al. 1999) and BAG-3 belongs to the family of BAG-like proteins (TAKAYAMA et al. 1999).

The common structural feature of these proteins is the presence of the WW (or WWP) domain (ROTIN 1998), one of several domains known to allow physical interactions between proteins. The WW domains contain about 40 amino acids characterized by a high proportion of hydrophobic and aromatic residues, two highly conserved tryptophans and one invariant proline. Their partners in interacting proteins are short poly-proline

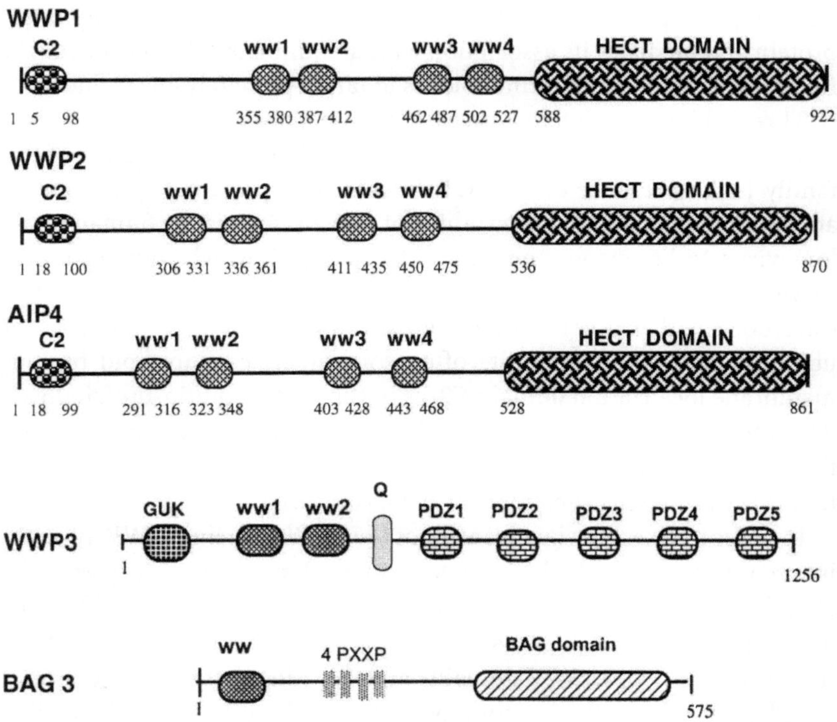

Fig. 3. Cellular partners of penton base protein

motifs that are different from those of the ligands of the SH3 domain. Three distinct poly-proline ligands of WW domains have been identified: xPPxY (Chen and Sudol 1995), xPPLP (Bedford et al. 1997) and proline, glycine, methionine-rich motif, PGM (Bedford et al. 1998). As stated above the base protein contains two xPPxY motifs in its N-terminal part, which are conserved in all sequenced serotypes.

3.1 Ubiquitin-Protein Ligases

Protein ubiquitination is involved in selective degradation of intracellular proteins in eukaryotic cells: conjugation of multiple ubiquitin residues in the majority of cases targets protein for degradation by proteasomes. In addition, protein ubiquitination is also involved in receptor-mediated endocytosis followed by lysosomal degradation. A cascade of three classes of enzymes performs the conjugation of ubiquitin, a small protein, to the target protein: ubiquitin-activating enzyme E1; ubiquitin-conjugating enzyme E2 and ubiquitin-protein ligase E3. E3 seems to recognize specifically target protein and mediates its association with ubiquitinated E2, permitting the final transfer of the ubiquitin moieties to target protein lysine residues.

Three ubiquitin-protein ligases found to interact with penton base belong to the HECT domain family (Fig. 3). A characteristic feature of this family is the presence of a HECT domain of approximately 350 amino acids, *h*omologous to the *C* *t*erminus of *E*6AP (E6AP is the human papillomavirus 16 E6 oncoprotein-associated protein), which contains a conserved catalytic cysteine residue indispensable to the formation of a thiol ester bond with a ubiquitin moiety (Staub et al. 1996). In addition, the ubiquitin-protein ligases quite often contain the calcium/lipid binding membrane localization domain C2, as in the first ubiquitin-protein ligase described, Nedd4 (Kumar et al. 1997). Domains of E3 enzymes provide a flexible means of linking a conserved ubiquitination function to potentially very numerous substrates in eukaryotic cells.

In order to assess the involvement of four WW domains of WWP1 in the interaction with the Ad base protein, these domains were expressed as glutathione S-transferase-tagged proteins in *Escherichia coli*, purified, and used separately in the blot overlay assay with different adenovirus structural proteins (Fig. 4). WW domains of WWP1 interact with base protein of Ad2, with dodecahedra composed of Ad3 base but not of Ad fiber protein. All four WW domains are able to interact with base proteins.

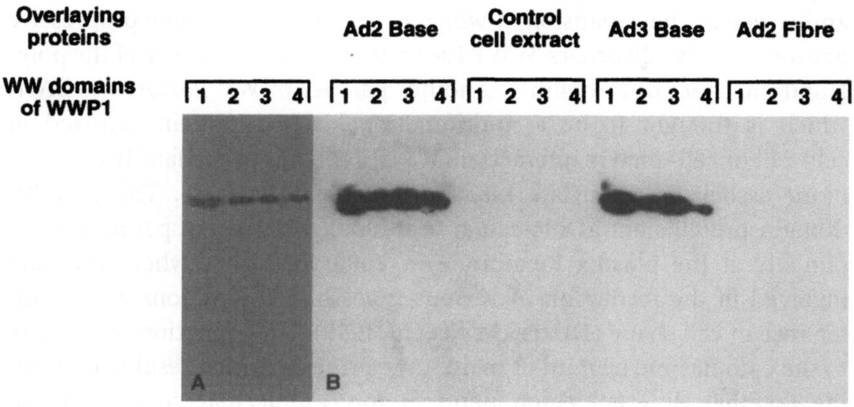

Fig. 4a, b. Interaction of Ad proteins with four WW domains of WWP1. The same amount of each of four WW domains of WWP1 was applied to each membrane. **a** Protein staining with Coomassie brilliant blue. **b** Blot overlay was performed with indicated ligands and revealed with the appropriate antibodies

Of the three kinds of motif that WW domains are known to recognize in partner proteins, the Ad base contains two xPPxY motifs. To determine whether both of these are required for the interaction with WW domains, we constructed appropriate mutants of the base protein. All three ubiquitin-protein ligases were able to interact in vitro with the wild-type adenovirus base protein. However, removal of the first xPPxY abolished the interaction. In contrast mutation of the essential proline (Pirozzi et al. 1997) in the second xPPxY motif did not alter the interaction. These results show that the first xPPxY motif in the Ad2 penton base protein is indispensable for physical contact of the penton base protein with ubiquitin ligases. These interactions were confirmed by immunoprecipitation of co-transfected HeLa cells (results not shown).

3.2 MAGUK-Like WWP3/BAP1 Protein

This protein was first described simultaneously by Wood et al. (1998) and Shiratsuchi et al. (1998). It was identified as an atrophin-1-binding protein and as a partner of brain-specific angiogenesis inhibitor 1 (BAI1). Atrophin-1, a brain protein of unknown function, contains a poly-glutamine repeat, the expansion of which is responsible for some progressive neurogenerative disorders. WWP3 seem to bind in the vicinity of the poly-glutamine tract of atrophin-1. Another partner of WWP3/BAP1 is BAI1, which is thought to be a multifunctional signal protein involved in cell–cell or cell–matrix interaction. WWP3 belongs to the family of membrane-associated guanylate kinase homologs (MAGUK). These multi-domain proteins act as molecular scaffolds for signaling pathway components at the plasma membrane of eukaryotic cells where they are involved in the formation of various types of cell junctions and maintenance of cell shape (Dimitratos et al. 1999). They function by binding to the cytoplasmic termini of membrane proteins, which results in bringing together different components of multi-molecular complexes and clustering them at the plasma membrane. All MAGUK homologs contain a GUK domain with a high degree of sequence similarity to guanylate kinase, an enzyme that converts GMP to GDT, using ATP as a phosphate donor. However, some of the MAGUK family members lost this enzymatic activity.

WWP3 protein, in addition to a GUK domain and two WW domains, contains a poly-glutamine stretch that is able to interact with other proteins (Waragai et al. 1999) and five GLGF-like repeats that are characteristic of the PDZ domain, another multi-interactional domain (Fanning and Anderson, 1999; see Fig. 3). WWP3 PDZ domains (Shiratsuchi et al. 1998) mediate an interaction of WWP3 with BAI1. Contrary to that, interaction of adenovirus base protein with WWP3 (Fig. 5) is mediated by WW domains, leaving the PDZ domain and the poly-Q stretch of WWP3 free to interact with other components of the eukaryotic cell.

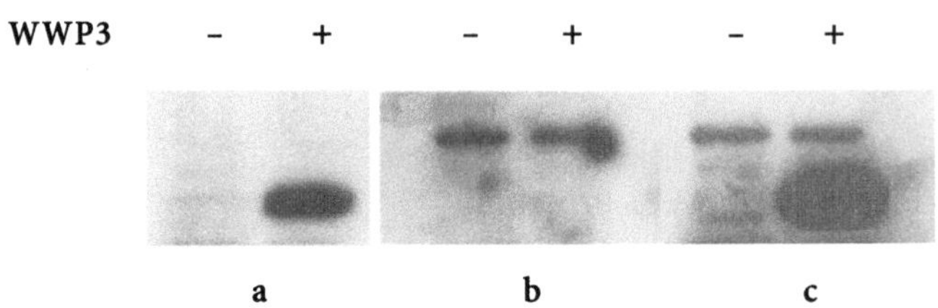

Fig. 5a–c. In vitro interaction of adenovirus base protein with WWP3. **a** Expression of WWP3 seen in transfected cells (+) as compared with non-transfected control cells (–). **b** a similar blot treated with anti-base antibody. **c** The same blot overlayed with fiber dodecahedron and revealed with anti-base antibody

3.3 BAG3 of the BAG Family of Hsp70/Hsc70 Regulators

The BAG3 protein belongs to the family of Hsp70/Hsc70 (molecular chaperones) regulators (TAKAYAMA et al. 1999). These proteins contain an approximately 45 amino acid BAG domain at their C termini that binds to the ATPase domain of Hsp70/Hsc70, resulting in the inhibition of Hsp70/Hsc70 chaperone activity. Ubiquitously expressed BAG3 is a protein that contains multiple protein interaction motifs and phosphorylation sites; it contains a WW domain at N terminus and four PXXP domains in the middle of its polypeptide chain (Fig. 3). BAG3 was also described as Bcl-2 interacting protein (with the name bis; LEE et al. 1999) and as the phospholipase Cγ subunit-interacting protein (with the name CAIR-1; DOONG et al. 2000). In the first case BAG3 was found to exert a weak anti-apoptotic activity, which was however synergistic with Bcl-2, in preventing Bax-induced and Fas-mediated apoptosis. The Bcl-2-binding domain on BAG3 was not precisely identified, but it seems that this interaction is not mediated by the WW domain (LEE et al. 1999). In the second study BAG3 was described as a partner of phospholipase C-γ able to form a tertiary complex with it and with Hsp70. It should be noted that phospholipase C-γ, which is involved in transmembrane signal transduction, has been linked to malignant transformation, invasive potential and angiogenesis. It has been shown that BAG3 interacts with the SH3 domain of phospholipase C-γ through BAG3 PXXP domains which allows simultaneous interaction of BAG3 BAG domain with Hsp70 (DOONG et al. 2000).

The interaction of the Ad penton base protein with BAG3 was found by library screening and was confirmed by ligand overlay analysis (Fig. 6). Interestingly, BAG3 devoid of the C-terminal Hsp70/Hsc70-binding BAG domain is still able to interact with base protein. This substantiates our supposition that BAG3 interacts with the PPXY of the base protein through its WW domain localized in the N-terminal part of the protein. It shows that this interaction leaves the C-terminal part of BAG3 with its BAG domain, free to interact with another partner, for example Hsp70.

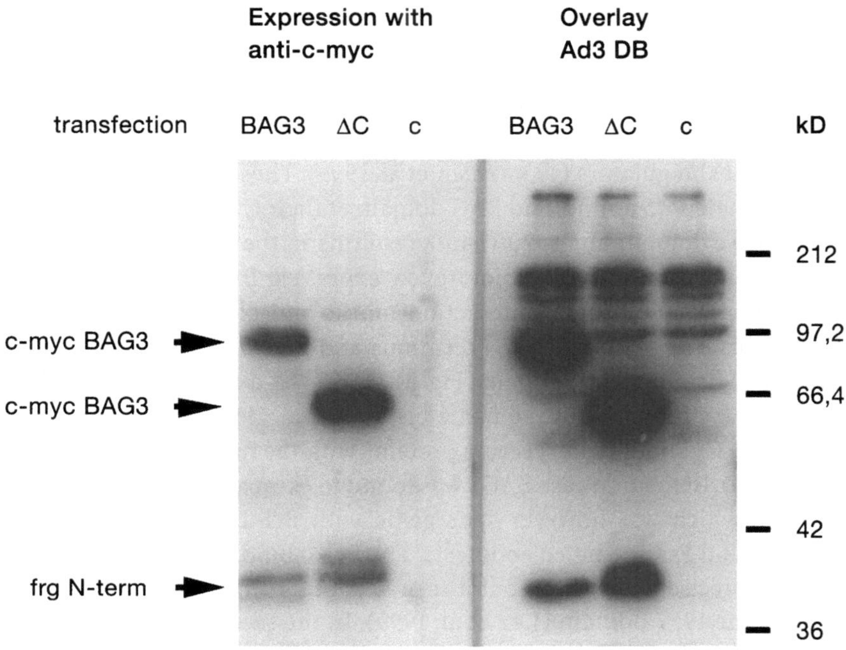

Fig. 6. Interaction of Ad3 penton base with BAG3. HeLa cells were transfected with c-myc tagged BAG3 and ΔCBAG3. The extract of the transfected cells was separated by polyacrylamide gel electrophoresis and the expression is shown by Western blot with anti-c-myc (*left panel*). A similar membrane was overlayed with Ad3 dodecahedron base and analyzed with anti-dodecahedron antibody (*right panel*). C-contra nontransfected cells

4
Novel Partner Proteins of Fiber

The enteric adenoviruses of serotype 40 and 41 (subgroup F) contain two fibers of different primary sequences and different length, each encoded by a separate gene (Kidd et al. 1993; Yeh et al. 1994). It was shown that the longer fiber, but not the shorter one, interacts with CAR, a protein that is a cell receptor for several human serotypes (Roelvink et al. 1998). However, the enteric tropism of Ad40 and Ad41 cannot be explained by the interaction of the long fiber with ubiquitously expressed CAR. It is possible that one of the reasons for the narrow and specific tropism of enteric serotypes is the presence in the gastrointestinal tract of a receptor or some other protein that interacts specifically with the short fiber.

To search for the partners of the short Ad41 fiber we used the two-hybrid approach. The expression library of the Ad41 permissive HEK kidney cells was used with the 'bait' containing the head domain of the short fiber. This approach produced a dozen positive clones which were confirmed as bona fide partners of Ad41 short fiber by gel overlay with the ligand, performed on yeast extracts expressing each of these proteins.

Ad41 short fiber partners can be divided roughly in three major classes (Table 1). The first one contains proteins involved in signal transduction which are localized in the cytoplasm but are capable of interaction with cell membranes (HAX-1, RhoGDI 1-α, C53). The second class contains the proteins containing in their sequences the transmembrane domains (APHC, etc.) and are thus capable of more intimate interactions with cell membranes. In the third class are the proteins able to interact with nucleic acids (CIRP, mBx, TAFII55, Meis3, ribosomal proteins). Several proteins on this list are only recently identified and their function is largely unknown.

With the exception of ribosomal proteins, all interactions have been confirmed by membrane overlay technique. As an example, interaction with RhoGDI 1a is shown in Fig. 7.

When using fibers of different serotypes for membrane overlay we found that the majority of partners of Ad41 short fiber are able to interact also with the Ad41 long fiber as well as with fibers of Ad3 and Ad2. At present we are studying the specificity of these interactions with the aim of identifying cellular partners specific for the Ad41 short fiber, most probably involved in the determination of enteric tropism of this serotype.

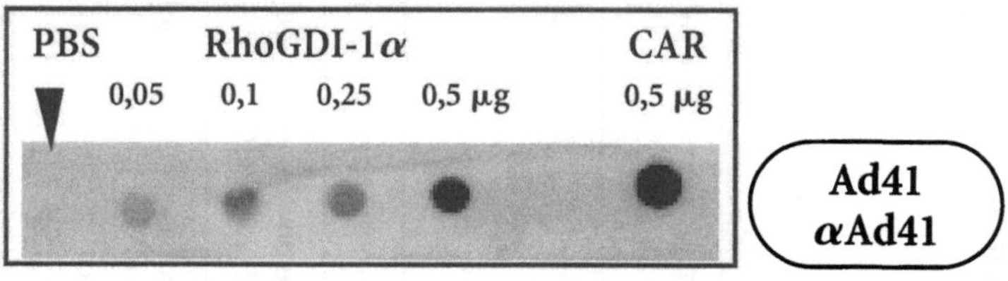

Fig. 7. Interaction of RhoGDI-1α with Ad41. Increasing amounts of purified RhGDI-1α were allowed to interact with 1 μg Ad41. Immobilized virus was detected by specific antibody and revealed by electrochemical luminescence. The negative control is phosphate-buffered saline and the positive control is CAR that is known to interact with the Ad41 long fiber

Table 1. Protein partners of the head domain of the Ad41 short fiber

Clone	Identification	Accession number	Amino acid fragment	Function	References
27	C53	gb|AF343090|[a]	1M-434D	Nck5a-binding protein of unknown function. Nck5a is the activating subunit of neuronal Cdc2-like kinase catalytic subunit Cdk5	Ching 2000
63a	RhoGDI-1α	sp|X69550|	144Y-204D	Inhibits Rho family proteins by sequestering them in cytoplasm	Keep 1997
133	Hax-1	sp|O00165|	179P-279R	Ubiquitous partner of several proteins, some involved in signal transduction. May play a role in motility and/or cell adhesion	Suzuki 1997
28	Unknown protein	gb|AF343342|[a]	1M-149G	Unknown	
44	Alkaline phyto-ceramidase (aPHC)	gb|AF214454|	1M-267H	Hydrolyzes phytoceramide. Localized in endoplasmic reticulum and Golgi	Mao 2001
94	ATP synthase subunit 9 precursor	gb|U09813|	1M-142M	Hydrophobic pore-forming subunit of ATP synthase, mediating proton translocation	Yan 1994
133	Hax-1	sp|O00165|	179P-279R	See above	Suzuki 1997

102	Unknown protein	gb|AF069072|	2711pb-3299pb	Unknown	
60	Unknown protein			Unknown	Tran 1999
53	Cold-inducible RNA-binding protein (CIRP)	sp|Q14011|	87D-172E	Involved in cold-inducible suppression of cell proliferation	Nishiyama 1997
58	mBx1 (rat)	embl|Y16641|	131G-177G	hnRNP component of A/B type. Involved in RNA metabolism	Plomaritoglou 2000
69	TAFII55	sp|P21675|	289K-349K	Subunit of transcription factor TFIID	Chiang 1995
127	MRG2 or Meis3, human	gb|AY091593|	1G-234L	Cofactor of Hox family transcription factors	Steelman 1997
57, 124	S14 with an N-terminal alternative exon RSALSHSLFPV-WSLETTCS	gb|M13934|	1M-138D	Component of 40S ribosome. Interacts with 18S rRNA	Rhoads 1986
74	L39	gb|D79205|	1M-52L	Component of 60 S ribosome. Involved in translational accuracy	Tsui 1996
118a	L41	sp|P28751|	1M-25K	Component of 60 S ribosome Stimulates protein kinase CKII by binding to its β subunit	Go 1998

[a] Sequence submitted by authors.

5
Conclusions

The initial interactions of the adenovirus penton with the primary and secondary receptors, both membrane proteins, are well described. In contrast with that, the role of the penton components downstream of the initial cell contact is not known. This work shows for the first time that two adenovirus structural proteins, fiber and base, are able to interact intimately with different classes of cellular targets. In the case of penton base, a protein responsible for virus internalization, the partners include three ubiquitin-protein ligases that are possibly involved in degradation and endocytosis. This is probably a part of the defensive mechanism of a cell attacked by virus. Another base protein partner, BAG3, is involved in modulating Bcl2 anti-apoptotic activity and in controlling Hsc70 chaperone activity. Some of the Ad41 short fiber partners are involved in signal transduction and cell growth. Further work will illustrate the implications of these interactions for both the viral and cellular life cycles.

Acknowledgements. We thank P. Boulanger (Faculté de Médecine, Lyon, France), M.-C. Dagher (CENG de Grenoble), R. Longnecker (Northwestern University Medical School, Chicago Ill, USA), G. Pirozzi (Cytogen Corp. Princeton NJ, USA), J. C. Reed and S. Takayama (Burnham Institute, La Jolla CA, USA) and J. Wood (John Hopkins University School of Medicine, Baltimore MD, USA) for generous gifts of plasmids, proteins and antibodies.

References

Arnberg N, Edlund K, Kidd AH, Wadell G (2000) Adenovirus type 37 uses sialic acid as a cellular receptor. J Virol 74:42–8

Arnberg N, Kidd AH, Edlund K, Olfat F, Wadell G (2000) Initial interactions of subgenus D adenoviruses with A549 cellular receptors: sialic acid versus alpha(v) integrins. J Virol 74:7691–3

Arnberg N, Pring-Akerblom P, Wadell G (2002) Adenovirus type 37 uses sialic acid as a cellular receptor on Chang C Cells. J Virol 76:8834–41

Albinsson B and AH Kidd (1999) Adenovirus type 41 lacks an RGD alpha(v) integrin binding motif on the penton base and undergoes delayed uptake in A549 cells. Virus. Res 64:125–36

Bal HP, Chroboczek J, Schoehn G, Ruigrok RW, Dewhurst S (2000) Adenovirus type 7 penton purification of soluble pentamers from Escherichia coli and development of an integrin-dependent gene delivery system. Eur J Biochem 267: 6074–81

BEDFORD MT, CHAN DC, LEDER P (1997) FBP WW domains and the AblSH3 domain bind to a specific class of proline rich ligand. EMBO J 9:2376–2383

BEDFORD MT, REED R and LEDER P (1998) WW domain-mediated interactions reveal a spliceosome-associated protein that binds a third class of proline rich motif: The proline, glycine a nd methionine rich motif. PNAS USA 95:10602-10607

BERGELSON JM, CUNNINGHAM JA, DROGUETT G, KURT-JONES EA, KRITHIVAS A, HONG JS, HORWITZ MS, CROWELL RL, FINBERG RW (1997) Isolation of a common receptor for Coxsackie B viruses and adenoviruses 2 and 5. Science 275: 1320–3

BROWN D, BRETON S (2000) Sorting proteins to their target membranes. Kidney Int 57:816–24

CHEN HI and SUDOL M (2000) The WW domain of Yes-associated protein binds a proline rich ligand that differs from the consensus established for Src homology 3-binding modules. PNAS USA 9: 7819–7823

DAVISON E, DIAZ RM, HART IR, SANTIS G, MARSHALL JF (1997) Integrin alpha5beta1-mediated adenovirus infection is enhanced by the integrin-activating antibody TS2/16. J Virol 71:6204–7

DAVISON E, KIRBY I, ELLIOTT T, SANTIS G (1999) The human HLA-A*0201 allele, expressed in hamster cells, is not a high-affinity receptor for adenovirus type 5 fiber. J Virol 73:4513–7

DELAHAYE L, MOTHE-SATNEY I, MYERS MG, WHITE MF, VAN OBBERGHEN E (1998) Interaction of insulin receptor substrate-1 (IRS-1) with phosphatidylinositol 3-kinase: effect of substitution of serine for alanine in potential IRS-1 serine phosphorylation sites. Endocrinology 139:4911–9

DESCHAMBEAULT J, LALONDE JP, CERVANTES-ACOSTA G, LODGE R, COHEN EA, LEMAY G (1998) Polarized human immunodeficiency virus budding in lymphocytes involves a tyrosine-based signal and favors cell-to-cell viral transmission. J Virol 73:5010–7

DI GIOVINE M, SALONE B, MARTINA Y, AMATI V, ZAMBRUNO G, CUNDARI E, FAILLA CM, SAGGIO I (2000) Binding properties, cell delivery, and gene transfer of adenoviral penton base displaying bacteriophage. Virology 282:102–12

DIMITRATOS SD, WOODS DF, STATHAKIS DG, BRYANT PJ (1999) Signaling pathways are focused at specialized regions of the plasma membrane by scaffolding proteins of the MAGUK family. Bioessays 21:912–21

DOONG H, PRICE J, KIM YS, GASBARRE C, PROBST J, LIOTTA LA, BLANCHETTE J, RIZZO K, KOHN E (2000) CAIR-1/BAG-3 forms an EGF-regulated ternary complex with phospholipase C-gamma and Hsp70/Hsc70. Oncogene 19:4385–95

FANNING AS, ANDERSON JM (1999) Protein modules as organizers of membrane structure. Curr Opin Cell Biol 11:432–9

FENDER P, RUIGROK RW, GOUT E, BUFFET S, CHROBOCZEK J (1997) Adenovirus dodecahedron, a new vector for human gene transfer. Nat Biotechnol 15:52–6

HONG JS, ENGLER JA (1991) The amino terminus of the adenovirus fiber protein encodes the nuclear localization signal. Virology 185:758–67

HONG SS, GAY B, KARAYAN L, DABAUVALLE MC, BOULANGER P (1999) Cellular uptake and nuclear delivery of recombinant adenovirus penton base. Virology 262: 163–77

Hong SS, Karayan L, Tournier J, Curiel DT, Boulanger PA (1997) Adenovirus type 5 fiber knob binds to MHC class I alpha2 domain at the surface of human epithelial and B lymphoblastoid cells. EMBO J 16:2294–306
Kibbey RG, Rizo J, Gierasch LM and Anderson RGW (1998) The LDL receptor clustering motif interacts with the clathrin terminal domain in a reverse turn conformation. J Cell Biol 142: 59–67
Kidd AH, Chroboczek J, Cusack S, Ruigrok RW (1993) Adenovirus type 40 virions contain two distinct fibers. Virology 192:73–84
Kumar S, Harvey KF, Kinoshita M, Copeland NG, Noda M, Jenkins NA (1997) cDNA cloning, expression analysis, and mapping of the mouse Nedd4 gene. Genomics 199, 40:435–43
Li Y, Paz Marzolo M, van Kerkhof P, Strous GJ, Bu G (2000) The YXXL motif, but not the two NPXY motifs, serves as the dominant endocytosis signal for low density lipoprotein receptor-related protein. J Biol Chem 275:17187–94
Mathias P, Wickham T, Moore M, Nemerow G (1994) Multiple adenovirus serotypes use alpha v integrins for infection. J Virol 68:6811–4
McDonald D, Stockwin L, Matzow T, Zajdel MB, Blair G (1999) Coxsackie and adenovirus receptor (CAR)-dependen t and major histocompatibility complex (MHC) class I-independent uptake of recombinant adenoviruses into human tumour cells. Gene Ther 6:1512–9
Medina-Kauwe LK, Kasahara N, Kedes L (2001) 3PO, a novel nonviral gene delivery system using engineered Ad5 penton proteins. Gene Ther 8:795–803
Pirozzi G, McConnell SJ, Uveges AJ, Carter JM, Sparks AB, Kay BK, Fowlkes DM (1997) Identification of novel human WW domain-containing proteins by cloning of ligand targets. J Biol Chem 272:14611–6
Roelvink PW, Lizonova A, Lee JG, Li Y, Bergelson JM, Finberg RW, Brough DE, Kovesdi I, Wickham TJ (1998) The coxsackievirus-adenovirus receptor protein can function as a cellular attachment protein for adenovirus serotypes from subgroups A, C, D, E and F. J Virol 72:7909–15
Rotin D (2000) WW (WWP) domains: from structure to function. Curr Top Microbiol Immunol 228:115–33
Schoehn G, Fender P, Chroboczek J, Hewat EA (1996) Adenovirus 3 penton dodecahedron exhibits structural changes of the base on fibre binding. EMBO J 15:6841–6
Seth P (1994) Adenovirus-dependent release of choline from plasma membrane vesicles at an acidic pH is mediated by the penton base protein. J Virol 68:1204–6
Shiratsuchi T, Futamura M, Oda K, Nishimori H, Nakamura Y, Tokino T (1998) Cloning and characterization of BAI-associated protein 1: a PDZ domain-containing protein that interacts with BAI1. Biochem Biophys Res Commun 247:597–604
Staub O, Dho S, Henry PC, Correa J, Ishikawa T, McGlade J, Rotin D (1996) WW domains of Nedd4 bind to the proline-rich PY motifs in the epithelial Na^+ channel deleted in Liddle's syndrom. EMBO J 15:2371–2380
Stewart PL, Chiu CY, Huang S, Muir T, Zhao Y, Chait B, Mathias P, Nemerow GR (1997) Cryo-EM visualization of an exposed RGD epitope on adenovirus that escapes antibodyneutralization. EMBO J 16:1189–98
Stevenson SC, Rollence M, White B, Weaver L, McClelland A (1995) Human adenovirus serotypes 3 and 5 bind to two different cellular receptors via the fiber head domain. J Virol 69:2850–7

Sun XJ, Pons S, Asano T, Myers MG Jr, Glasheen E, White MF (1996) The Fyn tyrosine kinase binds Irs-1 and forms a distinct signaling complex during insulin stimulation. J Biol Chem 271:10583–7

Takayama S, Xie Z, Reed JC (1999) An evolutionarily conserved family of Hsp70/Hsc70 molecular chaperone regulators. J Biol Chem 274:781–6

Tirabassi RS, Enquist LW (1999) Mutation of the YXXL endocytosis motif in the cytoplasmic tail of pseudorabies virus gE. J Virol 73:2717–28

Lee JH, Takahashi T, Yasuhara N, Inazawa J, Kamada S, Tsujimoto Y (1999) Bis, a Bcl-2-binding protein that synergizes with Bcl-2 in preventing cell death. Oncogene 18:6183–90

Waragai M, Lammers CH, Takeuchi S, Imafuku I, Udagawa Y, Kanazawa I, Kawabata M, Mouradian MM, Okazawa H (1999) PQBP-1, a novel poly-glutamine tract-binding protein, inhibits transcription activation by Brn-2 and affects cell survival. Hum Mol Genet 8:977–87

Wood JD, Yuan J, Margolis RL, Colomer V, Duan K, Kushi J, Kaminsky Z, Kleiderlein JJ, Sharp AH, Ross CA (1998) Atrophin-1, the DRPLA gene product, interacts with two families of WW domain-containing proteins. Mol Cell Neurosci 11:149–60

Wu E, Fernandez J, Fleck SK, Von Seggern DJ, Huang S, Nemerow GR (2001) A 50-kDa membrane protein mediates sialic acid-independent binding and infection of conjunctival cells by adenovirus type 37. Virology 279:78–89

Yeh HY, Pieniazek N, Pieniazek D, Gelderblom H, Luftig RB (1994) Human adenovirus type 41 contains two fibers. Virus. Res 33:179–183

Zhang F, Andreassen P, Fender P, Geissler E, Hernandez JF, Chroboczek J (1999) A transfecting peptide derived from adenovirus fiber protein. Gene Ther 6:171–81

CHAPTER 3

Structural Studies on Adenoviruses

C. SAN MARTÍN, R. M. BURNETT

Abstract. The adenovirus genome encodes more than 40 proteins, of which 11 combine with the viral DNA to form an icosahedral capsid of ~150 MDa molecular weight and ~900 Å in diameter. This chapter reviews the information that structural biology techniques have provided about the adenovirus proteins and capsid. The structures of two capsid proteins (hexon and fiber) and two non-structural polypeptides (DNA-binding protein and protease) have been solved by X-ray crystallography. Fiber and its knob have been the focus of the latest structural studies, due to

C. SAN MARTÍN, R. M. BURNETT
The Wistar Institute, 3601 Spruce Street, Philadelphia, PA 19104, USA
e-mail: carmen@cnb.vam.es
e-mail: BURNETT@WISTAR.UPENN.EDU

their role in host recognition and consequently in virus targeting for human gene therapy. The current model for the large capsid comes from a combination of electron microscopy and crystallography. The resultant images have revealed a surprising similarity between adenovirus and a bacterial virus, which suggests their common evolutionary origin.

1 Introduction

This chapter reviews the role that structural techniques have played in improving our understanding of the adenovirus proteins and their functions. Since their discovery 50 years ago (Rowe et al. 1953), much effort has been devoted to characterizing the adenoviruses, especially those infecting humans. From the structural point of view, this has resulted in a stoichiometric and architectural description of the adenovirus virion that is fairly complete considering its large size and complexity (Burnett 1997). In recent years, the emphasis on adenovirus studies seemed to shift from their basic biology to their use as gene therapy vectors (Russell 2000). However, this tendency has also motivated new structural studies, especially those related to viral tropism and receptor binding.

Of more than 40 proteins encoded by the adenovirus genome (Sussenbach 1984), only four have molecular models at atomic resolution (Table 1). This chapter starts with a description of the virion and the structural polypeptides. Because of the fiber knob's importance for targeting adenoviral vectors, its structure and its receptor-interactions have received the most recent attention (Xia et al. 1994; Bewley et al. 1999; van Raaij et al. 1999 a; Durmort et al. 2001). Progress on this topic has been considerable, and is the subject of a separate chapter in this book (see chapter by Freimuth, this volume). The other protagonist in the first stages of infection, penton base, still remains unconquered, but electron microscopy (EM) has shed some light on its interaction with fiber and integrin receptors (Stewart et al. 1993, 1997; Schoehn et al. 1996; Chiu et al. 1999, 2001; Von Seggern et al. 1999). The other most extensively characterized virion protein structure is that of hexon (Athappilly et al. 1994; Rux and Burnett 2000). While no high-resolution data exist for any of the minor capsid proteins (or those in the core), their shapes and general disposition in the virion are known from EM and biochemistry (Stewart and Burnett 1995). The next section of the chapter addresses the non-structural proteins, for which only the E2A DNA-binding protein (Tucker

Table 1. Proteins encoded by the adenovirus genome[a]

Proteins[b]	Transcription class	Structure data (Protein Data Bank codes)
13, 27, 32	E1A	–
16, 21, 55	E1B	–
DNA-binding protein (DBP)	E2A	1ADT, 1ADU, 1ADV, 1ANV (DNA-binding, C-terminal fragment) (TUCKER et al. 1994; KANELLOPOULOS et al. 1996)
DNA polymerase	E2B	–
Terminal protein[c]	E2B	–
4, 7, 8, 10, 12, 13, 15, 15, 19,	E3	–
7, 13, 13, 14, 15, 17	E4	–
p52/55 (scaffolding)	L1	–
IIIa (capsid cement)[c]	L1	–
pX (cleaved into X and μ; core proteins)[c]	L2	–
VII (core protein)[c]	L2	–
V (core)[c]	L2	–
III (penton base)[c]	L2	–
Protease	L3	1AVP (DING et al. 1996)
VI (capsid cement)[c]	L3	–
II (hexon)[c]	L3	1DHX (Ad2) (ATHAPPILLY et al. 1994), 1RUX (Ad5) (RUX and BURNETT 2000)
25	L4	–
VIII (capsid cement)[c]	L4	–
100 kDa	L4	–
IV (fiber)[c]	L5	Knob: 1NOB (Ad12) (BEWLEY et al. 1999), 1QHV (Ad2) (VAN RAAIJ et al. 1999a), 1H7Z (Ad3) (DURMORT et al. 2001), 1KNB (Ad5) (XIA et al. 1994) Knob + CAR: 1KAC (Ad12) (BEWLEY et al. 1999) Knob + shaft: 1QIU (Ad2) (VAN RAAIJ et al. 1999b)
IX (capsid cement)[c]	Intermediate	–
IVa2	Intermediate	–

[a] Modified from (BENKÖ et al. 2000).

[b] Proteins are denoted by their molecular weight in kDa, or by their common name (Roman numerals or function).

[c] Structural proteins.

et al. 1994; Kanellopoulos et al. 1996) and the protease (Ding et al. 1996) have yielded molecular structures.

A particularly interesting aspect of adenoviruses to be revealed by structural studies is an evolutionary link to PRD1, a membrane-containing bacterial virus (Butcher et al. 1995; Benson et al. 1999; San Martín et al. 2001); this is the focus of another section. The chapter ends with an overview of future expectations.

2 Structure of the Virion and Its Components

The adenovirus structural proteins combine to form an icosahedral non-enveloped particle that contains the dsDNA–protein core complex (Fig. 1 a). The composition and structure of the virion have been revealed by a combination of electrophoretic analyses (Maizel et al. 1968; van Oostrum and Burnett 1985), controlled disruption experiments (Smith et al. 1965; Laver et al. 1969; Prage et al. 1970), and EM (Horne et al. 1959; Furcinitti et al. 1989; Stewart et al. 1991, 1993).

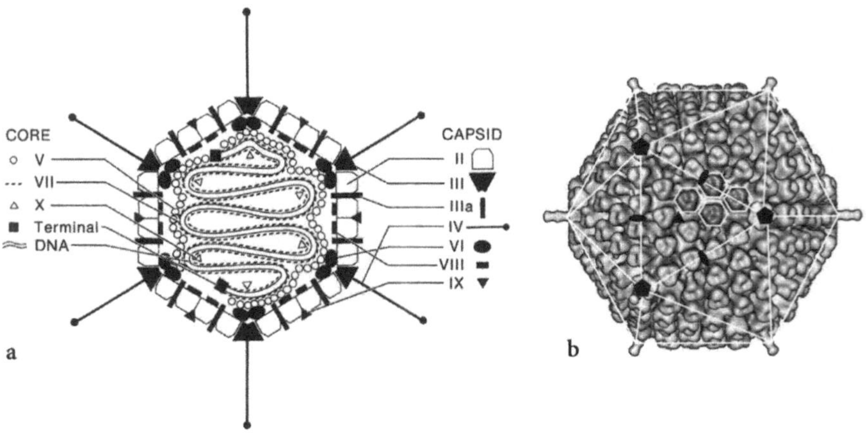

Fig. 1. **a** Schematic view of the adenovirus particle summarizing the current model of its architecture. **b** Ad2 cryo-EM reconstruction at 35-Å resolution overlaid by an ideal icosahedron in white. The icosahedral fivefold, threefold and twofold symmetry axes are labeled with *black pentagons*, *triangle* and *ovals*, respectively. The four hexon trimers in the icosahedral asymmetric unit are *highlighted with white hexagons*. (**a** Reprinted from Stewart and Burnett 1993, by permission of the Japanese Journal of Applied Physics)

The polypeptides present in the mature adenovirus virion are designated by Roman numerals (II–XII) in order of their decreasing apparent molecular weight on SDS gels (MAIZEL et al. 1968). Exceptions are polypeptide X in the core, also called μ, and the DNA-bound terminal protein. In addition, the two major coat proteins are commonly referred to as hexon (polypeptide II) and penton base (polypeptide III), according to the number of their surrounding neighbors in the virion. Penton base forms a complex with the long projecting fiber (polypeptide IV) at the vertex, called penton. Table 2 shows a list of the best characterized structural polypeptides. Very little is known about polypeptides XI and XII, and it is thought that they may be degradation products resulting from cleavage of other virion proteins, as a large number of these undergo proteolytic

Table 2. Adenovirus structural proteins[a]

Polypeptide	Molecular mass (Da)[b]	Number of residues[b]	Biochemical copy number[b,c]	Copy number in current virion model[d]
II (hexon)	109,077	967	720 ± 7	240 trimers
III (penton base)	63,296	571	56 ± 1	12 pentamers
IIIa (capsid cement)	63,535	570	68 ± 2	60 monomers
IV (fiber)	61,960	582	35 ± 1	12 trimers
V (core)	41,631	368	157 ± 1	–[e]
Terminal protein (core)[f]	~55,000	~500	2	–
VI (capsid cement)	22,118	206	342 ± 4	60 hexamers
VII (core)	19,412	174	833 ± 19	–
VIII (capsid cement)	15,390	140	127 ± 3	–
IX (capsid cement)	14,339	139	247 ± 2	80 trimers
μ (core)[g]	~4,000	~36	~104	–

[a] Modified from STEWART and BURNETT (1995).
[b] Values described for protein monomers in human adenovirus type 2.
[c] Data from VAN OOSTRUM and BURNETT (1985) considering the cleavage sites from ANDERSON (1990).
[d] STEWART et al. (1993).
[e] Dashes indicate proteins that have not yet been imaged.
[f] REKOSH (1981).
[g] HOSOKAWA and SUNG (1976).

processing by the viral protease during virus maturation (PETTERSON 1984). Hexon and fiber are the only two structural polypeptides for which crystallographic models have been obtained.

Most of the structural data on adenovirus come from human adenovirus type 2 (Ad2). The Ad2 virion has a diameter of 914 Å from vertex to vertex (STEWART et al. 1991) and a particle mass of 150×10^6 Da (VAN OOSTRUM and BURNETT 1985), of which 22.6×10^6 Da is DNA. The great size and complexity of adenovirus present a major challenge for structural biology methods. The first EM images showed 252 projections on an icosahedral particle, with 12 long fibers projecting from the vertices (HORNE et al. 1959). A three-dimensional reconstruction from cryo-EM images provided a more complete description of the Ad2 virion at 35 Å resolution (Fig. 1b) (STEWART et al. 1991). Later, this reconstruction was used as a reference frame to position a crystallographic hexon model in the quasi-equivalent positions of the capsid (STEWART et al. 1993). Subtraction of the resulting model from the EM reconstruction revealed the positions and shapes of the minor capsid proteins. Subsequent correlation of the difference map features with the known volumes and copy numbers of each protein (VAN OOSTRUM and BURNETT 1985) permitted the assignment of minor proteins IIIa and VI. Similarly, the positions of polypeptide IX in the capsid were assigned by a combined EM/X-ray study using two-dimensional images (FURCINITTI et al. 1989).

2.1 Hexon (Polypeptide II)

The homotrimeric hexon is the adenovirus major coat protein, accounting for approximately 63% of the virion mass (VAN OOSTRUM and BURNETT 1985). The structures of both human adenovirus type 2 and 5 hexons have been solved by X-ray crystallography (ATHAPPILLY et al. 1994; RUX and BURNETT 2000). Each hexon monomer is formed by two eight-stranded β-barrels and three extended loops (Fig. 2a). In the trimer, the six β-barrels generate a pseudo-hexagonal base with a central cavity, while the loops intertwine to form three 'towers' at the top of the molecule (Fig. 2b). The β-barrels in hexon have the same folding topology as those in the coat proteins of most spherical viruses (CHELVANAYAGAM et al. 1992). However, in adenovirus the β-barrel axes are perpendicular to the capsid surface, in contrast with the mostly in-plane orientation of a typical viral β-barrel.

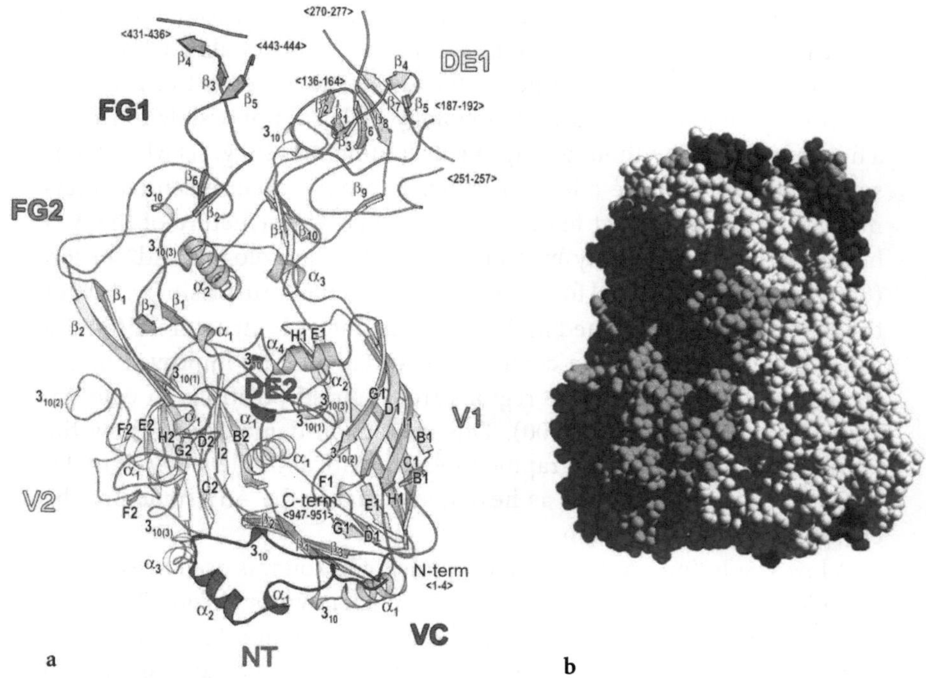

Fig. 2. **a** Ribbon diagram of the Ad5 hexon monomer. The loops at the top of the molecule form the outer surface of the virion. The *labels* indicate the individual secondary structure elements. **b** Space filling model of the Ad5 hexon trimer. Each monomer is depicted in a *different shade*, showing how each subunit clasps its neighbor to achieve a very stable structure. (**a, b** Reprinted from Rux and Burnett 2000, by permission of Academic Press)

The loops at the top of the hexon molecule form the outer surface of the virion. These regions of the polypeptide have the highest sequence variability between the different serotypes, and form the type-specific epitopes (Toogood et al. 1992; Rux and Burnett 2000). Loop DE1 (Fig. 2a) contains an acidic stretch of residues specific for subgroup C (Rux and Burnett 2000). It has recently been proposed that the interaction of this region with histone H1 at the nuclear pore brings about partial capsid disassembly and subsequent DNA release into the nucleus of the infected cell (Trotman et al. 2001). The way the loops interlock to form the towers, with a loop from each monomer participating in each of the three towers, explains the high stability of the trimer, which is resistant to conditions as harsh as 8 M urea (Shortridge and Biddle 1970). More stability is added by the conformation of the N-terminal region, which clasps the β-barrels at the bottom of the adjacent monomer. The effect is to stabilize the interface of two subunits of the trimer by the tower, and to tie the third subunit by an N terminus. This elaborate topology makes it impossible to isolate a hexon monomer without disrupting its folding, and suggests why a transient complex with the adenovirus 100 K protein is an essential intermediate ensuring correct folding of the trimer. The structure of the Ad5 hexon was solved recently by using the highly homologous Ad2 hexon (86% identity) as a model for molecular replacement. Surprisingly, 25% of the sequence was reassigned in the more accurate Ad5 structure, although the overall topology of the chain was maintained. The improved hexon model relocated two epitope regions from buried positions to external sites (Rux and Burnett 2000). This correction demonstrates how the latest advances in crystallographic tools are critical in the study of molecules as large and complex as hexon, and makes the Ad5 structure the current paradigm for hexon.

A low-resolution crystallographic model of hexon was used to establish a model for the overall architecture of the adenovirus virion (Burnett 1985). Each one of the 20 facets of the capsid contains 12 copies of the hexon trimer. The geometry of the adenovirus capsid corresponds to that for a T = 25 icosahedral particle, in the nomenclature introduced by Caspar and Klug (1962). With this geometry, a total of 25 × 60, that is, 1,500 identical subunits, and 25 different chemical environments would be expected for the adenovirus capsid. However, the use of the trimeric hexon as a stable building block leads to a drastic reduction in both numbers, with only 252 subunits in the capsid (240 hexons + 12 pentons) and only four different hexon environments in the icosahedral asymmetric unit (Fig. 1b). This economical arrangement is advantageous for accurate assembly, and the adenovirus particle is generally denoted as having 'pseudo' T = 25 geometry. Thanks to the pseudo-hexagonal shape of its base, the hexon trimer can occupy the sixfold coordinated positions in the capsid, while closely packing in a slightly curved $p3$ net (Stewart et al. 1991). The curvature brings the hexons along the edges into as close a contact as those within a facet. Thus, close packing of hexons occurs throughout the capsid despite the completely different chemical nature of the hexon–hexon interfaces at the edge and within the facet (Burnett 1985).

2.2
Penton Base (Polypeptide III)

Pentons lie at the 12 virion vertices, and are formed by the penton base (a pentamer of polypeptide III) and a long fiber (a trimer of polypeptide IV). These are the viral proteins responsible for the attachment (fiber) and internalization (penton base) of adenovirus in the host cell. Interaction of penton base with α_v integrins in the cell surface triggers virus internalization and membrane permeabilization (WICKHAM et al. 1993, 1994; MATHIAS et al. 1994). Binding to integrins is mediated by an ArgGlyAsp (RGD) sequence that is conserved within an otherwise highly variable region in the penton base protein of four different adenovirus subgroups (A, B, C and E) (BAI et al. 1993; MATHIAS et al. 1994).

Polypeptide III has so far resisted all attempts at crystallization, and most of the information on its three-dimensional structure comes from moderate resolution cryo-EM reconstructions. The difference map obtained from the Ad2 reconstruction at 35 Å resolution (STEWART et al. 1991, 1993) gave a first three-dimensional image of the penton base, its boundaries and dimensions. A later 25-Å three-dimensional reconstruction of dodecameric particles formed by adenovirus type 3 (Ad3) pentons established the penton base boundaries with more clarity (SCHOEHN et al. 1996; FENDER et al. 1997). The penton base has a spade-like profile, with a height of 124 Å, and maximum and minimum diameters of 112 and 50 Å (Fig. 3). A 30-Å diameter cavity which allows communication between the hollow interior of the penton base and the viral core is closed on the external side. The peripentonal hexon V1 β-barrels fit into grooves on the side of the penton base.

Five small (~22 Å) lobes of density protrude from the outer rim of the penton base (STEWART et al. 1993; SCHOEHN et al. 1996). As the EM density of these lobes is weaker than that for the rest of the penton base, they are thought to correspond to flexible loops. Based on similarities with the foot-and-mouth disease virus, which contains an integrin binding RGD sequence in a mobile loop of viral protein 1 (VERDAGUER et al. 1995), the five lobules could be correlated with the penton base sequence. This part contains the RGD sequence mediating recognition of the adenovirus internalization receptor, integrin (SCHOEHN et al. 1996). The assignment was confirmed by images of Ad2 bound to a Fab fragment of the monoclonal antibody DAV-1, which recognizes a linear epitope including the RGD sequence (STEWART et al. 1997). Comparison of Ad2 reconstructions,

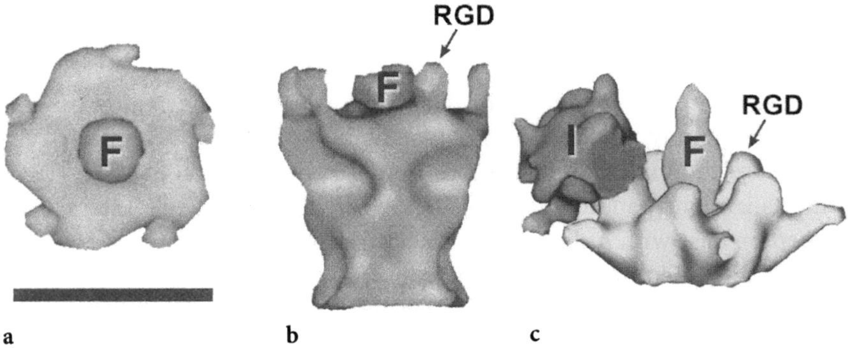

Fig. 3. The Ad2 penton complex as seen from outside the virion (**a**) and in a side view (**b**). Note that only a small portion of the fiber (*F*) appears in the cryo-EM reconstructions. Most of the trimeric fiber does not follow icosahedral symmetry and is averaged out. **c** Ad12 penton bound to soluble recombinant integrin $\alpha_v\beta_5$ (*I*). Only a single integrin heterodimer is shown. One of the lobes assigned to the RGD loop is labeled in **b** and **c**. The bar represents 100 Å. (**a, b** Reprinted from Stewart et al. 1993, by permission of Oxford University Press; **c** reprinted from Chiu et al. 1999, by permission of the American Society for Microbiology)

with and without DAV-1 Fab, showed that the sites for DAV-1 binding corresponded to lobules of weak density extending a further 24 Å from the previously observed 22 Å protrusions. The weakness of both the penton base protrusion and the Fab density was attributed to the high flexibility of the RGD epitope. Interestingly, adenovirus infection is inhibited in the presence of the Fab fragment, but not in the presence of the complete DAV-1 IgG molecule. Although the spacing between penton base protrusions (~60 Å) is in principle enough to allow IgG binding to all of them, the presence of the fiber increases the steric hindrance and so prevents binding of antibodies to all five RGD sites on the penton base. This leaves some sites accessible to integrins and allows the virion to escape neutralization.

Cryo-EM reconstructions were also used to structurally characterize the penton base–integrin complex, by imaging Ad2 and Ad12 virions bound to soluble recombinant integrin $\alpha_v\beta_5$ (Chiu et al. 1999). The Ad12 RGD loop is shorter (~20 residues) than that of Ad2 (~80 residues) and could be expected to be less flexible. Accordingly, the Ad12 reconstruction gave a better-defined picture of the integrin density (Fig. 3c). Biosensor experiments gave a stoichiometry of 4.2 integrin molecules bound per penton base (Chiu et al. 1999). These appeared as a ring-shaped density above the penton base in the EM reconstructions. The ring was segmented into five regions, each of them containing the two domains in the integrin heterodimer: a globular domain with a ~20-Å RGD binding cleft covering the penton base protrusions, and a distal domain with extended, flexible tails. The ring observed in the reconstruction would be formed by interactions between the proximal domains. The recently reported structure of soluble integrin $\alpha_v\beta_3$ in complex with an RGD peptide shows that the binding cleft is at the interface between the α_v and β_3 subunits in the head domain (Xiong et al. 2002). The aggregation of several integrins in a ring over the penton base might reflect clustering in the cell membrane upon interaction with multivalent ligands. This kind of clustering could in turn play a role in activation of integrin-mediated signaling pathways, and so could also be involved in inducing the signals necessary for virus endocytosis. This hypothesis is also supported by the finding that the penton base, but not an RGD peptide derived from the penton base sequence, activates the p72syk kinase and promotes B-lymphoblastoid cell adhesion (Nemerow and Stewart 1999). Also, the spacing of the integrin-binding sites on adenovirus is virtually identical to that of the unrelated foot-and-mouth disease virus (Acharya et al. 1989; Jackson et al. 1997), which also uses integrins for infection. Nevertheless, the IgG experiments (Stewart et al. 1997) would seem to indicate that integrin clustering is not essential for infection.

The reconstructions of human Ad3 penton dodecahedra with and without fiber revealed that the five penton base protuberances undergo a conformational change upon fiber binding, moving away from the fivefold axis by ~15 Å when the fiber is present (Schoehn et al. 1996). A similar change was seen when comparing reconstructions of wild-type Ad5 and the fiberless mutant Ad5.βgal.ΔF (von Seggern et al. 1999). Following the assignment of these features to the RGD integrin-binding loop, it can be hypothesized that this change, and in general the flexibility of the RGD loop, has a functional role in the process of virus entry into the cell. Host cell specificity would be enhanced by a sequential process whereby recognition through the fiber primes the integrin-mediated uptake mechanism.

Negatively stained EM images of isolated pentons, and the Ad2 reconstruction, initially suggested that the fiber was inserted in the central cavity of the penton base (Ruigrok et al. 1990; Stewart et al. 1993). More recent reconstructions show a hollow penton base, with the fiber not inserted, but attached to its external surface (Schoehn et al. 1996; von Seggern et al. 1999). Correct penton base assembly and incorporation into the virion occur without fiber, but fiberless virions are less stable than wild-type, contain less core protein VII, and suffer DNA leakage (von Seggern et al. 1999). This suggests that the fiber helps to seal the vertex region after DNA packaging. Unfortunately, structural studies on the interaction of penton base and fiber are especially challenging, due to the symmetry mismatch between a pentameric III and a trimeric IV. The problem is even more complicated in the adenoviruses with two different fibers: in the human enteric adenoviruses (Ad40 and Ad41) either of the two types of fibers can be present at a particular vertex (Kidd et al. 1993; Favier et al. 2002), while in the avian adenovirus CELO one fiber of each type binds simultaneously to a penton base (Hess et al. 1995).

2.3
Fiber (Polypeptide IV)

The trimeric fibers protruding from the vertices of the adenovirus virion have three structural domains (Green et al. 1983). The N-terminal tail attaches non-covalently to the penton base (polypeptide III), and the C-terminal segment folds into a globular 'knob' domain. This knob is necessary and sufficient for virion binding to host cells (Henry et al. 1994; Louis et al. 1994). The region between the two terminal domains, which is characterized by repeating motifs of ~15 residues, forms a shaft whose length varies among serotypes, from six repeating units in Ad3 (Signäs et al. 1985) to 22 in Ad2 and Ad5 (Green et al. 1983).

The structure of the Ad5 knob gave the first glimpse of the fiber structure (Fig. 4a) (Xia et al. 1994). It showed that each knob monomer is formed by an eight-stranded antiparallel β-sandwich structure, and that the functional state of the fiber knob is a propeller-shaped trimer. Two prominent surface features of the trimeric knob are the deep depression centered on the threefold symmetry axis, and the three symmetry-related valleys between subunits. These features were thought to be the receptor-binding sites, as they contain the main conserved regions between the Ad2 and Ad5 serotypes. These share the same cellular receptor, the coxsackie-virus–adenovirus receptor protein, CAR (Bergelson et al. 1997). It was proposed that adenovirus used a similar strategy to picornaviruses, where the receptor binding sites lie in deep crevices on the capsid surfaces, where they are shielded from the immune reaction (Bella et al. 1998). The Ad2 knob structure (van Raaij et al. 1999a) is very similar to that for the Ad5 knob, except for some of the flexible loops between β-strands that probably contribute to the serological differences.

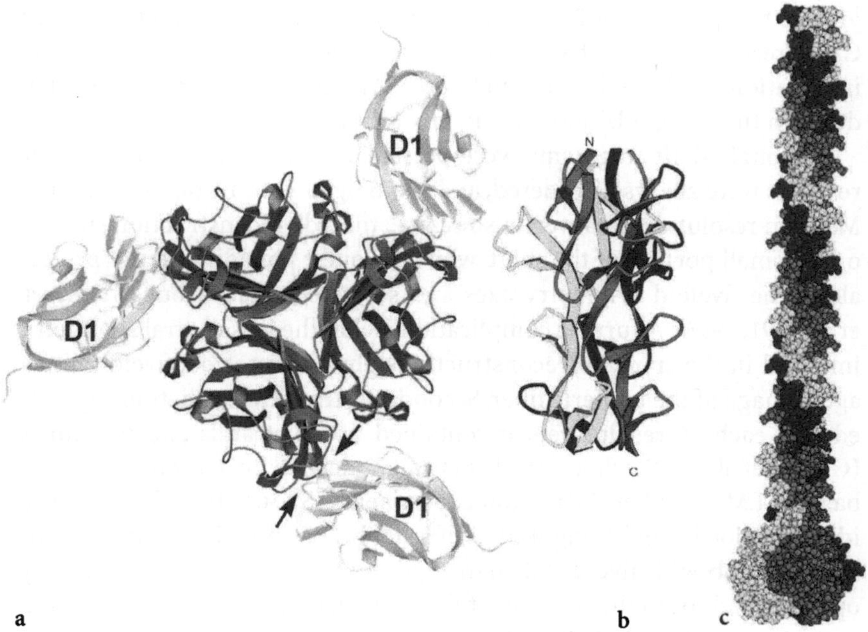

Fig. 4. **a** Structure of the Ad12 knob (*darker shades*) bound to the extracellular domain 1 of CAR (D1, *lighter shades*) viewed down the fiber. *Arrows* indicate the CAR–knob interaction region in one of the monomers. **b** Structure of four repeats from the Ad2 fiber shaft showing the triple β-spiral fold. The N and C termini of one monomer are labeled. **c** Model of a 21-repeat fiber showing the shaft and C-terminal knob. Each monomer is represented in a *different shade* in **b** and **c**. **a** Created with MOLSCRIPT (Kraulis 1991). (**b**, **c** Reprinted from van Raaij et al. 1999b, by permission of *Nature*)

The surprise came when the structure of a third CAR binding knob, that of Ad12, was solved in the presence of the extracellular domain 1 of CAR (Fig. 4a) (Bewley et al. 1999). This showed that the binding site was neither at the central knob depression nor in the three valleys, but at the periphery of the knob trimer, with the CAR–knob interface formed by several loops also involved in the intramolecular contacts between knob monomers. Thus, instead of following the canyon strategy, adenovirus uses surface loops, exposed to immunoselective pressure, to bind its primary receptor. The structure of a non-CAR binding knob, that of Ad3, showed a very similar fold to those solved previously (Durmort et al. 2001). Only the Glu at position 140 of Ad3, where CAR binding adenoviruses have Ser, Thr or Pro, seemed significant for preventing binding, as CAR presents residues that are mainly negatively charged. The functional implications of these findings on knob–CAR interactions are discussed in detail in the chapter by FREIMUTH, this volume.

Although shaft fragments were present in some knob crystals, their residues were always disordered, and so no light was cast on its structure. Medium resolution data were also scarce: the Ad2 reconstruction showed only a small portion of the shaft, while the outer portions did not project along the fivefold symmetry axes and so were averaged out (STEWART et al. 1991, 1993). A further complication is that the icosahedral symmetry imposed in the cryo-EM reconstructions gives a distorted, fivefold averaged, image of the trimeric fiber. Secondary structure prediction suggested that each 15 residue repeat contained two β-strands and two turns (GREEN et al. 1983), while a triple helical model had been proposed on the basis of EM and fiber diffraction experiments (STOUTEN et al. 1992). The identification by unfolding studies of a stable C-terminal domain containing the knob with five distal shaft repeats (MITRAKI et al. 1999) finally opened the door to the solution of the shaft structure. The stable domain, spanning residues 319–582 of the Ad2 fiber (knob plus four repeats), was cloned in *Escherichia coli*, crystallized, and its structure solved by molecular replacement using the Ad2 knob as a model (VAN RAAIJ et al. 1999a, b).

The shaft structure did not conform to either of the two models previously proposed. Instead, it revealed a unique new fold, named a triple β-spiral (Fig. 4b). The first seven residues of each repeat form an extended β-strand that runs parallel to the fiber axis. A type 2 β-turn, containing a conserved glycine, connects to another β strand, four residues long, which runs backwards at an angle of 45° to the shaft axis. The repeats are connected by a solvent-exposed variable loop. Each repeat extends over 13 Å of shaft length and twists by ~50°, so that seven repeats provide a complete 360° rotation around the shaft axis. The average diameter of the shaft is 15 Å, but the surface loops extend to a radius of 22 Å (Fig. 4c). Three conserved residues in each repeat line a longitudinal hydrophobic core. This core opens to form a solvent-filled cavity at the point where shaft and knob connect through a flexible six residue linker region.

The shaft structure showed that the trimer is highly cross-linked, and that one-third of the monomer solvent accessible area is buried on trimerization. This explains the high rigidity and stability of the shaft. However, some flexibility is tolerated, as the threefold axes of knob and shaft are tilted by 2° in the crystal. Likewise, the third repeat in the shaft of most human adenovirus subgroups does not correspond to the consensus sequence (Chroboczek et al. 1995), and correlates with the kink in the shaft that is observed in EM images of human and some animal adenoviruses (Ruigrok et al. 1990, 1994; Hess et al. 1995). These observations suggest that some conformational leeway in the fiber is probably required to mediate binding of the virion to its receptor and its subsequent entry. This idea has led to proposals that the usage of a particular receptor, and consequently the cellular tropism of different adenovirus serotypes, depends on the flexibility and length of the fiber (Roelvink et al. 1998; Shayakhmetov and Lieber 2000; Chiu et al. 2001; Wu et al. 2001; Seki et al. 2002). Structural evidence supporting this hypothesis comes from cryo-EM studies on pseudotyped Ad5 virions containing the Ad37 fiber (Chiu et al. 2001). Ad37 (subgroup D) does not infect cells via CAR, although its fiber knob binds to CAR molecules (Roelvink et al. 1998; Arnberg et al. 2000; Wu et al. 2001). As the Ad37 fiber does not contain the third non-consensus shaft repeat, and is very short (eight repeats) and rigid, geometrical constraints prevent CAR binding. Attachment of the side surface of the fiber knob to CAR would create collisions between the viral capsid and the host cell membrane that cannot be relieved by bending the shaft. Also, the short length of the fiber places the CAR binding surfaces within the region where the integrin ring binds to the penton base (Chiu et al. 1999), precluding simultaneous interaction with both receptors. Interestingly, Ad3, which has the shortest fiber described so far, is also unable to use CAR as a receptor, in common with the short fiber of Ad40 (Roelvink et al. 1998). In view of these results, it can be predicted that the non-CAR binding serotypes will use residues in the top surface of the knob to bind to their receptors (Chiu et al. 2001).

2.4 Minor Capsid Proteins

The adenovirus capsid presents a high level of complexity, with at least seven different polypeptides in its icosahedral coat (Table 2). In the current model, the minor proteins IIIa, VI, VIII and IX act as capsid cement and play the double role of stabilizing the virion structure while allowing the flexibility necessary for its disassembly during infection. These small proteins facilitate the use of the same hexon building block in four different chemical environments.

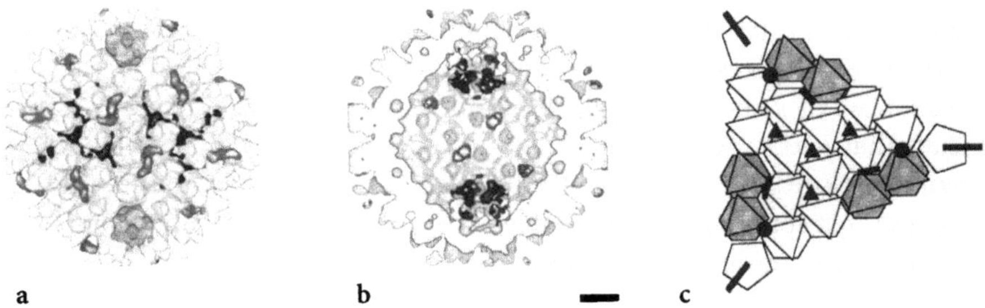

Fig. 5a–c. Location and shape of the minor proteins, as seen in the 35-Å resolution difference imaging. **a** A view from outside the virion. Different shades indicate, from *light* to *dark*: hexon, penton, IIIa, IX. **b** View of the internal surface showing, from *light* to *dark*: hexon and penton (in the same *gray shade*), internal portion of IIIa, VI. The bar represents 100 Å. **c** Facet schematic showing the location of the capsid components in the current model. The 12 hexon trimers in the facet are represented as *white triangles superimposed on hexagons.* Hexons belonging to adjacent facets are shaded *gray. Pentagons* and *bars* represent penton bases and fibers. The minor proteins are represented as *black rectangles* (IIIa), *circles* (VI) and *triangles* (IX). (Reprinted from STEWART et al. 1993, by permission of Oxford University Press)

Atomic resolution structural information is not presently available for any of the adenovirus minor coat proteins. Visualizing them by EM is also difficult, as the moderate resolution does not allow clear definition of protein boundaries in the virion. Nevertheless, a combination of an Ad2 cryo-EM reconstruction with the crystallographic structure of hexon allowed the first, and so far only, picture of the minor proteins and their position in the capsid (Fig. 5) (STEWART et al. 1993).

2.4.1
Polypeptide IIIa

The three-dimensional difference image of Ad2 showed that polypeptide IIIa is present as a large, elongated monomeric component (Fig. 5a, b). Most of the density attributed to IIIa is on the outer surface of the virion, contacting four different hexons at the icosahedral edge. The IIIa envelope narrows at the capsid surface and reappears in the inside and accounts for 97% of the expected volume. The 60 copies of IIIa in the virion, two at each edge, seem to rivet the virion facets together.

2.4.2
Polypeptide VI

Polypeptide VI has been assigned to a position on the inner capsid surface, where it plays the dual role of connecting the bases of two adjacent peripentonal hexons and tethering the highly ordered capsid to the less ordered core region (Fig. 5b). The molecule has a three-armed shape, with each one of its three 29-Å diameter lobes separated by 46 Å from the next. The volume of each lobe is consistent with the volume of a monomer, but stoichiometric analysis indicates that VI is organized in hexamers (6×60 monomers in the capsid). This discrepancy was resolved by proposing that half of each polypeptide is disordered, particularly a 115 residue stretch in the middle containing 13% basic residues and 15% prolines. Crystallographic studies have shown that similar domains in other viral capsids, which are presumed to interact with the non-icosahedral core, are frequently disordered (Harrison et al. 1978; Abad-Zapatero et al. 1980; Choi et al. 1991).

2.4.3
Polypeptide VIII

An early model for adenovirus placed polypeptide VIII in the interior of the capsid (Everitt et al. 1975). This position is consistent with the fact that there was no unassigned difference density on the exterior of the Ad2 virion (Stewart et al. 1993). Except for VI, the difference densities on the interior surface of the capsid were not interpretable, and VIII could not be identified. As with VI, VIII has a relatively high proline content (8%) and is rich in the basic residues arginine and lysine (11%), which suggests that it could be largely disordered.

2.4.4
Polypeptide IX

Disruption of adenovirus capsids under mild conditions results in a progressive loss of the vertex pentons and adjacent peripentonal hexons, and finally the remaining planar groups-of-nine hexons (GONs) (Smith et al. 1965; Laver et al. 1969; Prage et al. 1970). The GONs, which have an intriguing non-random dissociation pattern (Pereira and Wrigley 1974), were the subject of much early EM work. The result is that polypeptide IX, a minor component of the GONs, is the best characterized of all the adenovirus minor proteins. After stoichiometric analysis revealed that every facet in the virion contains 12 copies of IX (van Oostrum and Burnett 1985), a model was proposed in which four polypeptide IX trimers bind to the nine hexons in the GON, and act as capsid cement.

Observation of negatively stained GONs with the scanning-transmission electron microscope and subtraction of the crystallographic hexon density from an average image confirmed the model and gave a two-dimensional picture of the disposition of IX in the GON (Furcinitti et al. 1989). Polypeptide IX trimers were present at one of the two types of local threefold axes in the GON, with each elongated monomer extending along a hexon–hexon interface (Fig. 5a). The monomer size was estimated as 64 Å long and 18 Å in diameter. Based on the shape and dimensions from EM, and secondary structure predictions, a coiled-coil structure has been proposed for IX (Lupas et al. 1991; Burnett 1997; Rosa-Calatrava et al. 2001). Later, the three-dimensional difference imaging confirmed that IX lies in the positions within the facet found from the two-dimensional analysis, and shows that IX is on the outer surface of the capsid (Stewart et al. 1993). There is no IX at the local threefold axis adjacent to the peripentonal hexons. These positions explain the dissociation patterns of both the viral particle and the GON. The Ad5 deletion mutant *dl313*, lacking IX, assembles into wild-type like particles, but these have lower thermal stability and do not form GONs upon disruption (Colby and Shenk 1981). This provides the clearest evidence that IX is a capsid cement, as it adds stability to the virion but is not required for assembly.

Polypeptide IX is different from all other adenovirus structural proteins in its early expression in the infectious cycle, shortly after the start of viral DNA replication (Table 1). It is also notable that IX has a far more extensive role than its purely structural function suggests. Deletion mutants lacking IX and the early E1 and E3 regions of the genome were prepared with the expectation that they could have a larger DNA packaging capacity than the wild-type virus, and so be advantageous as gene vectors. On the contrary, they were even unable to package the complete viral genome (Ghosh-Choudhury et al. 1987). Packaging capacity was restored when the mutants were grown on cells supplying IX. A possible explanation of this phenomenon is that the particles are smaller in the absence of IX. However, cryo-EM studies of an Ad5 mutant lacking IX show that it is the same size as wild-type (C. San Martín, S.D. Fuller and R.M. Burnett, unpublished results). A recent computer simulation study showed that the internal pressure of a viral capsid increases more than 10-fold during the final third of the DNA packaging process (KINDT et al. 2001). Another possibility is, then, that the unstable IX-less particles cannot cope with the internal pressure built during the last stages of DNA packaging. In any case, IX carries out a critical, as yet unexplained, function in controlling DNA packaging. In addition, Ad5 polypeptide IX acts as a transcriptional activator, stimulating the activity of the adenovirus major late promoter, as well as other viral and cellular TATA-containing promoters (LUTZ et al. 1997). The physiological role of this transcriptional activity is not clear at present, given that mutants lacking IX propagate at almost wild-type levels (COLBY and SHENK 1981) and that IX is completely absent in some animal adenoviruses (BENKÖ et al. 2000).

The latest work on polypeptide IX has focused on dissecting the functional domains of the protein. Immunoelectron microscopy studies of negatively stained Ad3 capsids indicate that the C-terminal domain of IX, which is highly variable between different subgroups, is exposed on the capsid surface, while the more conserved N terminus is buried (AKALU et al. 1999). Site-directed mutagenesis of Ad5 polypeptide IX (ROSA-CALATRAVA et al. 2001) shows that N-terminal residues 22–26 are essential for incorporation of the protein in the virion, while transcriptional activity depends on the integrity of a C-terminal leucine zipper. A 10 residue, alanine-rich domain in the center of the sequence is present only in the human serotypes, and was proposed to act as a flexible hinge between the C- and N-terminal domains. Deletion of this region also abolished the transcriptional function of IX. Integrity of the leucine zipper and the alanine stretch was also required for protein oligomerization, suggesting that correct self-assembly and transcription activation are linked. The fact that the C-terminal half of the molecule can be modified without apparently affecting viral assembly opens the possibility to use C-terminal modifications of IX in adenovirus retargeting for gene transfer (DMITRIEV et al. 2002; ROSA-CALATRAVA et al. 2001).

2.5
Core and Core Proteins

The adenovirus core contains a single copy of linear double-stranded DNA and four viral proteins: polypeptides V, VII, μ and the terminal protein. The core composition in avian adenoviruses is different from the mammalian, as they lack V (BENKÖ et al. 2000). Two copies of the terminal protein are covalently linked to the 5′ genome ends, circularizing it and acting as a primer for DNA replication (REKOSH et al. 1977). Polypeptides V, VII and μ bind DNA (CHATTERJEE et al. 1986), and their high proportion of basic residues makes them similar to histones. Polypeptide VII is the major core protein, with over 800 copies per virion (Table 2). It complexes tightly with DNA to form compact repeating structures, termed 'adenosomes' by analogy with the nucleosomes observed in the nuclei of mammalian cells (TATE and PHILIPSON 1979; MIRZA and WEBER 1982; DEVAUX et al. 1983; VAYDA et al. 1983; VAYDA and FLINT 1987). Polypeptide V can also bind to penton base and to polypeptide VI, and so could link the core to the capsid (EVERITT et al. 1975; CHATTERJEE et al. 1986). The function of μ is unknown (HOSOKAWA and SUNG 1976).

The disposition of DNA and proteins in the core is still unclear. A model in which the viral DNA is organized into 8–12 large spherical domains was proposed on the basis of cross-linking studies (WONG and HSU 1989), EM of negatively stained preparations of sarcosyl cores (BROWN et al. 1975), and ion etching experiments (NEWCOMB et al. 1984). However, no sign of order was seen in the Ad2 cryo-EM reconstruction at 35 Å resolution (STEWART et al. 1991), nor in an Ad5 reconstruction at 15 Å (C. San Martín, S.D. Fuller and R.M. Burnett, unpublished results). In contrast, small-angle neutron and X-ray scattering experiments showed evidence of some ordering in the Ad2 core, with a maximum in the scattering pattern at 1/29 $Å^{-1}$ (DEVAUX et al. 1983).

The absence of order in the Ad2 and Ad5 reconstruction cores is surprising, as cryo-EM studies on bacteriophages and animal viruses are revealing an increasing number of images of viral DNA in concentric layers with a separation of ~25 Å (EARNSHAW and HARRISON 1977; CERRITELLI et al. 1997; EARNSHAW et al. 1978; ZHOU et al. 1999; ZHANG et al. 2000; SAN MARTÍN et al. 2001). Moreover, parallel lines with a spacing of 25 Å were observed in negatively stained EM images of collapsed fowl adenovirus particles (RUIGROK and SCHÜLLER 1993), which lack the core protein analog to human polypeptide V, and show no evidence of nucleosomal organization (LI et al. 1984). Rings in the EM images are consistent with DNA packing in liquid crystalline arrays.

The contrasting results for avian and mammalian adenoviruses can be reconciled if the avian cores are organized in a layered manner similar to those of bacteriophages, while other adenoviruses have a different arrangement, more like that found in nucleosomes. Both forms of organization would probably show some order in diffraction experiments. It is notable that bacteriophage PRD1, which resembles adenovirus, but lacks histone-like core proteins, presents a concentric ring pattern in cryo-EM reconstructions (San Martín et al. 2001; see Sect. 4). These tantalizing glimpses of the adenovirus core show that its structural organization remains an open and interesting question for further study.

3 Non-Structural Proteins

About 30 adenovirus non-structural proteins have been described so far (Table 1) (Petterson 1984; Benkö et al. 2000). Presumably all of these proteins carry out catalytic and regulatory functions during the viral life cycle, but the precise role for most of them is still unknown. The fact that they are generally produced in small quantities has hindered their study. The structures of two of the non-structural proteins have been solved at atomic resolution: the DNA binding protein encoded by the E2 region; and the viral protease.

3.1 DNA-Binding Protein

The early region E2 of the adenovirus genome encodes three proteins involved in viral DNA replication: the terminal protein; the viral DNA polymerase; and a ~ 50-kDa single-strand DNA binding protein, DBP (Table 1) (Ramachandra and Padmanabhan 1999). DBP (473–529 residues, depending on the serotype) is a multifunctional phosphoprotein that participates in adenovirus and adeno-associated virus DNA replication (Ward et al. 1998), as well as in other viral DNA metabolic processes. DBP binds cooperatively to single-stranded DNA, protecting it against nuclease digestion (van der Vliet et al. 1978), and destabilizing the double helix during the elongation phase of DNA replication (Zijderveld and van der Vliet 1994). As is usual in single-stranded DNA binding proteins, DBP has no sequence specificity. Cleavage experiments with chymotrypsin suggest that DBP has two domains (Klein et al. 1979; Linne and Philipson 1980; Schechter et al. 1980): an N-terminal region that contains the nuclear localization signal and all phosphorylation sites (Morin et al. 1989); and a highly conserved C-terminal domain that binds DNA and is active in DNA replication (Tsernoglou et al. 1985).

The structure of the DNA binding domain of Ad5 DBP has been solved by X-ray crystallography (Tsernoglou et al. 1984; Tucker et al. 1994).

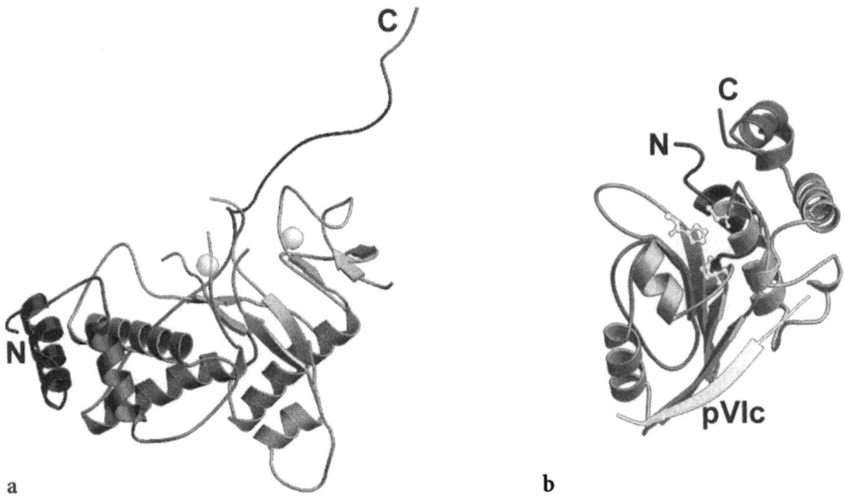

a b

Fig. 6. **a** Ribbon diagram of the Ad5 DBP DNA binding domain. The N and C termini are labeled, and two Zn atoms are shown as *white spheres*. **b** Ribbon diagram of the Ad2 proteinase bound to its cofactor (pVIc). The N and C termini of the protein, and the pVI 11-peptide cofactor are labeled. The residues forming the active site are indicated by *ball and stick* representations. Figure created with MOLSCRIPT (KRAULIS 1991)

The molecule forms a roughly globular (51 × 42 × 27 Å) single domain (Fig. 6a), with an α-helical base topped by three small β-sheets. A 40-Å long C-terminal extension protrudes from the main body of the molecule. This appendage hooks up one molecule to its neighbor, to form a protein chain, by binding to the cleft between two helices in the next DBP molecule in the crystal. This intermolecular interaction involves hydrophobic residues that are conserved throughout serotypes. Two zinc atoms, in a novel coordination different from all known zinc binding motifs for DNA binding proteins, stabilize the structure. The formation of a chain via the C-terminal arm suggested that this arm would be required to maintain the cooperativity of ssDNA binding. This hypothesis was confirmed by analysis of deletion mutants. Deleting the 17 C-terminal residues did not reduce DNA binding, but greatly decreased cooperativity (TUCKER et al. 1994). Other mutation studies on the C-terminal region and one of the helices in the binding cleft also support the importance of oligomerization for the correct function of DBP (VOS et al. 1989; DEKKER et al. 1997, 1998).

Although crystallization of the DBP DNA-binding domain bound to ssDNA has been reported (KANELLOPOULOS et al. 1995), no high resolution structure of the complex is yet available. However, some ideas about the possible DNA binding mechanism could be deduced from the structure of DBP alone. It was observed that the edge of one of the DBP β-sheets has a region of high positive potential, composed of several conserved arginine and lysine residues, which could interact with the DNA phosphate backbone. In the DBP chain, these regions would concatenate to form a highly positively charged ridge that winds around the protein multimer. Based on these observations, a model was proposed in which DBP would oligomerize as a protein filament, around which the DNA strand would wind during chain elongation (Tucker et al. 1994).

A second crystal form of the DBP DNA binding domain yielded an essentially identical structure for the base of the molecule (Kanellopoulos et al. 1996). The C-terminal arm, however, protruded from the base in a very different orientation, and the relative angle between DBP molecules in the chain was altered by 90°. The link between the C-terminal arm flexibility and the function of the protein has yet to be described, but it would presumably be advantageous for correct oligomerization.

3.2
Protease

Like many other DNA and RNA viruses, adenovirus encodes an endopeptidase that plays a crucial role in the assembly of infectious particles (Weber 1995, 1999). Between 10 and 30 molecules of the protease (204 residues, 25 kDa in Ad2) are packaged in each adenovirus virion (Chatterjee and Flint 1987; Anderson 1990). Mutants that do not package the protease cannot uncoat and so are unable to carry out a productive infection (Greber et al. 1996). An 11 residue peptide, the product of the cleavage of polypeptide VI (pVIc) by the protease, binds to the same protease via a disulfide bond and increases its activity by a factor of 300. The viral DNA itself increases the catalytic rate of the protease 6000-fold (Mangel et al. 1993, 1996; Baniecki et al. 2001), and is thought to drive protease packaging. The adenoviral enzyme is the only protease known to require DNA for maximum activity. Most of the minor capsid and the core polypeptides, including the terminal protein, are cleaved by the viral protease during viral maturation. The substantial effect that these cleavages have on particle stability can be appreciated by the example of polypeptide VI, where its binding to hexon increases by 40-fold when it is cleaved by the protease (Matthews and Russell 1995). The various cleavage sites are highly conserved, but the exact order and consequences of the cleavages are not known, although it seems that these take place only after particle assembly. It is possible that the adenoviral protease also digests cellular proteins in the last stages of infection.

The atomic structure of the Ad2 protease in complex with its activating peptide revealed an ovoid-shaped particle with α-helical regions at both poles and a central β-sheet (Fig. 6b) (DING et al. 1996; MCGRATH et al. 1996). The pVIc peptide is bound to the β-sheet as an additional strand. This location allows the peptide to interact with two non-contiguous regions of the protease, but it is far from the active site. Four large positively-charged patches on the protein surface might be involved in its association with DNA, which acts both as a cofactor and as a packaging vehicle. Neither the sequence nor the structure of the adenovirus protease are related to those of any other described protein. Like the use of DNA as a cofactor, these features emphasize the uniqueness of the enzyme. However, the disposition of the residues in the active site matches that of papain (POLGAR 1974), suggesting that both enzymes use the catalytic mechanism common to cysteine proteinases. The adenovirus protease has recently been crystallized in the absence of the pVIc cofactor (BANIECKI et al. 2002). A comparison of this structure with that of the protease-pVIc complex promises to show how peptide binding activates the enzyme.

4
Distant Relatives – Not So Distant?

One notable and unexpected characteristic of adenovirus revealed by structural studies is its similarity to a bacterial virus, PRD1 (BAMFORD et al. 1995). PRD1 is the prototype organism of the *Tectiviridae*, which are membrane-containing viruses infecting a broad range of Gram-negative bacteria. The parallels between adenovirus and PRD1 extend from their DNA replication mechanisms, to their capsid architecture and the folding of their major coat proteins. As in adenovirus, the phage genome consists of a single linear dsDNA molecule (15 kilo base pairs) with a protein covalently linked to each of its 5′ ends, and identical inverted terminal repeat sequences at both ends. The terminal protein is used as a primer in the initiation of DNA replication, which takes place through a sliding-back mechanism (CALDENTEY et al. 1993), as in adenovirus (KING and VAN DER VLIET 1994). Both virions are organized as a pseudo T = 25 lattice of trimeric major coat proteins, with secondary coat proteins and spikes at

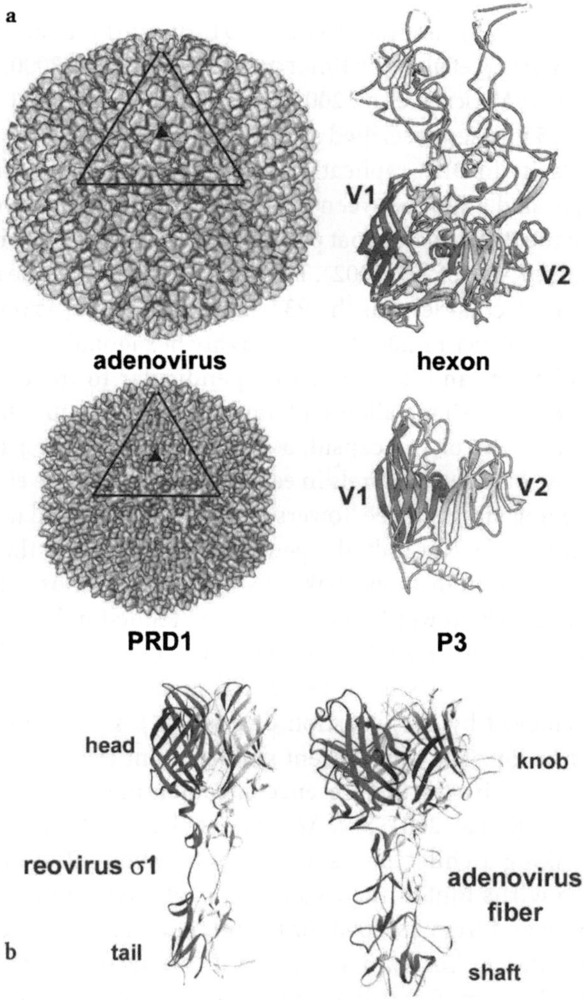

Fig. 7 a, b. Evolutionary links revealed by structural studies. **a** Comparison of Ad5 and bacteriophage PRD1 cryo-EM reconstructions (*left*) and major coat protein fold (*right*). A facet and its threefold axis are indicated in the cryo-EM reconstructions. The viral jellyrolls are labeled as V1 and V2. **b** Comparison of reovirus σ1 and adenovirus fiber. (**b** Reprinted from Chappell et al. 2002, by permission of Oxford University Press)

the vertices (Fig. 7a) (Stewart et al. 1991; Butcher et al. 1995; Grahn et al. 1999; Rydman et al. 1999; Bamford and Bamford 2000; Caldentey et al. 2000; San Martín et al. 2001). Adenovirus and PRD1 are the only pseudo T = 25 viruses described so far.

The similarity in DNA replication and general capsid organization had already suggested a link between adenovirus and PRD1 when the atomic structure of the PRD1 major coat protein, P3, was solved by X-ray crystallography (Benson et al. 1999, 2002). This study showed that the resemblance reached the molecular level as the P3 trimer, like hexon (Athappilly et al. 1994; Rux and Burnett 2000) has a roughly hexagonal base composed of six jellyroll barrels and is oriented perpendicular to the capsid surface. As with hexon, this shape allows a trimer of P3 to occupy the sixfold coordinated positions of the capsid, a characteristic of the pseudo T = 25 geometry. One of the two barrels in each monomer is more elongated than the other, giving rise to three 'towers' and a more triangular shape at the top of the molecule. Not only the general shape, but also the folding and topology of the protein show striking similarities to those of adenovirus hexon (Fig. 7a). The tower loops are more extended in hexon than in P3, probably due to the evolutionary pressure of dealing with the host immune system (Benson et al. 1999). The greater size of the adenovirus capsid is achieved by the insertion of a small β-sheet between the two jellyrolls of hexon. The most recent studies point at yet another parallel between the two viruses, the presence of cementing proteins in the PRD1 capsid (Rydman et al. 2001; San Martín et al. 2001, 2002).

These striking architectural similarities between PRD1 and the mammalian adenovirus hint at their common evolutionary origin (Hendrix 1999; Belnap and Steven 2000; Bamford et al. 2002a). The two viruses infect very different hosts (mammals/birds versus bacteria), using a very different mechanism: PRD1 has an internal membrane that forms a tail-like device used to inject its DNA into the bacterial cell (Bamford et al. 1995), whereas adenovirus enters the cell via endocytosis. And yet, adenovirus and PRD1 share a common structural design. This suggests that the architecture was already established in the early stages of evolution, before the branches of the evolutionary tree diverged into the three kingdoms known today. Furthermore, as the design has been conserved throughout time, even after all traces of sequence similarity have almost disappeared (for a comparison of hexon and P3 sequences, see chapter by Benkö, this volume), it must be a highly efficient arrangement. All of these results have contributed strong evidence towards the theory of an evolutionary relationship between animal and bacterial viruses. Adenovirus and PRD1 are not the only examples that give credence to this theory: other parallels have been found between herpesvirus and tailed bacteriophages, and between the dsRNA bacteriophage $\phi 6$ and reoviruses (HENDRIX 1999; BAMFORD et al. 2002a).

Interestingly, an evolutionary link also exists between reoviruses and adenoviruses. The recently solved structure of the reovirus attachment protein $\sigma 1$ exhibits a remarkable similarity to that of the adenovirus fiber (Fig. 7b) (CHAPPELL et al. 2002). Reovirus $\sigma 1$ fibers protrude from the 12 vertices of the virion. They are composed of an elongated tail and a globular head. The head binds to the reovirus receptor JAM (junction adhesion molecule) (BARTON et al. 2001). The shaft of adenovirus fiber and the tail of reovirus $\sigma 1$ are the only known examples of the 'triple β spiral' folding motif (VAN RAAIJ et al. 1999b). The folding of the $\sigma 1$ head is also very similar to that of the adenovirus knob (XIA et al. 1994). There is a similar match in function: both $\sigma 1$ and fiber bind to viral receptors in the cell surface; both receptors are in the immunoglobulin superfamily, have two extracellular immunoglobulin-like domains, and are present in cell tight junctions (BERGELSON et al. 1997; BARTON et al. 2001; COHEN et al. 2001). As in adenovirus fiber, the $\sigma 1$ tail allows a degree of flexibility that has probably arisen to optimize its interaction with the viral receptor. While the PRD1-adenovirus similarities spanned the evolutionary tree by linking prokaryotic and eukaryotic viruses, this newly described parallel between adenovirus and reovirus links the realm of dsDNA-viruses with that of dsRNA-containing viruses. The question whether this further link is due to host-specific requirements, or to innate characteristics of the viral ancestors (BAMFORD et al. 2002a) remains to be answered.

5
Conclusions and Perspectives

In the 50 years since their discovery (ROWE et al. 1953), the adenoviruses have been extremely useful as experimental systems in which to unravel many complicated processes in eukaryotic cell life, such as splicing and apoptosis. They have more recently shown promise as vectors for gene transfer into mammalian cells, particularly for vaccine delivery where only transient expression of the transgene is required (IMLER 1995; BABIUK and TIKOO 2000; SHIVER et al. 2002). However, many questions about the basic biology of adenoviruses remain open. In particular, a detailed picture of the penton base structure and its interaction with integrin is not available; there is no well-established model for the core structure and the function of the core proteins; there are no molecular structures for the minor capsid proteins, and few for the non-structural proteins.

Advances in the structural genomics field (STEVENS et al. 2001) will probably shed light on the structures of some more adenoviral proteins, either by direct experimental resolution or by modeling by analogy with homologous folds (BAKER and SALI 2001). The structural genomics approach, however, is most suitable for small proteins or isolated domains, but is not expected to offer results on larger entities, such as the 300-kDa penton base or the viral particle itself. Other methodologies will be needed to solve these structures.

The latest progress in the cryo-EM and image reconstruction field make this technique a good candidate to study structures of non-crystallized, large macromolecular complexes at a medium resolution range (6–10 Å) (BAUMEISTER and STEVEN 2000). Cryo-EM derived images of large viruses at resolutions better than 10 Å are starting to be obtained routinely (MANCINI et al. 2000; ZHOU et al. 2000, 2001). The combination of EM maps of virions with atomic resolution structures of viral components allows the construction of quasi-atomic resolution models of complete viral capsids, which delineate the interactions between capsomeres at the residue level (GRIMES et al. 1997; SAN MARTÍN et al. 2001; KUHN et al. 2002). Quasi-atomic models have been used to describe the large morphological changes occurring in the HK97 bacteriophage capsid during its maturation (CONWAY et al. 2001). The EM/X-Ray combined approach that was used with an Ad2 map at 35-Å resolution provided the most complete and detailed image of the virus particle available so far (STEWART et al. 1991, 1993). It can be predicted that the study of adenovirus will also benefit from the latest advances in the EM field. As an example, comparison of a 15-Å resolution map of Ad5 with a map at 35 Å shows the dramatic increase in available detail (Fig. 8) (C. San Martín, S.D. Fuller and R.M. Burnett, unpublished results).

Finally, it is also expected that the adenovirus particle will eventually yield to analysis by X-ray crystallography. Crystalline inclusions have been reported in cell nuclei at late stages of infection (BOULANGER et al. 1970; HENRY et al. 1971). However, three-dimensional crystallization of isolated virions has so far remained elusive. The main reason for this failure is that production and purification of the amount of virus needed for crystallization is not easy to achieve. In addition, the long and flexible fibers of

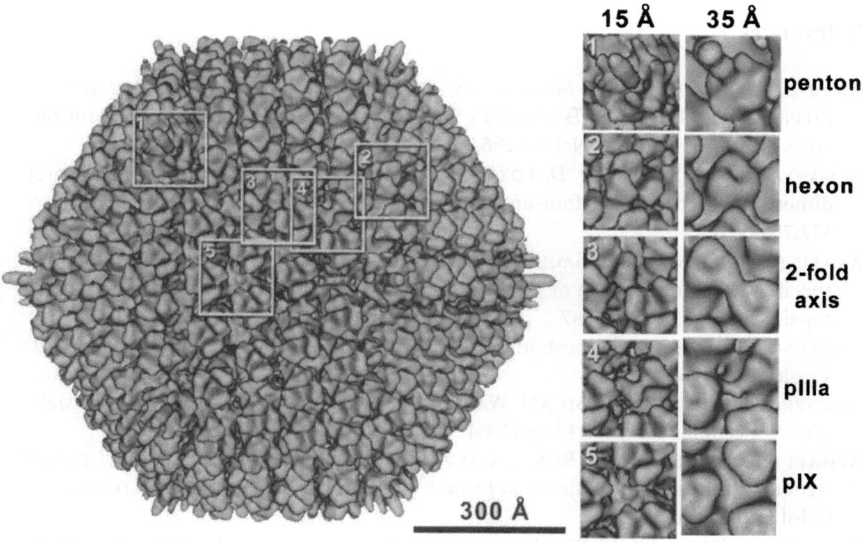

Fig. 8. Cryo-EM reconstruction of Ad5 at 15-Å resolution. The features inside the *boxes* are compared with the same features of the Ad2 reconstruction at 35 Å in the *panels* (*right*)

some adenoviruses could prevent formation of a regular lattice. Furthermore, the large size of the capsid would have made structure solution an almost intractable computational problem until a few years ago, even if diffraction data could be collected. These and similar problems are being overcome, and progress in crystallography of large viruses is steadily advancing (Grimes et al. 1998; Reinisch et al. 2000), including the very recent report of the crystallization of the adenovirus-like bacteriophage PRD1, complete with its internal membrane (Bamford et al. 2002b). In summary, there is a well-founded hope that structural techniques will produce new and exciting results in the coming years to increase our understanding of adenovirus biology and will help to develop improved vector systems for gene therapy and vaccine delivery.

Acknowledgements. This short review cannot do justice to the many people who have contributed to structural studies on adenovirus, in particular the early pioneers who made all subsequent work possible. We gratefully acknowledge their contribution. We also thank J. J. Rux for his careful reading of the manuscript. This work was supported by grants from the National Science Foundation (MCB 95-07102 and MCB 00-94577), the National Institutes of Health (AI-17270), the Human Frontiers Science Program (RG320/2001), and by the Wistar Cancer Center (CA 09171). C. S. M. was partially supported by a PNFPI Fellowship from the Spanish Ministerio de Educación, Cultura y Deporte.

References

Abad-Zapatero C, Abdel-Meguid SS, Johnson JE, Leslie AGW, Rayment I, Rossmann MG, Suck D, Tsukihara T (1980) Structure of southern bean mosaic virus at 2.8 Å resolution. Nature 286:33–39

Acharya R, Fry E, Stuart D, Fox G, Rowlands D, Brown F (1989) The three dimensional structure of foot-and-mouth disease virus at 2.9 Å resolution. Nature 337:709–716

Akalu A, Liebermann H, Bauer U, Granzow H, Seidel W (1999) The subgenus-specific C-terminal region of protein IX is located on the surface of the adenovirus capsid. J Virol 73:6182–6187

Anderson CW (1990) The proteinase polypeptide of adenovirus serotype 2 virions. Virology 177:259–272

Arnberg N, Edlund K, Kidd AH, Wadell G (2000) Adenovirus type 37 uses sialic acid as a cellular receptor. J Virol 74:42–48

Athappilly FK, Murali R, Rux JJ, Cai Z, Burnett RM (1994) The refined crystal structure of hexon, the major coat protein of adenovirus type 2, at 2.9 Å resolution. J Mol Biol 242:430–455

Babiuk LA, Tikoo SK (2000) Adenoviruses as vectors for delivering vaccines to mucosal surfaces. J Biotechnol 83:105–113

Bai M, Harfe B, Freimuth P (1993) Mutations that alter an Arg-Gly-Asp (RGD) sequence in the adenovirus type 2 penton base protein abolish its cell-rounding activity and delay virus reproduction in flat cells. J Virol 67:5198–5205

Baker D, Sali A (2001) Protein structure prediction and structural genomics. Science 294:93–96

Bamford DH, Caldentey J, Bamford JKH (1995) Bacteriophage PRD1: a broad host range dsDNA tectivirus with an internal membrane. Adv Virus Res 45:281–319

Bamford DH, Burnett RM, Stuart DI (2002a) Evolution of viral structure. Theoretical Population Biology 61:461–470

Bamford JKH, Bamford DH (2000) A new mutant class, made by targeted mutagenesis, of phage PRD1 reveals that protein P5 connects the receptor binding protein to the vertex. J Virol 74:7781–7786

Bamford JKH, Cockburn JJB, Diprose J, Grimes JM, Sutton G, Stuart DI, Bamford DH (2002b) Diffraction quality crystals of PRD1, a 66-MDa dsDNA virus with an internal membrane. J Struct Biol 139:103–112

Baniecki ML, McGrath WJ, McWhirter SM, Li C, Toledo DL, Pellicena P, Barnard DL, Thorn KS, Mangel WF (2001) Interaction of the human adenovirus proteinase with its 11-amino acid cofactor pVIc. Biochemistry 40:12349–12356

Baniecki ML, McGrath WJ, Dauter Z, Mangel WF (2002) Adenovirus proteinase: crystallization and preliminary X-ray diffraction studies to atomic resolution. Acta Crystallogr D58:1462–1464

Barton ES, Forrest JC, Connolly JL, Chappell JD, Liu Y, Schnell FJ, Nusrat A, Parkos CA, Dermody TS (2001) Junction adhesion molecule is a receptor for reovirus. Cell 104:441–451

Baumeister W, Steven AC (2000) Macromolecular electron microscopy in the era of structural genomics. Trends Biochem Sci 25:624–631

Bella J, Kolatkar PR, Marlor CW, Greve JM, Rossmann MG (1998) The structure of the two amino-terminal domains of human ICAM-1 suggests how it functions as a rhinovirus receptor and as an LFA-1 integrin ligand. Proc Natl Acad Sci USA 95:4140–4145

Belnap DM, Steven AC (2000) 'Déjà vu all over again': the similar structures of bacteriophage PRD1 and adenovirus. Trends Microbiol 8:91–93

Benkö M, Harrach B, Russell WC (2000) *Adenoviridae*. In: Van Regenmortel MHV, Fauquet CM, Bishop DHL, Carstens EB, Estes MK, Lemon SM, Maniloff J, Mayo MA, McGeoch DJ, Pringle CR, Wickner RB (eds) Virus Taxonomy. Academic Press, San Diego, pp 227–238

Benson SD, Bamford JKH, Bamford DH, Burnett RM (1999) Viral evolution revealed by bacteriophage PRD1 and human adenovirus coat protein structures. Cell 98:825–833

Benson SD, Bamford JKH, Bamford DH, Burnett RM (2002) The X-ray crystal structure of P3, the major coat protein of the lipid-containing bacteriophage PRD1, at 1.65 Å resolution. Acta Crystallogr D58:39–59

Bergelson JM, Cunningham JA, Droguett G, Kurt-Jones EA, Krithivas A, Hong JS, Horwitz MS, Crowell RL, Finberg RW (1997) Isolation of a common receptor for Coxsackie B viruses and adenoviruses 2 and 5. Science 275:1320–1323

Bewley MC, Springer K, Zhang Y-B, Freimuth P, Flanagan JM (1999) Structural analysis of the mechanism of adenovirus binding to its human cellular receptor, CAR. Science 286:1579–1583

Boulanger PA, Torpier G, Biserte G (1970) Investigation on intranuclear paracrystalline inclusions induced by adenovirus 5 in KB cells. J Gen Virol 6:329–332

Brown DT, Westphal M, Burlingham BT, Winterhoff U, Doerfler W (1975) Structure and composition of the adenovirus type 2 core. J Virol 16:366–387

Burnett RM (1985) The structure of the adenovirus capsid. II. The packing symmetry of hexon and its implications for viral architecture. J Mol Biol 185:125–143

Burnett RM (1997) The structure of adenovirus. In: Chiu W, Burnett RM, Garcea RL (eds) Structural biology of viruses. Oxford University Press, New York, pp 209–238

Butcher SJ, Bamford DH, Fuller SD (1995) DNA packaging orders the membrane of bacteriophage PRD1. EMBO J 14:6078–6086

Caldentey J, Tuma R, Bamford DH (2000) Assembly of bacteriophage PRD1 spike complex: role of the multidomain protein P5. Biochemistry 39:10566–10573

Caldentey J, Blanco L, Bamford DH, Salas M (1993) *In vitro* replication of bacteriophage PRD1 DNA. Characterization of the protein-primed initiation site. Nucleic Acids Res 21:3725–3730

Caspar DLD, Klug A (1962) Physical principles in the construction of regular viruses. Cold Spring Harb Symp Quant Biol 27:1–24

Cerritelli ME, Cheng N, Rosenberg AH, McPherson CE, Booy FP, Steven AC (1997) Encapsidated conformation of bacteriophage T7 DNA. Cell 91:271–280

Chappell JD, Prota AE, Dermody TS, Stehle T (2002) Crystal structure of reovirus attachment protein σ1 reveals evolutionary relationship to adenovirus fiber. EMBO J 21:1–11

Chatterjee PK, Flint SJ (1987) Adenovirus type 2 endopeptidase: an unusual phosphoprotein enzyme matured by autocatalysis. Proc Natl Acad Sci USA 84:714–718

Chatterjee PK, Vayda ME, Flint SJ (1986) Identification of proteins and protein domains that contact DNA within adenovirus nucleoprotein cores by ultraviolet light crosslinking of oligonucleotides ^{32}P-labelled *in vivo*. J Mol Biol 188:23–37

Chelvanayagam G, Heringa J, Argos P (1992) Anatomy and evolution of proteins displaying the viral capsid jellyroll topology. J Mol Biol 228:220–242

Chiu CY, Mathias P, Nemerow GR, Stewart PL (1999) Structure of adenovirus complexed with its internalization receptor, $\alpha_v\beta_5$ integrin. J Virol 73:6759–6768

Chiu CY, Wu E, Brown SL, von Seggern DJ, Nemerow GR, Stewart PL (2001) Structural analysis of a fiber-pseudotyped adenovirus with ocular tropism suggests differential modes of cell receptor interactions. J Virol 75:5375–5380

Choi H-K, Tong L, Minor W, Dumas P, Boege U, Rossmann MG, Wengler G (1991) Structure of Sindbis virus core protein reveals a chymotrypsin-like serine proteinase and the organization of the virion. Nature 354:37–43

Chroboczek J, Ruigrok RW, Cusack S (1995) Adenovirus fiber. Curr Top Microbiol Immunol 199:163–200

Cohen CJ, Shieh JTC, Pickles RJ, Okegawa T, Hsieh J-T, Bergelson JM (2001) The coxsackievirus and adenovirus receptor is a transmembrane component of the tight junction. Proc Natl Acad Sci USA 98:15191–15196

Colby WW, Shenk T (1981) Adenovirus type 5 virions can be assembled *in vivo* in the absence of detectable polypeptide IX. J Virol 39:977–980

Conway JF, Wikoff WR, Cheng N, Duda RL, Hendrix RW, Johnson JE, Steven AC (2001) Virus maturation involving large subunit rotations and local refolding. Science 292:744–748

Dekker J, Kanellopoulos PN, van Oosterhout JAWM, Stier G, Tucker PA, van der Vliet PC (1998) ATP-independent DNA unwinding by the adenovirus single-stranded DNA binding protein requires a flexible DNA binding loop. J Mol Biol 277:825–838

Dekker J, Kanellopoulos PN, Loonstra AK, van Oosterhout JAWM, Leonard K, Tucker PA, van der Vliet PC (1997) Multimerization of the adenovirus DNA-binding protein is the driving force for ATP-independent DNA unwinding during strand displacement synthesis. EMBO J 16:1455–1463

Devaux C, Timmins PA, Berthet-Colominas C (1983) Structural studies of adenovirus type 2 by neutron and X-ray scattering. J Mol Biol 167:119–132

Ding J, McGrath WJ, Sweet RM, Mangel WF (1996) Crystal structure of the human adenovirus proteinase with its 11 amino acid cofactor. EMBO J 15:1778–1783

Dmitriev IP, Kashentseva EA, Curiel DT (2002) Engineering of adenovirus vectors containing heterologous peptide sequences in the C terminus of capsid protein IX. J Virol 76:6893–6899

Durmort C, Stehlin C, Schoehn G, Mitraki A, Drouet E, Cusack S, Burmeister WP (2001) Structure of the fiber head of Ad3, a non-CAR-binding serotype of adenovirus. Virology 285:302–312

Earnshaw WC, Harrison SC (1977) DNA arrangement in isometric phage heads. Nature 268:598–602

Earnshaw WC, King J, Harrison SC, Eiserling FA (1978) The structural organization of DNA packaged within the heads of T4 wild-type, isometric and giant bacteriophages. Cell 14:559–568

Everitt E, Lutter L, Philipson L (1975) Structural proteins of adenoviruses. XII. Location and neighbor relationship among proteins of adenovirion type 2 as revealed by enzymatic iodination, immunoprecipitation and chemical cross-linking. Virology 67:197–208

Favier A-L, Schoehn G, Jaquinod M, Harsi C, Chroboczek J (2002) Structural studies of human enteric adenovirus type 41. Virology 293:75–85

Fender P, Ruigrok RWH, Gout E, Buffet S, Chroboczek J (1997) Adenovirus dodecahedron, a new vector for human gene transfer. Nature Biotechnol 15: 52–56

Furcinitti PS, van Oostrum J, Burnett RM (1989) Adenovirus polypeptide IX revealed as capsid cement by difference images from electron microscopy and crystallography. EMBO J 8:3563–3570

Ghosh-Choudhury G, Haj-Ahmad Y, Graham FL (1987) Protein IX, a minor component of the human adenovirus capsid, is essential for the packaging of full length genomes. EMBO J 6:1733–1739

Grahn AM, Caldentey J, Bamford JKH, Bamford DH (1999) Stable packaging of phage PRD1 DNA requires adsorption protein P2, which binds to the IncP plasmid-encoded conjugative transfer complex. J Bacteriol 181:6689–6696

Greber UF, Webster P, Weber J, Helenius A (1996) The role of the adenovirus protease in virus entry into cells. EMBO J 15:1766–1777

Green NM, Wrigley NG, Russell WC, Martin SR, McLachlan AD (1983) Evidence for a repeating cross-β sheet structure in the adenovirus fibre. EMBO J 2:1357–1365

Grimes JM, Jakana J, Ghosh M, Basak AK, Roy P, Chiu W, Stuart DI, Prasad BVV (1997) An atomic model of the outer layer of the bluetongue virus core derived from X-ray crystallography and electron cryomicroscopy. Structure 5: 885–893

Grimes JM, Burroughs JN, Gouet P, Diprose JM, Malby R, Ziéntara S, Mertens PPC, Stuart DI (1998) The atomic structure of the bluetongue virus core. Nature 395:470–478

Harrison SC, Olson AJ, Schutt CE, Winkler FK, Bricogne G (1978) Tomato bushy stunt virus at 2.9 Å resolution. Nature 276:368–373

Hendrix RW (1999) Evolution: the long evolutionary reach of viruses. Curr Biol 9:R914–917

Henry CJ, Slifkin M, Merkow LP, Pardo M (1971) The ultrastructure and nature of adenovirus type 2-induced paracrystalline formations. Virology 44:215–218

Henry LJ, Xia D, Wilke ME, Deisenhofer J, Gerard RD (1994) Characterization of the knob domain of the adenovirus type 5 fiber protein expressed in *Escherichia coli*. J Virol 68:5239–5246

Hess M, Cuzange A, Ruigrok RWH, Chroboczek J, Jacrot B (1995) The avian adenovirus penton: two fibres and one base. J Mol Biol 252:379–385

Horne RW, Brenner S, Waterson AP, Wildy P (1959) The icosahedral form of an adenovirus. J Mol Biol 1:84–86

Hosokawa K, Sung MT (1976) Isolation and characterization of an extremely basic protein from adenovirus type 5. J Virol 17:924–934

Imler J-L (1995) Adenovirus vectors as recombinant viral vaccines. Vaccine 13:1143–1151

Jackson T, Sharma A, Ghazaleh RA, Blakemore WE, Ellard FM, Simmons DL, Newman JWI, Stuart DI, King AMQ (1997) Arginine-glycine-aspartic acid-specific binding by foot-and-mouth disease viruses to the purified integrin $\alpha_v\beta_3$ in vitro. J Virol 71:8357–8361

Kanellopoulos PN, van der Zandt H, Tsernoglou D, van der Vliet PC, Tucker PA (1995) Crystallization and preliminary X-ray crystallographic studies on the adenovirus ssDNA binding protein in complex with ssDNA. J Struct Biol 115: 113–116
Kanellopoulos PN, Tsernoglou D, van der Vliet PC, Tucker PA (1996) Alternative arrangements of the protein chain are possible for the adenovirus single-stranded DNA binding protein. J Mol Biol 257:1–8
Kidd AH, Chroboczek J, Cusack S, Ruigrok RWH (1993) Adenovirus type 40 virions contain two distinct fibers. Virology 192:73–84
Kindt J, Tzlil S, Ben-Shaul A, Gelbart WM (2001) DNA packaging and ejection forces in bacteriophage. Proc Natl Acad Sci USA 98:13671–13674.
King AJ, van der Vliet PC (1994) A precursor terminal protein-trinucleotide intermediate during initiation of adenovirus DNA replication: regeneration of molecular ends *in vitro* by a jumping back mechanism. EMBO J 13:5786–5792
Klein H, Maltzman W, Levine AJ (1979) Structure-function relationships of the adenovirus DNA-binding protein. J Biol Chem 254:11051–11060
Kraulis PJ (1991) MOLSCRIPT: a program to produce both detailed and schematic plots of protein structures. J Appl Crystallogr 24:946–950
Kuhn RJ, Zhang W, Rossmann MG, Pletnev SV, Corver J, Lenches E, Jones CT, Mukhopadhyay S, Chipman PR, Strauss EG, Baker TS, Strauss JH (2002) Structure of dengue virus: implications for flavivirus organization, maturation, and fusion. Cell 108:717–725
Laver WG, Wrigley NG, Pereira HG (1969) Removal of pentons from particles of adenovirus type 2. Virology 39:599–605
Li P, Bellett AJD, Parish CR (1984) Structural organization and polypeptide composition of the avian adenovirus core. J Virol 52:638–649
Linne T, Philipson L (1980) Further characterization of the phosphate moiety of the adenovirus type 2 DNA-binding protein. Eur J Biochem 103:259–270
Louis N, Fender P, Barge A, Kitts P, Chroboczek J (1994) Cell-binding domain of adenovirus serotype 2 fiber. J Virol 68:4104–4106
Lupas A, van Dyke M, Stock J (1991) Predicting coiled coils from protein sequences. Science 252:1162–1164
Lutz P, Rosa-Calatrava M, Kedinger C (1997) The product of the adenovirus intermediate gene IX is a transcriptional activator. J Virol 71:5102–5109
Maizel JV, Jr, White DO, Scharff MD (1968) The polypeptides of adenovirus. I. Evidence for multiple protein components in the virion and a comparison of types 2, 7A, and 12. Virology 36:115–125
Mancini EJ, Clarke M, Gowen BE, Rutten T, Fuller SD (2000) Cryo-electron microscopy reveals the functional organization of an enveloped virus, Semliki Forest virus. Mol Cell 5:255–266
Mangel WF, McGrath WJ, Toledo DL, Anderson CW (1993) Viral DNA and a viral peptide can act as cofactors of adenovirus virion proteinase activity. Nature 361: 274–275
Mangel WF, Toledo DL, Brown MT, Martin JH, McGrath WJ (1996) Characterization of three components of human adenovirus proteinase activity *in vitro*. J Biol Chem 271:536–543
Mathias P, Wickham T, Moore M, Nemerow G (1994) Multiple adenovirus serotypes use α_v integrins for infection. J Virol 68:6811–6814

MATTHEWS DA, RUSSELL WC (1995) Adenovirus protein-protein interactions: molecular parameters governing the binding of protein VI to hexon and the activation of the adenovirus 23 K protease. J Gen Virol 76:1959–1969

MCGRATH WJ, DING J, SWEET RM, MANGEL WF (1996) Preparation and crystallization of a complex between human adenovirus serotype 2 proteinase and its 11-amino-acid cofactor pVIc. J Struct Biol 117:77–79

MIRZA MA, WEBER J (1982) Structure of adenovirus chromatin. Biochim Biophys Acta 696:76–86

MITRAKI A, BARGE A, CHROBOCZEK J, ANDRIEU J-P, GAGNON J, RUIGROK RWH (1999) Unfolding studies of human adenovirus type 2 fibre trimers. Evidence for a stable domain. Eur J Biochem 264:599–606

MORIN N, DELSERT C, KLESSIG DF (1989) Nuclear localization of the adenovirus DNA-binding protein: requirement for two signals and complementation during viral infection. Mol Cell Biol 9:4372–4380

NEMEROW GR, STEWART PL (1999) Role of α_v integrins in adenovirus cell entry and gene delivery. Microbiol Mol Biol Rev 63:725–734

NEWCOMB WW, BORING JW, BROWN JC (1984) Ion etching of human adenovirus 2: structure of the core. J Virol 51:52–56

PEREIRA HG, WRIGLEY NG (1974) *In vitro* reconstruction, hexon bonding and handedness of incomplete adenovirus capsid. J Mol Biol 85:617–631

PETTERSON U (1984) Structural and nonstructural adenovirus proteins. In: Ginsberg HS (ed) The Adenoviruses. Plenum Press, New York, pp 35–125

POLGAR L (1974) Mercaptide-imidazolium ion-pair: the reactive nucleophile in papain catalysis. FEBS Lett 47:15–18

PRAGE L, PETTERSSON U, HÖGLUND S, LONBERG-HOLM K, PHILIPSON L (1970) Structural proteins of adenoviruses. IV. Sequential degradation of the adenovirus type 2 virion. Virology 42:341–358

RAMACHANDRA M, PADMANABHAN R (1999) Adenovirus DNA replication. In: Seth P (ed) Adenoviruses: basic biology to gene therapy. RG Landes, Austin, Tex, USA, pp 59–68

REINISCH KM, NIBERT ML, HARRISON SC (2000) Structure of the reovirus core at 3.6 Å resolution. Nature 404:960–967

REKOSH DM, RUSSELL WC, BELLET AJ, ROBINSON AJ (1977) Identification of a protein linked to the ends of adenovirus DNA. Cell 11:283–295

ROELVINK PW, LIZONOVA A, LEE JGM, LI Y, BERGELSON JM, FINBERG RW, BROUGH DE, KOVESDI I, WICKHAM TJ (1998) The coxsackievirus-adenovirus receptor protein can function as a cellular attachment protein for adenovirus serotypes from subgroups A, C, D, E, and F. J Virol 72:7909–7915

ROSA-CALATRAVA M, GRAVE L, PUVION-DUTILLEUL F, CHATTON B, KEDINGER C (2001) Functional analysis of adenovirus protein IX identifies domains involved in capsid stability, transcriptional activity, and nuclear reorganization. J Virol 75: 7131–7141

ROWE WP, HUEBNER RJ, GILLMORE LK, PARROTT RH, WARD TG (1953) Isolation of a cytopathogenic agent from human adenoids undergoing spontaneous degeneration in tissue culture. Proc Soc Exp Biol Med 84:570–573

RUIGROK RWH, SCHÜLLER S (1993) Liquid crystalline DNA in fowl adenovirus. J Struct Biol 110:177–179

Ruigrok RWH, Barge A, Albiges-Rizo C, Dayan S (1990) Structure of adenovirus fibre. II. Morphology of single fibres. J Mol Biol 215:589–596

Ruigrok RWH, Barge A, Mittal SK, Jacrot B (1994) The fibre of bovine adenovirus type 3 is very long but bent. J Gen Virol 75:2069–2073

Russell WC (2000) Update on adenovirus and its vectors. J Gen Virol 81:2573–2604

Rux JJ, Burnett RM (2000) Type-specific epitope locations revealed by X-ray crystallographic study of adenovirus type 5 hexon. Mol Ther 1:18–30

Rydman PS, Caldentey J, Butcher SJ, Fuller SD, Rutten T, Bamford DH (1999) Bacteriophage PRD1 contains a labile receptor-binding structure at each vertex. J Mol Biol 291:575–587

Rydman PS, Bamford JKH, Bamford DH (2001) A minor capsid protein P30 is essential for bacteriophage PRD1 capsid assembly. J Mol Biol 313:785–795

San Martín C, Huiskonen JT, Bamford JKH, Butcher SJ, Fuller SD, Bamford DH, Burnett RM (2002) Minor proteins, mobile arms, and membrane-capsid interactions in the bacteriophage PRD1 capsid. Nature Struct Biol 9:756–762

San Martín C, Burnett RM, de Haas F, Heinkel R, Rutten T, Fuller SD, Butcher SJ, Bamford DH (2001) Combined EM/X-ray imaging yields a quasi-atomic model of the adenovirus-related bacteriophage PRD1 and shows key capsid and membrane interactions. Structure 9:917–930

Schechter NM, Davies W, Anderson CW (1980) Adenovirus coded deoxyribonucleic acid binding protein. Isolation, physical properties, and effects of proteolytic digestion. Biochemistry 19:2802–2810

Schoehn G, Fender P, Chroboczek J, Hewat EA (1996) Adenovirus 3 penton dodecahedron exhibits structural changes of the base on fibre binding. EMBO J 15:6841–6846

Seki T, Dmitriev I, Kashentseva E, Takayama K, Rots M, Suzuki K, Curiel DT (2002) Artificial extension of the adenovirus fiber shaft inhibits infectivity in coxsackievirus and adenovirus receptor-positive cell lines. J Virol 76:1100–1108.

Shayakhmetov DM, Lieber A (2000) Dependence of adenovirus infectivity on length of the fiber shaft domain. J Virol 74:10274–10286

Shiver JW, Fu T-M, Chen L, Casimiro DR, Davies M-E, Evans RK, Zhang Z-Q, Simon AJ, Trigona WL, Dubey SA, Huang L, Harris VA, Long RS, Liang X, Handt L, Schleif WA, Zhu L, Freed DC, Persaud NV, Guan L, Punt KS, Tang A, Chen M, Wilson KA, Collins KB, Heidecker GJ, Fernandez VR, Perry HC, Joyce JG, Grimm KM, Cook JC, Keller PM, Kresock DS, Mach H, Troutman RD, Isopi LA, Williams DM, Xu Z, Bohannon KE, Volkin DB, Montefiori DC, Miura A, Krivulka GR, Lifton MA, Kuroda MJ, Schmitz JE, Letvin NL, Caulfield MJ, Bett AJ, Youil R, Kaslow DC, Emini EA (2002) Replication-incompetent adenoviral vaccine vector elicits effective anti-immunodeficiency-virus immunity. Nature 415:331–335

Shortridge KF, Biddle F (1970) The proteins of adenovirus type 5. Arch Gesamte Virusforsch 29:1–24

Signäs C, Akusjärvi G, Pettersson U (1985) Adenovirus 3 fiber polypeptide gene: implications for the structure of the fiber protein. J Virol 53:672–678.

Smith KO, Gehle WD, Trousdale MD (1965) Architecture of the adenovirus capsid. J Bacteriol 90:254–261

Stevens RC, Yokoyama S, Wilson IA (2001) Global efforts in structural genomics. Science 294:89–92

Stewart PL, Burnett RM (1993) Adenovirus structure as revealed by X-ray crystallography, electron microscopy, and difference imaging. Jpn J Appl Phys 32:1342–1347

Stewart PL, Burnett RM (1995) Adenovirus structure by X-ray crystallography and electron microscopy. Curr Top Microbiol Immunol 199:25–38

Stewart PL, Burnett RM, Cyrklaff M, Fuller SD (1991) Image reconstruction reveals the complex molecular organization of adenovirus. Cell 67:145–154

Stewart PL, Fuller SD, Burnett RM (1993) Difference imaging of adenovirus: bridging the resolution gap between X-ray crystallography and electron microscopy. EMBO J 12:2589–2599

Stewart PL, Chiu CY, Huang S, Muir T, Zhao Y, Chait B, Mathias P, Nemerow GR (1997) Cryo-EM visualization of an exposed RGD epitope on adenovirus that escapes antibody neutralization. EMBO J 16:1189–1198

Stouten PFW, Sander C, Ruigrok RWH, Cusack S (1992) New triple-helical model for the shaft of the adenovirus fibre. J Mol Biol 226:1073–1084

Sussenbach JS (1984) The structure of the genome. In: Ginsberg HS (ed) The Adenoviruses. Plenum Press, New York, pp 35–125

Tate VE, Philipson L (1979) Parental adenovirus DNA accumulates in nucleosome-like structures in infected cells. Nucleic Acids Res 6:2769–2785

Toogood CIA, Crompton J, Hay RT (1992) Antipeptide antisera define neutralizing epitopes on the adenovirus hexon. J Gen Virol 73:1429–1435

Trotman LC, Mosberger N, Fornerod M, Stidwill RP, Greber UF (2001) Import of adenovirus DNA involves the nuclear pore complex receptor CAN/Nup214 and histone H1. Nat Cell Biol 3:1092–1100.

Tsernoglou D, Tucker AD, Van der Vliet PC (1984) Crystallization of a fragment of the adenovirus DNA binding protein. J Mol Biol 172:237–239

Tsernoglou D, Tsugita A, Tucker AD, van der Vliet PC (1985) Characterization of the chymotryptic core of the adenovirus DNA-binding protein. FEBS Lett 188: 248–252

Tucker PA, Tsernoglou D, Tucker AD, Coenjaerts FEJ, Leenders H, van der Vliet PC (1994) Crystal structure of the adenovirus DNA binding protein reveals a hook-on model for cooperative DNA binding. EMBO J 13:2994–3002

van der Vliet PC, Keegstra W, Jansz HS (1978) Complex formation between the adenovirus type 5 DNA-binding protein and single-stranded DNA. Eur J Biochem 86:389–398

van Oostrum J, Burnett RM (1985) Molecular composition of the adenovirus type 2 virion. J Virol 56:439–448

van Raaij MJ, Louis N, Chroboczek J, Cusack S (1999a) Structure of the human adenovirus serotype 2 fiber head domain at 1.5 Å resolution. Virology 262:333–343

van Raaij MJ, Mitraki A, Lavigne G, Cusack S (1999b) A triple β-spiral in the adenovirus fibre shaft reveals a new structural motif for a fibrous protein. Nature 401:935–938

Vayda ME, Flint SJ (1987) Isolation and characterization of adenovirus core nucleoprotein subunits. J Virol 61:3335–3339

Vayda ME, Rogers AE, Flint SJ (1983) The structure of nucleoprotein cores released from adenovirions. Nucleic Acids Res 11:441–460

Verdaguer N, Mateu MG, Andreu D, Giralt E, Domingo E, Fita I (1995) Structure of the major antigenic loop of foot-and-mouth disease virus complexed with a neutralizing antibody: direct involvement of the Arg-Gly-Asp motif in the interaction. EMBO J 14:1690–1696

Von Seggern DJ, Chiu CY, Fleck SK, Stewart PL, Nemerow GR (1999) A helper-independent adenovirus vector with E1, E3, and fiber deleted: structure and infectivity of fiberless particles. J Virol 73:1601–1608

Vos HL, Brough DE, van der Lee FM, Hoeben RC, Verheijden GFM, Dooijes D, Klessig DF, Sussenbach JS (1989) Characterization of adenovirus type 5 insertion and deletion mutants encoding altered DNA binding proteins. Virology 172: 634–642

Ward P, Dean FB, O'Donnell ME, Berns KI (1998) Role of the adenovirus DNA-binding protein in in vitro adeno-associated virus DNA replication. J Virol 72: 420–427

Weber JM (1995) Adenovirus endopeptidase and its role in virus infection. Curr Top Microbiol Immunol 199:227–235

Weber JM (1999) Role of endoprotease in adenovirus infection. In: Seth P (ed) Adenoviruses: basic biology to gene therapy. RG Landes, Austin, Tex, USA, pp 79–83

Wickham TJ, Mathias P, Cheresh DA, Nemerow GR (1993) Integrins $\alpha_v\beta_3$ and $\alpha_v\beta_5$ promote adenovirus internalization but not virus attachment. Cell 73:309–319

Wickham TJ, Filardo EJ, Cheresh DA, Nemerow GR (1994) Integrin $\alpha_v\beta_5$ selectively promotes adenovirus mediated cell membrane permeabilization. J Cell Biol 127:257–264

Wong M-L, Hsu M-T (1989) Linear adenovirus DNA is organized into supercoiled domains in virus particles. Nucleic Acids Res 17:3535–3550

Wu E, Fernandez J, Fleck SK, von Seggern DJ, Huang S, Nemerow GR (2001) A 50-kDa membrane protein mediates sialic acid-independent binding and infection of conjunctival cells by adenovirus type 37. Virology 279:78–89

Xia D, Henry LJ, Gerard RD, Deisenhofer J (1994) Crystal structure of the receptor-binding domain of adenovirus type 5 fiber protein at 1.7 Å resolution. Structure 2:1259–1270

Xiong J-P, Stehle T, Zhang R, Joachimiak A, Frech M, Goodman SL, Arnaout MA (2002) Crystal Structure of the Extracellular Segment of Integrin $\alpha V\beta 3$ in Complex with an Arg-Gly-Asp Ligand. Science 296:151–155

Zhang Z, Greene B, Thuman-Commike PA, Jakana J, Prevelige PE, Jr, King J, Chiu W (2000) Visualization of the maturation transition in bacteriophage P22 by electron cryomicroscopy. J Mol Biol 297:615–626

Zhou ZH, Chen DH, Jakana J, Rixon FJ, Chiu W (1999) Visualization of tegument-capsid interactions and DNA in intact herpes simplex virus type 1 virions. J Virol 73:3210–3218

Zhou ZH, Dougherty M, Jakana J, He J, Rixon FJ, Chiu W (2000) Seeing the herpesvirus capsid at 8.5 Å. Science 288:877–880

Zhou ZH, Baker ML, Jiang W, Dougherty M, Jakana J, Dong G, Lu G, Chiu W (2001) Electron cryomicroscopy and bioinformatics suggest protein fold models for rice dwarf virus. Nat Struct Biol 8:868–873

Zijderveld DC, van der Vliet PC (1994) Helix-destabilizing properties of the adenovirus DNA-binding protein. J Virol 68:1158–1164

2
Viral Replication

CHAPTER 4

The Multifunctional Role of E1A in the Transcriptional Regulation of CREB/CBP-Dependent Target Genes

D. BROCKMANN, H. ESCHE

Abstract. Oncoproteins encoded by the early region 1A (E1A) of adenoviruses (Ads) have been shown to be powerful tools to study gene regulatory mechanisms. As E1A proteins lack a sequence-specific DNA-binding activity, they modulate viral and cellular gene expression by interacting directly with a diverse array of cellular factors, among them sequence-specific transcription factors, proteins of the general transcription machinery, co-activators and chromatin-modifying enzymes. By making

D. BROCKMANN, H. ESCHE
Institute of Molecular Biology (Cancer Research), University of Essen Medical School, Hufelandstraße 55, 45122 Essen, Germany
e-mail: helmut.esche@uni-essen.de

use of these factors, E1A affects major cellular events such as cell cycle control, differentiation, apoptosis, and oncogenic transformation. In this review we will focus on the interaction of E1A with cellular components involved in the cAMP/PKA signal transduction pathway and we will discuss the consequences of these interactions in respect to the activation of CREB/CBP-dependent target genes.

1 Introduction

Adenovirus (Ad) early region 1A (E1A) proteins were among the first eukaryotic transcriptional regulators identified (BERK et al. 1979; JONES and SHENK 1979; LEWIS and MATHEWS 1980). They generally induce the transcription of all viral and specific cellular genes and act at multiple steps in the transcription process. For instance, E1A proteins stimulate transcription by stabilizing the preinitiation complex (PIC) (for a review see FLINT and SHENK 1997), by recruiting co-factors to respective promoters and by interfering with the nuclear proteasomal degradation machinery (TURNELL et al. 2000). Moreover, these viral proteins also inhibit the transcription of viral and cellular genes, thereby modulating viral gene expression at defined stages of the viral replication cycle and antagonizing expression of cellular genes that inhibit viral propagation (ROCHETTE-EGLY et al. 1990; BAYLEY and MYMRYK 1994; SHENK 1996). Thus, E1A proteins are excellent models for analysing mechanisms of eukaryotic transcriptional modulation.

Virus encoded transcriptional regulatory proteins can be divided into two groups based on whether or not they bind sequence-specifically to DNA. The Ad E1A proteins belong to the group of transcriptional regulators which do not bind to DNA in a sequence-specific manner (FERGUSON et al. 1985; ZU et al. 1992). Rather, they modulate transcription by interacting with cellular factors, e.g. components of the basal transcription machinery, transcription factors which bind to upstream promoter and enhancer elements, co-factors, and regulatory proteins that control the activity of DNA-binding factors. In this review we will focus on the functional relationship between E1A gene products and the cAMP/PKA signal transduction pathway. Especially, we will discuss the multifunctional role of E1A proteins in the transcriptional regulation of CREB/CBP-dependent target genes.

For reasons of clarity, a respective prefix is added to differentiate between E1A gene products of Ad2, Ad5 and Ad12 in the following text, which are the best characterized serotypes. If the prefix is missing, the data and conclusions refer to all three serotypes.

2 Structure and General Properties of Ad12-E1A Gene Products

The Ad12-E1A transcription unit, located at the extreme left end of the Ad12 genome, is the first to be expressed after the viral DNA enters the nucleus of the infected cell (for a review see SHENK 1996). By alternative splicing of a common Ad12-E1A RNA precursor, which is transcribed from a constitutively active promoter, six different mRNAs are generated (13S, 12S, 11S, 10S, 9.5S and 9S according to their sedimentation coefficients), which give rise to five distinct proteins of 266 amino acid (aa) residues (13S mRNA), 235aa (12S mRNA), 106aa (11S and 10S mRNA), 52aa (9.5S mRNA), and 52aa (9S mRNA) (Fig. 1; PERRICAUDET et al. 1980; SAWADA and FUJINAGA 1980; BROCKMANN et al. 1990; for a review see BROCKMANN and ESCHE 1995). These proteins are referred to as Ad12-$E1A_{13S}$, Ad12-$E1A_{12S}$, Ad12-$E1A_{11S/10S}$, Ad12-$E1A_{9.5S}$, and Ad12-$E1A_{9S}$ in this review. The E1A proteins translated from the two major differentially spliced mRNAs (13S and 12S) share the same N- and C-terminal aa sequences but differ in a 31aa internal, cystine-rich segment, unique to the larger E1A protein (Fig. 1B). This segment represents one out of three E1A domains, which are highly conserved among all human Ads (KIMELMANN et al. 1985; MORAN and MATHEWS 1987), termed conserved regions 1, 2, and 3 (CR1, CR2, CR3). As it will be discussed later, CR1–3 and the N-terminal region express most of the trans-regulatory functions of both larger proteins of Ad12-E1A. Ad12-$E1A_{11S/10S}$, Ad12-$E1A_{9.5S}$, and Ad12-$E1A_{9S}$ do not carry any of the CRs due to differential splicing, which causes frame shifts after the first splice (BROCKMANN et al. 1990; Fig. 1A). They share only the N terminus (aa 1–29) with Ad12-$E1A_{13S}$ and Ad12-$E1A_{12S}$ (Fig. 1B).

$E1A_{13S}$ and $E1A_{12S}$ are potent transcriptional regulators which interfere with a variety of cellular processes such as modulation of gene expression, inhibition of cellular differentiation, promotion of cell cycle progression and transformation. They are essential for the trans-activation of all other adenoviral transcription units (JONES and SHENK 1979, NEVINS 1981), induce the expression of specific cellular genes (for a review see SHENK 1996) and can repress enhancer-mediated gene expression (ROCHETTE-EGLY et al. 1990). In contrast, very little is known about the biological

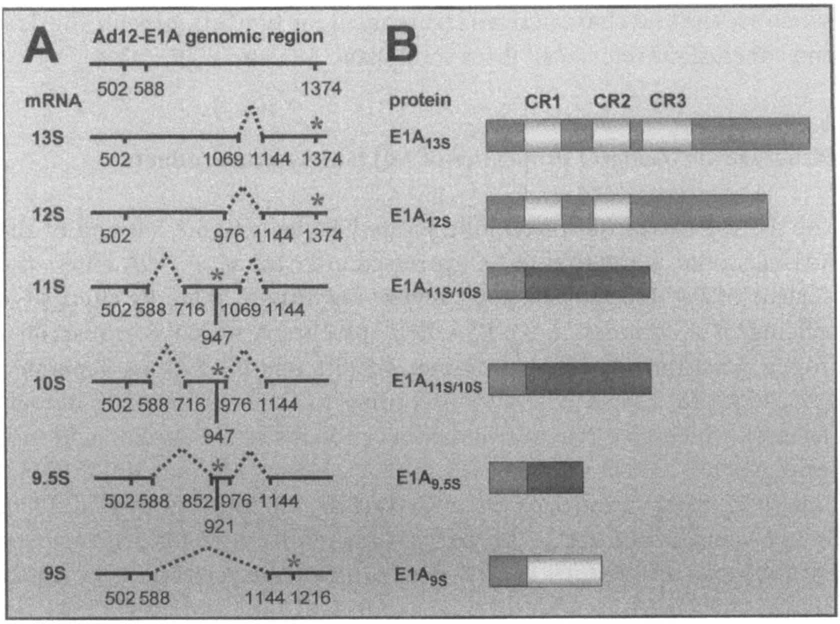

Fig. 1. Schematic representation of the Ad12-E1A mRNAs (**A**) and proteins (**B**). The *numbers below the lines* refer to the positions of nucleotides in the Ad12 genome, indicating the common start codon, the splice junctions, and the first nucleotide of the respective stop codon (indicated also by *red asterisks above the lines* in **A**. Reading frames and frame shifts of the Ad12-E1A proteins are indicated by different *colours* (**B**). The conserved regions 1–3 (CR1, CR2, CR3) are shown in *grey*

functions of the other Ad12-E1A proteins. Ad12-E1A$_{9.5S}$ represses activator protein 1 activated gene expression, which correlates with a physical interaction between E1A$_{9.5S}$ and the DNA-binding region and the leucine zipper structure of the activator protein 1 constituent Jun (BROCKMANN et al. 1995). In the case of Ad2 and Ad5, E1A$_{11S}$ and E1A$_{10S}$ are dispensable for lytic growth of the virus (STEPHENS and HARLOW 1987; ULFENDAHL et al. 1987) although it has been shown that Ad2-E1A$_{11S}$ constitutes an efficient transcriptional activator in transient expression assays (ULFENDAHL et al. 1987). Of note, neither Ad2/Ad5-E1A$_{11S}$ nor Ad2/Ad5-E1A$_{10S}$ have any homology with Ad12-E1A$_{11S/10S}$ or Ad12-E1A$_{9.5S}$ (for a review see BROCKMANN and ESCHE 1995). As Ad2/Ad5-E1A$_{11S}$ and Ad2/Ad5-E1A$_{10S}$ are translated in the same reading frame as Ad2/Ad5-E1A$_{13S}$ and Ad2/Ad5-E1A$_{12S}$ (STEPHENS and HARLOW 1987; ULFENDAHL et al. 1987) they, differ from both larger Ad2/Ad5 proteins only in that the functional domain CR1 is removed by splicing. Due to their poor functional characterization the smaller E1A proteins will not be discussed further in this review.

3
E1A Proteins as Transcription Factors

In contrast with classical transcription factors, E1A proteins lack a sequence-specific DNA-binding activity (Ferguson et al. 1985; Zu et al. 1992). The question therefore arises of how E1A might then modulate gene expression. The breakthrough in the understanding of the regulatory functions of E1A came with the identification of a great variety of cellular regulatory factors with which E1A proteins interact. These factors can be subdivided into four main classes.

3.1
General Transcription Factors

E1A binds to a variety of general transcription factors. For example, several reports have demonstrated that CR3, which is unique for the $E1A_{13S}$ protein, interacts with subunits of the $TF_{II}D$ complex, namely TBP (Horikoshi et al. 1991; Lee et al. 1991; Lipinski et al. 1998), and the TBP-associated factors TAF250 and TAF110 (Geisberg et al. 1995; Marzzarelli et al. 1995, 1997). Moreover, the N terminus of Ad12-E1A associates with the $TF_{II}F$ subunit RAP30 (Lipinski et al 1998). The ability to bind the components of $TF_{II}D$ and $TF_{II}F$ most probably enables E1A to act as a molecular bridge, connecting sequence-specific transcription factors to general transcription factors, thereby helping to stabilize the PIC.

3.2
Sequence-Specific Transcription Factors

The $E1A_{13S}$ protein can activate transcription through a CR3-mediated interaction with cellular transcription factors such as c-Jun and activating transcription factor 2 (ATF-2; Lillie and Green 1989; Liu and Green 1990, 1994). Amino acid sequences of Ad12-E1A spanning the non-conserved N terminus and CR1 (aa 1–79) interact with the cAMP-response element (CRE) binding protein (CREB) and with ATF-1 (Fax et al. 2000a;

see also Sect. 4.1.1). All but one of the early Ad promoters contain CREs, and E1A activates transcription through these sites partly because it can bind to the above-mentioned cellular transcription factors.

Other sequence-specific transcription factors interacting with CR3 and the N -terminal domain of E1A are YY1 (LEE et al. 1995; LEWIS et al. 1995) and Oct-4 (SCHÖLER et al. 1991; BREHM et al. 1999). YY1, which represses certain promoters, interacts with histone deacetylases (HDACs) in vitro and in vivo, indicating that it might repress transcription by targeting HDACs to the promoter (for a review see THOMAS and SETO 1999). As the HDAC binding site partially overlaps with the E1A binding site in YY1, it seems possible that E1A relieves repression by displacing HDACs from YY1.

3.3
Repressors of General or Sequence-Specific Transcription Factors

This class of E1A-interacting proteins includes the pocket proteins pRB, p107 and p130 (WHYTE et al. 1988; LI et al. 1993), which are inhibitors of the sequence-specific transcription factor E2F (BAGCHI et al. 1991; BANDARA and LATHANGUE 1991), and the Dr1/DRAP1 complex, which binds to TBP, thereby precluding the entry of $TF_{II}A$ and/or $TF_{II}B$ to the PIC (INOSTROZA et al. 1992; KRAUS et al. 1994; MERMELSTEIN et al. 1996). CR2 of E1A is primarily responsible for the interaction with the pocket proteins whereas the N-terminal portion of CR1 plays an auxiliary role, apparently stabilizing the interaction (DYSON et al. 1992; FATTAEY et al. 1993). The interaction of E2F with pRB family members is normally regulated by phosphorylation mediated by G1 cyclin-dependent kinases (BANDARA et al. 1991; HAMEL et al. 1992). By binding to hypophosphorylated pRB, the viral protein competes with E2F for binding to the pocket domain, dissociating the tumour suppressor protein from E2F (BAGCHI et al. 1991; BANDARA et al. 1991; CHELLAPPAN et al. 1992). Released from the pocket proteins, E2F can activate target promoters involved in the expression of S-phase specific genes, thereby promoting the cell into S-phase (HIEBERT et al. 1989, 1991; SIDLE et al. 1996).

The TBP inhibitor Dr1/DRAP1 is bound by the N terminus of the larger E1A proteins (KRAUS et al. 1994, MERMELSTEIN et al. 1996). This interaction leads to the dissociation of the DR1/DRAP1-TBP complex, releasing TBP which in turn is able to form an active transcription initiation complex (KRAUS et al. 1994).

3.4 Co-activators

E1A proteins also interact with a variety of co-activators. For example, the N terminus and CR1 bind to acetyltransferases (ATs) such as p300, the CREB-binding protein (CBP) and the p300/CBP associated factor (PCAF) (Eckner et al. 1994; Arany et al. 1995; Dorsman et al. 1997; Reid et al. 1998; Lipinski et al. 1999; see also Sect. 4.1.2). In addition, CR3 associates via the SUR-2 protein with the human Srb/Mediator complex (Boyer et al. 1999).

In summary, E1A proteins have evolved the property of binding to a huge variety of transcriptional regulatory proteins; most interactions lead to an activation whereas others lead to an inhibition of target gene expression. Physical and functional interactions of the E1A proteins with all these cellular factors allows the virus to activate the expression of its own genome and to deregulate the expression of cellular genes, with the consequence that the host cell is driven into the S-phase of the cell cycle, thereby providing an environment that is efficient for viral propagation. In the next chapters we will discuss in detail the multifunctional role that Ad12-E1A proteins play in the transcriptional regulation of the CREB/CBP-dependent target genes, for example in the regulation of the expression of the Ad12 E2 ($E2_{Ad12}$) transcription unit.

4 Cross-Talk of E1A with Factors of the cAMP/PKA Signal Transduction Pathway

The cAMP/protein kinase A (PKA) signal transduction pathway regulates a variety of different cellular processes such as gene transcription, proliferation, differentiation, neuronal activity and memory (for reviews see Shaywitz and Greenberg 1999; Edwards and Scott 2000). It is organized as a cascade of regulatory components among which the second messenger adenosine 3′,5′-cyclic monophosphate (cAMP) constitutes a key factor. The cAMP concentration is increased in response to a number of extracellular stimuli such as hormones, neurotransmitters or other signal substances via specific extracellular receptors linked by a heterotrimeric G-protein complex to an intracellular effector adenylyl cyclase. Two cAMP molecules in turn bind co-operatively to each regulatory (R) subunit of the inactive PKA holoenzyme, which consists of two catalytic (C) subunits bound to an R dimer. Upon binding, the PKA holoenzyme dissociates, giving rise to two active C subunits which translocate into the nucleus and phosphorylate target proteins such as CREB on Ser133 (Gonzales and Montminy 1989; Hagiwara et al. 1993). Phosphorylated CREB is then able to recruit CBP to the CREs of specific target promoters (Chrivia et al. 1993), giving rise to a transcriptionally active transcription factor/co-activator complex.

There are two main classes of PKA, type I and type II, which both share a common C subunit but contain different R-isoforms (RIa, RIb, RIIa, RIIb, respectively), which differ in their cAMP-binding affinities and in their localization in the cell (for a review see Taskén et al. 1997). Three distinct C subunits (Ca, Cb, and Cg) have also been isolated. Although there are subtle differences in the kinetic characteristics and cAMP sensitivities of Ca- and Cb-containing holoenzymes, the two isoforms are virtually indistinguishable with respect to substrate specificity and interaction with R subunits.

A large portion of PKA-II is targeted to subcellular components through association of the RII subunits with cellular adapter proteins known as A-kinase anchoring proteins (AKAPs) (for a review see Dell'Acqua and Scott 1997). In contrast, the PKA-I holoenzyme is predominantly soluble in the cytosol although growing evidence suggests that PKA-I might also be localized to specific subcellular components through dual-specificity AKAPs. It is assumed that this compartmentalization of PKA to distinct subcellular loci as well as the composition of the PKA holoenzyme contribute to the specificity in cellular responses to extracellular signals.

The expression of viral and cellular genes, whose transcriptional regulation is controlled by the cAMP/PKA signal transduction pathway, can be modulated by E1A either positively or negatively. For instance, the $E2_{Ad12}$ transcription unit that encodes proteins essential for viral DNA replication (the DNA-binding protein, the terminal-protein precursor and the DNA-polymerase; Swaminathan and Thimmapaya 1995) is activated by the Ad12-$E1A_{12S}$ protein (Fax et al. 2000a), whereas the CRE-containing rat somatostatin promoter is repressed by $E1A_{12S}$ (Arany et al. 1995; Lundblad et al. 1995). These observations underscore the high potential and flexibility with which E1A is able to interfere in a given pathway. On the other hand they give rise to two questions: (a) what are the target molecules for E1A in the cAMP/PKA signal transduction cascade, and (b) how can an interaction of E1A with a limited set of target molecules of the same pathway lead either to a positive or a negative response at the level of transcriptional regulation?

4.1 The $E2_{Ad12}$-Promoter – A Powerful Tool to Study the Interaction of E1A with Different Components of the cAMP/PKA-Pathway

As briefly mentioned above, expression of transcription unit E2 is essential for viral propagation. The $E2_{Ad12}$-promoter, which controls E2 expression, constitutes a simply organized promoter carrying a CRE (nt –99 to –92 relative to the transcriptional start site), a single E2F recognition sequence (nt –89 to –82) and a TATA box (nt –29 to –24; FAX et al. 2000 a). Mutational analyses revealed that the CRE and the TATA box are essential cis-regulatory elements for promoter activation whereas the E2F site is dispensable at least for the cell systems analysed and for the transcriptional activation mediated through Ad12-$E1A_{12S}$. The CRE is bound by CREB and ATF-1 and several experiments confirmed that the $E2_{Ad12}$-promoter is driven by CREB: (a) it is strongly activated by PKA, (b) it is inhibited by the PKA-specific inhibitor PKI. (c) Activation of the $E2_{Ad12}$-promoter is inhibited by A-CREB, which is a DNA-binding defective, dominant-negative inhibitor of CREB (AHN et al. 1998).

Interestingly, the $E2_{Ad12}$-promoter differs significantly from the respective promoter of non-oncogenic Ad2 (Ad2-E2e). The latter represents a classical model system for studying the regulation of E2F-dependent gene expression (for a review see SWAMINATHAN and THIMMAPAYA 1995). In contrast to the $E2_{Ad12}$-promoter, Ad2-E2e contains two inverted E2F and one ATF transcription factor binding site. All three cis-regulatory elements are critical for Ad2-E1A-mediated trans-activation. In addition, transcriptional activation is enhanced by the open reading frame 6/7 protein of the viral region E4, which enables the co-operative and stable binding of two E2F transcription factors. However, the factors binding to the Ad2-E2e ATF site are not yet clearly defined, and cAMP inducibility of the Ad2-E2e promoter seems to be dependent on the cell system analysed (SASSONE-CORSI 1988; LEE et al. 1989; LOEKEN 1993). In contrast, the $E2_{Ad12}$-promoter is inducible by PKA in all cell systems studied until now including KB, HeLa and F9 cells, indicating a regulation mechanism independent of the cell system (P. Fax, H. Esche, D. Brockmann, unpublished results). These results lead to the surprising conclusion that various Ad serotypes make use of different cellular transcription factors coupled to different cellular pathways to activate homologous genes.

4.1.1 *E1A Physically Interacts with CREB and ATF-1 – The Promoter Recruitment Model*

As Ad12-E1A proteins lack a sequence-specific DNA binding activity (Ferguson et al. 1985; Zu et al. 1992), the question arises how could they then activate the $E2_{Ad12}$ promoter. An attractive model predicts that the viral proteins might be recruited to the promoter by directly binding to CREB and/or ATF-1.

CREB and ATF-1 belong to the so-called CRE subfamily of the bZIP class of transcription factors, which bind as dimers to CRE elements (5′TGACGTCA-3′) localized in target promoters (for a review see Shaywitz and Greenberg 1999). Dimerization of CREB is facilitated through a leucine zipper structure located at the very C-terminal end of the transcription factor. A basic region immediately in front of the leucine zipper is responsible for binding to the CRE. The N-terminal portion of CREB harbours the trans-activation domain consisting of the kinase-inducible-domain (KID; aa 100–160) flanked by two glutamine-rich regions (Q1 and Q2). Among these the KID is both necessary and sufficient for signal-induced activation of CREB (Gonzales et al. 1991; Brindle et al. 1993; Quinn 1993). With respect to the cAMP/PKA signal transduction pathway, activation of CREB is mediated by phosphorylation of Ser133 within the KID through PKA. This phosphorylation allows phospho-CREB to interact with CBP, which is a prerequisite for transcriptional activation.

Biochemical analyses have shown that Ad12-$E1A_{12S}$ binds to both CREB and ATF-1 in vitro and in vivo (Fax et al. 2000a). Interaction is facilitated through the N terminus and CR1 of Ad12-$E1A_{12S}$. Most interestingly, usage of point mutants demonstrated that Ad12-$E1A_{12S}$ modules essential for binding to CREB/ATF-1 are also essential for the $E2_{Ad12}$ promoter activation. Several lines of evidence suggest that CREB as well as ATF-1 recruit the Ad12-$E1A_{12S}$ protein to the $E2_{Ad12}$-CRE indicating that the mechanism of $E2_{Ad12}$ promoter activation comprises the assembly of an Ad12-$E1A_{12S}$/CREB/ATF-1 complex on the $E2_{Ad12}$-CRE.

In addition to the N terminus and CR1, CR3 which is unique for the $E1A_{13S}$ protein, is also able to bind at least to ATF-1 (Liu and Green 1994). CR3 consists of two subdomains: (a) a promoter targeting region in the C-terminal part, and (b) an activation domain including a zinc finger structure of the type $CX_2CX_{13}CX_2C$ (C = cysteine, X = any aa; Glenn and

Ricciardi 1987; Culp et al. 1988; Lillie and Green 1989; Pei and Berk 1989; Martin et al. 1990; Webster and Ricciardi 1991). The promoter targeting subdomain adapts $E1A_{13S}$ to the bZIP domain of ATF-1 and therefore directs the viral protein to respective cis-regulatory promoter elements (Liu and Green 1994). Interestingly, the CR3 promoter-targeting domain interacts with the bZIP domains of other transcription factors of this family as well, such as ATF-2, ATF-3 and c-Jun (Liu and Green 1994). Moreover, this domain is also able to associate with the DNA-binding domains of a variety of other sequence-specific transcription factors, among them the zinc-finger DNA-binding domain of Sp1, the basic/helix-loop-helix domain of upstream stimulatory factor (Liu and Green 1994) and the POU domain of the germline transcription factor Oct-4 (Schöler et al. 1991; Brehm et al. 1999). On the other hand, it is not yet clear whether CR3 also binds directly to CREB. Taken together, these results indicate a general recruitment mechanism through a variety of different cellular transcription factors that enables E1A to activate the expression of a huge variety of target genes.

However, there are several experiments arguing against a simple targeting model in which the only function of sequence-specific transcription factors is to adapt the E1A trans-activation domains to respective promoters. For example, using a series of ATF-2 mutants, Liu and Green (1990, 1994) have shown that although promoter targeting is mediated through the bZIP domain of ATF-2, E1A-dependent transcriptional activation requires the trans-activation domain of ATF-2 located in the N terminus. Moreover, an Oct-4 mutant consisting only of the POU DNA-binding domain binds $E1A_{13S}$ but failed to activate target gene expression in the Ad5-E1A expressing cell line 293 (Brehm et al. 1999). Activation strictly depends on the N-terminus-located trans-activation domain of Oct-4. These experiments suggest: (a) a synergism between the trans-activation domains of E1A with those of sequence-specific transcription factors, and/or (b) that other factors with co-activator activity might be involved in the activation process, too.

4.1.2
P300/CBP – Important Check-Points for E1A Functions

Co-activators such as p300 and CBP have been the subjects of intensive research since it was recognized that both proteins play important roles in transcriptional processes. P300 was identified as an E1A-associated factor more than two decades ago (YEE and BRANTON 1985; EGAN et al. 1988; WHYTE et al. 1988; WANG et al. 1993a, 1993b; ARANY et al. 1994; ECKNER et al. 1994, DORSMAN et al. 1997; LIPINSKI et al. 1999). It quickly became clear that p300 constitutes a key factor in a variety of E1A functions such as modulation of viral and cellular gene expression, regulation of the cell cycle, inhibition of cellular differentiation, and transformation (for reviews see BAYLEY and MYMRYK 1994; GOODMAN and SMOLIK 2000). P300 is closely related to CBP the latter of which was identified independently of p300 as an essential co-activator of phospho-CREB (CHRIVIA et al. 1993). Although p300 and CBP have partially different biological properties, they are here referred to as p300/CBP unless otherwise specified due to their indistinguishable function concerning the activation and repression of some promoters discussed in this chapter (GOLDMAN et al. 1997; KAWASAKI et al. 1998; YAO et al. 1998; GIORDANO and AVANTAGGIATI 1999; KUNG et al. 2000).

P300 and CBP are large nuclear-localized proteins of 2,414 and 2,441 aa, respectively. Both bind to a large variety of cellular and viral transcription factors including E1A, SV40 large T, nuclear hormone receptors, Stat1, Stat2, CREB, c-Jun, c-Fos, c-Myc, c-Myb, p53, MyoD, E2F, and p65/NF-kB (for reviews see GIORDANO and AVANTAGGIATI 1999; GOODMAN and SMOLIK 2000). Furthermore, p300 and CBP were shown to bind to factors of the general transcriptional machinery such as TBP, $TF_{II}B$, RNA helicase A, and RNA PolII (ABRAHAM et al. 1993; KEE et al. 1996; KWOK et al. 1994; SWOPE et al. 1996; DALLAS et al. 1997; NAKAJIMA et al. 1997; CHO et al. 1998) and to form complexes with other co-factors, among them PCAF, SRC-1, or p/CIP (YANG et al. 1996; YAO et al. 1996; TORCHIA et al. 1997). Interactions are mediated through several protein motifs of p300/CBP such as the bromodomain, the KID interaction (KIX) domain, and three regions rich in Cys and His aa (C/H domains; GOODMAN and SMOLIK 2000).

In principal, p300/CBP activates gene expression through three mechanisms. (a) It serves as a molecular scaffold bridging sequence-specific transcription factors to other co-activators, such as SRC1 or PCAF, and to the RNA PolII holoenzyme (NAKAJIMA et al. 1997; for review see GOODMAN and SMOLIK 2000). (b) P300/CBP carries intrinsic AT activity which acetylates nucleosomal histone proteins (BANISTER and KOUZARIDES 1996; OGRYZKO et al. 1996). However, even today it is not clear how histone acetylation supports transcription. As nucleosomes are efficient repressors of transcription in vitro and in vivo, it is commonly thought that acetylation of the N termini of histones neutralizes their positively charged lysine residues, thereby decreasing their affinity for DNA (for a review see GRUNSTEIN 1997). This leads to an increased accessibility of the chromatin template to the transcriptional machinery. A second model, which might not be mutually exclusive, proposes that acetylation of histones destabilizes the higher-order folding of chromatin (for a review see FRY and PETERSON 2001). This inability to gain a higher-order folding correlates directly with a large increase in transcription initiation and elongation as tested for RNA PolIII. Consequently, these results suggest that p300/CBP might enhance transcription by acetylating histones, thereby disrupting the nucleosome–nucleosome interactions that stabilize chromatin folding. (c) P300/CBP regulates inherent activities of transcriptional activators such as p53, GATA-1 or E1A and basal transcription factors like $TF_{II}F$ and $TF_{II}E$ by acetylation (GU and ROEDER 1997; IMHOF et al. 1997; BOYES et al. 1998; ZHANG et al. 2000). These acetylations either increase DNA-binding activity (GU and ROEDER 1997; BOYES et al. 1998), decrease DNA binding-activity (MUNSHI et al. 1998) or prevent protein–protein interactions (WALTZER and BIENZ 1998; ZHANG et al. 2000).

Several interacting surfaces for E1A of different serotypes are identified in the p300/CBP molecule. Ad12-$E1A_{12S}$ binds at least three domains in p300/CBP (LIPINSKI et al. 1999; FAX et al. 2000a): (a) aa 461–688 carrying the KIX domain necessary to interact with the KID of CREB (PARKER et al. 1996; RADHAKRISHNAN et al. 1997); (b) aa 1519–1850 spanning C/H3 and a part of the AT domain; (c) the C terminus (aa 1999–2200). The C/H3 interaction domain is particularly well characterized. O'CONNOR and co-workers (1999) identified a short stretch of 12 aa (aa 1811–1822 in CBP) in C/H3 which is necessary for binding $E1A_{12S}$ as well as some cellular transcription factors. This region was termed the "transcriptional adapter motif" (TRAM).

As expected, Ad2/Ad5-$E1A_{12S}$ associates with comparable p300/CBP domains, namely aa 1459–1891 and aa 2058–2163 (KUROKAWA et al. 1998). Additionally, Ad2/Ad5-$E1A_{12S}$ was shown to bind to the N terminus of p300/CBP (aa 1–450) and both Ad2/Ad5-$E1A_{12S}$ and Ad2/Ad5-$E1A_{13S}$ bind to the minimal AT-domain of p300 lacking the C/H3 domain (CHAKRAVARTI et al. 1999). Latter interaction, which is not mediated via the classical E1A interaction motifs (the N terminus and CR1), but rather via CR2 and/or CR3, leads to an inhibition of AT activity.

The E1A domains that interact with p300 and CBP are also well defined. Interestingly, extensive mutational analyses showed subtle differences in the interaction profile of oncogenic and non-oncogenic E1A. In case of non-oncogenic Ad2/Ad5-E1A, the N terminus and CR1 cooperatively bind p300. WANG and co-workers (1993a) have shown that at least three N-terminal residues within Ad5-E1A are important for p300 interaction: Arg2, His3 and Leu20. Mutation of these aa leads either to a complete loss (Arg2) or to a strong reduction (His3, Leu20) of the Ad5-E1A/p300 binding. In addition, three regions in CR1 were identified to be involved in p300 binding: aa 30–49, aa 70–81 (EGAN et al. 1988; WANG et al. 1993a), and aa 65–71 (O'CONNOR et al. 1999). Latter domain, which spans the consensus sequence FXE/DXXXL (single letter code, X = any aa), was shown to bind the TRAM motif located in the C/H3 domain of p300/CBP. Most interestingly, some cellular factors including p53 and E2F-1, which also bind to the C/H3 domain, carry this consensus sequence too (O'CONNOR et al. 1999). These results suggest a common mechanism underlying the association of different transcription factors with the C/H3 domain of p300/CBP. Moreover, these data give rise to a molecular key of how E1A might repress the activity of some cellular transcription factors like p53.

Several experiments indicate that different aa of $E1A_{12S}$ are necessary to bind to the different p300/CBP subfragments. For example, a mutant in which aa 3 was changed from His to Asp was still able to bind to aa 1–450 and aa 2,058–2,163 of p300/CBP but failed to interact with the C/H3 domain (KUROKAWA et al. 1998).

In contrast to Ad2/Ad5-E1A, the N terminus of oncogenic Ad12-E1A (aa 1–29) is sufficient to bind the C/H3 domain of p300 (LIPINSKI et al. 1999). Three aa in the N terminus are essential for the association of p300/CBP with the Ad12-$E1A_{12S}$ protein (Asp16, Ile18, Leu19). Most surprisingly, Arg2 which constitutes an essential aa in the Ad2/Ad5 system (WANG et al. 1993a) is dispensable for binding to p300/CBP in the Ad12 system (LIPINSKI et al. 1999). Whether these subtle differences in binding result in different serotype-specific functional E1A–p300/CBP complexes or play a role in the different oncogenic behaviour remains to be elucidated.

Has the binding of $E1A_{12S}$ to different p300 and CBP subdomains any physiological impact? Published data indicate that $E1A_{12S}$ interferes with the functional interaction of p300/CBP with sequence-specific transcription factors or co-activators by binding to different p300/CBP subdomains, thereby modulating the expression of specific target genes. For instance, $E1A_{12S}$ may inhibit the activities of c-Fos (BANNISTER and KOUZARIDES 1995) and Stat1 (ZHANG et al. 1996) by competing for their binding to the C/H3 domain. The interaction between the N terminus of Ad2/Ad5-E1A and the C/H3 domain of CBP was shown to disrupt the complex consisting of MyoD, CBP, and PCAF, thereby preventing myogenic differentiation (Puri et al. 1997). Furthermore, E1A$_{12S}$ inhibits the association between p300/CBP and p/CIP, most probably by a competitive binding mechanism to the p300/CBP C terminus binding region for p/CIP (aa 2,058–2,163 in CBP; Kurokawa et al. 1998). This results in a repression of retinoic acid receptor functions. However, do these repressive effects result from a simple displacement mechanism? Indeed, several groups have shown that E1A inhibits the enzymatic activity of p300/CBP (Chakravarti et al. 1999; Hamamori et al. 1999). Data of Chakravarti and co-workers (1999) suggest a two-step repression model involving multiple, independent but mutually co-operative interactions: In the first step, the N-terminal 76 aa of E1A mediate direct binding to the C/H3 domain of p300/CBP. In the second step, the C terminus of E1A including CR2 and CR3 acts to inhibit the AT activity of p300/CBP directly. It is noteworthy that, although CR2 and CR3 seem to be independently involved in the repression of the AT activity of p300/CBP, both are necessary for complete repression. These results indicate that Ad2/Ad5-E1A$_{12S}$ and Ad2/Ad5-E1A$_{13S}$ are able to inhibit AT activity of p300/CBP but that Ad2/Ad5-E1A$_{13S}$ is a more potent repressor. Interestingly, a comparable mechanism is discussed for the repression of the AT activity of PCAF through E1A (Chakravarti et al. 1999).

However, the modulation of the p300/CBP AT activity by E1A is still a matter of controversial discussion. In contrast with the data described above, other researchers have demonstrated an activation of the p300/CBP AT activity. For instance, Ait-Si-Ali et al. (1998) have shown that Ad2/Ad5-E1A activates the AT activity of CBP in G1/S phase transition. Activation depends on the physical interaction between the C/H3 domain of CBP located immediately adjacent to the AT domain. The authors speculate that this interaction induces a conformational change in the AT domain which leads to an increased catalytic activity comparable to the activation mediated by phosphorylation through cyclin E-Cdk2.

Data from our laboratory might solve this controversy. We have shown that Ad12-E1A$_{12S}$ modulates the AT activity of p300/CBP in a dose-dependent manner (Fax et al. 2000b). Incubation of CBP$_{1-1890}$, which carries the N terminus of CBP including the AT- and the C/H3-domain with low concentrations of Ad12-E1A$_{12S}$ leads to an increase of the AT activity whereas a large molar excess of Ad12-E1A$_{12S}$ represses the AT activity. Interestingly, the AT activity of a CBP mutant lacking the C/H3 domain (CBP_{1-1690}) was not regulated by Ad12-$E1A_{12S}$, indicating that a physical interaction between C/H3 and Ad12-$E1A_{12S}$ is necessary for the modulation of the enzymatic activity. The reason for this dose-dependent modulation of the AT activity of CBP is not yet clear. One explanation might be that a high concentration of Ad12-$E1A_{12S}$ competes with the histone substrate for the AT activity of p300/CBP. This idea is supported by findings of Chakravarti et al. (1999) and Zhang et al. (2000) who have shown that E1A is efficiently acetylated by p300/CBP in vitro. Most interestingly, this acetylation blocks the interaction of E1A with the C-terminal binding protein (CtBP; Zhang et al. 2000). The complex consisting of E1A and CtBP inhibits CREB-dependent gene expression probably due to the ability of CtBP to interact with HDACs (Sundqvist et al. 1998). Taken together the data published up to now show no clear-cut functional consequences of an interaction between E1A and p300/CBP with respect to the modulation of the AT activity. The reasons for the discrepancies of the published data are not yet known. But they might be due to different E1A protein subforms ($E1A_{12S}$ vs. $E1A_{13S}$), E1A proteins from different serotypes or different in vitro assay conditions. Clearly, additional experiments are necessary to clarify this controversy.

Several CREB-p300/CBP-dependent promoters like the rat somatostatin CRE in a heterologous promoter context (Arany et al. 1995), a truncated somatostatin promoter (Lundblad et al. 1995), and the interleukin-6 promoter (Arany et al. 1995) are repressed by $E1A_{12S}$. Mechanistically, it is presumed that binding of E1A to p300/CBP leads to the dissociation of the co-activator–transcription factor complex. Due to the data summarized above, inhibition of the AT activity might also be part of the repression mechanism. In contrast with these results, the $E2_{Ad12}$-promoter is activated through Ad12-$E1A_{12S}$ (Fax et al. 2000a). This observation raises two closely-linked questions: (a) how does Ad12-$E1A_{12S}$ activate the $E2_{Ad12}$-promoter, and (b) how can E1A on the one hand activate, and on the other hand repress promoters whose expression is driven by the same transcription factor/co-activator sets? In case of the $E2_{Ad12}$-promoter, DNA-dependent protein–protein interaction assays have demonstrated the assembly of a stable ternary complex consisting of CREB, Ad12-$E1A_{12S}$ and CBP on the $E2_{Ad12}$-CRE (Fax et al. 2000a). In this complex, CREB functions as an adapter on the $E2_{Ad12}$-CRE to recruit CBP and Ad12-$E1A_{12S}$, whereas CBP fulfils important co-activator activity(ies). The role of the adenoviral protein is less clear. However, there are at least three possible functions. (a) CREB requires phosphorylation on Ser133 through PKA before it is able to recruit p300/CBP. As the cAMP level does not seem to be affected by E1A (Florin-Christensen et al. 1993), most CREB molecules should be non-phosphorylated and consequently unable to recruit p300/CBP. Therefore, one function of Ad12-E1A_{12S} might be to allow the formation of a stable CREB/CBP complex, thereby circumventing phosphorylation of Ser133. This idea is supported by the finding that Ad12-E1A_{12S} binds to the KIX domain of CBP which in turn is necessary for the binding to the KID domain of CREB (Parker et al. 1996; Radhakrishnan et al. 1997; Lipinski et al. 1999; Fax et al. 2000 a). By binding to CREB on the one hand and to CBP on the other hand, Ad12-E1A_{12S} might function as a molecular glue, fixing both factors on the $E2_{Ad12}$-CRE. A similar model is proposed for the Tax protein of HTLV-1. In this case, recruitment of CBP to the CREB/Tax complex bound to CRE-like sequences in the viral long terminal repeat is independent of the CREB trans-activation domain and therefore independent of CREB phosphorylation (Kwok et al. 1996; Giebler et al. 1997; Harrod et al. 1998). Rather, the C terminus of Tax, which binds to p300/CBP, tethers the co-activator to the promoter and allows transcriptional activation. However, several findings argue against such a mechanism for the $E2_{Ad12}$-promoter. (b) The $E2_{Ad12}$-promoter is strictly dependent on CREB phosphorylation by PKA as shown by the usage of specific PKA inhibitors (Fax et al. 2000 a, 2001). (c) Ad12-E1A_{12S} interacts with the PKA-R_2C_2 complex, which might result in an increase of phospho-CREB in infected and transformed cells (Fax et al. 2001; see Sect. 4.1.3).

An alternative function of Ad12-E1A_{12S} might be to support the assembly of the PIC on the $E2_{Ad12}$-TATA-box. Ad12-E1A_{12S} binds to the TATA-box binding protein TBP and the $TF_{II}F$ subunit RAP30 in vitro (Lipinski et al. 1998). The latter factor plays important roles in the recruitment of PolII to the PIC as well as in transcriptional initiation (Tan et al. 1994; Fang et al. 1996; Lei et al. 1999). These observations lead to an attractive model in which Ad12-E1A_{12S} promotes TBP binding to the TATA-box and/or helps to recruit PolII to the PIC through an interaction with RAP30.

Finally, after recruitment to the $E2_{Ad12}$-promoter, Ad12-E1A_{12S} might be necessary to activate the AT activity of CBP. This hypothesis, which is in agreement with findings of Korzus et al. (1998) that CREB-dependent gene expression requires AT activity of p300/CBP, is supported by several findings. (a) The $E2_{Ad12}$-promoter is regulated in an acetylation/deacetylation-dependent mechanism (Fax et al. 2000 b): it is repressed by HDAC1, and p300/CBP counteracts the HDAC1-mediated repression in an AT-domain-dependent manner. Moreover, in contrast with a point-mutated, inactive $E2_{Ad12}$-promoter, the active $E2_{Ad12}$-promoter is associated with acetylated histones. (b) Ad12-E1A_{12S} activates the AT activity of p300/CBP. (c) The Ad12-E1A_{12S}-mediated activation of the $E2_{Ad12}$-promoter correlates well with the ability of the viral protein to associate with the AT activity of p300/CBP in vivo. Which of these models holds true for the activation of the $E2_{Ad12}$-promoter is not yet clear. However, it seems probably that combinatorial effects are responsible for the strong $E2_{Ad12}$-promoter activation mediated by the Ad12-E1A_{12S} protein.

With respect to the regulation through E1A, the promoter of the proliferating cell nuclear antigen (PCNA), which is cAMP-responsive, is of particular interest as it can be activated or repressed by the viral protein which depends on the cell system. The Ad2/Ad5-E1A_{12S} protein activates the PCNA promoter through two elements in HeLa and Saos-2 cells (MORRIS and MATHEWS 1991; KARUPPAYIL et al. 1998): it relieves the repression mediated by RFX1/p107 (LEE et al. 1998; LIU et al. 1999) and it activates the promoter through a cis-regulatory element called PERE (PCNA-E1A responsive element; LABRIE et al. 1993). The PERE is bound by CREB/ATF-1 heterodimers (LABRIE et al. 1995; LEE and MATHEWS 1997). Ad2/Ad5-E1A_{12S} mediates transcriptional activation of a minimal PCNA-promoter through CREB in transient expression assays. Moreover, an antisense CBP plasmid suppresses Ad2/Ad5-E1A_{12S}-dependent activation of PCNA transcription, demonstrating the necessity of CBP in the process of the Ad2/Ad5-E1A_{12S}-mediated transcriptional activation of the PCNA promoter (KARUPPAYIL et al. 1998). As CBP associates with the PERE-bound transcription factor complex (LEE and MATHEWS 1997), it is obvious that the adenoviral protein activates the PCNA promoter via a CBP-CREB-PERE pathway.

In contrast with HeLa and Saos-2 cells, the PCNA promoter is repressed by Ad2/Ad5-E1A_{12S} in primary rat kidney and CREF (cloned rat embryo fibroblast) cells (KANNABIRAN et al. 1999). Repression is not mediated via the PERE but depends on a p53 binding site in the PCNA promoter (nt −213 to −234) and on the N terminus and CR1 of the adenoviral protein, indicating a model in which Ad2/Ad5-E1A_{12S} disrupts the functional interaction between p53 and p300/CBP. Consistent with this idea is the finding that over-expressed CBP can partially relieve Ad2/Ad5-E1A_{12S}-mediated repression of the PCNA transcriptional activation by p53 in Saos-2 cells (KARUPPAYIL et al. 1998).

The important question remains how E1A can differentially regulate CREB/CBP-dependent promoters. One important checkpoint seems to be the co-activator whose recruitment results in a promoter activation and whose E1A-driven dissociation from CREB leads to promoter repression. The mechanisms which decide for either recruitment or dissociation, are poorly understood. However, the studies characterizing the PCNA promoter identify the promoter context and the transcription factor composition of the respective cell type as decisive factors.

4.1.3 *Ad12-E1A$_{12S}$ – A Viral AKAP*

CREB-mediated E2$_{Ad12}$ promoter activation requires at least one phosphorylation step, namely the phosphorylation of CREB on Ser133 which is underscored by the findings that PKA inhibitors, such as PKI, which exports the C subunit from the nucleus (Wen et al. 1995), or the ATP analogue H89, which competes with endogenous ATP for binding to the C subunit, prevent promoter activation (Fax et al. 2000a). However, with the exception of the first 15–30 min after infection, during which Ads transiently induce the cAMP/PKA signal transduction pathway to allow efficient delivery of the viral genome to the nucleus (Suomalainen et al. 2001), the cAMP level remains unchanged in Ad-infected cells (Florin-Christensen et al. 1993). Consequently, the question arises whether Ad12-E1A$_{12S}$ interacts directly with the PKA holoenzyme to facilitate phosphorylation of CREB.

One way in which Ad12-E1A$_{12S}$ might interfere with the activity of the PKA holoenzyme is by influencing its subcellular localization. Indeed, anchoring inhibitor peptides such as Ht31$_{418-718}$ (which binds to the RII subunits of PKA; Lester et al. 1997) or AKAP$_{326-425}$ (which binds RI and RII subunits; Lin et al. 1995; Trendelenburg et al. 1996; Huang et al. 1997), which contain the R-binding domains but no targeting domain and therefore displace and solubilize PKA by competition with endogenous AKAPs, inhibit the Ad12-E1A$_{12S}$-induced activation of the E2$_{Ad12}$-promoter (Fax et al. 2001). Subsequent analyses have shown that Ad12-E1A$_{12S}$ strongly binds in vitro and in vivo to RIa and RIIa. In contrast, Ad12-E1A$_{12S}$ binds only weakly to RIIb and does not bind to RIb, demonstrating a selectivity for specific regulatory PKA subunits. The interaction with RIa and RIIa is direct, mediated via the N terminus/CR1 region of Ad12-E1A$_{12S}$ and the AKAP interaction domain of RIa or RIIa, and is inhibited by a Ht31 anchoring competitor peptide. Binding of Ad12-E1A$_{12S}$ to RIIa correlates with a redistribution of RIIa from the cytoplasm into the nucleus as shown in Ad12-infected, Ad12-transformed cells as well as cells transiently expressing Ad12-E1A_{12S}. Surprisingly RIa is not redistributed in the presence of the adenoviral oncoprotein for unknown reasons. These data indicate that Ad12-E1A_{12S} might be responsible for the selective translocation of RIIa from the cytoplasm into the nucleus. In this context, it is noteworthy that it is not the PKA-I holoenzyme, which is soluble in the cytosol, but the C subunit anchored via RII to membrane-bound AKAPs that is responsible for the amplification of the cAMP signalling to the nucleus, for the increase of phosphorylation of CREB at position Ser-133 in response to cAMP, and for an increase of CREB-dependent target gene expression (CASSANO et al. 1996; FELICIELLO et al. 1997). These results are consistent with our observation that co-transfected Ad12-E1A_{12S} and RIIa co-operatively activate the $E2_{Ad12}$-promoter which in turn is consistent with data of TORTORA and CHO-CHUNG (1990) showing that the RIIa and RIIb but not the RI genes restore somatostatin promoter activity in the cAMP-unresponsive mutant pheochromocytoma cell line A126–1B2. These results suggest that the RII subunits of PKA are essential for cAMP-induced gene transcription.

However, the molecular mechanism by which Ad12-E1A_{12S} and RIIa co-operate in the activation of the $E2_{Ad12}$-promoter is still unknown. The simplest model would be that Ad12-E1A_{12S}-recruited RIIa might function as co-activator on the $E2_{Ad12}$ promoter. Published data show that RIIb binds directly as well as in a complex with CREB to the CRE consensus sequence and that an artificial Gal-RIIb fusion protein activates a Gal-dependent reporter construct in transient expression assays in COS7 cells (SRIVASTAVA et al. 1998). However, the findings that RIIa failed to bind directly to the CRE in these experiments (SRIVASTAVA et al. 1998) and that RIIa strongly activates the $E2_{Ad12}$-promoter only in the presence of the Ad12-E1A_{12S} protein (FAX et al. 2001) argue against this hypothesis. A second model might be that the adenoviral protein recruits the regulatory subunit to the viral promoter, where it functions as an adapter for the C subunit. Once anchored to the promoter, the catalytic activity is needed for phosphorylation of factors involved in $E2_{Ad12}$ promoter activation. This hypothesis is in agreement with our findings that co-transfection of Ad12-E1A_{12S} and RIIa results in a strong reporter gene expression and that PKI inhibits Ad12-E1A_{12S}-mediated activation of the $E2_{Ad12}$ promoter. Moreover, it is consistent with our preliminary unpublished observations that: (a) binding of Ad12-E1A_{12S} to RIIa does not lead to a dissociation of the PKA holoenzyme in vitro (F. Herberg, P. Fax, D. Brockmann, unpublished results); (b) Ad12-E1A$_{12S}$ pulls-down PKA activity from cellular extracts; and (c) that the C subunit and Ad12-E1A$_{12S}$ co-precipitate from infected COS7 cells using an anti-Ca antibody, although we did not observe a direct interaction in glutathione S-transferase pull-down assays. On the other hand, it is well documented that the kinase activity of the C subunits of PKA is inhibited by binding to R subunits (for a review see Beebe 1994). Clearly, further studies are necessary to elucidate the mechanism by which RIIa participates in the activation of the E2$_{Ad12}$ promoter.

5 Regulation of the CREB/CBP-Dependent E2$_{Ad12}$-Promoter by Ad12-E1A$_{12S}$ – A Complex Model with a Couple of Open Questions

Combining the data sets described above leads to a complex working scheme of how Ad12-E1A$_{12S}$ might activate the E2$_{Ad12}$-promoter: The E2$_{Ad12}$-CRE is bound by CREB, which serves as an adapter to recruit p300/CBP and Ad12-E1A$_{12S}$. In this complex the role of the adenoviral protein is multifunctional (Fig. 2) including the recruitment of TBP and/or PolII via RAP30 as well as the recruitment and/or activation of the AT activity of p300CBP, which in turn is necessary to modify the nucleosomal structure, thereby increasing access of the general transcriptional machinery to the E2$_{Ad12}$-promoter. This concept immediately gives rise to the question of whether other chromatin-remodelling complexes like SWI/SNF participate in the process of E2$_{Ad12}$-promoter activation and whether Ad12-E1A$_{12S}$ also interferes with the function of such complexes. Both questions should be the subject of future experiments. Moreover, besides acetylation of histones, p300/CBP might also acetylate Ad12-E1A$_{12S}$. This acetylation might block the formation of an Ad12-E1A$_{12S}$/CtBP repressor complex which was shown to inhibit CREB/CBP-dependent gene expression (Zhang et al. 2000).

One of the major unsolved questions refers to the function of RIIa in the activation process of the E2$_{Ad12}$-promoter. Activation of the E2$_{Ad12}$-promoter requires at least one phosphorylation step mediated by PKA. However, the cAMP level is not increased in Ad-infected cells (Florin-Christensen et al. 1993). The interaction between Ad12-E1A$_{12S}$ and RIIa might unravel this paradoxical situation. Ad12-E1A$_{12S}$ binds strongly to RIIa and dissociates the PKA holoenzyme from cytoplasmically localized AKAPs which leads to a redistribution of RIIa into the nucleus. These observations give rise to the hypothesis that the adenoviral protein

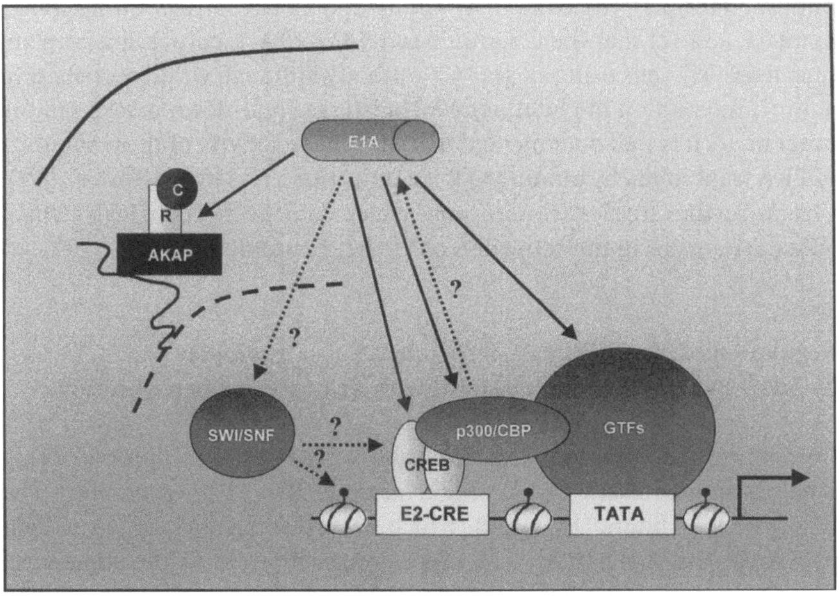

Fig. 2. Model of the complex factor network in the activation process of the $E2_{Ad12}$-promoter. The expression of the $E2_{Ad12}$-gene is driven by the cellular transcription factor CREB which in turn recruits p300/CBP to the E2-CRE. To facilitate promoter activation, Ad12-E1A interacts with key components of the cAMP/PKA signal transduction pathway: It binds to (indicated by an *arrow*) and dissociates the PKA-II holoenzyme from AKAPs, allowing at least the redistribution of RIIa into the nucleus where it functions in an as yet undefined mechanism as co-activator for the adenoviral protein. In addition, Ad12-E1A binds to CREB as well as to p300/CBP, the AT of which exhibits increased activity in response to its association with Ad12-E1A. P300/CBP in turn acetylates histones (symbolized by *black dots* on the *grey* nucleosomes) and might acetylate Ad12-E1A (indicated by a *dotted arrow* and a *question mark*) to prevent the formation of an inhibitory Ad12-E1A/CtBP complex. Moreover, Ad12-E1A binds directly to components of the general transcription machinery (*GTFs*) like TBP and the $TF_{II}F$ subunit RAP30. Finally, preliminary data from our laboratory suggest that Ad12-E1A also binds to the SWI/SNF subunit BRG-1, directing the chromatin remodelling complex to the $E2_{Ad12}$-promoter (R. Hashemi, H. Esche, D. Brockmann, unpublished results). The cell membrane is indicated by an *uninterrupted curve*, the nuclear membrane by *a dotted curve*. Expression of the $E2_{Ad12}$ gene is drawn schematically as an *arrow* on the adenoviral E2 DNA

recruits the R subunit to the viral promoter, where it functions either as an essential co-activator for Ad12-E1A_{12S} or as an adapter for the C subunit.

If PKA is recruited to the $E2_{Ad12}$-promoter, which target protein(s) is/are phosphorylated by its kinase activity? An obvious factor might be CREB, as only CREB phosphorylated on Ser133 interacts with CBP (Chrivia et al. 1993). Surprisingly, we did not observe an increase of CREB phosphorylated on Ser133 in the nuclei of Ad12-E1A_{12S}-expressing cells compared to mock-transfected cells. However, the fraction of CREB associated with Ad12-E1A_{12S} in these cells might be too small to detect changes in the phosphorylation status. An alternative candidate for phosphorylation might be the co-activator CBP. Consistent with this idea is the finding that the trans-activation potential of CBP can be regulated by PKA in transient expression assays in F9 cells using GAL4 fusion constructs (Chrivia et al. 1993). In addition, a recent report dealing with a constitutively active CREB shows that the recruitment of CBP is sufficient for a significant portion of CREB activity, but full activation of reporter gene expression requires a PKA-mediated activation step downstream from CREB phosphorylation and CREB recruitment (Cardinaux et al. 2000). Moreover, it has recently been shown in the cortical neuron and the AtT20 mouse pituitary cell system that the CBP recruitment step (phosphorylation of CREB on Ser133) can be uncoupled from CREB/CBP-activated transcription (Chawla et al. 1998; Hu et al. 1999). Taken together, these data indicate that recruitment of CBP to the promoter is not sufficient for full promoter trans-activation, and that CBP itself has also to be activated through phosphorylation by, e.g. PKA. These findings are in agreement with data showing that CaM-kinases phosphorylate CREB on Ser133 in PC12 cells which lack functional PKA, but that active PKA is required to induce *c-fos* gene expression (Ginty et al. 1991). These results can be interpreted in the way that a PKA-dependent phosphorylation of a second protein in this pathway, such as CBP or a factor downstream of the CREB/CBP complex, is needed. Consequently, recruitment of the C subunit to the promoter through an Ad12-E1A_{12S}/RIIa complex might not be necessary for the phosphorylation of CREB but for the phosphorylation and activation of p300/CBP. However, the modulation of the phosphorylation pattern of p300/CBP by E1A is controversially discussed in the literature. For example, Banerjee and co-workers (1994) showed that Ad2/Ad5-E1A inhibits the phosphorylation of p300 in vitro. In contrast, Ad5-E1A seems to induce p300 phoshorylation in Ad5-infected F9 cells (Kitabayashi et al. 1995) and p300 phosphorylation is not inhibited by E1A in Ad2-

infected HeLa or Ad5-infected BRK cells (YACIUK and MORAN 1991; AVANTAGGIATI et al. 1996). Most interestingly, p300 seems to be differentially phosphorylated in Ad12-E1A and Ad5-E1A-transformed rat cells (PEREIRA et al. 1994). Due to these results, it is now important to analyse whether the phosphorylation pattern of p300/CBP is altered on the active $E2_{Ad12}$-promoter and whether this alteration is due to the expression of the adenoviral protein.

Acknowledgements. We wish to thank O. Lehmkühler for critical reading of the manuscript. Work in the laboratory is supported by the Deutsche Forschungsgemeinschaft, grants BR 1150/4-1 and Br 1150/6-1, as well as by the IFORES-programme of the University of Essen Medical School.

References

ABRAHAM SE, LOBO S, YACIUK P, WANG HG, MORAN E (1993) p300, and p300-associated proteins, are components of TATA-binding protein (TBP) complexes. Oncogene 8:1639–1647

AHN S, OLIVE M, AGGARWAL S, KRYLOV D, GINTY DD, VINSON C (1998) A dominant-negative inhibitor of CREB reveals that it is a general mediator of stimulus-dependent transcription of c-fos. Mol Cell Biol 18:967–977

AIT-SI-ALI S, RAMIREZ S, BARRE F-X, DKHISSE F, MAGNAGHI-JAULIN L, GIRAULT JA, ROBIN P, KNIBIEHLER M, PRITCHARD LL, DUCOMMUN B, TROUCHE D, HAREL-BELLAN A (1998) Histone acetyltransferase activity of CBP is controlled by cycle-dependent kinases and oncoprotein E1A. Nature 396:184–186

ARANY Z, SELLERS WR, LIVINGSTON D, ECKNER R (1994) E1A-associated p300 and CREB-associated CBP belong to a conserved family of coactivators. Cell 77:799–800

ARANY Z, NEWSOME D, OLDREAD E, LIVINGSTONE DM, ECKNER R (1995) A family of transcriptional adaptor proteins targeted by the E1A oncoprotein. Nature 374:81–84

AVANTAGGIATI ML, CARBONE M, GRAESSMANN A, NAKATANI Y, HOWARD B, LEVINE AS (1996) The SV40 large T antigen and Adenovirus E1A oncoproteins interact with distinct isoforms of the transcriptional co-activator p300. EMBO J 15:2236–2248

BAGCHI S, WEINMANN R, RAYCHAUDHURI P (1991) The retinoblastoma protein co-purifies with E2F-1, an E1A-regulated inhibitor of the transcription factor E2F. Cell 65:1063–1072

BANDARA LR, LATHANGUE NB (1991) Adenovirus E1A prevents the retinoblastoma gene product from complexing with a cellular transcription factor. Nature 351: 494–497

BANDARA LR, ADAMCZEWSKI JP, HUNT T, LATHANGUE NB (1991) Cyclin A and the retinoblastoma gene product complex with a common transcription factor. Nature 352:249–251

BANERJEE AC, RECUPERO AJ, MAL A, PIOTRKOWSKI AM, WANG D-M, HARTER M (1994) The Adenovirus E1A 289R and 243R proteins inhibit the phosphorylation of p300. Oncogene 9:1733–1737

Bannister AJ and Kouzarides (1995) CBP-induced stimulation of c-Fos activity is abrogated by E1A. EMBO J 14:4758–4762

Bannister AJ, Kouzarides T (1996) The CBP co-activator is a histone acetyltransferase. Nature 384:641–643

Bayley ST, Mymryk JS (1994) Adenovirus E1A proteins and transformation. Int J Oncol 5:425–444

Beebe SJ (1994) The cAMP-dependent protein kinases and cAMP signal transduction. Sem Cancer Biol 5:285–294

Berk AJ, Lee F, Harrison T, Williams J, Sharp PA (1979) Pre-early adenovirus 5 gene product regulates synthesis of early viral messenger RNAs. Cell 17:935–944

Boyer TG, Martin ME, Lees E, Ricciardi RP, Berk AJ (1999) Mammalian Srb/Mediator complex is targeted by adenovirus E1A protein. Nature 399:276–279

Boyes J, Byfield P, Nakatani Y, Ogryzko VV (1998) Regulation of activity of the transcription factor GATA-1 by acetylation. Nature 396:594–598

Brehm A, Ohbo K, Zwerschke W, Botquin V, Jansen-Dürr P, Schöler HR (1999) Synergism with germ line transcription factor Oct-4: viral oncoproteins share the ability to mimic a stem cell-specific activity. Mol Cell Biol 19:2635–2643

Brindle P, Linke S, Montminy M (1993) Protein-kinase-A-dependent activator in transcription factor CREB reveals new role for CREM repressors. Nature 364: 821–824

Brockmann D, Tries B, Esche H (1990) Isolation and characterisation of novel adenovirus type 12 E1A mRNAs by cDNA PCR technique. Virology 179:585–590

Brockmann D, Esche H (1995) Regulation of viral and cellular gene expression by E1A proteins encoded by the oncogenic adenovirus type 12. In: Doerfler W, Böhm P (eds) The Molecular repertoire of Adenoviruses III, Biology and pathogenesis. Springer, Berlin Heidelberg, pp 81–112

Brockmann D, Bury C, Kröner G, Kirch H-C, Esche H (1995) Repression of the c-Jun trans-activation function by the adenovirus type 12 E1A 52R protein correlates with the inhibition of phosphorylation of the c-Jun activation domain. J Biol Chem 270:10754–10763

Cardinaux J-R, Notis JC, Zhang Q, Vo N, Craig JC, Fass DM, Brennan RG, Goodman RH (2000) Recruitment of CREB binding protein is sufficient for CREB-mediated gene activation. Mol Cell Biol 20:1546–1552

Cassano S, Gallo A, Buccigrossi V, Porcellini A, Cerillo R, Gottesman ME, Avvedimento EV (1996) Membrane localization of cAMP-dependent protein kinase amplifies cAMP signaling to the nucleus in PC12 cells. J Biol Chem 271: 29870–29875

Chakravarti D, Ogryzko V, Kao H-Y, Nash A, Chen H, Nakatani Y, Evans RM (1999) A viral mechanism for inhibition of p300 and PCAF acetyltransferase activity. Cell 96: 393–403

Chawla S, Hardingham GE, Quinn DR, Bading H (1998) CBP: A signal-regulated transcriptional co-activator controlled by nuclear calcium and CaM kinase IV. Science 281:1505–1509

Chellappan SP, Kraus VB, Kroger R, Munger K, Howley PM (1992) Adenovirus E1A, simian virus 40 tumor antigen and human papillomavirus E7 protein share the capacity to disrupt the interaction between transcription factor E2F and the retionoblastoma gene product. Proc Natl Acad Sci USA 89:4549–4553

Cho H, Orphanides G, Sun X, Yang XJ, Ogryzko V, Lees E, Nakatani Y, Reinberg D (1998) A human RNA polymerase II complex containing factors that modify chromatin structure. Mol Cell Biol 18:5355–5363

Chrivia JC, Kwok RPS, Lamb N, Haniwawa M, Montminy MR, Goodman RH (1993) Phosphorylated CREB binds specifically to the nuclear protein CBP. Nature 365:855–859

Culp JS, Webster LC, Friedman DJ, Smith CL, Huang W-J, Wu FY-H, Rosenberg M, Ricciardi RP (1988) The 289-amino acid E1A protein of adenovirus binds zinc in a region that is important for trans-activation. Proc Natl Acad Sci USA 85:6450–6454

Dallas PB, Yaciuk P, Moran E (1997) Characterization of monoclonal antibodies raised against p300: both p300 and CBP are present in intracellular TBP complexes. J Virol 71:1726–1731

Dell'Acqua ML, Scott JD (1997) Protein kinase A anchoring. J Biol Chem 272:12881–12884

Dorsman JC, Teunisse AFAS, Zantema A, van der Eb AJ (1997) The adenovirus 12 E1A proteins can bind directly to proteins of the p300 transcription co-activator family, including the CREB-binding protein CBP and p300. J Gen Virol 78:423–426

Dyson N, Guida P, McCall C, Harlow E (1992) Adenovirus E1A makes two distinct contacts with the retinoblastoma protein. J Virol 66:4606–4611

Eckner R, Ewen ME, Newsome D, Gerdes M, DeCaprio JA, Lawrence JB, Livingston DM (1994) Molecular cloning and functional analysis of the adenovirus E1A-associated 300-kD protein (p300) reveals a protein with properties of a transcriptional adaptor. Genes Dev 8:869–884

Edwards AS, Scott JD (2000) A-kinase anchoring proteins: protein kinase A and beyond. Curr Opin in Cell Biol 12:217–221.

Egan C, Jelsma TN, Howe JA, Bayley ST, Ferguson B, Branton PE (1988) Mapping of cellular protein-binding sites on the products of early-region 1A of human adenovirus type 5. Mol Cell Biol 8:3955–3959

Fang SM, Burton ZF (1996) RNA polymerase II-associated protein (RAP) 74 binds transcription factor (TF) IIB and blocks TFIIB-RAP30 binding. J Biol Chem 271: 11703–11709

Fattaey AR, Harlow E, Helin K (1993) Independent regions of adenovirus E1A are required for binding to and dissociation of E2F-protein complexes. Mol Cell Biol 13:7267–7277

Fax P, Lipinski KS, Esche H, Brockmann D (2000a) cAMP-independent activation of the Adenovirus type 12 E2 promoter correlates with the recruitment of CREB-1/ATF-1, E1A$_{12S}$, and CBP to the E2-CRE. J Biol Chem 275:8911–8920

Fax P, Lehmkühler O, Kühn C, Esche H, Brockmann D (2000b) E1A$_{12S}$-mediated activation of the adenovirus type 12 E2 promoter depends on the histone acetyltransferase activity of p300/CBP. J Biol Chem 275:40554–40560

Fax P, Carlson CR, Collas P, Taskén K, Esche H, Brockmann D (2001) Binding of PKA-RIIá to the Adenovirus E1A$_{12S}$ oncoprotein correlates with its nuclear translocation and an increase in PKA-dependent promoter activity. Virology 285:30–41

Feliciello A, Li Y, Avvedimento EV, Gottesman ME, Rubin CS (1997) A-kinase anchor protein 75 increases the rate and magnitude of cAMP signaling to the nucleus. Curr Biol 7:1011–1014

Ferguson B, Krippl B, Andrisani O, Jones N, Westphal H, Rosenberg M (1985) E1A 13S and 12S mRNA products made in Escherichia coli both function as nuclear-localized transcription activators but do not directly bind DNA. Mol Cell Biol 5:2653–2661

Flint J, Shenk T (1997) Viral transactivating proteins. Annu Rev Genet 31:177–212

Florin-Christensen M, Missero C, Florin-Christensen J, Tranque P, Ramon Y Cajal S, Dotto GP (1993) Counteracting effects of E1a transformation on cAMP growth inhibition. Exp Cell Res 207:57–61

Fry CJ, Peterson CL (2001) Chromatin remodelling enzymes: who's on first? Curr Biol 11:185–197

Geisberg VJ, Chen JL, Ricciardi R (1995) Subregions of the adenovirus E1A transactivation domain target multiple components of the TFIID complex. Mol Cell Biol 15:6283–6290

Giebler HA, Loring JE, van Orden K, Colgin MA, Garrus JE, Escudero KW, Brauweiler A, Nyborg JK (1997) Anchoring of CREB binding protein to the human T-cell leukemia virus type 1 promoter: a molecular mechanism of Tax transactivation. Mol Cell Biol 17:5156–64

Ginty DD, Glowacka D, Bader DS, Hidaka H, Wagner JD (1991) Induction of immediate early genes by Ca^{2+} influx requires cAMP-dependent protein kinase in PC12 cells. J Biol Chem 266:17454–17458

Giordano A, Avantaggiati ML (1999) p300 and CBP: partners for life and death. J Cell Physiol 181:218–230

Glenn GM, Ricciardi RP (1987) An adenovirus type 5 E1A protein with a single amino acid substitution blocks wild-type E1A transactivation. Mol Cell Biol 7:1004–1001

Goldman PS, van Tran K, Goodman RH (1997) The multifunctional role of the co-activator CBP in transcriptional regulation. Recent Prog Horm Res 52:103–119

Gonzales GA, Montminy MR (1989) Cyclic AMP stimulates somatostatin gene transcription by phosphorylation of CREB at Serine 133. Cell 59:675–680

Gonzalez GA, Menzel P, Leonard J, Fischer WH, Montminy MR (1991) Characterization of motifs which are critical for activity of the cyclic AMP-responsive transcription factor CREB. Mol Cell Biol 11:1306–1312

Goodman RH, Smolik S (2000) CBP/p300 in cell growth, transformation, and development. Genes Dev 14:1553–1577

Grunstein M (1997) Histone acetylation in chromatin structure and transcription. Nature 389:349–352

Gu W, Roeder RG (1997) Activation of p53 sequence-specific DNA binding by acetylation of the p53 C-terminal domain. Cell 90:595–606

Hagiwara M, Brindle P, Harotunian A, Armstrong R, Rivier J, Vale W, Tsien R, Montminy MR (1993) Coupling of hormonal stimulation and transcription via the cyclic AMP-responsive factor CREB is rate limited by nuclear entry of protein kinase A. Mol Cell Biol 13:4852–4859

Hamamori Y, Sartorelli V, Ogryzko V, Puri PL, Wu H-Y, Wang JYJ, Nakatani Y, Kedes L (1999) Regulation of histone acetyltransferases p300 and PCAF by the bHLH protein twist and adenoviral oncoprotein E1A. Cell 96:405–413

Hamel PA, Gill RM, Phillips RA, Gallie BL (1992) Regions controlling hyperphosphorylation and conformation of the retinoblastoma gene product are independent of domains required for transcriptional repression. Oncogene 7:693–701

Harrod R, Tang Y, Nicot C, Lu HS, Vassilev A, Nakatani Y, Giam C-Z (1998) An exposed KID-like domain in human T-cell lymphotropic virus type 1 Tax is responsible for the recruitment of coactivators CBP/p300. Mol Cell Biol 18:5052–5061

Hiebert SW, Lipp M, Nevins JR (1989) E1A-dependent trans-activation of the human MYC promoter is mediated by the E2F factor. Proc Natl Acad Sci USA 86:3594–3598

Hiebert SW, Blake M, Azizkhan J, Nevins JR (1991) Role of E2F transcription factor in E1A-mediated trans-activation of cellular genes. J Virol 65:3547–3552

Horikoshi N, Maguire K, Kralli A, Maldonado E, Reinberg D, Weinmann R (1991) Direct interaction between adenovirus E1A protein and the TATA box binding transcription factor IID. Proc Natl Acad Sci USA 88:5124–5128

Hu S, Chrivia J, Ghosh A (1999) Regulation of CBP-mediated transcription by neuronal calcium signalling. Neuron 22:799–808

Huang LJ, Durick K, Weiner JA, Chun J, Taylor SS (1997) Identification of a novel protein kinase A anchoring protein that binds both type I and type II regulatory subunits. J Biol Chem 272:8057–8064

Imhof A, Yang XJ, Ogryzko VV, Nakatani Y, Wolffe AP, Ge H (1997) Acetylation of general transcription factors by histone acetyltransferases. Curr Biol 7:689–692

Inostroza JA, Mermelstein FH, Ha I, Lane WS, Reinberg D (1992) Dr1, a TATA-binding protein-associated phosphoprotein and inhibitor of class II gene transcription. Cell 70:477–489

Jones N, Shenk T (1979) An adenovirus type 5 early gene function regulates expression of other early genes. Proc Natl Acad Sci USA 76:3665–3669

Kannabiran C, Morris GF, Mathews MB (1999) Dual action of the adenovirus E1A 243R oncoprotein on the human proliferating cell nuclear antigen promoter: repression of transcriptional activation by p53. Oncogene 18:7825–7833

Karuppayil SM, Moran E, Das G (1998) Differential regulation of p53-dependent and -independent proliferating cell nuclear antigen gene transcription by 12 S E1A oncoprotein requires CBP. J Biol Chem 273:17303–17306

Kawasaki H, Eckner R, Yao TP, Taira K, Chiu R, Livingston DM, Yokoyma KK (1998) Distinct roles of the co-activators p300 and CBP in retinoic-acid-induced F9-cell differentiation. Nature 393:284–289

Kee B, Arias J and Montminy M (1996) Adaptor-mediated recruitment of RNA polymerase II to a signal-dependent activator. J Biol Chem 271:2373–2375

Kimelmann D, Miller SJ, Porter D, Roberts BE (1985) E1A regions of the human adenoviruses and of the highly oncogenic simian adenovirus 7 are closely related. J Virol 53:399–409

Kitabayashi I, Eckner R, Arany Z, Chiu R, Gachelin G, Livingston DM, Yokoyama KK (1995) Phosphorylation of the Adenovirus E1A-associated 300 kDa protein in response to retinoic acid and E1A during differentiation of F9 cells. EMBO J 14:3496–3509

Korzus E, Torchia J, Rose DW, Xu L, Kurokawa R, McInerney EM, Mullen T-M, Glass CK, Rosenfeld MG (1998) Transcription factor-specific requirements for coactivators and their acetyltransferase functions. Science 279:703–707

Kraus VB, Inostroza JA, Yeung K, Reinberg D, Nevins JR (1994) Interaction of the Dr1 inhibitory factor with the TATA binding protein is disrupted by adenovirus E1A. Proc Natl Acad Sci USA 91:6279–6282

Kung AL, Rebel VI, Bronson RT, Ch'ng LE, Sieff CA, Livingston DM, Yao TP (2000) Gene dose-dependent control of hematopoiesis and hematologic tumor suppression by CBP. Genes Dev 14:272–277

Kurokawa R, Kalafus D, Ogliastro M-H, Kioussi C, Xu L, Torchia J, Rosenfeld MG, Glass CK (1998) Differential use of CREB binding protein–coactivator complexes. Science 279:700–703

Kwok RP, Lundblad JR, Chrivia JC, Richards JP, Bachinger HP, Brennan RG, Roberts SG, Green MR, Goodman RH (1994) Nuclear protein CBP is a coactivator for the transcription factor CREB. Nature 370:223–226

Kwok RP, Laurance ME, Lundblad JR, Goldman PS, Shih H, Connor LM, Marriott SJ, Goodman RH (1996) Control of cAMP-regulated enhancers by the viral transactivator Tax through CREB and the co-activator CBP. Nature 380: 642–646

Labrie C, Morris GF, Mathews M (1993) A complex promoter element mediates transactivation of the human proliferating cell nuclear antigen promoter by the 243-residue adenovirus E1A oncoprotein. Mol Cell Biol 13:1697–1707

Labrie C, Lee BH, Mathews MB (1995) Transcription factors RFX1/EF-C and ATF-1 associate with the adenovirus E1A-responsive element of the human proliferating cell nuclear antigen promoter. Nucleic Acids Res 23:3732–3741

Lee KAW, Fink JS, Goodman RH and Green MR (1989) Distinguishable promoter elements are involved in transcriptional activation by E1A and cyclic AMP. Mol Cell Biol 9:4390–4397

Lee WS, Kao CC, Bryant GO, Liu X, Berk AJ (1991) Adenovirus E1A activation domain binds the basic repeat in the TATA box transcription factor. Cell 67:365–376

Lee JS, See RH, Galvin KM, Wang J, Shi Y (1995) Functional interactions between YY1 and adenovirus E1A. Nucleic Acids Res 23:925–931

Lee B, Mathews MB (1997) Transcriptional coactivator cAMP response element binding protein mediates induction of the human proliferating cell nuclear antigen promoter by the adenovirus E1A oncoprotein. Proc Natl Acad Sci USA 94: 4481–4486

Lee BH, Liu M, Mathews MB (1998) Regulation of the human proliferating cell nuclear antigen promoter by the adenovirus E1A-associated protein p107. J Virol 72:1138–1145

Lei L, Ren D, Burton ZF (1999) The RAP74 subunit of human transcription factor IIF has similar roles in initiation and elongation. Mol Cell Biol 19:8372–8382

Lester LB, Langeberg LK, Scott JD (1997) Anchoring of protein kinase A facilitates hormone-mediated insulin secretion. Proc Natl Acad Sci USA 94:14942–14947

Lewis JB, Mathews MB (1980) Control of adenovirus gene expression: a class of immediate early products. Cell 21:303–313

Lewis BA, Tullis G, Seto E, Horikoshi N, Weinmann R, Shenk T (1995) Adenovirus E1A proteins interact with the cellular YY1 transcription factor. J Virol 69:1628–1636

Li Y, Graham C, Lacy S, Duncan AM, Whyte P (1993) The adenovirus E1A associated 130 kD protein is encoded by a member of the retinoblastoma gene family and physically interacts with cyclin A and E. Genes Dev 7:2366–2377

Lillie JW, Green MR (1989) Transcription activation by the adenovirus E1a protein. Nature 338:39–44

Lin R-Y, Moss SB, Rubin CS (1995) Characterization of S-AKAP84, a novel developmentally regulated A kinase anchor protein of male germ cells. J Biol Chem 270: 27804–27811

Lipinski KS, Esche H, Brockmann D (1998) Amino acids 1–29 of the adenovirus serotypes 12 and 2 E1A proteins interact with rap30 ($TF_{II}F$) and TBP in vitro. Virus Res 54:99–106

Lipinski KS, Fax P, Wilker B, Hennemann H, Brockmann D, Esche H (1999) Differences in the interactions of oncogenic adenovirus 12 early region 1A and nononcogenic adenovirus 2 early region 1A with the cellular coactivators p300 and CBP. Virology 255:94–105

Liu F, Green MR (1990) A specific member of the ATF transcription factor family can mediate transcription activation by the Adenovirus E1A protein. Cell 61:1217–1224

Liu F, Green MR (1994) Promoter targeting by Adenovirus E1A through interaction with different cellular DNA-binding domains. Nature 368:520–525

Liu M, Lee BH, Mathews MB (1999) Involvement of RFX1 protein in the regulation of the human proliferating cell nuclear antigen promoter. J Biol Chem 274:15433–15439

Loeken MR (1993) Effects of mutation of the CREB binding site of the somatostatin promoter on cyclic AMP responsiveness in CV-1 cells. Gene Expr 3:253–264

Lundblad JR, Kwok RPS, Laurance ME, Harter ML and Goodman RH (1995) Adenoviral E1A-associated protein p300 as a functional homologue of the transcriptional co-activator CBP. Nature 374:85–88

Martin KJ, Lillie JW, Green MR (1990) Evidence for interaction of different eukaryotic transcriptional activators with distinct cellular targets. Nature 346:147–152

Mazzarelli JM, Aktins GB, Geisberg JV, Ricciardi RP (1995) The viral oncoproteins Ad5 E1a, HPV16 E7 and SV40 Tag bind a common region of the TBP-associated factor-110. Oncogene 11:1859–1864

Mazzarelli JM, Mengus G, Davidson I, Ricciardi RP (1997) The transactivation domain of adenovirus E1A interacts with the C terminus of human TAF(II)135. J Virol 71:7978–7983

Mermelstein F, Yeung K, Cao J, Inostroza JA, Erdjument-Bromage H, Eagelson K, Landsman D, Levitt P, Tempst P, Reinberg D (1996) Requirement of a corepressor for Dr1-mediated repression of transcription. Genes Dev 10:1033–1048

Moran E, Mathews MB (1987) Multiple functional domains in the adenovirus E1A gene. Cell 48:177–187

Morris GF, Mathews MB (1991) The adenovirus E1A transforming protein activates the proliferating cell nuclear antigen promoter via an activating transcription factor site. J Virol 12:6397–6406

Munshi N, Merika M, Yie J, Senger K, Chen G, Thanos D (1998) Acetylation of HMG I(Y) by CBP turns off IFN beta expression by disrupting the enhanceosome. Mol Cell 4: 457–467

Nakajima T, Uchida C, Anderson SF, Lee CG, Hurwitz J, Parvin JD, Montminy M (1997) RNA helicase A mediates association of CBP with RNA polymerase II. Cell 90:1107–1112

Nevins JR (1981) Mechanism of activation of early viral transcription by the adenovirus E1A gene product. Cell 26:213–220

O'Connor MJ, Zimmermann H, Nielsen S, Bernard H-U, Kouzarides T (1999) Characterization of an E1A-CBP interaction defines a novel transcriptional adapter motif (TRAM) in CBP/p300. J Virol 73:3574–3581

Ogryzko VV, Schilz RL, Russanova V, Howard BH, Nakatani Y (1996) The transcriptional coactivators p300 and CBP are histone acetyltransferases. Cell 87:953–959

PARKER D, FERRERI K, NAKAJIMA T, LAMORTE VJ, EVANS R, KOERBER SC, HOEGER C, MONTMINY M (1996) Phosphorylation of CREB at Ser133 induces complex formation with CBP via a direct mechanism. Mol Cell Biol 16:694–703

PEI R, BERK AJ (1989) Multiple transcription factor binding sites mediate adenovirus E1A transactivation. J Virol 63:3499–3506

PEREIRA DS, JELINEK T, GRAHAM FL (1994) The Adenovirus E1A-associated p300 protein is differentially phosphorylated in Ad12 E1A- compared to Ad5 E1A-transformed cells. Int J Oncol 5:1197–1205

PERRICAUDET M, LEMOULLEC JM, TIOLAIS P, PETTERSSON U (1980) Structure of two adeno-virus type 12 transforming polypeptides and their evolutionary implications. Nature 288:174–176

PURI P, SARTORELLI V, YANG XJ, HAMAMORI Y, OGRYZKO VV, HOWARD B, KEDES L, WANG JY, GRAESSMANN A, NAKATANI Y, LEVRERO M (1997) Differential roles of p300 and PCAF acetyltransferases in muscle differentiation. Mol Cell 1:35–45

QUINN PG (1993) Distinct activation domains within cAMP response element-binding protein (CREB) mediate basal and cAMP-stimulated transcription. J Biol Chem 268: 16999–17009

RADHAKRISHNAN I, PÉREZ-ALVARADO GC, PARKER D, DYSON HJ, MONTMINY MR, WRIGHT PE (1997) Solution structure of the KIX domain of CBP bound to the transactivation domain of CREB: A model for activator:coactivator interaction. Cell 91:741–752

REID JL, BANNISTER AJ, ZEGERMAN P, MARTINEZ-BALBAS MA, KOUZARIDES T (1998) E1A directly binds and regulates the P/CAF acetyltransferase. EMBO J 17:4469–4477

ROCHETTE-EGLY C, FROMENTAL C, CHAMBON P (1990) General repression of enhancer activity by the adenovirus 2 E1A proteins. Genes Dev 4:137–150

SASSONE-CORSI P (1988) Cyclic AMP induction of early Adenovirus promoters involves sequences required for E1A transactivation. Proc Natl Acad Sci USA 85:7192–7196

SAWADA Y, FUJINAGA K (1980) Mapping of adenovirus 12 mRNA's transcribed from the transforming region. J Virol 36:639–651

SCHÖLER HR, CIESIOLKA T, GRUSS P (1991) A nexus between Oct-4 and E1A: implications for gene regulation on embryonic stem cells. Cell 66:291–304

SHAYWITZ AJ, GREENBERG ME (1999) CREB: A stimulus-induced transcription factor activated by a diverse array of extracellular signals. Annu Rev Biochem 68: 821–861

SHENK T (1996) Adenoviridae and their replication. In: Fields B, Howley P, Knipe D (eds) Virology. Raven Press, New York, pp 2111–2148

SIDLE A, PALATY C, DIRKS P, WIGGAN O, KIESS M, GILL RM, WONG AK, HAMEL PA (1996) Activity of the retinoblastoma family proteins, pRB, p107, and p130, during cellular proliferation and differentiation. Crit Rev Biochem Mol Biol 31:237–271

SRIVASTAVA RK, LEE YN, NOGUCHI K, PARK YG, ELLIS MJC, JEONG J-S, KIM SN, CHO-CHUNG YS (1998) The RIIâ regulatory subunit of protein kinase A binds to cAMP response element: An alternative cAMP signalling pathway. Proc Natl Acad Sci USA 95:6687–6692

STEPHENS C, HARLOW E (1987) Differential splicing yields novel adenovirus 5 E1A mRNAs that encode 30 kd and 35 kd proteins. EMBO J 6:2027–2035

Sundqvist A, Sollerbrant K, Svensson C (1998) The C-terminal region of Adenovirus E1A activates transcription through targeting of a C-terminal binding protein-histone deacetylase complex. FEBS Lett 429:183–188

Suomalainen M, Nakano MY, Boucke K, Keller S, Greber UF (2001) Adenovirus-activated PKA and p38/MAPK pathways boost microtubule-mediated nuclear targeting of virus. EMBO J 20:1310–1319

Swaminathan S, Thimmapaya B (1995) Regulation of Adenovirus E2 transcription unit. In: Doerfler W, Böhm P (eds) The Molecular repertoire of Adenoviruses III, Biology and pathogenesis. Springer, Berlin Heidelberg, pp 177–194

Swope DL, Mueller CL, Chrivia JC (1996) CREB-binding protein activates transcription through multiple domains. J Biol Chem 271:28138–28145

Tan S, Aso T, Conaway RC and Conaway JW (1994) Roles for both the RAP30 and RAP74 subunits of transcription factor IIF in transcription initiation and elongation by RNA polymerase II. J Biol Chem 269:25684–25691

Taskén K, Skålhegg BS, Taskén KA, Solberg R, Knutsen HK, Levy FO, Sandberg M, Orstavik S, Larsen T, Johansen AK, Vang T, Schrader HP, Reinton NT, Torgersen KM, Hansson V, Jahnsen T (1997) Structure, function, and regulation of human cAMP-dependent protein kinases. In: Corbin J, Francis S (eds) Signal Transduction in Health and Disease, Advances in Second Messenger and Phosphoprotein Research. Lippincott-Raven Publishers, Philadelphia, pp 191–204

Thomas MJ, Seto E (1999) Unlocking the mechanisms of transcription factor YY1: are chromatin modifying enzymes the key? Gene 236:197–208

Torchia J, Rose DW, Inostroza J, Kamei Y, Westin S, Glass CK, Rosenfeld G (1997) The transcriptional co-activator p/CIP binds CBP and mediates nuclear-receptor function. Nature 387:677–684

Tortora G, Cho-Chung YS (1990) Type II regulatory subunit of protein kinase restores cAMP-dependent transcription in a cAMP-unresponsive cell line. J Biol Chem 265: 18067–18070

Trendelenburg G, Hummel M, Riecken EO, Hanski C (1996) Molecular characterization of AKAP149, a novel kinase anchor protein with a KH domain. Biochim Biophys Res Commun 225:313–319

Turnell AS, Grand RJ, Gorbea C, Zhang X, Wang W, Mymryk SJ, Gallimore PH (2000) Regulation of the 26 S proteasome by adenovirus E1A. EMBO 19:4759–4773

Ulfendahl PJ, Linder S, Kreivi JP, Nordqvist K, Sevensson C, Hultberg H, Akusjarvi G (1987) A novel adenovirus-2 E1A mRNA encoding a protein with transcription activation properties. EMBO J 6:2037–2044

Waltzer L, Bienz M (1998) Drosophila CBP represses the transcription factor TCF to antagonize Wingless signalling. Nature 395:521–525

Wang H-GH, Rikitake Y, Carter MC, Yaciuk P, Abraham SE, Zerler B, Moran E (1993a) Identification of specific adenovirus E1A N-terminal residues critical to the binding of cellular proteins and to the control of cell growth. J Virol 67:476–488

Wang H-GH, Yaciuk P, Ricciardi RP, Green M, Yokoyama K, Moran E (1993b) The E1A products of oncogenic adenovirus serotype 12 include amino-terminally modified forms able to bind the retinoblastoma protein but not p300. J Virol 67:4804–4813

WEBSTER LC, RICCIARDI RP (1991) Trans-dominant mutants of E1A provide genetic evidence that the zinc finger of the trans-activating domain binds a transcription factor. Mol Cell Biol 11:4287–4296

WEN W, MEINKOTH JL, TSIEN RY, TAYLOR SS (1995) Identification of a signal for rapid export of proteins from the nucleus. Cell 82:463–473

WHYTE P, BUCHKOVICH KJ, HOROWITZ JM, FRIEND SH, RAYBUCK M, WEINBERG RA, HARLOW E (1988) Association between an oncogene and an anti-oncogene: the adenovirus E1A proteins bind to the retinoblastoma gene product. Nature 334: 124–129

YACIUK P, MORAN E (1991) Analysis with specific polyclonal antiserum indicates that the E1A-associated 300-kDa product is a stable nuclear phosphoprotein that undergoes cell cycle phase-specific modification. Mol Cell Biol 11:5389–5397

YANG XL, OGRYZKO VV, NISHIKAWA J, HOWARD BH, NAKATANI Y (1996) A p300/CBP-associated factor that competes with the adenoviral oncoprotein E1A. Nature 382: 319–324

YAO TP, KU G, ZHOU N, SCULLY R, LIVINGSTON DM (1996) The nuclear hormone receptor coactivator SRC-1 is a specific target of p300. Proc Natl Acad Sci USA 93: 10626–10631

YAO TP, OH SP, FUCHS M, ZHOU ND, CH'NG LE, NEWSOME D, BRONSON T, LI E, LIVINGSTON DM, ECKNER R (1998) Gene dosage-dependent embryonic development and proliferation defects in mice lacking the transcriptional integrator p300. Cell 93:361–372

YEE SP and BRANTON PE (1985) Detection of cellular proteins associated with human adenovirus type 5 early region 1A polypeptides. Virology 147:142–153

ZHANG JJ, VINKEMEIER U, GU W, CHAKRAVARTI D, HORVATH CM, DARNELL JE Jr (1996) Two contact regions between Stat1 and CBP/p300 in interferon gamma signaling. Proc Natl Acad Sci USA 93:15092–15096

ZHANG Q, YAO H, VO N, GOODMAN RH (2000) Acetylation of Adenovirus E1A regulates binding of the transcriptional corepressor CtBP. Proc Natl Acad Sci USA 97: 14323–14328

ZU YL, TAKAMATSU Y, ZHAO MJ, MAEKAWA T, HANDA H, ISHII S (1992) Transcriptional regulation by a point mutant of adenovirus-2 E1A product lacking DNA binding activity. J Biol Chem 267:20161–20187

Chapter 5

Adenovirus DNA Replication

H. Liu, J. H. Naismith, R. T. Hay

Abstract. Replication of the adenovirus genome is catalysed by adenovirus DNA polymerase in which the adenovirus preterminal protein acts as a protein primer. DNA polymerase and preterminal protein form a heterodimer which, in the presence of the cellular transcription factors NFI/CTFI and NFIII/Oct-1, binds to the origin of DNA replication. DNA replication is initiated by DNA polymerase mediated transfer of dCMP onto preterminal protein. Further DNA synthesis is catalysed by DNA polymerase in a strand displacement mechanism which also requires adenovirus DNA binding protein. Here, we discuss the role of individual proteins in this process as revealed by biochemical analysis, mutagenesis and molecular modelling.

H. Liu, J. H. Naismith, R. T. Hay
Centre for Biomolecular Science, Biomolecular Science Building, The University of St. Andrews, North Haugh, St. Andrews, KY16 9ST, UK
e-mail: rth@st-andrews.ac.uk

1 Introduction

Adenoviruses, of which there are more than 80 serologically distinct types, are similar in morphology and genome organization. The viruses have a linear double-stranded DNA (ds DNA) genome of 25–36 kb with inverted terminal repeats (ITRs) of various sizes depending on the virus or serotype. The virus genome is maintained inside the virion in a highly condensed form by small basic proteins but after infection of susceptible cells, the virion is uncoated and viral DNA is released into the nucleus. The DNA genome of the frequently studied adenovirus serotype 2 (Ad2) comprises 35,937 base pairs (bp) with ITRs of 10 bp. Located within the ITRs are *cis*-acting sequences which define the origin of DNA replication (*ori*). Covalently attached to each 5′ end of the genome is a terminal protein (TP) which is likely to constitute an additional *cis*-acting component of *ori*. In vivo, as E2 gene products accumulate and the infected cell enters the S phase of the cell cycle, viral DNA replication begins (Challberg and Kelly 1989; Hay and Russell 1989). Genome replication is accomplished by the combined action of three viral proteins encoded by the E2 transcription unit. These are precursor terminal protein (pTP), adenovirus DNA polymerase (Adpol) and DNA binding protein (DBP). Adpol and pTP are present in the infected cells as a heterodimer (Lichy et al. 1982). These three proteins with the core origin can sustain a low level of replication. With the development of in vitro systems (Challberg and Kelly 1979), it was shown that cellular nuclear factor I (NFI), or CAAT transcription factor (CTF1) and nuclear factor III (NFIII) or octamer-binding protein (Oct-1) were required for an efficient replication in addition to the three viral proteins (de Jong and Van der Vliet 1999). Finally, another cellular factor, named nuclear factor II (NFII) that shows topoisomerase activity, is required for complete elongation of the viral genome (Nagata et al. 1983a). The proteins involved in adenovirus DNA replication and their functions are listed in Table 1.

Replication of the adenovirus genome is catalysed by Adpol via a protein-priming mechanism (reviewed in de Jong and Van der Vliet 1999; Hay 1996; Van der Vliet 1995), in which pTP acts as the protein primer. Following binding of pTP–Adpol heterodimer (pTP/pol) to the core origin of replication (Mul and Van der Vliet 1992; Temperley and Hay 1992), DNA replication initiates at either end of the linear viral genome with formation of a covalent linkage between the α-phosphoryl group of the

Table 1. Properties of proteins involved in Ad2 or Ad5 DNA replication

Proteins	Size (kDa)	Domain required for DNA	Interacting protein(s)	Functions in DNA replication
Adpol	140	All	Ad pTP,	DNA polymerase, 3′-5′ proof-reading exonuclease; binding to the core of *ori* as pTP/pol; transferring dCMP onto pTP during initiation of DNA replication; catalysing strand-displacement synthesis during elongation phase of DNA replication
Ad pTP	80	All	Adpol, NFIII	Protein primer for initiation of DNA replication; binding onto the core of *ori* as pTP/pol; substrate for the Ad protease to release TP
Ad DBP	59	DNA-binding domain aa 174–529	-	Enhance initiation; required for elongation; ss and ds DNA binding; stimulation of the NFI binding to *ori*; reducing dCTP K_m; increasing Ad pol processivity; helix-destabilizing
NFI/CTFI	55	DNA-binding domain aa 1–220	Adpol	Binding to the auxiliary region of *ori*; enhanced by DBP; enhancing initiation of DNA replication
NFII	30	All	-	Topoisomerase activity to release the topological stress during full-length genome synthesis
NFIII/ Oct-1	90	DNA-binding domain aa 280–439	Ad pTP	Binding to auxiliary region of *ori*; enhancing initiation of the DNA replication

terminal residue, dCTP, and the β-OH group of S580 in pTP. Synthesis of the nascent strand is then primed by the β-OH group of pTP–dCMP complex. Initiation and further DNA replication is enhanced by DBP and NFI and NFIII (DE JONG and VAN DER VLIET 1999; MONAGHAN et al. 1994; MUL et al. 1990; PRUIJN et al. 1986). The presence of adenovirus TP which is covalently attached to the 5′ end of viral DNA, also stimulates initiation of DNA replication in vitro (PRONK et al. 1992; PRONK and VAN DER VLIET 1993). After initiation, DNA synthesis is accompanied by displacement of the non-template strand which can form a partial duplex by base-pairing of the ITRs on which a second round of DNA synthesis can be initiated (HAY et al. 1984; STOW 1982). Alternatively, the displaced strands may simply reanneal to form ds DNAs which are subsequently processed and encapsidated into virions. This process is probably enhanced by DBP, which promotes intermolecular reannealing but inhibits intramolecular annealing in vitro (ZIJDERVELD et al. 1993).

The progress on adenovirus DNA replication has been reviewed extensively (DE JONG and VAN DER VLIET 1999; HAY 1995, 1996; HAY and RUSSELL 1989; VAN DER VLIET 1995, 1996). However, new knowledge in this field accumulates continuously. This overview summarizes the latest developments regarding adenovirus replication and the roles of the replication proteins.

2 Adenovirus Origins of DNA Replication

Efficient initiation of adenovirus DNA replication requires the presence of specific terminal nucleotide sequences (WIDES et al. 1987). This *cis*-acting DNA sequences which define *ori*, are located at the ends of viral genome within the ITRs (RAWLINS et al. 1984). TP, which is covalently attached to each 5′ end of DNA, may well be an additional *cis*-acting component of *ori* (PRONK and VAN DER VLIET 1993). Sequence comparison of *oris* from a wide range of adenoviruses and mutational analysis of *ori* of the Ad2 genome has identified four regions within the terminal 50 bp that contribute to *ori* activity in vitro and in vivo (HAY 1985b). Similar to many other *oris*, *ori* of Ad2 consists of an essential core region and auxiliary regions (RAWLINS et al. 1984; Fig. 1). The terminal 18 bp of ITRs, designated as the core region, represent the minimal *ori* (CHALLBERG and RAWLINS 1984). Sequence 5′-ATAATATACC-3′ (bp 9–18, Fig. 1), which is conserved in all human adenoviruses (STILLMAN et al.

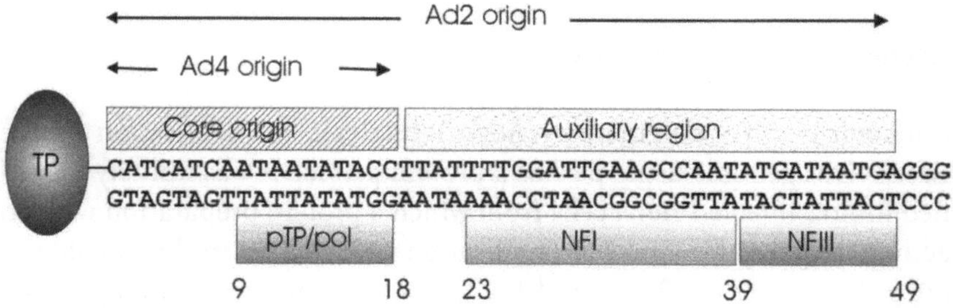

Fig. 1. Structure and organization of Ad2 and Ad4 origins of DNA replication. The binding sites for pTP/pol, NFI and NFIII/Oct-1 are indicated. (Modified from HAY 1996)

1982b), is the pTP/pol binding site. Although the terminal 18-bp of the ITR can function as *ori*, in isolation it supports only a limited, basal level of initiation of DNA replication (BAILEY and MAUTNER 1994; VAN BERGEN et al. 1983). Separated from the minimal *ori* by a spacer region of defined length (ADHYA et al. 1986; BOSHER et al. 1990; WIDES et al. 1987) is the NFI/CTF1 DNA-binding site positioned between bp 23 and 38 within the auxiliary region (ADHYA et al. 1986; JONES et al. 1987; NAGATA et al. 1982, 1983b). Distal to the NFI-binding site is the DNA recognition site (bp 39–49) for another cellular protein, NFIII/Oct-1. This recognition site includes the partially conserved nucleotides 5′-TATGATAATGA-3′ (O'NEILL and KELLY 1988; PRUIJN et al. 1986; ROSENFELD et al. 1987; WIDES et al. 1987). In contrast with Ad2, Ad4 requires only an 18-bp terminal sequence of the viral genome to effectively initiate DNA replication in vivo (HAY 1985a) and in vitro (HARRIS and HAY 1988; TEMPERLEY et al. 1991; TEMPERLEY and HAY 1991). The ITR of Ad4 does not contain an NFI recognition site, and although it does have a binding site for NFIII/Oct-1, neither factor is required for DNA replication in vivo or is capable of stimulating DNA replication in vitro (HARRIS and HAY 1988). Ad4 therefore appears to have circumvented the need for the host factors NFI/CTF1 and NFIII/Oct-1. The reasons for the different requirements of Ad2 and Ad4 DNA replication are not obvious, and DNA-binding assays have indicated that the affinity of baculovirus-expressed Ad4 pTP/pol for the core region is not substantially different from that of the Ad2 pTP/pol.

3
Adenovirus DNA Polymerase

Adenovirus DNA polymerase (Adpol) catalyses the replication of the adenovirus genome using pTP as a protein primer. It was initially identified in Ad2-infected HeLa cells from which a protein preparation with an activity required for viral DNA replication yielded a heterodimer containing polypeptides of 80 kDa and 140 kDa with a unique DNA polymerase activity (Enomoto et al. 1981; Ikeda et al. 1981). The adenovirus origin of these two proteins was demonstrated using complementation experiments in vitro with temperature-sensitive mutants of Ad5 (group N mutants) defective in both the initiation and elongation of DNA replication (Stillman et al. 1982a; Van Bergen et al. 1983). Nuclear extract prepared from cells infected with these mutant viruses was unable to support initiation of DNA replication, but activity was restored by addition the of wild-type Adpol (Stillman et al. 1982a). The biochemical defects associated with the temperature-sensitive phenotypes are as a result of altered Adpol–DNA binding (Chen et al. 1994). Early studies of Adpol in adenovirus replication were limited because of the difficulty of over-expressing Adpol in *Escherichia coli*. High expression of functional protein was achieved in HeLa cells (Shu et al. 1988), COS cells (Pettit et al. 1988; Shu et al. 1987) using recombinant vaccinia viruses containing Adpol cDNA (Stunnenberg et al. 1988) and in insect cells using recombinant baculovirus (Watson and Hay 1990). The role of Adpol in adenovirus DNA replication has been analysed by mutagenesis (Brenkman et al. 2001, 2002; Joung and Engler 1992; Joung et al. 1991; Liu et al. 2000) and sequence comparison with other DNA polymerases (Pols; Knopf 1998; Liu et al. 2000).

The in vitro systems which replicate adenovirus genome (Challberg and Kelly 1979; Stillman 1983) have been used to establish the role of Adpol in virus DNA synthesis. Prior to DNA synthesis, a large pre-initiation complex is formed, which contains three viral proteins pTP, Adpol and DBP and two cellular transcription factors NFI (Bosher et al. 1990; Chen et al. 1990; Mul and Van der Vliet 1992) and NFIII (Mul et al. 1990) held in a precise spatial arrangement on *ori* by a combination of DNA–protein and protein–protein interactions (Chen et al. 1990). The pTP/pol heterodimer recognizes bp 9–18 in the core of *ori* (Brenkman et al. 2002; Temperley et al. 1991, Temperley and Hay 1992). This binding is stabilized in the case of Ad2 and Ad5 by specific interactions with NFI and NFIII that bind to adjacent auxiliary regions of *ori* (Armentero et al. 1994; Bosher et al. 1990; Chen et al. 1990; Coenjaerts et al. 1994; Mul and Van der Vliet 1992). After initia-tion, this initiation complex dissociates, enhancing the elongation and proofreading activity of Adpol (Coenjaerts and Van der Vliet 1994; King et al. 1997). Finally, Adpol catalyses synthesis of the full genome with the participation of NFII (Nagata et al. 1983a; Wong and Hsu 1990).

Adpol is one of several protein-primed DNA Pols included in the Pol α family (Field et al. 1984; Hay 1996). Like all other Pols, Adpol has intrinsic 3′–5′ proofreading exonuclease activity and contains sequence motifs (EXO I, II, and III) characteristic of exonucleases in the amino-terminal portion. Conserved motifs associated with DNA polymerase activity (designated Pol I–VII) are located in the carboxy-terminal portion of Adpol (Hwang et al. 1992; Knopf 1998; Liu et al. 2000). Structural analysis of Pols from diverse sources showed that these molecules share a similar domain arrangement that has been compared with the shape of a hand with subdomains consisting of 'thumb', 'palm', and 'fingers' (Joyce and Steitz 1994). In RB69 Pol, a member of Pol α family, these domains form a molecule with the overall shape of a disc with a hole in the centre. Residues surrounding the central hole are directly involved in catalytic activity and three deep grooves, formed by five domains, converge on this point (Wang et al. 1997). Comparison of domain organization and the conserved residues of Adpol with that of phage ϕ29 Pol and RB69 Pol are shown in Fig. 2. Although the sizes of these DNA polymerases are different, the domain organization and apparent spatial relationship of these residues has been uncovered (Liu et al. 2000) suggesting a similar catalytic mechanism for these DNA polymerases. Analysis of the spatial relationship of domains in Adpol using oligonucleotides with a translocation block showed that the distance between the enzyme surface and the exonuclease active site was approximately 5 nucleotides, whereas the distance between the entrance of the primer binding groove and the polymerase active site was 9–10 nucleotides (Brenkman et al. 2002). This is rather similar to that of RB69 Pol calculated based on the structure (Franklin et al. 2001; Wang et al. 1997). Adpol has similar EXO domain organization to all other DNA Pols. Mutation of conserved residues D283 and E285 in EXO I, N417, F421 and D422 in EXO II and Y580 and D584 in EXO III alter or abolish exonuclease activity (Brenkman et al. 2002; H. Liu, J.H. Naismith and R.T. Hay, unpublished results).

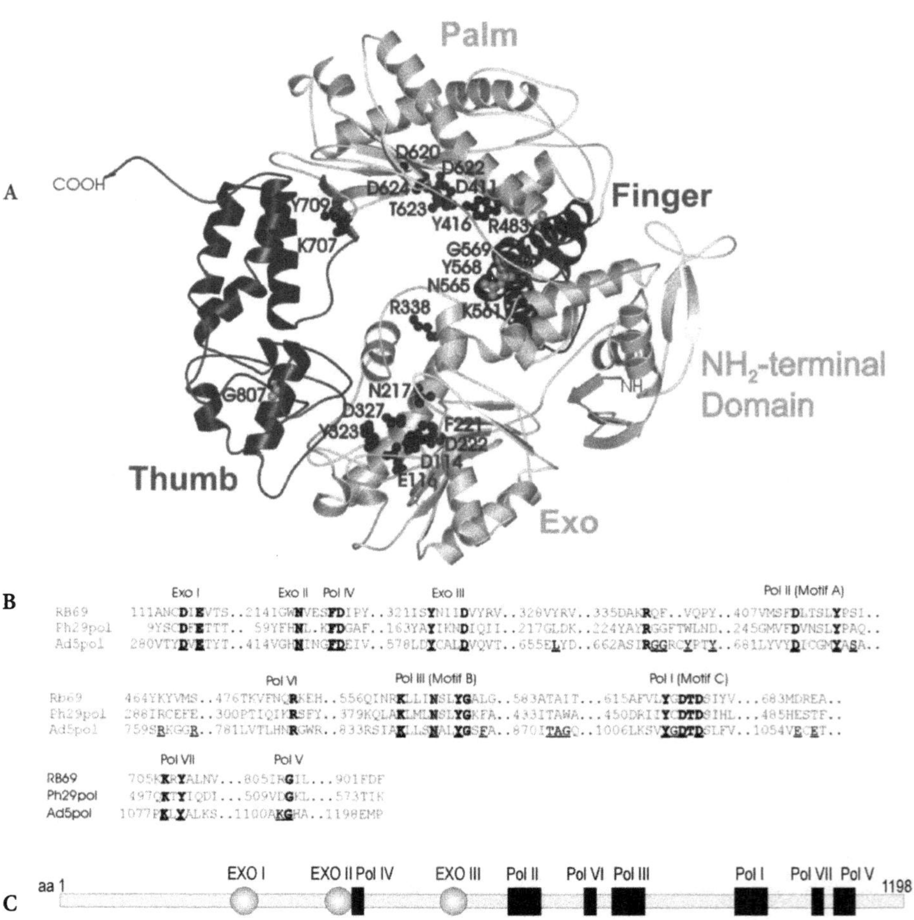

Fig. 2A–C. Sequence alignment of the conserved domains in Adpol with those in phage 29 and RB69 Pols. **A** Ribbon diagram of RB69 DNA Pol from WANG et al. showing the relative locations of five domains (EXO, amino-terminus, fingers, palm and thumb) and the conserved residues in these domains. Each residue is marked with *a letter* (one-letter amino acid code) and *a number* indicating its position in the amino acid sequence. **B** Multiple-sequence alignment of conserved domain in RB69 Pol, phage 29 Pol and Adpol. The conserved residues in each domain indicated in **A** are in *boldface*. Residues mutated in Adpol are *underlined*. **C** Locations of EXO (*filled circles*) and Pol (*filled rectangles*) in the Adpol amino acid sequence. *Numbers* at each end indicate the length of the Adpol peptide chain. (Modified from LIU et al. 2000)

The polymerase catalytic site of Adpol is composed of the conserved motifs YGDTD in Pol I, Dx_2SLYP in Pol II, and Kx_3NSx_2YG in Pol III similar to that of all other DNA Pols. Residues within or between the conserved regions of Adpol are also involved in DNA polymerase activity (Chen and Horwitz 1989; Joung and Engler 1992; Joung et al. 1991; Liu et al. 2000). The functions of these conserved residues in adenovirus DNA replication are summarized in Table 2. Similar to all other polynucleotide Pols, Adpol-catalysed nucleotide addition appears to proceed by a common 'two-metal-ion' mechanism. The dNTP is associated with two magnesium ions which are bound to the phosphates of the nucleotide and to two completely conserved aspartic acid residues D685 and D1014 (Liu et al. 2000). It is proposed that the two metal ions serve both to correctly position the dNTP in the active site and participate in the chemistry of phosphoryl transfer. While one metal ion activates the 3′ hydroxyl group of the primer, the other helps the departure of pyrophosphate and both stabilize the charge and structure of the predicted transition state (Steitz 1998). This suggestion has been further confirmed in the structure of the replication complex of RB69 Pol (Franklin et al. 2001). It is supposed that in Adpol, residue D685 in Pol II ligates Mg^{2+}, binds dNTP and participates in catalysis (Joung et al. 1991). D1014, the second aspartic acid in the DTD sequence of motif I, is involved in metal-specific catalysis and metal-induced infidelity in DNA synthesis. Mutation of these residues abolished DNA polymerase and initiation activity (Joung et al. 1991; Liu et al. 2000). Three chemically distinct clusters of amino acids are expected to interact with the primer terminus and the incoming dNTP. These conserved clusters consist of exposed aromatic residues Y670, Y690, Y844, F847, Y1010, and Y1080; negatively charged residues D685, D1012, D1014, E1055 and E1057 and positively charged residues R787, R791, K837 and K1078. These residues play important roles in initiation of adenovirus DNA replication and mutation of these residues affects Adpol activity (Liu et al. 2000). Two aromatic residues Y690 and Y844 in Adpol are highly conserved between the members of Pol α family, and appear to be important in maintaining the fidelity of DNA replication. In each case, these residues are involved in selecting between ribo- and deoxyribonucleotide triphosphates. Similar function of equivalent residues was identified in ϕ29 Pol (Saturno et al. 1998) and RB69 pol (Franklin et al. 2001).

Distinct regions of Adpol contribute to interact with pTP and NFI (Armentero et al. 1994; Bosher et al. 1990; Chen et al. 1990) to form the

Table 2. Phenotypes of mutations in Adpol[a] (from Liu et al. 2000)

Adpol mutants	DNA Binding (0.2 M)[b]		Adpol-pTP complex[c]	Pol activity[d] (%)	Initiation[e] (%)
	ss DNA	ds DNA			
Adpol-WT	++++	++++	++++	++++	++++
656L/A	+++	+++	+++	++++	+++
666GG/AA	+++	++	+++	++++	+++
667G/D	+++	+++	+++	++++	+++
670Y/A	++++	++	+++	+++	+
673Y/A	+	+	+	++	+
685D/A	++++	+++	++++	–	–
690Y/A	+++	+++	++++	+	++
692S/A	+++	++	+++	+++	+++
760R/A	++++	++	++++	++++	++
764R/A	+++	++	+++	++++	++++
837K/A	++++	++++	+++	++++	++++
841N/A	++++	++++	++++	++	+
844Y/A	+	++	++	+	+
847F/A	+	+	+++	++	+
871TAG/AAA	+	+	++	++	+
1010YG/AA	+	+	+++	+	+
1012DTD/ATA	+	+	++	–	–
1055E/A	+	+	++++	++++	++
1057E/A	–	–	–	+	+
1078K/A	+	+	+++	++	++
1080Y/A	–	–	+	++	+
1101KG/AA	+	+	+++	++	++

[a] Activities of wild-type Adpol and the mutants in the assays. ++++, Activity of the wild-type Adpol or the mutants showing 76%–100% activity of the wild-type; +++, ++ and +, mutants showing 51%–75%, 25%–50% and <25% activity of the wild-type respectively; –, no activity detected.

[b] Ability of Adpol mutants to bind ss or ds DNA at 0.2 M NaCl. Adpol bound to ss or ds DNA-Sepharose beads was determined by Western blotting with an Adpol-specific antiserum. The scanned image of the electrochemiluminescence blot was used to determine binding of each mutant Adpol relative to that of wild-type.

[c] Ability of Adpol mutants to form the Adpol–pTP complex. The amount of Adpol bound by pTP was determined by immunoprecipitation of the Adpol–pTP complex using an anti-pTP monoclonal antibody (16H1) and detection of Adpol by Western blotting.

[d] Catalytic activity of Adpol on activated DNA was determined in a DNA polymerase assay and is based on the dAMP incorporation to wild-type Adpol.

[e] Ability of Adpol mutants to participate in initiation of DNA replication was determined based on the incorporation of pTP–dCMP complex.

pre-initiation complex. Domain structure analysis indicated at least four domains present within Adpol (Parker et al. 1998). Adenovirus pTP interacts with three of the four proteolytically derived domains indicating that like other members of Pol α family, Adpol has a multi-domain structure and interaction with pTP takes place with non-contiguous regions of the polypeptide chain over a large surface area (Parker et al. 1998). Mutagenesis and biochemical analysis showed that residues Y670, Y673, Y844, Y1080 or E1057 are involved in the pTP/pol interaction. They also play roles in protein primer selection and template interaction (Liu et al. 2000). Although these residues are dispersed in the linear Adpol sequence, they might have close spatial relationship. It is reported that an YxGG/A motif in phage ϕ29 Pol are required for protein-primed replication (Truniger et al. 1999). A similar motif is found in Adpol (Fig. 2) and mutation of these residues affects template DNA binding and the transition from initiation to elongation (Brenkman et al. 2001). Two potential zinc-binding motifs in Adpol are important for DNA binding and replication initiation (Joung and Engler 1992). Mutation of these zinc finger domains resulted in the loss of the ability to bind *ori* but maintained the ability to interact with pTP, indicating that domains of Adpol involved in protein–protein and protein–DNA interaction function independently (Chen et al. 1990). Although no direct interaction of DBP with Adpol has yet been found, DBP was shown to have a profound effect on both DNA polymerase and exonuclease activities of Adpol (Field et al. 1984). DNA synthesis on poly (dT):oligo (rA) was stimulated 10- to 100-fold by the presence of DBP due to an increase in processivity of Adpol. It is thought that DBP may act by stabilizing the interaction between Adpol and the DNA template. However, this function of DBP is different from that of other single-strand DNA binding protein (SSBs), such as *E. coli* SSB, which was unable to substitute for Ad DBP. Likewise, Ad DBP could not substitute for human SSB in stimulating the activity of human Pol α. Binding of DBP to the displaced strand drives the ATP-independent unwinding of the template during adenovirus DNA replication (Dekker et al. 1997). In contrast with DNA polymerase activity, the proofreading exonuclease activity of Adpol is inhibited by DBP. Inhibition by DBP is a direct effect on the exonuclease activity of Adpol rather than binding to and protecting DNA ends (Lindenbaum et al. 1986). Although DBP increases the thermostability of Adpol, it has not yet been possible to demonstrate a direct physical association between the two proteins. Mutagenesis and DNA binding analysis showed that interaction of Adpol with NFI targets and stabilizes Adpol onto *ori* (MUL and VAN DER VLIET 1992). Although the interaction site on Adpol is not clear, the site on NFI bound by Adpol is between amino acids 68 and 150, which overlap the DNA binding and replication activation domain of this factor.

Intriguingly, phosphorylation of Adpol was reported and a role for phosphorylation in the activities of Adpol has been suggested (RAMACHANDRA and PADMANABHAN 1993). Phosphorylation occurs on S67 in Adpol by a stably associated 'cdc2-like' histone H1 kinase. Although the precise function of the phosphorylation is unclear, substitution of alanine for serine at this position abolishes DNA replication initiation activity of Adpol. Dephosphorylation of Adpol with calf intestinal alkaline phosphatase resulted in significant decrease of Adpol activity in vitro (RAMACHANDRA et al. 1993).

4 Adenovirus DNA Binding Protein

Adenovirus DNA binding protein (DBP) is a viral protein abundantly accumulated in adenovirus-infected cells. It is the first identified and purified viral non-structural protein with high affinity for single-stranded DNA (ss DNA) and was shown to also bind ds DNA and RNA (VAN DER VLIET and LEVINE 1973). The involvement of DBP in adenovirus replication and its function are well documented and have been reviewed extensively (HAY 1996; VAN DER VLIET 1995). The role of DBP in viral infectivity was first demonstrated genetically by isolation of the temperature-sensitive mutations H5ts125 and H5ts107 in the DBP gene. Nuclear extracts from cells infected with these mutants are defective for adenovirus DNA replication with both exogenous and endogenous templates (FRIEFELD et al. 1983). In both cases, activity was restored by addition of the purified wild-type DBP.

DBP is not essential for the initiation of adenovirus DNA replication but is required for chain elongation. The role of DBP in the elongation of DNA replication, both in vivo and in vitro, has been well documented. However, its specific role in the initiation of adenovirus DNA replication is contentious. DBP stimulates the initiation in vitro using purified components (CLEAT and HAY 1989; KENNY and HURWITZ 1988; MUL and VAN DER VLIET 1993). Since NFI-mediated stimulation of initiation of DNA replication in vitro is dependent on the concentration of DBP (DE VRIES et al. 1985) and DBP increases the affinity of NFI for its recognition site in the adenovirus *ori* (CLEAT and HAY 1989; STUIVER and VAN DER VLIET 1990), it is suggested that this stimulation was a consequence of a functional interaction between DBP and NFI. However, no direct interaction between these two proteins has been detected and the observed effect is likely to be a consequence of altering the structure of bound DNA by DBP. In fact, binding of DBP to ss DNA is stronger and DBP can remove the tertiary structure of ds DNA fragments upon binding (STUIVER et al. 1992). Therefore, it is likely that in the DBP/DNA complex, the relative positions of hydrogen-bound donor and acceptor groups in the major grooves are altered, causing a fine turning of the contacts between NFI and DNA that leads to a higher affinity interaction. This fine turning may increase the dNTP accessibility to the initiation complex as DBP can stimulate initiation of DNA replication by decreasing the K_m of the polymerase for the initiator nucleotide dCTP (MUL and VAN DER VLIET 1993).

Sequence comparisons and proteolytic digestion of DBP showed that DBP has two distinguishable domains, the amino-terminal domain, involved in host-range determination (HARFST and LEPPARD 1999; KITCHINGMAN 1995; Vos et al. 1988) and the carboxyl-terminal domain, involved in DNA replication and DNA binding. By partial chymotrypic digestion, DBP yields a carboxyl-terminal fragment around 40 kDa and a highly phosphorylated 27-kDa amino-terminal fragment (ARIGA et al. 1980; CLEGHON et al. 1993; KLEIN et al. 1979). The amino-terminal fragment is dispensable for DNA replication and the carboxyl-terminal domain of the molecule can fully substitute for the intact molecule during the adenovirus DNA replication in vitro. Within the carboxyl-terminal region are three highly conserved motifs (EAGLE and KLESSIG 1992; KITCHINGMAN 1985; Vos et al. 1988) that are important for binding to ss DNA (NEALE and KITCHINGMAN 1989, 1990). In a photo cross-linking assay, two residues M229 and F418 in the carboxyl-terminal domain were identified to contact DNA. Mutation of residues N322, R323, and K470 also lowered the affinity of DBP for ss DNA binding (NEALE and KITCHINGMAN 1989) suggesting that residues within different regions of the carboxyl-terminal domain also play role in the interaction with DNA. Structure determination of the carboxyl-terminal domain has revealed that the conserved motifs described are important elements in the architecture of the DBP molecule. Two zinc atoms inside the protein appear to stabilize the overall protein folding rather than to contact directly with the DNA (TUCKER et al. 1994). A positively charged region incorporating the cross-linked F418 is thought to mediate interactions with DNA. A mechanism for the cooperative binding of the carboxyl-terminal region of DBP to DNA was suggested by the crystal structure (KANELLOPOULOS et al. 1995). A carboxyl-terminal extension of 17 amino acids from one molecule hooks onto a second molecule to form a protein chain around which the ss DNA is proposed to wind (TUCKER et al. 1994).

DBP binds cooperatively to ss DNA in a non-sequence specific manner. It possess the properties of a helix-destabilizing protein which enhances transcription by specifically increasing the DNA binding activity of a transcription factor to a promoter region through subtle changes in DNA structure, similar to the stimulation of DBP in the initiation of DNA replication (MONAGHAN et al. 1994; ZIJDERVELD et al. 1994). Like other helix-destabilizing proteins, DBP-catalysed DNA unwinding requires neither ATP nor $MgCl_2$. In fact the latter, similar to other agents such as NaCl, inhibits the unwinding process by increasing the stability of duplex DNA molecules (MONAGHAN et al. 1994). The mechanism of DBP-catalysed unwinding of ds DNA has not been established but one possibility is that DBP first binds to ds DNA then invades the DNA duplex and binds to the exposed ss DNA in a stable fashion (CLEAT and HAY 1989; STUIVER and VAN DER VLIET 1990). Alternatively, DBP could bind transiently to single-stranded region of DNA exposed during 'breathing' of short ds DNA molecules. In both cases, a rigid DNA structure is finally formed by interwinding DBP protein chains around the molecule (STUIVER et al. 1992). DBP is unable to unwind the DNA bound by either NFI or NFII/Oct-1 at their cognate recognition sites suggesting that the bound proteins stabilize the DNA duplex, or that the bound proteins may interfere with the ability of DBP to form a continuous protein chain on the DNA. Unlike DNA helicases that translocate unidirectionally on DNA, DBP-catalysed DNA unwinding does not appear to be strictly directional. This unwinding is ATP-independent and driven by DBP protein chain formation and high affinity binding to the displaced strand during adenovirus DNA replication (DEKKER et al. 1997). In the DBP carboxyl-terminal domain, the presence of a flexible loop makes the multimerized protein chain flexible (VAN BREUKELEN et al. 2000) to obtain high affinity for ss DNA binding and subsequent local destabilization of the replication fork (DEKKER et al. 1998). The ability of DBP to impose a rigid structure on DNA is responsible for the ability of DBP to inhibit intramolecular renaturation but promote intermolecular renaturation (ZIJDERVELD et al. 1993). While destabilization of ds DNA is inhibited by high concentrations of monovalent and divalent cations, the renaturation reaction is highly resistant to their presence (ZIJDERVELD et al. 1993).

While Ad2 DBP decreases the K_m for transfer of dCTP onto pTP it increases binding of NFI to *ori*, Ad4 DBP stimulates initiation of replication, but in this case the extent of the stimulation is independent of the concentrations of NFI and dCTP (TEMPERLEY and HAY 1991). In this case, DBP may participate with other replication proteins in unwinding the DNA duplex at the termini of the viral genome prior to initiation. In fact, TP (PRONK et al. 1992; PRONK and VAN DER VLIET 1993), NFI (DE VRIES et al. 1987; MUL and VAN DER VLIET 1992), and NFIII/Oct-1 (VERRIJZER et al. 1991) can all modify the structure of viral DNA. In combination with the ability of DBP to distort and unwind short, fully duplex DNA, this could allow adenovirus proteins to destabilize and unwind *ori* and function together in initiation of DNA replication. The role of DBP in elongation is likely to be a consequence of the ability of DBP to stabilize displaced single strands and alter the properties of Adpol as described previously. DBP may also stimulate initiation of DNA replication by additional activities. In fact, DBP displays multiple biochemical activities that are required for effective viral DNA synthesis (BROUGH et al. 1993) and was found in a large complex associated with cellular proteins (RICIGLIANO et al. 1994). However, the functional significance of the interaction with other cellular proteins has yet to be determined.

5
Adenovirus Protease, pTP and TP

Adenovirus DNA contains the 55-kDa TP covalently linked to its both 5′ ends but the E2-encoded 80-kDa pTP is the active form in initiation of viral DNA replication (CHALLBERG et al. 1980; ENOMOTO et al. 1981). At late times in infection pTP is processed by proteolysis to generate a TP, a 55-kDa carboxyl-terminal fragment of pTP. Ad2 pTP is cleaved to TP via a 62-kDa intermediate (iTP), and because the cleavage sites are conserved in all identified pTPs, it is likely that the cleavage of pTP to TP, via iTP, plays a role in the infectious cycle (WEBSTER et al. 1989). Processing of pTP was first studied with an Ad2 mutant (H2ts1) encoding a temperature sensitive protease (Ad protease). At the restrictive temperature, the mutant replicates efficiently, but the progeny virions are not infectious. Analysis of those virions revealed that in contrast with a wild-type virus infection, mutant H2ts1 virions contain only the intact unprocessed form of pTP

(Stillman et al. 1981). Thus pTP is processed by the H2ts1-specific protease (Ad protease; Ruzindana-Umunyana et al. 2000), which is essential for viral infectivity and accumulates to high levels late in the infection cycle. Ad protease is an L3 gene product with a molecular mass of 23 kDa. Site-directed mutagenesis and sequence comparison of viral proteases indicate that the enzyme may represent a new subclass of cysteine proteases (Rancourt et al. 1994, 1996). Structure determination reveals that Ad protease has an active site made up of H54, E71, C122 and a papain-like fold (Ding et al. 1996; Sircar et al. 1998). The protein is synthesized as an inactive form and activated by an 11-amino acid disulfide-linked peptide (Mangel et al. 1993; Tihanyi et al. 1993; Webster et al. 1993) derived from the carboxy terminus of the pVI virion protein (pVIc; Webster et al. 1993). Viral DNA was reported to stimulate protease activity (Mangel et al. 1993), but is not absolutely required for protease activity (Webster et al. 1993, 1994). Ad protease is involved in processing of pTP to TP (Webster et al. 1994) and other structural proteins (Sung et al. 1983) to generate an infectious virus capsid. Proteolytic activity is required for adenovirus maturation (Matthews and Russell 1995) and the ability of mature virus capsids to uncoat the viral DNA (Ruzindana-Umunyana et al. 2000) and enhance gene delivery (Cotten and Weber 1995). Inhibition of this proteolytic activity reduced or abolished the viral infectivity (Greber et al. 1996).

The roles of pTP in adenovirus replication are to function as a protein primer and then to produce the terminal protein linked to virus genome. Adenovirus DNA replication requires formation of the initiation complex, and elongation after disassociation of the complex (King et al. 1997). In the first step the incoming pTP/pol is positioned at *ori* by protein–DNA and protein–protein interactions. In adenovirus-infected cells, pTP exists as a heterodimer in which it binds tightly to Adpol (Lichy et al. 1982; Stillman et al. 1982a). Studies on pTP have been facilitated by cloning the cDNA copies of the gene (Pettit et al. 1988) into vaccinia virus (Stunnenberg et al. 1988) and in baculovirus (Bosher et al. 1990) for high-level expression in human and insect cells. This has allowed its function and the functional domains to be defined. Genetic and biochemical analysis of pTP demonstrated that the amino-terminal region, removed during processing by Ad protease, is required for specific DNA binding, interaction with Adpol and initiation of DNA replication (Fredman et al. 1991; Freimuth and Ginsberg 1986; Pettit et al. 1989; Roovers et al. 1993; Webster et al. 1994). A nuclear localization sequence, located in the central region of pTP (or the amino-terminal region of TP), mediates the nuclear transport of pTP and also enhances the nuclear transport of pTP/pol (ZHAO and PADMANABHAN 1988). It was initially demonstrated that the sequence-specific binding of pTP to DNA allows the pTP/pol to recognize *ori* (TEMPERLEY and HAY 1992). It is thought that the phosphorylation state pTP (KUSUKAWA et al. 1994) may modulate its DNA binding activity. Partial proteolysis of pTP/pol or cross-linked pTP–DNA complex also indicated a polypeptide of 20 kDa in the amino-terminal region contacting DNA while multiple regions within the carboxyl-terminal two-thirds of pTP interact with Adpol (WEBSTER et al. 1997a). Interestingly, a region of about 60 amino acids within the carboxyl terminus appears to regulate the DNA binding ability of pTP. Analysis of the pTP/pol complex using primer/template elongation with biotin showed that pTP bound to the primer-binding groove of Adpol (BRENKMAN et al. 2002). However, the precise interaction of pTP with Adpol is still unclear although the interaction is through non-contiguous regions of the polypeptide chain over a large surface area of Adpol (PARKER et al. 1998). In addition to Adpol, pTP also interacts with NFIII via the POU homeodomain in pre-initiation complex (COENJAERTS et al. 1994; VAN LEEUWEN et al. 1997). Co-precipitation of POU with individual domains of pTP indicated that POU contacts multiple sites on pTP. These results were confirmed by partial proteolysis of pTP in the presence or absence of POU homeodomain (POUhd). A number of sites accessible to proteases in free pTP are resistant to cleavage in the presence of POUhd suggesting that the POUhd induces a conformational change in pTP which may influence the initiation of adenovirus DNA replication (BOTTING and HAY 1999). Further analysis showed that the region of POUhd bound by pTP overlaps with DNA binding domain of POUhd (DE JONG et al. 2002). Mutation analysis of pTP showed that the role of pTP in initiation is complex with each of different activities of pTP contributing to its role in the initiation of DNA replication (BOTTING and HAY 2001).

It is reported that in adenovirus-infected Hep-2 cells, TP–DNA (SCHAACK et al. 1990 a), pTP (FREDMAN and ENGLER 1993), viral structural polypeptides V, PVII, 11 kDa (core proteins) and capsids are intimately associated with nuclear matrix (NM). The presence of structural proteins in NM during the infection suggests that mostly newly formed virions and viral cores might be bound to this structure (KHITTOO et al. 1986). As Ad protease activity was associated with NM, it was proposed that NM is the site of adenovirus assembly and mature virions are released from the matrix by the viral endoproteinase. However, packaging of TP–DNA complex during Ad2 H2ts1 infection at the non-permissive temperature would argue against this proposition (Weber 1990). It has also been suggested that the association of TP–DNA or pTP with NM may be involved in targeting adenovirus replication complexes to a specific subnuclear domain (Fredman and Engler 1993; Webster et al. 1997b). Release of pTP from NM seems to be controlled by a tyrosine kinase, thus phosphorylation may play a key role in the modulation of pTP binding sites on the NM (Angeletti and Engler 1996). Although the domains of pTP which interact with NM have not been identified, three pTP linker insertion mutants, which were lethal or partially defective with respect to virion formation but active in DNA replication in vitro, are thought to be defective for binding to NM (Fredman et al. 1991). NM associated with the pTP interacts with number of cellular proteins (Angeletti and Engler 1998), although their function in adenovirus replication has yet to be determined.

Processing of pTP by Ad protease into TP via iTP seems to be a highly regulated process during adenovirus infection (Webster et al. 1994). In considering the role of this processing in adenovirus DNA replication, free pTP and pTP/TP covalently attached to the 5′ end of viral DNA should be regarded as distinct functional entities. Whereas the role of free pTP is to act as the protein primer for DNA replication, the significance of the covalently attached pTP/TP remains to be elucidated. In transfection experiments, the infectivity of TP–DNA is orders of magnitude greater than that of protein-free DNA (Tamanoi and Stillman 1982; Van Bergen et al. 1983). TP–DNA is also a much more efficient template than protein-free DNA in in vitro replication assay with a difference of 20-fold under certain conditions (Pronk and Van der Vliet 1993). It has been proposed that genome-bound pTP/TP may protect viral DNA from 5′–3′ exonuclease degradation, prevent the inhibitory effect of end-binding proteins such as the NFIV/Ku (De Vries et al. 1989), allow attachment of viral DNA to the nuclear matrix (Schaack et al. 1990a) and participate in unwinding of the DNA duplex at *ori*. In addition, genome-bound TP serves to enhance the DNA binding of pTP/pol by stabilizing binding of the incoming pTP/pol at *ori* (Pronk and Van der Vliet 1993) and induces conformational changes in *ori*. However, the relationship between these two phenomena has not yet been established. It should be emphasized that in vivo the template for early transcription and the first round of DNA replication is TP–DNA, whereas the template for subsequent rounds of replication is likely to be pTP–DNA as processing of pTP to TP takes place late in the infection cycle. It is also possible that the NM-binding properties of TP–DNA and pTP–DNA are distinct and that this has some bearing on the specific transcription/replication sites of viral DNA. Since pTP binds to DNA and Adpol, and TP does not (WEBSTER et al. 1994), it would not be surprising to find differences between the properties of TP–DNA and pTP–DNA as templates for DNA replication.

6 Cellular Nuclear Factors

During DNA replication in vitro, only limited initiation of replication occurs in the presence of three virus-encoded proteins and *ori*, but the frequency of initiation was increased by the addition of a crude nuclear extract of uninfected HeLa cells. This stimulatory effect is due to the presence of two cellular transcription factors NFI/CTF1 and NFIII/Oct-1. In addition, topoisomerase or NFII is also required for complete adenovirus genome replication. The roles of these factors in adenovirus DNA replication have been reviewed (DE JONG and VAN DER VLIET 1999; VAN DER VLIET 1995, 1996).

7 NFI

In adenovirus DNA replication assays in vitro with pTP/pol and DBP, addition of nuclear extract from uninfected HeLa cells can stimulate the efficiency of initiation and elongation of DNA replication (IKEDA et al. 1982; LICHY et al. 1982). This stimulatory activity of the nuclear extracts was associated with a protein termed NFI (NAGATA et al. 1982). In DNA binding assays, NFI bound specifically to a DNA sequence 5′-TGGC(N6)-GCCAA-3′ present in a number of adenovirus *oris* (LEEGWATER et al. 1985) and the interaction of NFI with this sequence increases the efficiency of initiation of viral DNA replication both in vivo and in vitro (DE VRIES et al. 1985; GUGGENHEIMER et al. 1984; HAY 1985a, NAGATA et al. 1983b, RAWLINS et al. 1984). Since the NFI recognition sequence is similar to the CCAAT box and NFI also binds to monomeric CCAAT sites, NFI protein is also called CCAAT binding factor (CTF), a family of proteins involved in cellular gene transcription (JONES et al. 1987). Thus, NFI/CTF can serve as transcription selectivity factor for RNA polymerase II and as an initiation factor for adenovirus DNA replication. Further analysis showed that NFI protein is indistinguishable from CTF transcription factors in polypeptide composition, DNA-binding properties, immunological cross-reactivity, and in vitro stimulation of DNA replication and transcription initiation (JONES et al. 1987). Molecular analysis of cDNAs from various human NFI/CTF mRNAs revealed that they all originate from a single gene, but give rise to multiple mRNA transcripts by differentiated splicing of a precursor molecule (MEISTERERNST et al. 1989; PAONESSA et al. 1988; SANTORO et al. 1988; WENZELIDES et al. 1996). NFI/CTF proteins formed by differential splicing display a broad range of transcriptional activities. Mutational analysis of the cDNA of the largest NFI (CTF-1) protein identified two functional domains. A highly conserved amino-terminal domain (NFI-DBD), stretching from amino acids 1 to approximately 220, is sufficient for DNA binding, dimerization, and DNA replication (BANDYOPADHYAY and GRONOSTAJSKI 1994; BOSHER et al. 1991; GOUNARI et al. 1990; MERMOD et al. 1989; SANTORO et al. 1988) while a less conserved, carboxyl-terminal domain has a transcriptional activation function (ALTMANN et al. 1994; MERMOD et al. 1989).

The stimulation of NFI in Ad DNA replication in vitro is complex. Both NFI and its binding site in *ori* affect the stimulatory function. The orientation and spacing between the NFI site and the core origin sequence is critical in Ad DNA replication. Insertion of additional sequence between these two regions abolished NFI-mediated stimulation of DNA replication in vitro (ADHYA et al. 1986; COENJAERTS et al. 1991; WIDES et al. 1987) and in vivo (BOSHER et al. 1990). Stimulation is also strongly dependent on the concentration of DBP (DE VRIES et al. 1985) and pTP/pol (MUL et al. 1990). At low concentrations of pTP/pol, NFI or NFI-DBD stimulated replication up to 50-fold, whereas at high concentrations, stimulation was less than twofold (MUL et al. 1990). Since NFI, its recognition site and the core origin sequence are critical in adenovirus DNA replication, a strict constraint on spatial arrangement between the core origin sequence and the NFI site is necessary to allow specific interactions between NFI and other replication proteins. A direct interaction between NFI-DBD and Adpol was subsequently demonstrated (BOSHER et al. 1990; CHEN et al. 1990; MUL et al. 1990) and the interaction site on NFI-DBD is between amino acids 68 and 150, which overlap the DNA binding and replication activation domain (CHEN et al. 1990). The interaction of NFI/Adpol stabilizes the relatively weak interaction between the pTP/pol and its recognition site in the adenovirus *ori*. Although dimerization, DNA binding, and activation of

DNA replication are co-localized within the N-terminal portion of the protein, mutational analysis has demonstrated that the DNA-binding and Adpol interaction of NFI can be separated (Armentero et al. 1994). Further analysis indicated that in isolation a highly basic amino-terminal subdomain bound non-specifically to DNA while a carboxyl-terminal subdomain bound specifically but with very low affinity to the NFI recognition site. Both of these subdomains stimulate DNA replication in a pTP/pol-dependent manner, although not to the same extent as the intact NFI–DBD. The carboxy-terminal domain could interact with the pTP/pol complex and recruit pTP/pol to DNA, while the amino-terminal subdomain, which was alpha-helical enhanced binding of pTP/pol to DNA by inducing structural changes in the DNA (Dekker et al. 1996).

8 NFIII

Cellular factor NFIII was initially identified in HeLa cell nuclear extracts through its ability to stimulate the initiation of Ad DNA replication in vitro in the presence of all three viral proteins and NFI (Pruijn et al. 1986). The protein consists of 743 amino acids with a molecular mass of 92 kDa and stimulates the level of initiation in vitro three- to sevenfold when bound to its recognition site in the adenovirus *ori* (O'Neill and Kelly 1988). In Ad2, the NFIII core binding site, 5′-TATGATAAT-3′, is situated between nucleotides 39 and 49 in the Ad2 ITR. NFIII is functionally identical to transcription factor Oct-1 (OTF-1; O'Neill et al. 1988). Unlike in vitro, the role of NFIII/Oct-1 in vivo is unclear as genomes of an Ad2 mutant with a deletion of the NFIII/Oct-1 site in *ori* are as infectious as wild-type virus with an intact NFIII/Oct-1 site (Hay and McDougall 1986), although the site may be functionally redundant with sequences adjacent to the ITR (Hatfield and Hearing 1993). However, transfection assays with plasmids containing only core origin of replication linked to an NFIII/Oct-1 site (no NFI site) indicated that these templates could replicate more efficiently than plasmids containing only core origin of replication (Hay et al. 1988). Similar results were reported in Adenovirus DNA replication in vitro (Mul et al. 1990). Although both the NFIII/Oct-1 and NFI-binding sites were needed for optimal stimulation of replication, the enhancing effect of the two transcription factors is independent and non-synergetic. NFIII/Oct-1 stimulates DNA replication in the absence of NFI or NFI–BD and vice versa. It is possible that NFIII/Oct-1 induces DNA bending and that *ori* stimulates DNA replication by promoting interactions between the various components in the pre-initiation complex (Verrijzer et al. 1991).

Oct-1 belongs to the POU-protein superfamily of transcription factors, of which the prototypes Pit-1, Oct-1 and Unc-86 are characterized by a common DNA-binding domain called the POU-domain (Herr et al. 1988; Sturm et al. 1988). Similar to other Oct proteins, NFIII/Oct-1 contains a POU domain that is sufficient for stimulation of Ad2 DNA replication in vitro (Verrijzer et al. 1990). The POU domain is a 160-amino acid bipartite DNA-binding domain containing a POU-specific subdomain (POUs, approximately 75 amino acids) linked to a homeo subdomain (POUhd) of 60 amino acids. For stimulation of both Ad DNA replication and site-specific DNA binding, the intact POU domain is required (Verrijzer et al. 1990). The POUs domain is responsible for site-specific DNA binding of NFIII, mutations within this domain which strongly reduced DNA binding were incapable of stimulating DNA replication (Van Leeuwen et al. 1995). Structure determination of the DNA–POUhd complex revealed that The POU homeodomain contacts the 3′ half of 5′-TATGCAAAT-3′ (Klemm et al. 1994). Mutation analysis also showed that this domain interacts with pTP (Brenkman et al. 2002; Coenjaerts et al. 1994). The POUhd domain contacts a number of sites on pTP to induce a conformational change which may influence the initiation of Ad DNA replication (Botting and Hay 1999) or interact with pTP/pol to stabilize the pre-initiation complex formed at *ori* (Coenjaerts et al. 1994). It is thought that POUhd interacts with pTP and the POUs domain binds the specific site to recruit the pTP/pol to the correct position (Brenkman et al. 2002). The POUhd domain alone inhibits adenovirus DNA replication (Verrijzer et al. 1990) and this could be a consequence of the failure to recognize the replication origin (Brenkman et al. 2002).

9 NFII

Although three viral proteins, together with NFI and NFIII are sufficient to sustain initiation and elongation of DNA replication in vitro, the combination of these proteins is not sufficient for replication of the whole viral genome unless a third cellular protein is present (Nagata et al. 1983a). This protein, isolated from nuclear extracts of uninfected HeLa cells, has been designated as NF II with a molecular mass of 30 kDa and topoisomerase activity. Treatment of infected cells with topoisomerase-specific inhibitors prevented efficient replication of adenovirus DNA. Inhibition of topoisomerase I activity led to an immediate cessation of adenovirus DNA replication, while inhibition of topoisomerase II blocked replication only after completion of approximately one additional round (SCHAACK et al. 1990b). It is supposed that topological stress accumulates during movement of the replication forks, suggesting that unwinding is limited despite the presence of free molecular ends. Since viral DNA is attached to the nuclear matrix through TP (FREDMAN and ENGLER 1993; SCHAACK et al. 1990a) in vivo, this may restrict the free rotation of the template. However, this cannot explain the need for topoisomerase in vitro. It was suggested that TP-DNA circularizes by interactions between the two terminal proteins, thereby restricting the rotational freedom of the templates. This suggestion is supported by early electron microscopic observations of a circular DNA-protein complex from adenoviruses, which led to the discovery of TP (ROBINSON et al. 1973). However, subsequent analysis using *Xba* I digested 1,339 bp and 34,596 bp fragments containing TP at only one end showed that topoisomerase is still required for complete replication of the larger fragment (VAN DER VLIET 1995). As these templates initiate at the TP-containing origins and thus are TP-free at other end, there is no apparent reason for these molecules not to rotate freely. Therefore, the blockage is not caused by TP-TP interactions but by the length of the DNA. It is possible that the proteins in the replication fork disassociate when positive supercoils accumulate and are not relieved in time. Thus in the absence of topoisomerase it is possible that the release of topological stress is too slow compared to the rate of fork movement, despite the presence of a free molecular end. In this respect it is worth noting that several bacteriophages containing linear genome DNA such as T7 also require DNA gyrase action during their DNA replication (ITOH and TOMIZAWA 1977).

10
A Model for Adenovirus DNA Replication

Based on the observations described above, it is possible to suggest a model for the initiation of Ad DNA replication. After infection of susceptible cells, the adenovirus virion is uncoated in the cytosol and the core enters the nucleus to release the viral DNA. DNA replication is initiated after the transcription and expression of the E2 genes which encode Adpol, pTP and DBP. Adpol and pTP form a heterodimer and DBP, produced in large amounts, binds ds DNA to coat the entire Ad DNA genome. The coating of DBP on Ad DNA template increases the affinity of NFI for its recognition site in the Ad2 *ori* thus ensuring that the NFI-binding site in *ori* is fully occupied (Cleat and Hay 1989; Stuiver and Van der Vliet 1990). A direct interaction between NFIII PUOhd with pTP and POUs domain with *ori* (bp 39–49) recruits the pTP/pol onto its binding site (bp 9–18) in *ori* (Coenjaerts et al. 1994; de Jong et al. 2002). A pre-initiation complex is then formed and stabilized by direct interaction of Adpol with NFI DBD bound to its recognition site (bp 23–38) in *ori* (Armentero et al. 1994; Bosher et al. 1990; Chen et al. 1990; Mul et al. 1990). A further stabilization of the complex may be accomplished by interactions between the incoming pTP/pol and genome-bound TP (Pronk and Van der Vliet 1993) after which the origin may unwind to expose the ss DNA that is the template for DNA synthesis. The ATP-independent helix-destabilizing property of DBP may be particularly important in catalysing this reaction (Pronk et al. 1994). Unlike most other systems, this unwinding reaction does not appear to require the activity of bona fide helicase. The heterodimer is then repositioned for initiation such that the acceptor serine in pTP is positioned opposite the GTA at positions 4–8 in the genome for formation of the covalent bond between the α-phosphoryl group of the terminal residue, dCTP, and the β-OH group of a serine of pTP. Addition of the second and third nucleotides is then primed by the 3′-OH group of pTP bound dCMP to form a pTP–trinucleotide complex (pTP–CAT) catalysed by Adpol. pTP–CAT and the initiation complex then jump back three bases to the first GTA at position 1–3 in *ori* (King and Van der Vliet 1994) and the disassociation of pre-initiation complex begins (King et al. 1997). Synthesis of the nascent strand is then primed by the 3′-OH group of TMP in the pTP–CAT complex and Adpol becomes a more efficient enzyme after release of the pTP primer and its processivity is dramatically increased by DNA-bound DBP. DNA synthesis is accompanied by displacement of the non-template strand, which can form a partial duplex by base-pairing of the ITRs on which a second round of DNA synthesis may be initiated (Hay et al. 1984; Stow 1982). The displaced strands may simply reanneal to form double-stranded products which are subsequently processed and encapsidated into virions.

Although great strides have been made in our understanding of adenovirus DNA replication a complete description of the process awaits structure determination of the pre-initiation complex at the viral origin of DNA replication. This represents a formidable challenge but should be achievable given the relative simplicity of the adenovirus system.

References

Adhya S, Shneidman PS, Hurwitz J (1986) Reconstruction of adenovirus replication origins with a human nuclear factor I binding site. J Biol Chem 261:3339–46

Altmann H, Wendler W, Winnacker EL (1994) Transcriptional activation by CTF proteins is mediated by a bipartite low-proline domain. Proc Natl Acad Sci USA 91:3901–5

Angeletti PC, Engler JA (1996) Tyrosine kinase-dependent release of an adenovirus preterminal protein complex from the nuclear matrix. J Virol 70:3060–7

Angeletti PC, Engler JA (1998) Adenovirus preterminal protein binds to the CAD enzyme at active sites of viral DNA replication on the nuclear matrix. J Virol 72: 2896–904

Ariga H, Klein H, Levine AJ, Horwitz MS (1980) A cleavage product of the adenovirus DNA binding protein is active in DNA replication in vitro. Virology 101: 307–10

Armentero MT, Horwitz M, Mermod N (1994) Targeting of DNA polymerase to the adenovirus origin of DNA replication by interaction with nuclear factor I. Proc Natl Acad Sci USA 91:11537–41

Bailey A, Mautner V (1994) Phylogenetic relationships among adenovirus serotypes. Virology 205:438–52

Bandyopadhyay S, Gronostajski RM (1994) Identification of a conserved oxidation-sensitive cysteine residue in the NFI family of DNA-binding proteins. J Biol Chem 269:29949–55

Bosher J, Leith IR, Temperley SM, Wells M, Hay RT (1991) The DNA-binding domain of nuclear factor I is sufficient to cooperate with the adenovirus type 2 DNA-binding protein in viral DNA replication. J Gen Virol 72:2975–80

Bosher J, Robinson EC, Hay RT (1990) Interactions between the adenovirus type 2 DNA polymerase and the DNA binding domain of nuclear factor I. New Biol 2: 1083–90

Botting CH, Hay RT (1999) Characterisation of the adenovirus preterminal protein and its interaction with the POU homeodomain of NFIII (Oct-1). Nucleic Acids Res 27:2799–805

Botting CH, Hay RT (2001) Role of conserved residues in the activity of adenovirus preterminal protein. J Gen Virol 82:1917–27

Brenkman AB, Breure EC, Van Der Vliet PC (2002) Molecular architecture of adenovirus DNA polymerase and location of the protein primer. J Virol 76:8200–7

Brenkman AB, Heideman MR, Truniger V, Salas M, van der Vliet PC (2001) The (I/Y)XGG motif of adenovirus DNA polymerase affects template DNA binding and the transition from initiation to elongation. J Biol Chem 276:29846–53

Brough DE, Droguett G, Horwitz MS, Klessig DF (1993) Multiple functions of the adenovirus DNA-binding protein are required for efficient viral DNA synthesis. Virology 196:269–81

Challberg MD, Desiderio SV, Kelly TJ (1980) Adenovirus DNA replication in vitro: Characterisation of a protein covalently linked to nascent DNA strands. Proc Natl Acad Sci USA 77:5105–9

Challberg MD, Kelly TJ (1979) Adenovirus DNA replication in vitro: origin and direction of daughter strand synthesis. J Mol Biol 135:999–1012

Challberg MD, Kelly TJ (1989) Animal virus replication. Annual Reviews in Biochemistry 58:671–717

Challberg MD, Rawlins DR (1984) Template requirements for the initiation of adenovirus DNA replication. Proc Natl Acad Sci USA 81:100–4

Chen H, Ramachandra M, Padmanabhan R (1994) Biochemical characterization of a temperature-sensitive adenovirus DNA polymerase. Virology 205:364–70

Chen M, Horwitz MS (1989) Dissection of functional domains of adenovirus DNA polymerase by linker- insertion mutagenesis. Proc Natl Acad Sci USA 86: 6116–20

Chen M, Mermod N, Horwitz MS (1990) Protein-protein interactions between adenovirus DNA polymerase and nuclear factor I mediate formation of the DNA replication preinitiation complex. J Biol Chem 265:18634–42

Cleat PH, Hay RT (1989) Co-operative interactions between NFI and the adenovirus DNA binding protein at the adenovirus origin of replication. EMBO J 8:1841–8

Cleghon V, Piderit A, Brough DE, Klessig DF (1993) Phosphorylation of the adenovirus DNA-binding protein and epitope mapping of monoclonal antibodies against it. Virology 197:564–75

Coenjaerts FE, De Vries E, Pruijn GJ, Van Driel W, Bloemers SM et al. (1991) Enhancement of DNA replication by transcription factors NFI and NFIII/Oct-1 depends critically on the positions of their binding sites in the adenovirus origin of replication. Biochim Biophys Acta 1090:61–9

Coenjaerts FE, Van der Vliet PC (1994) Early dissociation of nuclear factor I from the origin during initiation of adenovirus DNA replication studied by origin immobilization. Nucleic Acids Res 22:5235–40

Coenjaerts FE, Van Oosterhout JA, Van der Vliet PC (1994) The Oct-1 POU domain stimulates adenovirus DNA replication by a direct interaction between the viral precursor terminal protein-DNA polymerase complex and the POU homeodomain. EMBO J 13:5401–9

Cotten M, Weber JM (1995) The adenovirus protease is required for virus entry into host cells. Virology 213:494–502

De Jong RN, Mysiak ME, Meijer LA, Van der Linden M, Van der Vliet PC (2002) Recruitment of the priming protein pTP and DNA binding occur by overlapping Oct-1 POU homeodomain surfaces. EMBO J 21:725–35

De Jong RN, Van der Vliet PC (1999) Mechanism of DNA replication in eukaryotic cells:cellular host factors stimulating adenovirus DNA replication. Gene 236: 1–12

De Vries E, Bloemers SM, Van der Vliet PC (1987) Incorporation of 5-bromodeoxycytidine in the adenovirus 2 replication origin interferes with nuclear factor 1 binding. Nucleic Acids Res 15:7223–34

De Vries E, Van Driel W, Tromp M, Van Boom J, Van der Vliet PC (1985) Adenovirus DNA replication in vitro: site-directed mutagenesis of the nuclear factor I binding site of the Ad2 origin. Nucleic Acids Res 13:4935–52

De Vries EW, Van Driel W, Bergsma WG, Arnberg AC, Van der Vliet PC (1989) Hela nuclear protein recognising DNA termini and translocating on DNA forming a regular DNA-multimeric protein complex. J Mol Biol 208:65–78

Dekker J, Kanellopoulos PN, Loonstra AK, Van Oosterhout JA, Leonard K et al. (1997) Multimerization of the adenovirus DNA-binding protein is the driving force for ATP-independent DNA unwinding during strand displacement synthesis. EMBO J 16:1455–63

Dekker J, Kanellopoulos PN, Van Oosterhout JA, Stier G, Tucker PA, Van der Vliet PC (1998) ATP-independent DNA unwinding by the adenovirus single-stranded DNA binding protein requires a flexible DNA binding loop. J Mol Biol 277:825–38

Dekker J, Van Oosterhout JA, Van der Vliet PC (1996) Two regions within the DNA binding domain of nuclear factor I interact with DNA and stimulate adenovirus DNA replication independently. Mol Cell Biol 16:4073–80

Ding J, McGrath WJ, Sweet RM, Mangel WF (1996) Crystal structure of the human adenovirus proteinase with its 11 amino acid cofactor. EMBO J 15:1778–83

Eagle PA, Klessig DF (1992) A zinc-binding motif located between amino acids 273 and 286 in the adenovirus DNA-binding protein is necessary for ssDNA binding. Virology 187:777–87

Enomoto T, Lichy JH, Ikeda JE, Hurwitz J (1981) Adenovirus DNA replication in vitro: purification of the terminal protein in a functional form. Proc Natl Acad Sci USA 78:6779–83

Field J, Gronostajski RM, Hurwitz J (1984) Properties of the adenovirus DNA polymerase. J Biol Chem 259:9487–95

Franklin MC, Wang J, Steitz TA (2001) Structure of the replicating complex of a pol alpha family DNA polymerase. Cell 105:657–67

Fredman JN, Engler JA (1993) Adenovirus precursor to terminal protein interacts with the nuclear matrix in vivo and in vitro. J Virol 67:3384–95

Fredman JN, Pettit SC, Horwitz MS, Engler JA (1991) Linker insertion mutations in the adenovirus preterminal protein that affect DNA replication activity in vivo and in vitro. J Virol 65:4591–7

Freimuth PI, Ginsberg HS (1986) Codon insertion mutants of the adenovirus terminal protein. Proc Natl Acad Sci USA 83:7816–20

Friefeld BR, Krevolin MD, Horwitz MS (1983) Effects of the adenovirus H5ts125 and H5ts107 DNA binding proteins on DNA replication in vitro. Virology 124:380–9

Gounari F, De Francesco R, Schmitt J, van der Vliet P, Cortese R, Stunnenberg H (1990) Amino-terminal domain of NF1 binds to DNA as a dimer and activates adenovirus DNA replication. EMBO J 9:559–66

Greber UF, Webster P, Weber J, Helenius A (1996) The role of the adenovirus protease on virus entry into cells. EMBO J 15:1766–77

Guggenheimer RA, Stillman BW, Nagata K, Tamanoi F, Hurwitz J (1984) DNA sequences required for the in vitro replication of adenovirus DNA. Proc Natl Acad Sci USA 81:3069–73

Harfst E, Leppard KN (1999) A comparative analysis of the phosphorylation and biochemical properties of wild type and host range variant DNA binding proteins of human adenovirus 5. Virus Genes 18:97–106

Harris MP, Hay RT (1988) DNA sequences required for the initiation of adenovirus type 4 DNA replication in vitro. J Mol Biol 201:57–67
Hatfield L, Hearing P (1993) The NFIII/OCT-1 binding site stimulates adenovirus DNA replication in vivo and is functionally redundant with adjacent sequences. J Virol 67:3931–9
Hay RH (1996) Adenovirus DNA replication. In: DNA Deplication in Eukaryotic Cells, ed. ML DePamphilis, pp 699–719. Cold Spring Harbour, NY: Cold Spring Harbour Laboratory Press
Hay RT (1985a) Origin of adenovirus DNA replication. Role of the nuclear factor I binding site in vivo. J Mol Biol 186:129–36
Hay RT (1985b) The origin of adenovirus DNA replication: minimal DNA sequence requirement in vivo. EMBO J 4:421–6
Hay RT, Clark AM, Cleat PH, Harris MPG, Robertson EC, Watson CJ (1988) Requirements for the initiation of adenovirus type 2 and 4 DNA replication. Cancer Cells 6:71–5
Hay RT, Freeman A, Leith I, Monaghan A, Webster A (1995) Molecular interactions during adenovirus DNA replication. Curr Top Microbiol Immunol 199:31–48
Hay RT, McDougall IM (1986) Viable viruses with deletions in the left inverted terminal repeat define the adenovirus origin of DNA replication. J Gen Virol 67:321–32
Hay RT, Russell WC (1989) Recognition mechanisms in the synthesis of animal virus DNA. Biochem J 258:3–16
Hay RT, Stow ND, McDougall IM (1984) Replication of adenovirus mini-chromosomes. J Mol Biol 175:493–510
Herr W, Sturm RA, Clerc RG, Corcoran LM, Baltimore D et al. (1988) The POU domain: a large conserved region in the mammalian pit-1, oct-1, oct-2, and Caenorhabditis elegans unc-86 gene products. Genes Dev 2:1513–6
Hwang CB, Ruffner KL, Coen DM (1992) A point mutation within a distinct conserved region of the herpes simplex virus DNA polymerase gene confers drug resistance. J Virol 66:1774–6
Ikeda JE, Enomoto T, Hurwitz J (1981) Replication of adenovirus DNA-protein complex with purified proteins. Proc Natl Acad Sci USA 78:884–8
Ikeda JE, Enomoto T, Hurwitz J (1982) Adenoviral protein-primed initiation of DNA chains in vitro. Proc Natl Acad Sci USA 79:2442–6
Itoh T, Tomizawa JI (1977) Involvement of DNA gyrase in bacteriophage T7 DNA replication. Nature 270:78–80
Jones KA, Kadonaga JT, Rosenfeld PJ, Kelly TJ, Tjian R (1987) A cellular DNA-binding protein that activates eukaryotic transcription and DNA replication. Cell 48:79–89
Joung I, Engler JA (1992) Mutations in two cysteine-histidine-rich clusters in adenovirus type 2 DNA polymerase affect DNA binding. J Virol 66:5788–96
Joung I, Horwitz MS, Engler JA (1991) Mutagenesis of conserved region I in the DNA polymerase from human adenovirus serotype 2. Virology 184:235–41
Joyce CM, Steitz TA (1994) Function and structure relationships in DNA polymerases. Annu Rev Biochem 63:777–822
Kanellopoulos PN, Van der Zandt H, Tsernoglou D, Van der Vliet PC, Tucker PA (1995) Crystallization and preliminary X-ray crystallographic studies on the adenovirus ssDNA binding protein in complex with ssDNA. J Struct Biol 115:113–6

Kenny MK, Hurwitz J (1988) Initiation of adenovirus DNA replication. II. Structural requirements using synthetic oligonucleotide adenovirus templates. J Biol Chem 263:9809–17

Khittoo G, Delorme L, Dery CV, Tremblay ML, Weber JM et al. (1986) Role of the nuclear matrix in adenovirus maturation. Virus Res 4:391–403

King AJ, Teertstra WR, van der Vliet PC (1997) Dissociation of the protein primer and DNA polymerase after initiation of adenovirus DNA replication. J Biol Chem 272:24617–23

King AJ, van der Vliet PC (1994) A precursor terminal protein-trinucleotide intermediate during initiation of adenovirus DNA replication: regeneration of molecular ends in vitro by a jumping back mechanism. EMBO J 13:5786–92

Kitchingman GR (1985) Sequence of the DNA-binding protein of a human subgroup E adenovirus (type 4): comparisons with subgroup A (type 12), subgroup B (type 7), and subgroup C (type 5). Virology 146:90–101

Kitchingman GR (1995) Mutations in the adenovirus-encoded single-stranded DNA binding protein that result in altered accumulation of early and late viral RNAs. Virology 212:91–101

Klein H, Maltzman W, Levin AJ (1979) Structure function relationships of the adenovirus DNA-binding protein. Journal of Biological Chemistry 254:11051–60

Klemm JD, Rould MA, Aurora R, Herr W, Pabo CO (1994) Crystal structure of the Oct-1 POU domain bound to an octamer site: DNA recognition with tethered DNA-binding modules. Cell 77:21–32

Knopf CW (1998) Evolution of viral DNA-dependent DNA polymerases. Virus Genes 16:47–58

Kusukawa J, Ramachandra M, Nakano R, Padmanabhan R (1994) Phosphorylation-dependent interaction of adenovirus preterminal protein with the viral origin of DNA replication. J Biol Chem 269:2189–96

Leegwater PA, van Driel W, van der Vliet PC (1985) Recognition site of nuclear factor I, a sequence-specific DNA-binding protein from HeLa cells that stimulates adenovirus DNA replication. EMBO J 4:1515–21

Lichy JH, Field J, Horwitz MS, Hurwitz J (1982) Separation of the adenovirus terminal protein precursor from its associated DNA polymerase: role of both proteins in the initiation of adenovirus DNA replication. Proc Natl Acad Sci USA 79:5225–9

Lindenbaum JO, Field J, Hurwitz J (1986) The adenovirus DNA binding protein and adenovirus DNA polymerase interact to catalyze elongation of primed DNA templates. J Biol Chem 261:10218–27

Liu H, Naismith JH, Hay RT (2000) Identification of conserved residues contributing to the activities of adenovirus DNA polymerase. J Virol 74:11681–9

Mangel WF, McGrath WJ, Toledo DL, Anderson CW (1993) Viral DNA and a viral peptide can act as cofactors of adenovirus virion proteinase activity. Nature 361: 274–5

Matthews DA, Russell WC (1995) Adenovirus protein-protein interactions: molecular parameters governing the binding of protein VI to hexon and the activation of the adenovirus 23 K protease. J Gen Virol 76:1959–69

Meisterernst M, Rogge L, Foeckler R, Karaghiosoff M, Winnacker EL (1989) Structural and functional organization of a porcine gene coding for nuclear factor I. Biochemistry 28:8191–200

Mermod N, O'Neill EA, Kelly TJ, Tjian R (1989) The proline-rich transcriptional activator of CTF/NF-I is distinct from the replication and DNA binding domain. Cell 58:741–53

Monaghan A, Webster A, Hay RT (1994) Adenovirus DNA binding protein: helix destabilising properties. Nucleic Acids Res 22:742–8

Mul YM, Van der Vliet PC (1992) Nuclear factor I enhances adenovirus DNA replication by increasing the stability of a preinitiation complex. EMBO J 11:751–60

Mul YM, Van der Vliet PC (1993) The adenovirus DNA binding protein effects the kinetics of DNA replication by a mechanism distinct from NFI or Oct-1. Nucleic Acids Res 21:641–7

Mul YM, Verrijzer CP, Van der Vliet PC (1990) Transcription factors NFI and NFIII/oct-1 function independently, employing different mechanisms to enhance adenovirus DNA replication. J Virol 64:5510–8

Nagata K, Guggenheimer RA, Enomoto T, Lichy JH, Hurwitz J (1982) Adenovirus DNA replication in vitro: identification of a host factor that stimulates synthesis of the preterminal protein-dCMP complex. Proc Natl Acad Sci USA 79:6438–42

Nagata K, Guggenheimer RA, Hurwitz J (1983a) Adenovirus DNA replication in vitro: synthesis of full-length DNA with purified proteins. Proc Natl Acad Sci USA 80:4266–70

Nagata K, Guggenheimer RA, Hurwitz J (1983b) Specific binding of a cellular DNA replication protein to the origin of replication of adenovirus DNA. Proc Natl Acad Sci USA 80:6177–81

Neale GA, Kitchingman GR (1989) Biochemical analysis of adenovirus type 5 DNA-binding protein mutants. J Biol Chem 264:3153–9

Neale GA, Kitchingman GR (1990) Conserved region 3 of the adenovirus type 5 DNA-binding protein is important for interaction with single-stranded DNA. J Virol 64:630–8

O'Neill EA, Fletcher C, Burrow CR, Heintz N, Roeder RG, Kelly TJ (1988) Transcription factor OTF-1 is functionally identical to the DNA replication factor NF-III. Science 241:1210–3

O'Neill EA, Kelly TJ (1988) Purification and characterization of nuclear factor III (origin recognition protein C), a sequence-specific DNA binding protein required for efficient initiation of adenovirus DNA replication. J Biol Chem 263:931–7

Paonessa G, Gounari F, Frank R, Cortese R (1988) Purification of a NF1-like DNA-binding protein from rat liver and cloning of the corresponding cDNA. EMBO J 7:3115–23

Parker EJ, Botting CH, Webster A, Hay RT (1998) Adenovirus DNA polymerase: domain organisation and interaction with preterminal protein. Nucleic Acids Res 26:1240–7

Pettit SC, Horwitz MS, Engler JA (1988) Adenovirus preterminal protein synthesized in COS cells from cloned DNA is active in DNA replication in vitro. J Virol 62:496–500

Pettit SC, Horwitz MS, Engler JA (1989) Mutations of the precursor to the terminal protein of adenovirus serotypes 2 and 5. J Virol 63:5244–50

Pronk R, Stuiver MH, Van der Vliet PC (1992) Adenovirus DNA replication: the function of the covalently bound terminal protein. Chromosoma 102:S39–45

PRONK R, VAN DER VLIET PC (1993) The adenovirus terminal protein influences binding of replication proteins and changes the origin structure. Nucleic Acids Res 21:2293–300

PRONK R, VAN DRIEL W, VAN DER VLIET PC (1994) Replication of adenovirus DNA in vitro is ATP-independent. FEBS Lett 337:33–8

PRUIJN GJ, VAN DRIEL W, VAN DER VLIET PC (1986) Nuclear factor III, a novel sequence-specific DNA-binding protein from HeLa cells stimulating adenovirus DNA replication. Nature 322:656–9

RAMACHANDRA M, NAKANO R, MOHAN PM, RAWITCH AB, PADMANABHAN R (1993) Adenovirus DNA polymerase is a phosphoprotein. J Biol Chem 268:442–8

RAMACHANDRA M, PADMANABHAN R (1993) Adenovirus DNA polymerase is phosphorylated by a stably associated histone H1 kinase. J Biol Chem 268:17448–56

RANCOURT C, KEYVANI-AMINEH H, DIOURI M, WEBER JM (1996) Mutagenesis of conserved residues of the adenovirus protease. Virology 224:561–3

RANCOURT C, TIHANYI K, BOURBONNIERE M, WEBER JM (1994) Identification of active-site residues of the adenovirus endopeptidase. Proc Natl Acad Sci USA 91:844–7

RAWLINS DR, ROSENFELD PJ, WIDES RJ, CHALLBERG MD, KELLY TJ (1984) Structure and function of the adenovirus origin of replication. Cell 37:309–19

RICIGLIANO JW, BROUGH DE, KLESSIG DF (1994) Identification of a high-molecular-weight cellular protein complex containing the adenovirus DNA binding protein. Virology 202:715–23

ROBINSON AJ, YOUNGHUSBAND HB, BELLETT AJ (1973) A circular DNA-protein complex from adenoviruses. Virology 56:54–69

ROOVERS DJ, VAN DER LEE FM, VAN DER WEES J, SUSSENBACH JS (1993) Analysis of the adenovirus type 5 terminal protein precursor and DNA polymerase by linker insertion mutagenesis. J Virol 67:265–76

ROSENFELD PJ, O'NEILL EA, WIDES RJ, KELLY TJ (1987) Sequence-specific interactions between cellular DNA-binding proteins and the adenovirus origin of DNA replication. Mol Cell Biol 7:875–86

RUZINDANA-UMUNYANA A, SIRCAR S, WEBER JM (2000) The effect of mutant peptide cofactors on adenovirus protease activity and virus infection. Virology 270:173–9

SANTORO C, MERMOD N, ANDREWS PC, TJIAN R (1988) A family of human CCAAT-box-binding proteins active in transcription and DNA replication: cloning and expression of multiple cDNAs. Nature 334:218–24

SATURNO J, LAZARO JM, BLANCO L, SALAS M (1998) Role of the first aspartate residue of the „YxDTDS" motif of phi29 DNA polymerase as a metal ligand during both TP-primed and DNA-primed DNA synthesis. J Mol Biol 283:633–42

SCHAACK J, HO WY, FREIMUTH P, SHENK T (1990a) Adenovirus terminal protein mediates both nuclear matrix association and efficient transcription of adenovirus DNA. Genes Dev 4:1197–208

SCHAACK J, SCHEDL P, SHENK T (1990b) Topoisomerase I and II cleavage of adenovirus DNA in vivo: both topoisomerase activities appear to be required for adenovirus DNA replication. J Virol 64:78–85

SHU L, PETTIT SC, ENGLER JA (1988) The precise structure and coding capacity of mRNAs from early region 2B of human adenovirus serotype 2. Virology 165: 348–56

Shu LM, Horwitz MS, Engler JA (1987) Expression of enzymatically active adenovirus DNA polymerase from cloned DNA requires sequences upstream of the main open reading frame. Virology 161:520–6
Sircar S, Ruzindana-Umunyana A, Neugebauer W, Weber JM (1998) Adenovirus endopeptidase and papain are inhibited by the same agents. Antiviral Res 40:45–51
Steitz TA (1998) A mechanism for all polymerases. Nature 391:231–2
Stillman BW (1983) The replication of adenovirus DNA with purified proteins. Cell 35:7–9
Stillman BW, Lewis JB, Chow LT, Mathews MB, Smart JE (1981) Identification of the gene and mRNA for the adenovirus terminal protein precursor. Cell 23: 497–508
Stillman BW, Tamanoi F, Mathews MB (1982a) Purification of an adenovirus-coded DNA polymerase that is required for initiation of DNA replication. Cell 31:613–23
Stillman BW, Topp WC, Engler JA (1982b) Conserved sequences at the origin of adenovirus DNA replication. J Virol 44:530–7
Stow ND (1982) The infectivity of adenovirus genomes lacking DNA sequences from their left-hand termini. Nucleic Acids Res 10:5105–19
Stuiver MH, Bergsma WG, Arnberg AC, Van Amerongen H, Van Grondelle R, Van der Vliet PC (1992) Structural alterations of double-stranded DNA in complex with the adenovirus DNA-binding protein. Implications for its function in DNA replication. J Mol Biol 225:999–1011
Stuiver MH, Van der Vliet PC (1990) Adenovirus DNA-binding protein forms a multimeric protein complex with double-stranded DNA and enhances binding of nuclear factor I. J Virol 64:379–86
Stunnenberg HG, Lange H, Philipson L, Van Miltenburg RT, Van der Vliet PC (1988) High expression of functional adenovirus DNA polymerase and precursor terminal protein using recombinant vaccinia virus. Nucleic Acids Res 16: 2431–44
Sturm RA, Das G, Herr W (1988) The ubiquitous octamer-binding protein Oct-1 contains a POU domain with a homeo box subdomain. Genes Dev 2:1582–99
Sung MT, Cao TM, Lischwe MA, Coleman RT (1983) Molecular processing of adenovirus proteins. J Biol Chem 258:8266–72
Tamanoi F, Stillman BW (1982) Function of adenovirus terminal protein in the initiation of DNA replication. Proc Natl Acad Sci USA 79:2221–5
Temperley SM, Burrow CR, Kelly TJ, Hay RT (1991) Identification of two distinct regions within the adenovirus minimal origin of replication that are required for adenovirus type 4 DNA replication in vitro. J Virol 65:5037–44
Temperley SM, Hay RT (1991) Replication of adenovirus type 4 DNA by a purified fraction from infected cells. Nucleic Acids Res 19:3243–9
Temperley SM, Hay RT (1992) Recognition of the adenovirus type 2 origin of DNA replication by the virally encoded DNA polymerase and preterminal proteins. EMBO J 11:761–8
Tihanyi K, Bourbonniere M, Houde A, Rancourt C, Weber JM (1993) Isolation and properties of adenovirus type 2 proteinase. J Biol Chem 268:1780–5
Truniger V, Blanco L, Salas M (1999) Role of the "YxGG/A" motif of Phi29 DNA polymerase in protein-primed replication. J Mol Biol 286:57–69

TUCKER PA, TSERNOGLOU D, TUCKER AD, COENJAERTS FE, LEENDERS H, VAN DER VLIET PC (1994) Crystal structure of the adenovirus DNA binding protein reveals a hook- on model for cooperative DNA binding. EMBO J 13:2994–3002

VAN BERGEN BG, VAN DER LEY PA, VAN DRIEL W, VAN MANSFELD AD, VAN DER VLIET PC (1983) Replication of origin containing adenovirus DNA fragments that do not carry the terminal protein. Nucleic Acids Res 11:1975–89

VAN BREUKELEN B, KANELLOPOULOS PN, TUCKER PA, VAN DER VLIET PC (2000) The Formation of a Flexible DNA-binding Protein Chain Is Required for Efficient DNA Unwinding and Adenovirus DNA Chain Elongation. J Biol Chem 275: 40897–903

VAN DER VLIET PC (1995) Adenovirus DNA Replication. Current Topics in Microbiology and Immunology: 1030

VAN DER VLIET PC (1996) Roles of transcription factors in DNA replication. In: DNA Replication in Eukaryotic Cells, ed. ML DePamphilis, pp 87–118. Cold Spring Harbour, NY: Cold Spring Harbour Laboratory Press

VAN DER VLIET PC, LEVINE AJ (1973) DNA binding proteins specific for cells infected by adenovirus. Nat New Biol 246:170–4

VAN LEEUWEN HC, RENSEN M, VAN DER VLIET PC (1997) The Oct-1 POU homeodomain stabilizes the adenovirus preinitiation complex via a direct interaction with the priming protein and is displaced when the replication fork passes. J Biol Chem 272:3398–405

VAN LEEUWEN HC, STRATING MJ, COX M, KAPTEIN R, VAN DER VLIET PC (1995) Mutation of the Oct-1 POU-specific recognition helix leads to altered DNA binding and influences enhancement of adenovirus DNA replication. Nucleic Acids Res 23:3189–97

VERRIJZER CP, KAL AJ, VAN DER VLIET PC (1990) The DNA binding domain (POU domain) of transcription factor oct-1 suffices for stimulation of DNA replication. EMBO J 9:1883–8

VERRIJZER CP, VAN OOSTERHOUT JA, VAN WEPEREN WW, VAN DER VLIET PC (1991) POU proteins bend DNA via the POU-specific domain. Embo J 10:3007–14

VOS HL, VAN DER LEE FM, REEMST AM, VAN LOON AE, SUSSENBACH JS (1988) The genes encoding the DNA binding protein and the 23 K protease of adenovirus types 40 and 41. Virology 163:1–10

WANG J, SATTAR AK, WANG CC, KARAM JD, KONIGSBERG WH, STEITZ TA (1997) Crystal structure of a pol alpha family replication DNA polymerase from bacteriophage RB69. Cell 89:1087–99

WATSON CJ, HAY RT (1990) Expression of adenovirus type 2 DNA polymerase in insect cells infected with a recombinant baculovirus. Nucleic Acids Res 18:1167–73

WEBER J (1990) The adenovirus proteinases. Semin Virol 1:379–84

WEBSTER A, HAY RT, KEMP G (1993) The adenovirus protease is activated by a virus-coded disulphide-linked peptide. Cell 72:97–104

WEBSTER A, LEITH IR, HAY RT (1994) Activation of adenovirus-coded protease and processing of preterminal protein. J Virol 68:7292–300

WEBSTER A, LEITH IR, HAY RT (1997a) Domain organization of the adenovirus preterminal protein. J Virol 71:539–47

WEBSTER A, LEITH IR, NICHOLSON J, HOUNSELL J, HAY RT (1997b) Role of preterminal protein processing in adenovirus replication. J Virol 71:6381–9

Webster A, Russell S, Talbot P, Russell WC, Kemp GD (1989) Characterization of the adenovirus proteinase: substrate specificity. J Gen Virol 70:3225–34

Wenzelides S, Altmann H, Wendler W, Winnacker EL (1996) CTF5-a new transcriptional activator of the NFI/CTF family. Nucleic Acids Res 24:2416–21

Wides RJ, Challberg MD, Rawlins DR, Kelly TJ (1987) Adenovirus origin of DNA replication: sequence requirements for replication in vitro. Mol Cell Biol 7:864–74

Wong ML, Hsu MT (1990) Involvement of topoisomerases in replication, transcription, and packaging of the linear adenovirus genome. J Virol 64:691–9

Zhao LJ, Padmanabhan R (1988) Nuclear transport of adenovirus DNA polymerase is facilitated by interaction with preterminal protein. Cell 55:1005–15

Zijderveld DC, d'Adda di Fagagna F, Giacca M, Timmers HT, Van der Vliet PC (1994) Stimulation of the adenovirus major late promoter in vitro by transcription factor USF is enhanced by the adenovirus DNA binding protein. J Virol 68: 8288–95

Zijderveld DC, Stuiver MH, Van der Vliet PC (1993) The adenovirus DNA binding protein enhances intermolecular DNA renaturation but inhibits intramolecular DNA renaturation. Nucleic Acids Research 21:2591–8

CHAPTER 6

Regulation of Adenovirus Packaging

P. OSTAPCHUK, P. HEARING

Abstract. The application of fundamental concepts about the packaging of the adenovirus genome has contributed significantly to the development of therapeutic viral vectors for gene therapy. The packaging of adenovirus DNA into virus particles requires a cis-acting domain at the left end of the genome. This region contains a series of repeated sequences, termed A repeats due to their AT-rich character, that direct the packaging process. A repeats are believed to represent the binding sites for viral and cellular factors that mediate viral DNA packaging. This review will focus on fundamental aspects of adenovirus DNA packaging as well as how this information has been used and may be used to augment the selectivity of viral DNA packaging in applications pertaining to gene therapy vectors.

1 Adenovirus Assembly Intermediates

The assembly of adenovirus (Ad) particles has been probed using viral temperature-conditional and deletion mutants blocked at different stages

P. OSTAPCHUK, P. HEARING
Department of Molecular Genetics and Microbiology, Health Sciences Center, Stony Brook University, Stony Brook, NY 11794, USA
e-mail: phearing@ms.cc.sunysb.edu

of assembly, by pulse-chase kinetic analyses, and through the analysis of assembly intermediates found during a natural infection. Ad assembly appears to proceed through an ordered series of events analogous to those described for the assembly of bacteriophage where empty particles are produced first, viral DNA is then inserted, and final maturation of the particle takes place (reviewed in D'Halluin 1995). With adenovirus, light intermediate particles (buoyant density of 1.315 g/ml in a CsCl equilibrium gradient), heavy intermediate particles (1.37 g/ml buoyant density), and mature virus particles (1.34 g/ml buoyant density), are observed. Light particles contain the capsid structural components and no or very little viral DNA and associated core proteins. Additionally, light intermediate particles contain several proteins that exit the particle during maturation (50 kDa and 39 kDa polypeptides) that may represent the equivalent of phage scaffolding proteins. The light intermediate particles may convert to heavy intermediate particles upon the insertion of viral DNA, but this precursor–product relationship is not entirely certain based on pulse-chase experiments. The heavy intermediate particles contain viral DNA, lack the 50-kDa and 39-kDa polypeptides found in light particles, and exhibit the precursor forms of viral proteins that are proteolytically processed during virion maturation. The issue of separate or combined entry of DNA and core proteins into heavy particles remains uncertain. As the final step in maturation, the virus-encoded and encapsidated proteinase performs numerous cleavages of multiple viral proteins to generate the mature, infectious virion. When analyzed by electron microscopy, light, heavy and mature Ad particles have a similar morpho-logy although there is some evidence that light particles may have an indentation on one facet. Perhaps this corresponds to a portal vertex by analogy to phage assembly.

2
Packaging of Adenovirus DNA into Capsids

In addition to the three classes of virus particles described above, additional Ad particles are observed following infection of cells in culture with different human adenoviruses. These virus particles contain incomplete copies of the viral genome. The predominance of incomplete virus particles relative to complete virions is characteristic for each adenovirus serotype. Incomplete particles of lower density than mature virions were isolated from Ad2, Ad3 and Ad7 infected cells (Daniell 1976; Tibbetts 1977). The lightest incomplete particles seem to be devoid of viral DNA.

Among the other defective virions, a direct relationship between the length of the encapsidated viral DNA and the density of the incomplete particles was established. The size of the genome ranged from 15% of the length of the complete genome to full size. Analysis of the packaged subgenomic DNAs revealed that sequences derived from the left end were significantly over-represented, which suggested that DNA packaging occurs in a polar fashion from left to right (DANIELL 1976; TIBBETTS 1977). In support of this model, both ends of the genome were equally represented in the pool of subgenomic adenovirus DNA within infected cells.

The first indication for the presence of cis-acting sequences in the adenovirus genome that direct selective DNA encapsidation from the left end came from studies with naturally occurring evolutionary variants of Ad16 (HAMMARSKJOLD and WINBERG 1980). A duplication of the left-end 390, but not 290 base pairs (bp) at the right end of the viral genome allowed DNA packaging to initiate from both ends of the genome. It was suggested that the region between nucleotides (nt) 290 and 390 harbors an essential signal for polar viral DNA packaging. In separate studies, a number of Ad3 variant viruses were selected after repeated high multiplicity passage in HeLa cells (ROBBINSON and TIBBETTS 1984). Those variants carry various mutations in their left-end 750 bp, but retain full capacity for growth and polar packaging of their genomes. Mutations of sequences within nt 136–318 did not affect viral growth, whereas the maintenance of nt 319–390 in all the mutants suggested that this region is indispensable. This interval coincides with the aforementioned region between nt 290 and 390 in Ad16 which appeared to direct DNA encapsidation. Sequence comparisons of Ad3 with Ad5 and Ad12, as representatives of adenovirus subgroups B, C, and A, respectively, reveals that the interval between nt 237 and nt 491 is highly conserved.

3
Cis-Acting Packaging Elements

The DNA comprising the packaging domain of adenovirus include a series of at least seven AT-rich sequences. These AT-rich sequences were originally identified from results of deletion, insertion and site-directed mutagenesis of adenovirus nt 194–358 (GRABLE and HEARING 1990, 1992; HEARING and SHENK 1983; reviewed in FUJISAWA and HEARING 1994; SCHMID and HEARING 1995). The following consensus sequence of 5′-$GTN_{3-4}TTTG$-3′ was derived from comparisons of these related sequences

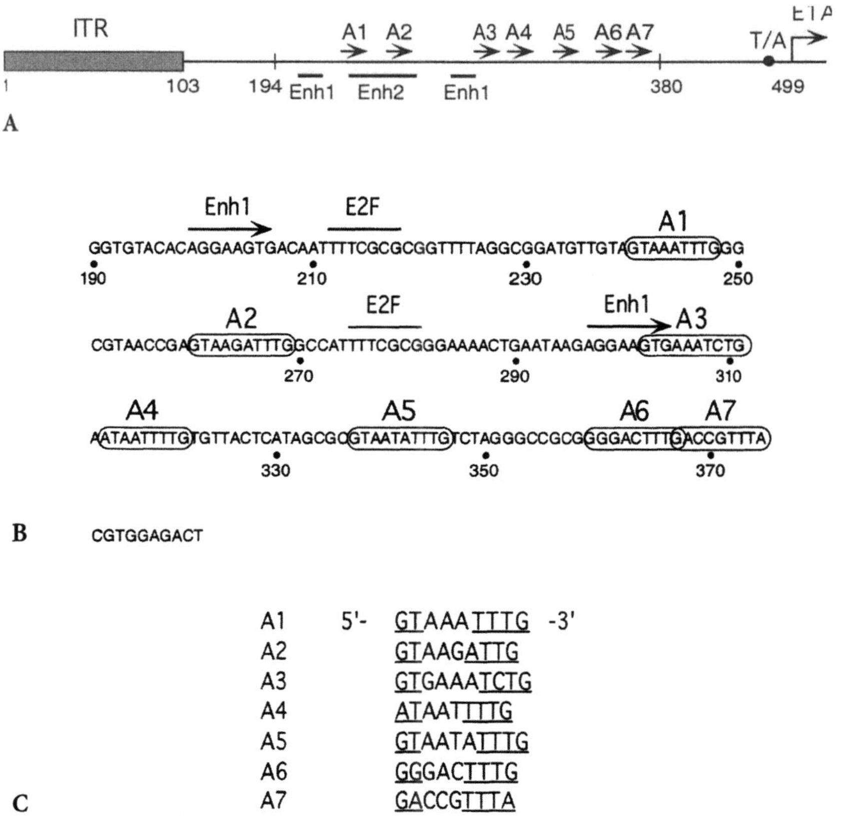

Fig. 1A–C. The adenovirus type 5 packaging domain. **A** Schematic representation of the left end of the Ad5 genome. Nucleotide positions, relative to the left terminus, are indicated by *numbers*. The inverted terminal repeat (*ITR*) is represented by a *gray box*. Viral packaging repeats, termed A repeats 1–7 (*arrows*), are located between nt 194 and nt 380. The E1A transcriptional start site is indicated by an *arrow* at nt 499 and the E1A TATA box is designated T/A. Viral enhancer elements 1 and 2 are designated as Enh1 and Enh2. **B** Nucleotide sequence of the Ad5 packaging domain and E1A enhancer region is shown. Packaging A repeats are *encircled* and indicated by a *number*, and E1A transcriptional elements (Enh1, E2F) are shown. Nucleotides numbers are indicated relative to the left terminus at nt 1. **C** Sequence of Ad5 packaging A repeats

and are referred to as A repeats 1–7. As we will see later, this consensus makes up only a portion of a functional 'A' repeat for packaging (Fig. 1).

A repeats are functionally redundant; however, there appears to be a hierarchy among the repeats with respect to efficiency of packaging as first suggested from the results of packaging assays of viruses with mutations within these sequences (GRABLE and HEARING 1990, 1992). A repeats 3 and 4 do not contribute significantly to packaging and perhaps more than circumstantially, show little homology to the consensus described above and to the further refined consensus discussed later. The more functionally significant A repeats are the two pairs of repeats 1 and 2 and A repeats 5 and 6. Figure 2A shows an alignment of the sequences of the pairs of repeats and flanking sequences. Interestingly, both of these pairs of repeats share in common a 21-bp spacing between the two TTTG elements within a pair. One can imagine that this alignment of the A repeats has been maintained for facilitating the binding of a packaging protein(s) on the same face of the helix as they are spaced apart by two turns of the helix. Results from more recent analyses of the nucleotide and spacing requirements of A repeat sequences has led to a refinement of the important sequence requirements for a functional A repeat (SCHMID and HEARING 1997). Site-directed and insertion mutagenesis was used to specifically target the region between A repeats within a pair; this has helped to distinguish the significance of the nucleotide sequences per se from spacing requirements among the sequences. SCHMID and HEARING have shown that there is a critical dinucleotide, CG, found 8 nt downstream of the TTTG. The 8-nt spacing between the consensus TTTG and the CG was critical for A repeat function in packaging. That is, the introduction of an additional 4 nt between the TTTG and the CG 8 bp downstream resulted in a significant reduction in packaging, whereas the insertion of 4 bp between adjacent A repeat units did not reduce packaging efficiency. Alignment of these four A repeats (Fig. 2B) shows conservation of the 5′-TTTGN_8CG-3′ sequences. This consensus includes the TTTG of the initial consensus described previously with the addition of the 8 nt and the critical CG. It may be that a single protein binds this bipartite sequence of TTTG and CG, or that several proteins bind, one to the TTTG region and another to the CG region and interact to enhance packaging of the viral genome. Interestingly, the 8 nt between the conserved sequences show little homology. Mutations within these sequences, at least in the context of assaying a pair of A repeats, appear to have little or no effect on the efficiency of packaging (SCHMID and HEARING 1997). Importantly, these

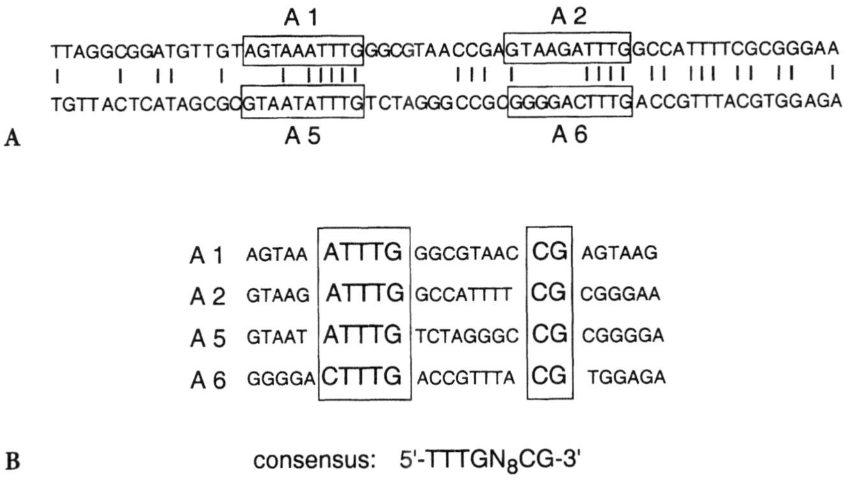

Fig. 2 A, B. Alignment of Ad5 packaging repeats. **A** The nucleotide sequences surrounding A repeats 1 and 2 and A repeats 5 and 6 are shown with the A repeats *boxed*. *Vertical lines* correspond to identical nucleotides in the alignment. **B** The packaging repeat consensus motif. Shown is an alignment of A repeats 1, 2, 5 and 6. Nucleotides comprising the bipartite consensus motif for A repeats 1, 2, 5 and 6 are *boxed and enlarged*. The consensus bipartite A repeat motif is shown at the *bottom*

results suggested that all the A repeats could bind the same or related protein(s) by virtue of the conserved TTTG and CG.

As A repeats are redundant for function, it seemed plausible that one could generate synthetic packaging sequences from any one A repeat by making multiple head-to-tail copies. Multiple copies seemed necessary as results from earlier mutagenesis analyses showed that more than one A repeat was needed for packaging of the viral genome to occur efficiently. If, in fact, such a synthetic packaging element functioned, having a single A repeat as a target for analyses would facilitate the characterization of the essential nucleotides needed for the packaging element and simplify the search for the putative trans-acting factors binding these elements that play a role in packaging. Indeed, several synthetic packaging elements have been generated based on this premise. Successful substitution of the authentic adenovirus packaging element with synthetic packaging domains demonstrates the significance of the A repeats as minimal DNA packaging domains (SCHMID and HEARING 1998). Measurements of packaging efficiencies for these synthetic packaging domains has reinforced the earlier observation that there is a functional hierarchy among the A repeats with regard to the efficiency of packaging. A minimal packaging domain that includes two tandem copies of nt 334–385 that include the pair of A repeats 5 and 6 as well as A repeat 7 functions at wild-type levels for packaging. In addition, viruses were generated that have in place of the authentic packaging element (deletion of nt 194–814 of Ad5) six head-to-tail copies of either A1, A2 or A6. Viruses with the different synthetic packaging elements grew, in single infections, to differing levels; part of that growth defect was due to their differences in packaging efficiencies as measured in a co-infection packaging assay (SCHMID and HEARING 1999). The analysis of viruses carrying synthetic packaging sequences in place of the entire packaging domain demonstrated that A repeats, although redundant for function, are not equivalent in efficiency of packaging. The results with these viruses demonstrated that: (1) a virus with a dimer of A5–A7 packaged at wild-type efficiency; (2) a virus with six copies of A1 was only slightly defective for packaging (twofold down); (3) a virus with six copies of A2 was reduced an additional fivefold for packaging; and (4) a virus with six copies of A6 was reduced approximately 100-fold in growth. Packaging of the virus with six copies of A6 was undetectable in the packaging assay (the lower limits of detection in this assay represents an approximately 50-fold reduction in packaging). The different synthetic packaging elements are valuable tools with which to identify proteins that interact with these sequences and that may play a role in packaging. This application is being used as described in the next section of this review.

The virus with six copies of A6 in place of the packaging domain was severely defective for packaging. Serial passage of this defective virus led to selection of revertants that grew to almost wild-type levels (SCHMID and HEARING 1998). These revertant viruses had altered packaging elements. In every case, there was an amplification in the number of copies of the A6 repeat. Virus were isolated with either 12, 16, 17 or 20 copies of A6. Results of co-infection experiments with these viruses showed that a virus with 20 copies of A6 packaged at wild-type levels and competed more effectively for packaging when in competition with virus that had 16 or fewer copies of A6. These results supported the idea that packaging factors are limiting and suggested that the A6 repeat is less effective for binding the necessary factors for packaging hence the addition of more sites to recover efficient packaging. Unfortunately, second site revertants elsewhere in the viral genome were not obtained. The identity of the packaging factor(s) is not known; the isolation of second site revertants might have targeted viral proteins involved in the process. Taken together, results from the analyses of viruses with synthetic packaging elements have shown that efficient packaging of these viruses occurs with multiple copies of single A repeats and demonstrate that an A repeat constitutes a minimal functional unit in the packaging domain. These repeats are probably binding sites for proteins involved in the packaging process.

Paradoxically, although these repeats share in common the consensus 5′-TTTG N_8CG-3′, they are not functionally equivalent. Therefore, the flanking sequences are probably influential in the packaging function; either by altering the binding affinity of a common factor(s) or by providing additional binding sites for factors that may differ from one repeat to another. Analyses of detailed mutations introduced into a simplified packaging domain – like the synthetic domains described that have a single functional A repeat – will be useful in determining essential sequence requirements.

Thus far, analyses of the cis-acting packaging element have been largely confined to *Human adenovirus type 5*. The location of packaging elements to the left ends of other Ad genomes is experimentally supported for human subtypes 2, 3, 7 and 16 (DANIELL 1976; HAMMARSKJOLD and WINBERG 1980; ROBBINSON and TIBBETTS 1984; TIBBETTS 1977) and for canine adenovirus 2 (SOUDAIS et al. 2001). In addition, the physical layout of the genes of the various adenovirus subtypes are quite similar, suggesting that the left-end DNA probably includes the cis-acting packaging sequences for members of *Adenoviridae*. Alignment of the DNA sequences from equivalent regions of other human adenovirus subtypes with the packaging region of Ad5 shows some interesting similarities. This is not entirely surprising as the packaging sequence is buried in a transcriptional enhancer of the adenovirus genome (HEARING and SHENK 1983). Some of the sequence similarities may be due to transcription factor binding sites that are conserved among adenoviruses and not just DNA sequences required for packaging. A search within the left end of adenovirus subtypes A, B, D, E and F for arrangements of the consensus sequence, 5′-TTTGN_8CG-3′, similar to the two pairs of repeats A1–A2 and A5–A6 of Ad5 yields provocative similarities (Fig. 3). The overall homologies identified among the various human subtypes of adenovirus may reflect a common mechanism for packaging of the genome which would not be surprising. But the detailed differences in the sequences may reflect the co-evolution of the packaging sequences along with viral proteins that interact with these domains. That there may be subtype specificity to the packaging of the viral genome will be discussed in the next section.

Repeats from serotypes:	A1			and	A2		
Ad5	TTTG	GGCGTAAC	CGAG	TAAGA-	TTTG	GCCATTTT--	CGCG
Ad3 and Ad7	TTTG	GGCGGATG	CAAG	TGAAA-	ATTC	TCCATTTT--	CGCG
Ad4	TTTG	GGCGGATG	CAAG	TGAAA-	ACGG	ACCATTTT--	CGCG
Ad9 and Ad17	TCTG	GGCGGATG	CAAG	TGAAA-	TTAG	GTCATTTTGG	CGCG
Ad40	ATTG	GGCGGGTT	TTTG	TAACT-	TTTG	GCCATTTTGG	CGCG
Ad41	ATTG	GGCGGGTT	TTTG	TAACTT	TTTG	GTTATTTTGG	CGCG
Ad12	TTTG	GGCGTTGT	TTGT	GCAAAT	TTTG	TGTTTTAGG-	CGCG

Repeats from serotypes:	A5			and	A6		
Ad5	TTTG	TCTAGGGC	CGCG	GGGAC	TTTG	-ACCGTTTA	CGTG
Ad3, 4,and 7	TTTG	CCGAGGGC	CGAG	TAGAC	TTTG	-ACCGTTTA	CGTG
Ad9 and 17	TTTA	CCGAGGGC	CGAG	-AGAC	TTTG	-ACCGATTA	CGTG
Ad40	ACTG	CTGATTAT	TACT	-GAAC	TTTG	-GCCCATGA	CGAA
Ad41	ACTG	CTGATCAT	TGCT	-GAAC	TTTG	GGGCTTTGA	CGTG
Ad12	TTTA	CCGAGGGC	AGAG	TGAAC	TCTG	-AGCCTCTA	CGTG

Fig. 3. The nucleotide sequence for pairs of Ad5 A repeats 1 and 2, and 5 and 6 are shown. Also shown are similar sequences found in the adenovirus genomes of the various serotypes indicated. The *gray boxes* outline the TTTG and CG regions of the A repeats and the *black boxes* outline the sequences that best fit the consensus: 5′-TTTGN$_8$CG-3′; "–" indicates the absence of a nucleotide

4
Trans-Acting Factors Involved in Ad Packaging

Little is know about the molecular mechanism for the selective, polar encapsidation of the adenovirus genome. A simplified concept of the initiating steps in the process is that the DNA packaging domain binds proteins that target the genome to a capsid competent for packaging, followed by additional events that allow the uptake of the viral DNA into the capsid. It has been shown that the efficient growth of a wild-type virus was impeded by the co-transfection of excess copies of the packaging domain of adenovirus (GRABLE and HEARING 1992). The growth defect was not at the level of replication or transcription. The results suggested that competition was occurring between the unlinked packaging sequences and the viral packaging sequences for a limiting factor. What the trans-acting proteins are for packaging is still unclear. However, there are some recent candidates of both viral and cellular origin to consider, including several that have been shown to bind the packaging sequences.

The synthetic minimal packaging elements described above have been used as tools to identify viral and cellular proteins that interact with these sequences. Although the minimal elements all function as packaging domains, they are not functionally equivalent. This could reflect a difference in the proteins that bind to these sequences; however, they do share in common the critical consensus 5′-TTTGN_8CGNG-3′. It seems likely that the same protein(s) is binding the different A repeats.

Gel mobility shift assays were used for identification of proteins binding the packaging domains. No qualitative differences were seen in binding activities to a DNA containing the A5–7 repeats utilizing extracts prepared from either uninfected or adenovirus infected 293 cells (SCHMID and HEARING 1998). Furthermore, there did not appear to be any changes in on-rates and off-rates of binding activities with uninfected verses infected extracts. The binding activities observed could be cellular proteins involved in the encapsidation process of the Ad genome. This is not surprising; an example for a role for cellular proteins in virus packaging had been demonstrated for SV40 and the transcription factors Sp1 and Ap-2 (DALYOT-HERMAN et al. 1996). The major binding activities to the A5–7 repeat sequences also bound the other minimal packaging sequences (A1, A2, A5 and A6). In addition, mutations of the TTTG nucleotides of the different A repeats resulted in a loss of binding of the complexes. TTTG is a critical part of the bipartite consensus sequence 5′-TTTGN_8CG-3′ common to all the A repeats. Mutation of the critical CG in the consensus, however, did not affect binding suggesting that there may be additional proteins that recognize this part of the consensus for packaging. The two significant binding activities, seen with both 293 and HeLa cell extracts, included transcription factor COUP-TF and an activity of unknown identity referred to as P complex (SCHMID and HEARING 1998). COUP-TF is an orphan receptor and a member of the thyroid, steroid hormone receptor super-family (BEATO et al. 1995). Depending upon the context of the COUP-TF binding site in a promoter, the protein functions as an activator or a repressor of transcription (KADOWAKE et al. 1995; LENG et al. 1996; POWER and CEREGHINI 1996).

It seems unlikely that COUP-TF binding to the packaging domain is required for encapsidation of the viral genome based on the results of comparative, quantitative binding studies of COUP-TF to the different A repeats. The binding efficiency of COUP-TF to the different minimal A repeats was inversely related to the efficiency of packaging activity associated with these elements. The A6 repeat which functions weakly as a minimal packaging element, bound COUP more efficiently than the A1 or A5–7 repeats. A1 and A5–7 minimal packaging elements function at efficiencies similar to that of the wild-type packaging element. Additionally, the high-affinity binding of COUP-TF to A repeat 6 and not to the other repeats is correlated with the presence in A6 of a highly conserved dimeric consensus binding site that happens to overlap the TTTG of the A repeat consensus. This conservation is not seen with the other A repeats.

P complex, on the other hand, displayed binding activities to the different minimal A repeats that directly correlated with the packaging efficiencies of the minimal A repeats. Repeats A1 and A5–7 constituted the strongest binding sites for P complex, A2 was intermediate in binding and A6 had the weakest binding activity to P complex. These binding activities correlated with the packaging efficiencies of the minimal A repeats described above. These results provided a direct correlation of P complex binding with packaging efficiency.

In addition to P complex binding of A repeats, P complex also bound to the first 13 nt of the inverted terminal repeat (ITR) of adenovirus. This binding was competed by an excess of A repeat sequences but not by A repeat sequences with TTTG mutated (Schmid and Hearing 1998). Interestingly, there is no obvious sequence similarity between the A repeat and the ITR binding sequences other than an AT-rich character. A cellular binding activity, ORP-A, that interacts with the first 12 nt of the ITR was previously identified (Rosenfeld et al. 1987). If this is a part of P complex or is identical to P complex awaits further characterization of the protein(s) involved in the binding. It is unknown whether the ITRs are required for packaging as the packaging assays used required replication and the ITRs are essential for viral DNA replication. Significantly, the packaging sequences must be located in close proximity to the ends of the adenovirus genome in order to function (Hearing et al. 1987). This spatial requirement may reflect an interaction of packaging factors bound to the packaging domain and the ITR for packaging to occur.

It seems likely that viral proteins are also required for targeting the genome to the capsid in the process of packaging of the adenovirus genome as this is a common theme in the packaging processes of other viruses and phage (reviewed in FUJISAWA and HEARING 1994). There are several recent observations that suggest a role for viral proteins in Ad packaging. It has been shown that co-infection of Ad5 with a virus of the same subtype (Ad2), as well as co-infection with adenoviruses of different subtypes, rescued a defective Ad5 virus for packaging. The source for the Ad5 genome was a temperature sensitive (ts) mutant of Ad5, ts147. ts147 is temperature sensitive for hexon and cannot package at the non-permissive temperature for lack of capsids. When relative packaging efficiencies for Ad5 ts147 compared to its co-infecting wild-type partner were measured, it became clear that packaging of the ts147 genome by a virus of its own subtype was more efficient than rescue by viruses of other subtypes (OSTAPCHUK and HEARING 2001). One possibility for this difference in relative packaging efficiencies is that viral proteins are involved in recognizing the DNA packaging domain and have co-evolved with the DNA so that there is a subtype preference.

The results of experiments with adenovirus mutants defective for the L1 protein, 52/55-kDa protein (52/55kD), suggests a role for this protein in packaging. Ad5 pm8001 is a genomic clone of Ad5 containing mutations in the *52/55kD* gene resulting in loss of its expression. Transfection of pm8001 DNA in 293 cells does not yield infectious particles. However, pm8001 virus was rescued by using a 293 cell line that expresses Ad5 52/55kD. The block in production of infectious particles was not at the level of replication or viral gene expression, but rather at the level of packaging of the viral genome. Although pm8001 was reduced for replication approximately fourfold in 293 cells and late protein mRNAs showed a lag in accumulation concomitant with a reduction in late proteins, virus particles could be isolated by CsCl density gradient centrifugation. The density of the pm8001 particles was similar to that of empty capsids. Consistent with that, no detectable portion of the adenovirus genome was found associated with the empty particles (GUSTIN and IMPERIALE 1998). Although this virus was rescued by co-infection with wild-type Ad5, attempts to rescue this virus by co-infection with viruses from other subtypes failed (ZHANG et al. 2001). Another 52/55kD mutant, temperature sensitive mutant H5ts369, was similarly blocked for lytic growth. This virus, however, accumulated particles of a density intermediate between the empty and mature particles when grown at the non-permissive temperature. A portion of the left end of the viral genome was detected in these intermediate particles as though the full uptake of the viral genome was defective (HASSON et al. 1989). The 52/55-kDa protein is found associated with empty particles but full-length protein is not detected in mature, infectious virus (HASSON et al. 1992). Taken together, these results suggest a role for the 52/55-kDa protein in encapsidation of the viral genome; perhaps interactions of 52/55kD with other viral proteins are subtype-specific. The exact role of the 52/55-kDa protein in Ad encapsidation is unclear. The protein shares characteristics of a scaffolding protein (DOKLAND 1999). However, the presence of empty particles with pm8001 suggests that it may be directly involved in initiating packaging. If indeed the 52/55-kDa protein is involved in the initiating events of packaging, that is the binding of the packaging domain of Ad by proteins that bridge the genome to a capsid competent for DNA uptake, there are probably other players in the process. Attempts to detect binding of the 52/55-kDa protein to the packaging domain have failed (ZHANG and IMPERIALE 2000). A possible candidate for contributing to this bridge is the viral late protein, IVa2, that has been shown to interact with 52/55kD as well as bind the packaging sequences.

ZHANG and IMPERIALE noted that there was DNA sequence homology between the DE element of the major late promoter (MLP) of adenovirus and the A repeats of the packaging domain. Results of mobility shift and competition assays suggested that the DE element and the A repeats of adenovirus bound some of the same proteins. Importantly, the binding was specific to Ad-infected cells (ZHANG and IMPERIALE 2000), suggesting that the binding activity might involve viral proteins. Buffer conditions used for mobility shift assays and the probe construction were different from conditions used by SCHMID and HEARING. These differences seem to explain the different complexes observed in mobility shift assays by these two groups (E. Erturk and P. Hearing, unpublished results). Additional results of in vitro binding studies showed that the introduction of mutations in either the TTTG or CG nucleotides critical to the consensus A repeat sequence were critical to the binding activities observed with the infected cell extracts. Two complexes, x and y, specific to infected cell extracts could be formed on DNA fragments containing repeats A1 and A2 and mutations in the TTTG disrupted x, the slower migrating complex, while mutations in the CG disrupted both the x and y complexes. Previously, it had been shown that IVa2 bound the DE element and acted as a transcriptional activator of the MLP (TRIBOULEY et al. 1994). Binding of IVa2 to the DE element yields two complexes in a mobility shift assay. The slower migrating complex is a heterodimer of IVa2 and an unknown protein, and the faster migrating complex is a homodimer of IVa2. Binding of the heterodimer at its site requires cooperation of binding of the homodimer at a nearby site. The x and y complexes formed on the A repeat may be similar to the binding complexes on DE. The only evidence that IVa2 is in the complexes x and y is that a monoclonal antibody directed against IVa2 added to a mobility shift assay of the A repeat packaging sites eliminated binding. Bacterially expressed IVa2 can bind DE in vitro but this has not been demonstrated for the A repeats (LUTZ and KEDINGER 1996).

Although IVa2 plays a transcriptional role, there are multiple reasons to consider IVa2 having a role in packaging of the viral genome as well. First, mutations of the A repeat sequences that disrupt IVa2 binding disrupt sequences critical for packaging function. Second, IVa2 is found in gradient purified viral particles – both empty particles and mature, infectious particles (WINTER and D'HALLUIN 1991). Its presence in the capsid does not appear to be necessary for early gene expression of the in-coming virus. Furthermore, IVa2 is made after early genes but prior to late genes, about 12 h post-infection (WINTER and D'HALLUIN 1991) at a time prior to packaging. Interestingly, IVa2 contains a protein domain that resembles a Walker box for ATP binding (KOONIN et al. 1993), a feature common to other phage proteins involved in packaging (reviewed in FUJISAWA and HEARING 1994). Although, it is not known whether IVa2 binds ATP. Thirdly, IVa2 has been shown to bind the protein product of L1, the 52/55-kDa protein (GUSTIN et al. 1996) whose possible role in encapsidation has been discussed. The interaction of IVa2 with the empty particles, however, appears independent of the 52/55-kDa protein (ZHANG et al. 2001). Lastly, provocative results with an Ad5/Ad7 mosaic virus support a role for IVa2 in packaging (ZHANG et al. 2001). A mosaic virus genome was constructed in a bacterial plasmid that replaced the Ad7 ITRs and packaging sequence with the Ad5 left end 384 nt and right-end 200 nt flanking Ad7 genomic sequences starting at nt 2711. The E1 early region was replaced with an expression cassette for green fluorescent protein (GFP) as a marker for the virus. DNA for this virus was obtained by propagating the plasmid in bacteria and later releasing the viral genome from the bacterial plasmid by restriction endonuclease cleavage. Transfection of the mosaic virus DNA into either 293 cells or into 293 cells expressing the Ad5 52/55-kDa protein resulted in cells expressing GFP. Time course analysis of the accumulation of viral DNAs suggested that replication of the mosaic genome was occurring but the virus was not spreading, and infectious particles were not recovered from either cell line. This result was somewhat unexpected, based on previous results it had been anticipated that the presence of the Ad5 52/55-kDa protein might rescue the mosaic virus for packaging. Perhaps an additional Ad5 protein was required. Indications of virus spread were seen when the mosaic virus DNA was co-transfected with an expression vector for Ad5 IVa2. In addition, lysates prepared after transfection of the mosaic genome into a 293 cell line expressing Ad5 IVa2 could be used to infect 293 cells to yield GFP-positive cells. The enticing possibility suggested from the results of these experiments is that IVa2 indeed is playing a role in packaging and that role is subtype specific. However, the results present a paradox. It is puzzling that the Ad5 IVa2 rescued infectivity of the mosaic virus but did not rescue pm8001 in the co-infection. The GFP assay is probably very sensitive and the actual quantity of mosaic virus rescued is unknown. It may be that the levels of the mosaic virus are very low, and that similarly low levels of virus were obtained in the co-infection experiments pm8001 but the assay used was not sensitive enough to detect the virus produced. Further characterization of the mosaic virus is necessary to understand the role of IVa2 and the 52/55-kDa protein with the phenotype observed.

The characterization of the proteins necessary for the packaging process is in its infancy. The exciting findings of proteins that bind the packaging sequences are aiding in the elucidation of the packaging process. Cellular protein(s), P complex, and viral proteins IVa2 and 52/55kD are important candidates as packaging factors. The binding of P complex or IVa2 does not necessarily preclude binding of the other and they may work in concert for packaging of the viral genome.

5 Packaging and Gene Therapy

The application of fundamental concepts about the packaging of the adenovirus genome has contributed significantly to the development of therapeutic viral vectors for gene therapy. Modifications have been made of adenovirus-based gene therapy vectors with the goal of developing safer and more effective reagents. Vectors have been developed that are deleted for many of the essential genes of the virus thus requiring helper function for growth. The vectors with deletions of all genes are referred to as the 'gutless', 'gutted', 'high capacity' or 'HD' (for helper-dependent) vectors. They include, as the only authentic adenovirus sequences, the

ITRs necessary for replication of the vector and the packaging sequence (nt 194–358) required for encapsidation of the vector genome. The remainder of the backbone of the vector includes sequences of the gene of interest to be expressed, sequences to promote the expression of the gene and, in addition, so-called stuffer sequences that contribute to a DNA backbone of an appropriate size for stable encapsidation into the adenovirus capsid (BETT et al. 1994; GHOSH-CHOUDHURY et al. 1987; MITANI et al. 1995; PARKS and GRAHAM 1997) (see chapter by KOCHANEK, this volume). Since the helper-dependent vectors no longer contain genes for replication or for the structural proteins of the virus they require complementation for growth. A helper virus provides in trans all viral proteins for growth of the helper-dependent vector virus. In addition, the therapeutic vector virus needs to be purified from the helper virus. In this section, we will focus only on the developments made in the design of the helper virus, and how information regarding the packaging domain has contributed to their development.

Based on earlier findings that defective Ad genomes could be packaged in the presence of co-infecting wild-type Ad virus (DEURING et al. 1981; GLUZMAN and DOREN 1983; HAMMARSKJOLD and WINBERG 1980; HEARING et al. 1987), a helper-dependent vector virus was rescued and propagated using wild-type Ad2 as the helper virus (MITANI et al. 1995). Unlike some of its predecessors (MANSOUR et al. 1985; YAMADA et al. 1985), this helper-dependent vector was rescued independent of selection. An approximation of the resulting ratio of vector/helper, however, appeared low. A recombination event between the Ad vector and the helper virus genomes resulted in size differences permitting a partial purification of the vector virus from the helper virus by density gradient centrifugation. These results were encouraging and suggested that yields of a therapeutic vector virus could be improved if provisions for selection against production of the helper virus could be provided. Furthermore, the results illustrated the importance of having a means to purify the Ad vector virus from its helper virus.

Several methods have been used to select against helper virus by specifically decreasing or preventing the encapsidation of the helper virus genome in favor of encapsidation of the vector. Viruses with previously characterized deletion mutations of multiple A repeats in the packaging domain (GRABLE and HEARING 1990) were used as helpers to rescue helper-dependent virus vectors (ALEMANY et al. 1997; KOCHANEK et al. 1996). These viruses were chosen for helper function because their packaging domains were disabled such that they did not compete effectively for packaging in a co-infection with a wild-type virus that had an authentic packaging domain. Helper-dependent vectors contain wild-type packaging domains and should have a competitive advantage over such helper viruses for packaging. Although the helper viruses with deletions in the packaging domain rescued helper dependent vectors, there was still substantial helper virus contamination in the final yields.

A novel method was developed that produced a helper virus regulated for packaging (PARKS et al. 1996). Recognition sites, Lox P, were introduced into the helper virus flanking the packaging domain (referred to as floxed). Lox P sites are DNA sequences that are recognized and cleaved by the bacteriophage recombinase, Cre (HOESS and ABREMSKI 1985). This virus grew to wild-type levels in 293 cells. Infection of the floxed virus into 293 cells that express the Cre recombinase in the nucleus resulted in virus genomes that were specifically deleted for the packaging domain. Replication and production of late gene products was unaffected, but due to the lack of packaging domains the floxed virus was not encapsidated. The Lox P virus has been used successfully to rescue various helper-dependent vector viruses (O'NEAL et al. 2000; OKA et al. 2001; SCHIEDNER et al. 1998; also see references in MORSY and CASKEY 1999). The levels of contamination of the vector virus with helper virus is effected by the efficiency of recombination resulting in deletion of the floxed packaging sequence (HARDY et al. 1997; PARKS et al. 1996). In addition, helper virus contamination can result when recombination occurs between the helper virus packaging sequences and the vector virus sequences resulting in the loss of the downstream lox site. Cre recombination requires the presence of the two Lox P sites. Improvements in this approach have been introduced targeting the various problems of recombination. A more efficient recombination enzyme for the yeast recombinase FLP was developed in vitro (BUCHHOLZ et al. 1998). FLPe was expressed in cells to remove Ad packaging sequences flanked by the FLPe-recognition sequences (UMANA et al. 2001). This system is close to 100% efficient for recombination compared to 80%–90% reported for the Cre–Lox system resulting in less contamination of the vector virus with helper virus. An additional improvement in the Cre–Lox system was the modification of the packaging domain to reduce homology with the packaging sequences of the vector virus with the goal of eliminating recombination between the packaging sequences of the therapeutic vector and helper virus (SANDIG et al. 2000). Modifications introduced into the packaging domain were based on information about the bipartite consensus sequence for the A repeats, 5′-ATTTGN$_8$CG-3′. The resulting packaging domain contained sequences with only 22 bp of contiguous homology with the vector virus packaging sequences reducing targets for recombination. The modified system resulted in lower helper virus contamination of the vector virus. Elimination of the helper virus from the therapeutic vector virus is necessary. Improvements in this regard are feasible as we learn more about the process of packaging of adenovirus. The finding that the packaging domain may function in a subtype specific manner opens other avenues for development and improvement of helper systems that will yield vector virus free of helper virus (OSTAPCHUK and HEARING 2001; ZHANG et al. 2001).

References

ALEMANY R, DAI Y, LOU YC, SETHI E, PROKOPENKO E, JOSEPHS SF and ZHANG WW (1997) Complementation of helper-dependent adenoviral vectors: size effects and titer fluctuations. J Virol Methods 68:147–159

BEATO M, HERRLICH P and SCHUTZ G (1995) Steroid hormone receptors: many actors in search of a plot. Cell 83:851–857

BETT AJ, HADDARA W, PREVEC L and GRAHAM FL (1994) An efficient and flexible system for construction of adenovirus vectors with insertions or deletions in early regions 1 and 3. Proc Natl Acad Sci USA 91:8802–8806

BUCHHOLZ F, ANGRAND PO and STEWART AF (1998) Improved properties of FLP recombinase evolved by cycling mutagenesis. Nat Biotechnol 16:657–662

D'HALLUIN JC (1995) Virus assembly. Curr Top Microbiol Immunol 199 (Pt 1) 47–66

DALYOT-HERMAN N, BEN-NUN-SHAUL O, GORDON-SHAAG A and OPPENHEIM A (1996) The simian virus 40 packaging signal ses is composed of redundant DNA elements which are partly interchangeable. J Mol Biol 259:69–80

DANIELL E (1976) Genome structure of incomplete particles of adenovirus. J Virol 19: 685–708

DEURING R, KLOTZ G and DOERFLER W (1981) An unusual symmetric recombinant between adenovirus type 12 DNA and human cell DNA. Proc Natl Acad Sci USA 78:3142–3146

DOKLAND T (1999) Scaffolding proteins and their role in viral assembly. Cell Mol Life Sci 56:580–603.

FUJISAWA H and HEARING P (1994) Structure, function and specificity of the DNA packaging signals in double-stranded DNA viruses. In: "Seminars in Virology", Vol 5, pp 5–13. Academic Press Ltd

GHOSH-CHOUDHURY G, HAJ-AHMAD Y and GRAHAM FL (1987) Protein IX, a minor component of the human adenovirus capsid, is essential for the packaging of full length genomes. EMBO J 6:1733–1739

GLUZMAN Y and DOREN KV (1983) Palindromic adenovirus type 5-simian virus 40 hybrid. J Virol 45:91–103

GRABLE M and HEARING P (1990) Adenovirus type 5 packaging domain is composed of a repeated element that is functionally redundant. J Virol 64:2047–5206.

GRABLE M and HEARING P (1992) cis and trans requirements for the selective packaging of adenovirus type 5 DNA. J Virol 66:723–731

GUSTIN KE and IMPERIALE MJ (1998) Encapsidation of viral DNA requires the adenovirus L1 52/55-kilodalton protein. J Virol 72:7860–7870

GUSTIN KE, LUTZ P and IMPERIALE MJ (1996) Interaction of the adenovirus L1 52/55-kilodalton protein with the IVa2 gene product during infection. J Virol 70:6463–6467

HAMMARSKJOLD ML and WINBERG G (1980) Encapsidation of adenovirus 16 DNA is directed by a small DNA sequence at the left end of the genome. Cell 20:787–795

HARDY S, KITAMURA M, HARRIS-STANSIL T, DAI Y and PHIPPS ML (1997) Construction of adenovirus vectors through Cre-lox recombination. J Virol 71:1842–1849

HASSON TB, ORNELLES DA and SHENK T (1992) Adenovirus L1 52- and 55-kilodalton proteins are present within assembling virions and colocalize with nuclear structures distinct from replication centers. J Virol 66:6133–6142

HASSON TB, SOLOWAY PD, ORNELLES DA, DOERFLER W and SHENK T (1989) Adenovirus L1 52- and 55-kilodalton proteins are required for assembly of virions. J Virol 63:3612–3621

HEARING P, SAMULSKI RJ, WISHART WL and SHENK T (1987) Identification of a repeated sequence element required for efficient encapsidation of the adenovirus type 5 chromosome. J Virol 61:2555–2558

HEARING P and SHENK T (1983) The adenovirus type 5 E1 A transcriptional control region contains a duplicated enhancer element. Cell 33:695–703

HOESS RH AND ABREMSKI K (1985) Mechanism of strand cleavage and exchange in the Cre-lox site-specific recombination system. J Mol Biol 181:351–362

KADOWAKE Y, TOYOSHIMA K and YAMAMOTO T (1995) Dual transcriptional control by Ear3/COUP: Negative regulation through the DR1 direct repeat and positive regulation through a sequence downstream of the transcriptional start site of the mouse mammary tumor virus promoter. Proc Natl Acad Sci USA 92:4432–4436

KOCHANEK S, CLEMENS PR, MITANI K, CHEN HH, CHAN S and CASKEY CT (1996) A new adenoviral vector: Replacement of all viral coding sequences with 28 kb of DNA independently expressing both full-length dystrophin and beta-galactosidase. Proc Natl Acad Sci USA 93:5731–5736

KOONIN E, SENKEVICH TG and CHERNOS VI (1993) Gene A32 product of vaccinia virus may be an ATPase involved in viral DNA packaging as indicated by sequence comparisons with other putative viral ATPases. Virus Genes 7:89–94

LENG X, COONEY AJ, TSAI SY and TSAI M-J (1996) Molecular mechanisms of COUP-TF-mediated transcriptional repression: evidence for transrepression and active repression. Mol Cell Biol 16:2332–2340

LUTZ P and KEDINGER C (1996) Properties of the adenovirus IVa2 gene product, an effector of late- phase-dependent activation of the major late promoter. J Virol 70:1396–1405

MANSOUR SL, GRODZICKER T and TJIAN R (1985) An adenovirus vector system used to express polyoma virus tumor antigens. Proc Natl Acad Sci USA 82:1359–1363

MITANI K, GRAHAM FL, CASKEY CT and KOCHANEK S (1995) Rescue, propagation and partial purification of a helper virus- dependent adenovirus vector. Proc Natl Acad Sci USA 92:3854–3858
MORSY MA AND CASKEY CT (1999) Expanded-capacity adenoviral vectors–the helper-dependent vectors. Mol Med Today 5:18–24
O'NEAL WK, ZHOU H, MORRAL N, LANGSTON C, PARKS RJ, GRAHAM FL, KOCHANEK S and BEAUDET AL (2000) Toxicity associated with repeated administration of first-generation adenovirus vectors does not occur with a helper-dependent vector. Mol Med 6:179–195
OKA K, PASTORE L, KIM IH, MERCHED A, NOMURA S, LEE HJ, MERCHED-SAUVAGE M, ARDEN-RILEY C, LEE B, FINEGOLD M, BEAUDET A and CHAN L (2001) Long-term stable correction of low-density lipoprotein receptor-deficient mice with a helper-dependent adenoviral vector expressing the very low-density lipoprotein receptor. Circulation 103: 1274–1281
OSTAPCHUK P and HEARING P (2001) Pseudopackaging of adenovirus type 5 genomes into capsids containing the hexon proteins of adenovirus serotypes B, D, or E. J Virol 75:45–51
PARKS RJ, CHEN L, ANTON M, SANKAR U, RUDNICKI MA and GRAHAM FL (1996) A helper-dependent adenovirus vector system: removal of helper virus by Cre-mediated excision of the viral packaging signal. Proc Natl Acad Sci USA 93:13565–13570
PARKS RJ and GRAHAM FL (1997) A helper-dependent system for adenovirus vector production helps define a lower limit for efficient DNA packaging. J Virol 71:3293–3298
POWER SC and CEREGHINI S (1996) Positive regulation of the vHNF1 promoter by the orphan receptors COUP-TF1/Ear3 and COUP TFII/Arp1. Mol Cell Biol 16:778–791
ROBBINSON CC and TIBBETTS C (1984) Polar encapsidation of adenovirus DNA: evolutionary variants reveal dispensable sequences near the left ends of Ad3 genomes. Virology 137:5851–5860
ROSENFELD PJ, O'NEILL EA, WIDES RJ and KELLY TJ (1987) Sequence-specific interactions between cellular DNA-binding proteins and the adenovirus origin of DNA replication. Mol Cell Biol 7:875–886
SANDIG V, YOUIL R, BETT AJ, FRANLIN LL, OSHIMA M, MAIONE D, WANG F, METZKER ML, SAVINO R and CASKEY CT (2000) Optimization of the helper-dependent adenovirus system for production and potency in vivo. Proc Natl Acad Sci USA 97:1002–1007
SCHIEDNER G, MORRAL N, PARKS RJ, WU Y, KOOPMANS SC, LANGSTON C, GRAHAM FL, BEAUDET AL and KOCHANEK S (1998) Genomic DNA transfer with a high-capacity adenovirus vector results in improved in vivo gene expression and decreased toxicity. Nat Genet 18:180–183
SCHMID SI and HEARING P (1995) Selective encapsidation of adenovirus DNA. Curr Top Microbiol Immunol 199 (Pt 1), pp 67–80
SCHMID SI and HEARING P (1997) Bipartite structure and functional independence of adenovirus type 5 packaging elements. Journal of Virology 71:3375–3384
SCHMID SI and HEARING P (1998) Cellular components interact with adenovirus type 5 minimal DNA packaging domains. J Virol 72:6339–6347

SCHMID SI and HEARING P (1999) Adenovirus DNA Packaging. In: "Adenovirus Methods and Protocols" (WSM Wold, Ed.), pp 47–59. Humana Press, Totowa

SOUDAIS C, BOUTIN S and KREMER EJ (2001) Characterization of cis-acting sequences involved in canine adenovirus packaging. Mol Ther 3: 631–640

TIBBETTS C (1977) Viral DNA sequences from incomplete particles of human adenovirus type 7. Cell 12:243–249

TRIBOULEY C, LUTZ P, STAUB A and KEDINGER C (1994) The product of the adenovirus intermediate gene IVa2 is a transcriptional activator of the major late promoter. J Virol 68:4450–4457

UMANA P, GERDES CA, STONE D, DAVIS JR, WARD D, CASTRO MG and LOWENSTEIN PR (2001) Efficient FLPe recombinase enables scalable production of helper- dependent adenoviral vectors with negligible helper-virus contamination. Nat Biotechnol 19:582–585

WINTER N and D'HALLUIN JC (1991) Regulation of the biosynthesis of subgroup C adenovirus protein IVa2. J Virol 65:5250–5259

YAMADA M, LEWIS JA and GRODZICKER T (1985) Overproduction of the protein product of a nonselected foreign gene carried by an adenovirus vector. Proc Natl Acad Sci USA 82:3567–3571.

ZHANG W and IMPERIALE MJ (2000) Interaction of the adenovirus IVa2 protein with viral packaging sequences. J Virol 74:2687–2693

ZHANG W, LOW JA, CHRISTENSEN JB and IMPERIALE MJ (2001) Role for the adenovirus IVa2 protein in packaging of viral DNA. J Virol 75:10446–10454

Chapter 7

Adenovirus DNA Replication: Protein Priming, Jumping Back and the Role of the DNA Binding Protein DBP

R. N. de Jong, P. C. van der Vliet, A. B. Brenkman

Abstract. The adenovirus (Ad) genome is a linear double-stranded (ds) molecule containing about 36 kilobase pairs. At each end of the genome an approximately 100 base pair (bp) inverted terminal repeat (ITR) is found, the exact length depending on the serotype. To the 5′-end of each ITR, a

---

R. N. de Jong, P. C. van der Vliet, A. B. Brenkman
University Medical Center, Department of Physiological Chemistry and Center for Biomedical Genetics, Utrecht, The Netherlands
e-mail: r.n.dejong@med.uu.nl

P. C. van der Vliet
Universiteitsweg 100, P. O. Box 85060, 3508 AB Utrecht, The Netherlands
e-mail: P.C.vanderVliet@med.uu.nl

55-kDa terminal protein (TP) is covalently coupled. The Ad DNA replication system was one of the first replication systems that could be reconstituted in vitro (CHALLBERG and KELLY 1979). The system requires three virally encoded proteins: precursor TP (pTP), DNA polymerase (Pol) and the DNA binding protein (DBP). In addition, three stimulating human cellular proteins have been identified. These are the transcription factors NFI (NAGATA et al. 1982) and Oct-1 (PRUIJN et al. 1986) and the type I topoisomerase NFII (NAGATA et al. 1983). Ad DNA replication uses a protein primer for replication initiation. The transition from initiation to elongation is marked by a jumping back mechanism (KING and VAN DER VLIET 1994), followed by elongation. In order to elongate DBP is required. In this review we discuss the roles of DBP during initiation and elongation and we relate biochemical data on the jumping back mechanism used by Ad Pol to the recently solved crystal structure of a Pol α-like replication complex (FRANKLIN et al. 2001). We comment on the conditions and possible functions of jumping back and propose a model to describe the jumping back mechanism.

1
Pre-initiation Complex Formation

In the process of characterizing the in vitro replication system, it was soon found that the combined action of two cellular transcription factors, NFI and Oct-1, could stimulate initiation of replication up to 200-fold depending on the pTP–Pol concentration (DE JONG and VAN DER VLIET 1999; and references therein). Conserved binding sites of NFI and Oct-1 were found in the ITRs of the Ad genome downstream of the core origin, in the so-called auxiliary region. NFI binds specifically to Pol whereas Oct-1 binds to pTP. By using these direct interactions, NFI and Oct-1 recruit the pTP–Pol complex to the core origin by increasing both binding affinity and specificity for this sequence. The binding of NFI to the replication origin was furthermore shown to be stimulated by DBP which increased the association rate and decreased the dissociation rate of the NFI–DNA complex (CLEAT and HAY 1989; STUIVER and VAN DER VLIET 1990). In addition, the genome-bound TP stabilizes core origin binding of the pTP–Pol complex and induces changes in the origin structure (PRONK and VAN DER VLIET 1993). Together these five proteins are involved in stabilizing the pre-initiation complex and correct positioning of pTP–Pol.

2
The Ad DNA Binding Protein

The first viral replication protein to be discovered was the Ad DNA binding protein. In hindsight this was not surprising, as DBP is a very abundant protein with at least 10 million copies per infected cell. Moreover, its strong interaction with single-stranded (ss) DNA makes it easy to detect. The DNA binding capacity and the genetically determined need for stoichiometric amounts immediately suggested a role in elongation. Indeed, mutant and electron microscopy (EM) studies as well as in vitro reconstitution confirmed its essential role in elongation, while an enhancing function in initiation was established later. DNA replication turned out to be only one of the processes in which DBP is involved. Others include transcription control, mRNA stability, transformation, recombination, virus assembly and determination of the host range (HAY et al. 1995; VAN DER VLIET 1995). It is also essential as a helper for adeno-associated virus replication (WARD et al. 1998), showing a high functional versatility. Here we restrict the discussion to DNA replication.

2.1
Structure of DBP

DBP is the product of the E2A gene. Its expression is controlled by two different promoters, active early and late in infection, respectively (SWAMINATHAN and THIMMAPAYA 1995). Ad2/5 DBP consists of 529 amino acids (aa) with an apparent mobility of 72 kDa in SDS-containing polyacrylamide gels. DBP can be separated into two domains by limited chymotrypsin treatment: a highly phosphorylated N-terminal domain of 173 aa and a C-terminal domain of 356 aa. The latter is non-phosphorylated, contains two zinc atoms and is sufficient for all DNA replication functions in vitro. The crystal structure of this domain (see Fig. 1) shows a mainly globular core consisting of seven alpha-helices, three beta sheets, and a remarkable protruding arm of 17 aa at the C-terminus (TUCKER et al. 1994). This arm contains a hook that fits perfectly into a hydrophobic cleft of a neighboring molecule, thereby enabling multimerization. Indeed, a protein chain can be observed by EM and chain formation is abolished upon deletion of the C-terminal arm, confirming the X-ray diffraction results. Interestingly, a second crystal structure has been found in which only the angle between the C-terminal arm and the core is changed

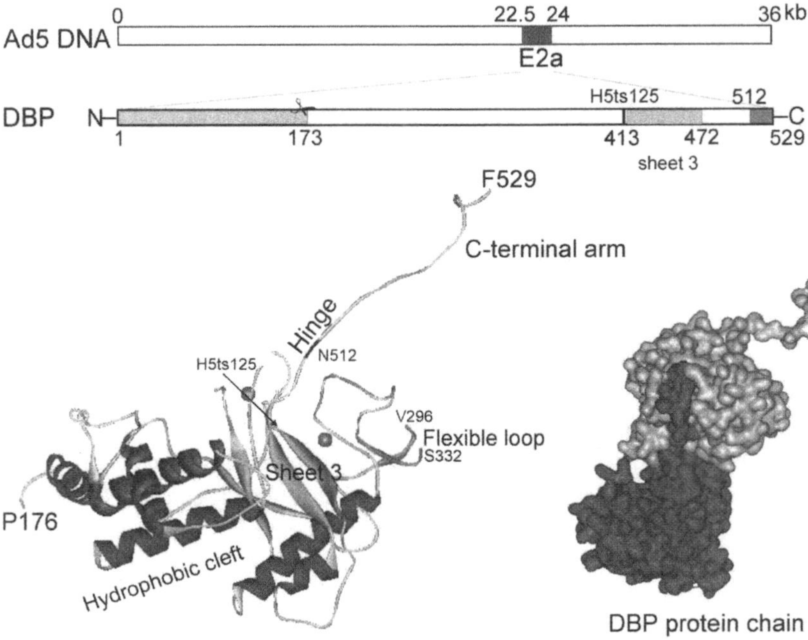

Fig. 1. The Ad DNA binding protein DBP. The crystal structure of the conserved DNA binding region of Ad5 DBP is shown (Kanellopoulos et al. 1996; Tucker et al. 1994). The positions of the two Zn atoms (dark spheres) are indicated. Sheet 3 and the flexible loop are involved in DNA binding. The 17-aa C-terminal arm fits in the hydrophobic cleft of another molecule, giving rise to a protein chain. A space-filling example is given for two molecules. DNA is wound around the protein surface of the chain (not shown). The chain is flexible around the hinge region, enabling adaptation to different DNA sequences. Chain formation is essential for unwinding of DNA in the replication fork

due to a different arrangement of residues 512–515, called the hinge region (Kanellopoulos et al. 1996). This suggests that the C-terminal arm has a high degree of rotational freedom, turning around the hinge region and in particular Asn512. Thus, the C-terminal arm can adopt different orientations leading to different conformations of the protein chain. The functional significance of this flexibility in DBP chains will be discussed below.

2.2 The Structure of DNA Is Altered by DBP

DBP binds ssDNA as well as dsDNA and RNA without apparent sequence specificity, but the modes of binding differ. Binding to ssDNA or RNA is stable and cooperative, with a cooperativity constant of 20–30 (VAN AMERONGEN et al. 1987). The length of the binding site varies between 10 and 15 bases. In contrast, dsDNA is bound in a non-cooperative fashion and binding is less stable (STUIVER et al. 1992). ssDNA is bound to the surface of the DBP chain, contacting the positively charged, three-stranded beta sheet 3, which is close to one of the zinc atoms. In this region also the mutant H5ts125 is located, which is temperature-sensitive for DNA binding (Fig. 1). Although the details of the DNA structure in the complex are not known, it is clear that DBP binding has a significant effect. In complex with DBP, ssDNA has an irregular, considerably extended structure with the bases unstacked as shown by spectroscopic and preliminary cocrystal data as well as model building (KANELLOPOULOS et al. 1995; TUCKER et al. 1994; VAN AMERONGEN et al. 1987). This DBP–ssDNA structure increases the rigidity of ssDNA as indicated by a reduced intramolecular renaturation. Intermolecular renaturation is enhanced, presumably by removal of secondary structures or shielding the electrostatic repulsion between the two strands (ZIJDERVELD et al. 1993). The dsDNA–DBP complex also forms a rigid nucleoprotein structure in which tertiary structures are removed as measured by hydroxyl radical foot printing and several other techniques (STUIVER et al. 1992). Thus DBP changes the DNA conformation drastically, favoring DNA replication.

2.3 Role of DBP in Chain Elongation

Together with the DNA polymerase, DBP is essential for elongation. How does DBP achieve this? Four modes of action have been observed. First of all, DBP enhances the rate and processivity of the DNA polymerase and modifies the sensitivity to nucleotide analogues (FIELD et al. 1984; LINDENBAUM et al. 1986; MUL and VAN DER VLIET 1993). These effects could involve modification of the active site of the polymerase, possibly indirectly through modification of the DNA structure. The collaboration between Pol and DBP is not understood in detail but it is specific, suggesting an interaction between the two proteins. Secondly, DBP can unwind short stretches of dsDNA or even longer stretches if short protruding ends are present (Georgaki et al. 1992; Monaghan et al. 1994; Zijderveld and van der Vliet 1994). The strong and cooperative binding to ssDNA unwinds dsDNA independently of ATP hydrolysis, in agreement with the notion that DNA replication in vitro does not need a helicase. Thirdly, DBP removes secondary structures that could act as a roadblock for the advancing polymerase during displacement synthesis as well as during duplication of the displaced strand. Removal of secondary structures could also be instrumental in the very efficient recombination process in adenoviruses and the renaturation of complementary strands originating from displacement (Zijderveld et al. 1993). Finally DBP protects ssDNA against nuclease attack, for instance in the vulnerable replication fork, a function common to many helix destabilizing proteins.

2.4 Multimerization Is the Driving Force

How does the ability of DBP to multimerize come in? Deletion of the C-terminal arm resulted in a monomeric protein which bound with greatly reduced affinity to DNA, but still stimulated initiation. In contrast, the mutant did not support elongation on a ds template and was defective in unwinding (Dekker et al. 1997). This indicates that protein chain formation coupled to high affinity binding to ssDNA is the driving force for unwinding during DNA chain elongation.

2.5 Flexibility Is the Key to Success

The different arrangement of the hinge region connecting the C-terminal arm to the core (Kanellopoulos et al. 1996) shows that the protein chain can adopt different conformations. What is the significance of this flexibility for the function of DBP? When proline residues were introduced in the hinge region to reduce flexibility, elongation was not possible and unwinding was severely impaired (van Breukelen et al. 2000). Still, DBP could bind DNA efficiently and cooperatively. This suggests that flexibility of the protein chain is an essential prerequisite for DNA chain elongation. One explanation for the need of flexibility is that, when bound in the replication fork, the position of DBP could lead to slightly different orientations of the C-terminal arm. Assuming that conformational changes are required to accommodate the transition of DBP from binding to dsDNA to shifting to ssDNA, such a flexibility ensures that the C-terminal arm still can hook into its neighbor, giving rise to a stable ssDNA–DBP complex in the displaced strand. If flexibility is lost, DBP will dissociate rapidly from the dsDNA in the fork, preventing unwinding and thus blocking elongation. Sequence differences in the fork could have an effect as well as some mutants in the hinge region are defective in binding polydA, which has an aberrant structure (VAN BREUKELEN et al. 2000). Thus, the ability of DBP to adopt more than one conformation by using a flexible C-terminal arm enables it to adapt to different DNA conformations thereby optimizing formation of the DBP–DNA chain and subsequent unwinding and elongation.

Another example of the need for flexibility has been observed. In the crystal, electron density for the region between aa 297 and 331 is not visible and this 34-aa long loop may well be flexible (Fig. 1). This region contains several residues that contact ssDNA based on mutagenesis and cross-linking (CLEGHON and KLESSIG 1992) and may fold after DNA binding, thereby stabilizing the complex. When the flexible loop was deleted, DBP could still enhance initiation but became defective in supporting elongation. Mixing experiments showed that the flexible loop and the C-terminal arm have distinct functions in unwinding during replication. Apparently multimerization via the C-terminal arm is required for the formation of a protein filament that saturates the displaced strand, whereas the flexible loop guarantees local destabilization of the fork by contributing to stability and high affinity, independent of multimerization (DEKKER et al. 1998).

2.6
Role of DBP in Initiation

The enhancing effect of DBP on initiation is strongly dependent on the pTP–Pol concentration, being highest at low pTP–Pol levels (DEKKER et al. 1997). As with the transcription factors NFI and Oct-1, this suggests recruitment of pTP–Pol to the origin and a specific interaction between DBP and the pTP–Pol complex. In agreement with an interaction, DBP protects Pol against thermal inactivation (LINDENBAUM et al. 1986) and binding between the two proteins has been observed using immobilized DBP (B. van Breukelen, unpublished results). However, most other common assays to demonstrate such an interaction, such as pull-down or immune precipitation, have been unsuccessful suggesting that the interaction is weak or unstable. Besides aiding in recruitment, DBP might also help to destabilize the origin, as initiation on a partially ss origin can not be stimulated by DBP anymore (B. van Breukelen, unpublished results). Mutants defective in unwinding can still stimulate initiation. Therefore, rather than ascribing the stimulation to simple unwinding of the origin, the effect could be more complex. DBP might stabilize pre-initiation complexes or prevent aberrant positioning of pTP–Pol, effects less dependent on DNA binding and more on protein–protein interactions.

DBP has also a direct effect on the kinetics of the initiation reaction, which is independent of the pTP–Pol concentration. It stimulates the formation of the pTP-CAT intermediate by lowering the K_m for the reaction, indicating that it can influence the Pol active center, either directly or by changing the template conformation (Mul and van der Vliet 1993). Finally, another way in which DBP enhances initiation is by enhancing the binding of NFI to the auxiliary origin (Cleat and Hay 1989; Stuiver and van der Vliet 1990). This is not due to a direct interaction between the two proteins but is based on a structural change leading to increased flexibility of the NFI binding site (Stuiver et al. 1992).

3
Ad pTP–Pol Uses a Jumping Back Mechanism to Start Replication

Ad Pol is a 140-kDa protein that belongs to the eukaryotic Pol α family (Ito and Braithwaite 1991). Besides a DNA polymerase activity, it also contains a proofreading $3' \rightarrow 5'$ exonuclease activity. In the early stages of infection, the protein is found as a stable heterodimer with the 80-kDa pTP (Enomoto et al. 1981). The biochemical properties and characterization of pTP and Pol are discussed in detail in the chapter by Liu and Hay, this volume.

The Ad origin template[1] starts with the repetitive sequence 3′-GTAGTA-5′ (see Table 1 for a summary of templates discussed). Ad pTP–Pol initiates DNA replication opposite nucleotide 4 and synthesizes a pTP-CAT trinucleotide intermediate. This intermediate pairs with template nucleotides 1 to 3 after jumping back, after which elongation starts as illustrated in Fig. 2

[1] Throughout this chapter, we will refer to the template strand as the strand that directs protein priming at the origin, in contrast to the displaced strand.

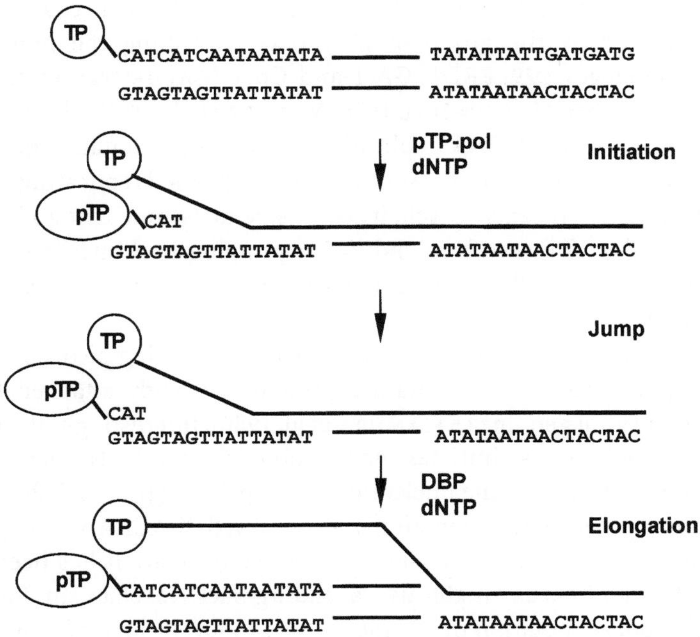

Fig. 2. Ad replication initiates opposite nucleotide 4, followed by jumping back. See text for details

Table 1. Oligonucleotides used as templates directing in vitro Ad DNA replication. The templates referred to in the text are compared with the wildtype Ad5 template (top). Initiation takes place opposite template nucleotide G4, followed by synthesis of a CAT trinucleotide intermediate directed by template nucleotides 4–6. pTP-CAT pairs with template nucleotides 1–3 after jumping back. The core origin, which is essential for Ad DNA replication, is shaded. Mutations in the templates are underlined

Template	Sequence
Wildtype	3´-GTAGTAGTTATTATATGGAATAAAACCTAA-5´
G4A	3´-GTAATAGTTATTATATGGAATAAAACCTAA-5´
G7C	3´-GTAGTACTTATTATATGGAATAAAACCTAA-5´
Δ1	3´-·TAGTAGTTATTATATGGAATAAAACCTAA-5´
Δ2	3´-··AGTAGTTATTATATGGAATAAAACCTAA-5´
Δ3	3´-···GTAGTTATTATATGGAATAAAACCTAA-5´
Δ3G7C	3´-···GTACTTATTATATGGAATAAAACCTAA-5´
GAGA	3´-GAGAGAGATATTATATGGAATAAAACCTAA-5´
Position	····v····|····v····|····v····|

(KING and VAN DER VLIET 1994). A similar sliding-back mechanism was first identified by the group of M. SALAS describing the replication of bacteriophages φ29, PRD1, GA-1 and Cp-1 (CALDENTEY et al. 1993; ILLANA et al. 1996; MARTIN et al. 1996; MENDEZ et al. 1992). These phages use a protein-primed DNA replication system and contain nucleotide repeats at the genome termini. As other protein-primed replicating phages contain sequence repeats as well, it seems safe to assume that an internal replication start followed by a sliding- or jumping-back mechanism is a common feature of all protein-primed DNA replication systems (SALAS 1991, 1996).

This could extend to a group of RNA viruses including picorna viruses such as poliovirus, which contain a protein covalently attached to the 5′ end of the viral genome (KITAMURA et al. 1980). Interestingly, the poliovirus RNA polymerase initiates replication of the viral RNA genome by the coupling of two uridine nucleotides to its priming protein VPg directed by a hairpin structure containing a conserved 3′-ACAAA-5′ sequence in the loop (PAUL et al. 1998, 2000; RIEDER et al. 2000). It has been proposed that poliovirus might use a sliding-back mechanism for VPg uridylylation, as mutation of the most 5′ A produced the most severe replication defect, while mutation of the 3′ preceding A resulted in vitro in the formation of a mono-uridylylated VPg. This suggested that replication could start at the most 5′ A, after which a putative sliding-back would be prohibited.

3.1
Which Genomic Features Make Jumping Back Possible?

Jumping back has been observed both on natural templates and on ss oligonucleotide templates, indicating that the process does not depend on the presence of the TP moiety or the ds structure of the genome terminus (KING and VAN DER VLIET 1994). Unwinding of the origin must precede the base-pairing of incoming nucleotides, because the template strand terminus bound by the polymerase will be ss during replication initiation of genomic DNA. Indeed, removal of 5′ terminal nucleotides of synthetic ds templates facilitates in vitro DNA replication by exposing a ss template region (KENNY and HURWITZ 1988).

Almost all adenoviruses sequenced to date contain di-, tri- or tetranucleotide repeats at their genome termini and we suspect that the exceptions to this rule are due to the difficulty of sequencing the genome termini. Such repeats are necessary to allow optimal base-pairing of internal replication intermediates after jumping back, but small mismatches and deletions of terminal template nucleotides did not block DNA replication (GRAHAM et al. 1989; KING and VAN DER VLIET 1994).

When replication was performed on a GAGA template (see Table 1), G5 was used as the start site and a pTP–tetranucleotide intermediate was formed (KING et al. 1997b). Ad pTP–Pol apparently possesses some flexibility in both the choice of the start site and the size of the jumping intermediate. A strong preference for coupling C to pTP was observed, since even on the G4A template (Table 1), a pTP-C product was formed, probably caused by initiation on G7 or G1 (KING and VAN DER VLIET 1994).

3.2
Template Strand Movement Is Essential During Jumping Back

During replication, the polymerase stays bound to its DNA template, as it can elongate primers using an M13 template to completion and is able to combine its exonuclease and polymerase activities in a processive manner (KING et al. 1997a). Flexibility in the binding to the template must allow the template to translocate along the catalytic site in order for jumping back to occur. While the template strand has to be retracted relative to the polymerase active center, the pTP-CAT primer still has to present its 3′-OH to the catalytic site to accept the incoming nucleotide. The movement of pTP-CAT relative to the Pol catalytic site is therefore restricted. The essential motion during jumping back is the movement of the template strand relative to the pTP-CAT/Pol complex.

This template strand movement causes a paradox, as pTP/Pol has been shown to bind the core origin (bp 9–18), a conserved sequence feature which is essential to DNA replication of both ss and ds templates (TEMPERLEY and HAY 1992). When the interaction between pTP–Pol and the core origin would remain intact, it would be hard to envisage a movement of the template strand of more than 10 Å relative to the polymerase. The paradox might be solved by dissociation of the pTP–Pol complex during the jump, changing its DNA binding affinity for the core origin and allowing the polymerase to proceed. An alternative explanation could be a flexibility of the pTP–DNA interaction that allows pTP to slide along the template during the early steps of replication. The interaction with specific downstream sequences like the core origin is not needed for elongation of a DNA primer by Pol, so core origin binding is important primarily during the initiation phase of DNA replication. When ds templates were mutated in their core nucleotides and assayed for their ability to support both initiation and elongation, mutations had similar effects in both phases (TEMPERLEY et al. 1991). No elongation specific defects were observed, suggesting that the interaction might be more important before the start of replication, than after initiation.

3.3 Why Jump Back During Genome Replication?

Jumping back has two major advantages: first, it enables the polymerase to correct mistakes during the early phase of replication and second, it allows for an internal start site of the replication of a linear genome.

When the proofreading exonuclease activity of φ29 and PRD1 was assayed using ss oligonucleotides, terminal protein severely inhibited exonucleolytic breakdown by its native DNA polymerase (CALDENTEY et al. 1993; ESTEBAN et al. 1993). This was verified for Ad DNA polymerase. When bound to pTP, the polymerization efficiency of Ad Pol is low, whereas the exonuclease activity is inhibited (KING et al. 1997a, b). By starting replication internally, mistakes in the synthesis of the first nucleotides which are not corrected by the impaired exonuclease activity will be corrected in the next round of replication, because these nucleotides will not serve as a template (see Fig. 3). Mismatched intermediates could in principle fail to base-pair to the genome terminus and thereby lead to abortive replication, although this was not observed when mutating the third nucleotide, inhibiting the intermediate to base-pair at position 3 (KING and VAN DER VLIET 1994). Finally, small deletions of the genome termini can be repaired using the jumping back mechanism (GRAHAM et al. 1989; KING and VAN DER VLIET 1994), providing a safeguard for the integrity of the replication start site.

Apart from being an error-correction mechanism, jumping and sliding back could also be a requirement of starting DNA replication of a linear genome internally. Jumping back will then prevent shortening of this genome at every replication cycle. Protein priming in general may depend on the ability to initiate replication internally, as not only adenoviruses, but also φ29, PRD1 and Cp-1 start replication internally. When DNA or RNA primers are used to initiate DNA replication, DNA polymerases are presented a matched primer–template hybrid to elongate. One could

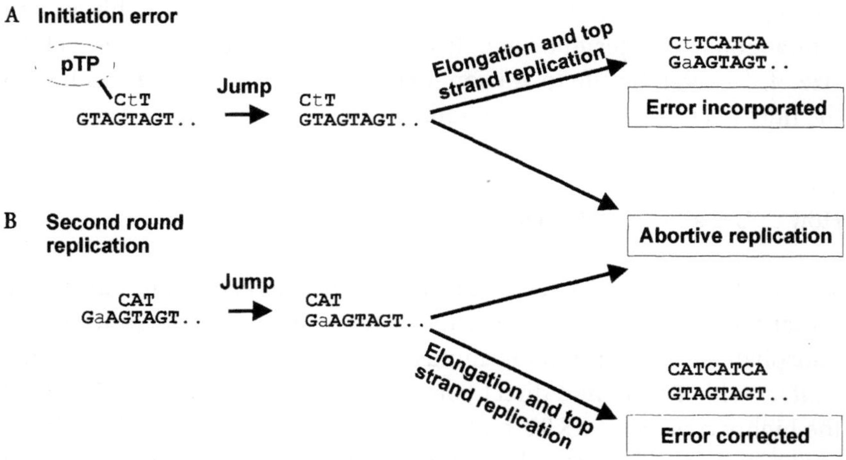

Fig. 3 A, B. Jumping back mechanism corrects initiation errors. pTP is schematically presented for only one initiation event and Pol is omitted for clarity. **A** When an error is made during initiation, it can either result in abortive replication after the jump as it will not base-pair correctly with the template or it will be incorporated into the Ad genome leading to a mutated genome when the new top strand is replicated. **B** When in the second round the bottom strand of the mutated genome is replicated, a correct initiation intermediate is formed that jumps back. This will result in either abortive replication if base-pairing is ineffective or it will result in elongation, thereby correcting the error after replication of the top strand

explain the need for an internal start by assuming that the protein primer has to interact with the preceding unpaired template nucleotide(s) to stably present its serine OH group to the incoming dNTP.

Interacting with a preceding template nucleotide might have yet another advantage. DNA polymerases do not select Watson–Crick basepairs by hydrogen bonding to (Doublie et al. 1998) and/or interacting hydrophobically with the base moieties of the nascent basepair (Franklin et al. 2001). Rather, this new basepair is held in position by numerous interactions of the polymerase with the phosphates of the incoming dNTP, the sugar of the dNTP and the phosphate backbone of the pairing template nucleotide. Moreover, the preceding primer–template basepair provides a stacking surface to the nascent basepair, further confining the dimensions of the active site to harbor only Watson–Crick basepairs. During protein-primed initiation, the preceding basepair is absent, but pTP has to present its serine to the active site like a DNA primer. Interestingly, the priming serine of many Ad pTPs is preceded by an asparagine or histidine that could partially substitute for the lost stacking interactions. In this way, an interaction with 3′ terminal template nucleotides could help to restrict the size of the catalytic site to Watson–Crick basepair dimensions, adding to the fidelity of the initiation reaction.

3.4 How to Use a Protein Primer

When pTP presents its serine to the polymerase active site during the initiation reaction, it will come close to the template strand. In RB69, the phosphate of a DNA primer terminus is held in place by a water-mediated contact with Tyr619 and direct contacts with Lys726 and with Tyr708 of the highly conserved KKRY motif (Franklin et al. 2001). RB69 residues Lys705 and Arg707 contact the template strand, while the conserved Lys706 and Tyr708 bind the primer. This draws the two strands together and forces the DNA to adopt a B-DNA structure. In almost all protein-primed DNA polymerases however, residues corresponding to Lys705 and Arg707 are absent (Blasco et al. 1995), suggesting that the use of a protein primer prohibits a compact primer–template hybrid during initiation. Mutation of Tyr1081 (Tyr708 of RB69) in Ad Pol resulted in a strong reduction of pTP interaction, initiation activity and DNA binding (Liu et al. 2000). When in φ29 this tyrosine was mutated to a serine, DNA polymerization was much more affected than initiation activity, in agreement with the notion that this region is involved in primer–template binding in both DNA- and protein-primed polymerases (Blasco et al. 1995).

The backbone contacts are normally supplemented by stacking on top of the penultimate basepair and hydrogen bonding to its template pairing nucleotide. The primer 3′-OH group should be liganded by one of the two metals in order to lower its affinity for the leaving hydrogen, facilitating its attack on the dNTP α-phosphate. In this framework of potentially 'lost' interactions, the pTP Ser580-OH still has to be positioned correctly. We consider it highly likely that interactions with the template strand nucleotides will contribute to proper positioning of pTP Ser580 in the polymerase active site (see Fig. 1). In agreement with this, initiation by bacteriophage PRD1 and Cp-1 polymerase is severely inhibited by mutations preceding the initiation start site (Caldentey et al. 1993; Martin et al. 1996).

As we discussed before, in DNA primed polymerization the preceding basepair may contribute to the size exclusion of non Watson–Crick basepairs by the active site structure. Moreover, preceding template (and primer) bases are proofread by polymerase contacts with universal minor groove hydrogen bond acceptors (FRANKLIN et al. 2001). Both types of interactions will be different during protein priming. It is interesting to note that the exonuclease activity of Pol is inhibited in the presence of pTP (KING et al. 1997a). This could prevent exonucleolytic degradation of the primer initiation product caused by non-Watson–Crick base-pairing of template nucleotides downstream of the active site. Exonuclease activity could be inhibited because the protein primer Ser580 region is too large to enter the exonuclease active site. Alternatively, pTP could sterically block entry to this active site by binding the exonuclease domain. Finally, pTP might restrict the movement of the thumb needed to transfer the primer strand from the polymerase to the exonuclease active site by binding in the dsDNA binding channel of the DNA polymerase, where the unpaired template strand is located.

4
The Mechanism of Jumping Back

A model of the mechanism of jumping back is illustrated in Fig. 4. Ad Pol is drawn schematically analogous to the structural data of RB69, a member of the eukaryotic DNA Pol α-like family to which Ad Pol belongs (FRANKLIN et al. 2001). RB69 DNA polymerase was the first eukaryotic DNA polymerase of which the crystal structures of the apo-enzyme, as well as the editing- and replicating complexes were solved. Three other structures of the Pol α family DNA polymerases display a high degree of structural similarity (HOPFNER et al. 1999; RODRIGUEZ et al. 2000; ZHAO et al. 1999), which makes it likely that Ad Pol adopts a similar structure. All polymerases of which the structure is known at present show a common architecture that resembles the shape of a right hand, with the palm, thumb and finger domains (STEITZ 1999).

In Fig. 4, the palm domain containing the polymerase active site and the thumb domain are shown. The finger domain lies to the back of the figure and is omitted for clarity. The exonuclease domain with its active site, and the N-terminal domain are indicated as well. Although not conserved, the presence of a separate N-terminal domain in Ad Pol will be discussed below. Initiation starts opposite the fourth nucleotide of the template strand of the origin of replication. Based on the ternary structure of the replicating complex of RB69 DNA polymerase, it was recently suggested

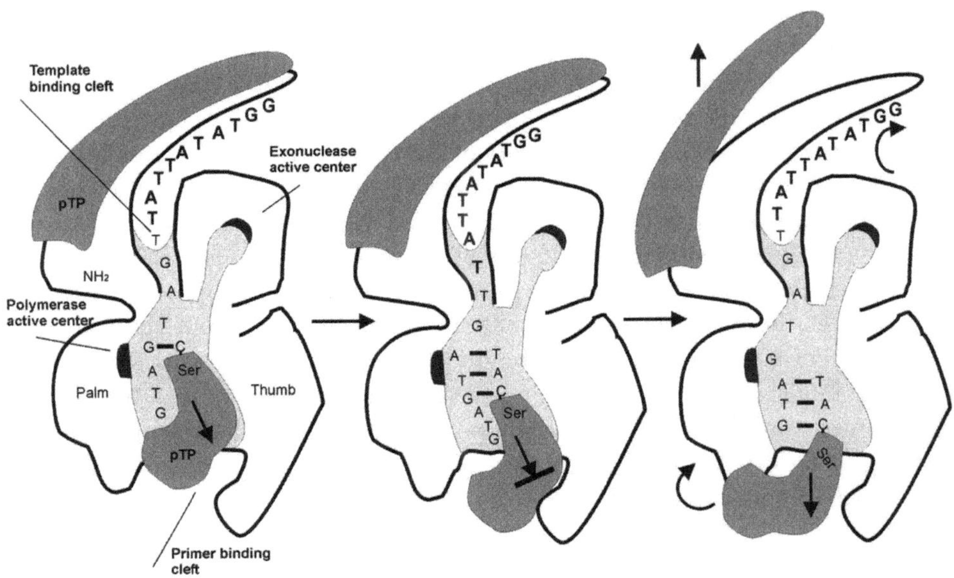

A B C

Fig. 4 A–C. A schematic model of the jumping back mechanism. The Ad polymerase is presented as a *cloverleaf* following the structural data of the ternary complex of RB69 DNA polymerase (FRANKLIN et al. 2001). The *lower part* of the *cloverleaf* shows the polymerase domain with the palm containing the polymerase active site, and the thumb. The fingers are omitted for clarity. The exonuclease domain containing its active site and a putative N-terminal domain (indicated as *NH₂*) are indicated at the *upper part* of the *cloverleaf*. pTP is displayed as two *gray* modules. One of the pTP modules binds the primer binding cleft, locating the priming serine hydroxyl group (*Ser*) at the Pol active site. The other module together with the N-terminal Pol domain contacts the core origin and lies close to the template binding cleft. The sequence of the core origin is marked in *bold*. *Arrows* mark the direction of DNA or protein movement. **A** After pre-initiation complex formation, pTP-C is formed. The primer will move the distance of one nucleotide out of the Pol active site. **B** When pTP-CAT is formed, the region of pTP fixed to Pol forms a roadblock and pTP-CAT cannot leave the Pol active site. **C** Repositioning of pTP releases the roadblock and results in dissociation of the pTP region fixed to Pol. The template strand has jumped back the distance of three bases and pTP-CAT has re-annealed with template strand residues 1–3. The resulting hybrid can move out of the Pol active site for subsequent elongation

that the template enters at a template binding cleft located between the N-terminal and exonuclease domains, as indicated in Fig. 4. Several biochemical studies of e.g. T4 DNA polymerase (Munn and Alberts 1991) and phage 29 DNA polymerase (de Vega et al. 1999) have shown that these polymerases cover a region of 10 bp, with a distance of 4 bases from the entrance of the template binding cleft to the polymerase active site. Therefore, during pre-initiation complex formation, the polymerase likely spans at least seven template bases. In addition, foot printing analysis of pTP, Ad Pol and the pTP–Pol complex on origin DNA demonstrated specific binding of the pTP–Pol complex to the core origin basepairs 9–18 (Mul and van der Vliet 1992; Temperley and Hay 1992). As the left border of the core origin would be located just before the entrance of the template binding cleft during initiation, it is possible that the N terminus of Ad Pol folds into a domain that binds to the core origin sequence as indicated in Fig. 4. Indeed, there are about 250 aa located N terminally that contain a putative Zn finger motif. Point and linker-insertion mutants in this motif retained DNA elongation, but had severely reduced initiation activity and lost the ability to bind the Ad core origin DNA, essential for Ad DNA replication (Joung and Engler 1992).

The protein primer for DNA replication, pTP, is indicated in Fig. 4 at the entrance of the primer binding cleft presenting the priming 3′ hydroxyl group of Ser580 to the polymerase active site. Part of the pTP is also shown at the entrance of the template binding cleft, contacting the core origin.

After pre-initiation complex formation, replication initiates with the covalent coupling of dCTP to pTP (Fig. 4A). After incorporation of the first and second dNMP residues, the protein primer and the template strand have to move one nucleotide out of the polymerase active site in order to make room for incorporation of the next nucleotide. However, after the incorporation of the third nucleotide (Fig. 4B), the template strand jumps back to base-pair with pTP-CAT on template residues 1–3 (Fig. 4C). The sequence of these events is at present unclear. We propose that the pTP region located at the primer binding cleft becomes strained by the movement of the primer–template after each nucleotide incorporation. At the formation of pTP-CAT, further elongation is blocked by the presence of pTP, which is fixed to the polymerase. This block allows the polymerase to melt three bases, after which the strain on pTP could be released by the repositioning of pTP which leads to dissociation. The template jumps back either actively or passively and pTP starts to dissociate from the polymerase occupying the space now available in the primer binding cleft.

Guided by DBP, the polymerase now further copies its genome with high fidelity and processivity.

4.1
pTP Might Trigger the Jump by Acting as a Lever

Instead of incorporating a fourth nucleotide, the template jumps back. To achieve this, an event must take place within the pTP–Pol complex that determines if and when the jump has to occur. During DNA primed replication, the newly formed duplex product binds in a groove formed by the palm and thumb domains, making contacts with the protein over one full turn of the DNA helix (Franklin et al. 2001). After incorporation of a correctly base-paired nucleotide, the newly synthesized duplex DNA moves one basepair out of the polymerase active site. This movement is not only a vertical translocation but a helical rotation as well. In protein-primed initiation, pTP probably moves in a similar fashion upon each nucleotide incorporated although pTP remains bound to the polymerase (King et al. 1997b). While pTP interacts over a large surface of the polymerase during initiation (Parker et al. 1998), a region must be present at the primer binding cleft to provide the hydroxyl group of Ser580 to the polymerase active site as discussed previously. pTP-CAT movement is thus restricted as it interacts stably with the polymerase. We therefore presume that there is a flexible region of pTP that binds in the primer binding cleft.

After formation of pTP-CAT, not only the template strand jumps back, but concomitantly the pTP–Pol complex starts to dissociate (King et al. 1997b; King and van der Vliet 1994). pTP might therefore trigger the jump and dissociation by acting as a lever. From each primer–template movement (both vertical and helical) strain will result in the pTP region that binds in the primer binding cleft while the rest of pTP remains fixed to the polymerase. When pTP-CAT is formed, the strain on pTP is maximal and is released by jumping back of the template strand and simultaneous dissociation of pTP, permitting further primer–template elongation. Alternatively, the jumping back mechanism could be an intrinsic property of the DNA polymerase that can only be observed when unwinding of the primer–template hybrid is energetically inexpensive. Since a self-imposed block would probably affect the processivity when encountering repeats that cannot be jumped, we favor the first explanation.

Early studies on the pTP–Pol complex demonstrated a clear difference in the kinetics of initiation and elongation. During initiation, the K_m for nucleotides is lower than during elongation (Mul and van der Vliet 1993). Moreover, initiation is resistant to ddNTPs and aphidicolin, whereas elongation is partially sensitive (Lichy et al. 1981). These results suggest that a major change within the pTP–Pol complex takes place upon the transition from initiation to elongation. Further experimental evidence for such a change comes from studies on the dissociation of the pTP–Pol complex. When an oligonucleotide template was used which cannot jump back because it lacks nucleotides 1–3 (oligonucleotide Δ3G7C, Table 1), about 60% of pTP dissociated from Pol after the formation of pTP-CAT, indicating a change within the pTP–Pol complex (King et al. 1997b). In addition, this result suggested that jumping back and dissociation are separate processes.

We consider it likely that the changes observed after pTP-CAT formation are a consequence of reorientation of pTP as described above. This would not only enable the polymerase to processively elongate the entire genome without inhibition of bound pTP (King et al. 1997b) but would also prevent another jump after the synthesis of pTP-CATCAT. We further propose that the size of the pTP intermediate is a determining factor for triggering dissociation and jumping back. Interestingly, when a GAGAGAGA repeat (Table 1) was used as template, initiation of replication started at G5. Instead of making a dinucleotide and jumping back twice, a pTP-tetranucleotide was synthesized (King et al. 1997b). This result is in agreement with the idea that a pTP-intermediate of a certain length is required to release the pTP lever.

4.2 How Does the Template Strand Move?

During the jump, three Watson–Crick basepairs are disrupted and the template strand moves. Members of the Pol α family of DNA polymerases have an intrinsic capacity to destabilize the dsDNA necessary for their editing function (Shamoo and Steitz 1999) and this may play a role during jumping back. The tertiary structure of the editing and replicating complexes of RB69 DNA polymerase show that when a mismatch is incorporated or when a primer-template remains too long at the polymerase active site without replicating, three basepairs are unwound and the primer strand is transferred to the exonuclease active site (Franklin et al. 2001; Shamoo and Steitz 1999). Although exonuclease activity is strongly decreased when pTP is complexed with the polymerase (King et al.

1997a), the capacity to melt three bases might still be intact. A fourth nucleotide cannot be incorporated into pTP-CAT and elongation is blocked. This may give the polymerase the opportunity to melt three bases similar to an exonucleolytic event. Instead of locating the primer strand at the exonuclease active site, the template strand jumps three bases and re-anneals with the primer at template residues 1–3. The template strand is contacted by the polymerase by multiple hydrogen bonds and charge–charge interactions (Franklin et al. 2001). At the polymerase active site, the I/YXGG motif in Ad DNA polymerase has recently been demonstrated to stabilize the template strand (Brenkman et al. 2001), in agreement with the ternary structure of the replicating RB69 DNA polymerase (Franklin et al. 2001). Mutation of both glycines to alanines (GG666/7AA) inhibited the transition from initiation to elongation, illustrating that template stabilization is important for jumping back (Brenkman et al. 2001).

When the template strand unwinds, it could move either simultaneously or just after pTP dissociation. To dissociate from the polymerase, pTP might be sterically hindered by the 3′-template overhang located at the primer–template binding groove. It is therefore possible that the tension released from pTP after formation of pTP-CAT, actively or passively pushes the template back the required distance. This could mean that the 3′-template overhang is an additional determining factor for the initiation start site and the jump of the template. Dissociation of pTP then allows it to leave the primer binding groove, explaining why the polymerase does not stall again after nucleotide 6.

4.3 Experimental Evidence for this Mechanism

Replication of template Δ3G7C (Table 1) resulted in dissociation after pTP-CAT formation in the absence of jumping back (described above; King et al. 1997b). Although this experiment suggests that dissociation of pTP and jumping back are independent processes, this template has no 3′-overhang and initiation is restricted to residue G1. When pTP-CAT is formed, reorientation of pTP within the pTP–Pol complex could therefore be facilitated as more space is available at the primer–template binding cleft, due to the lack of a 3′-template overhang. Dissociation of pTP then starts, without the need to jump back. In agreement with this, when Ad origin templates were used with deletions of one or two terminal 3′-

template residues, jumping back proceeds and recovers the terminal deletions (KING and VAN DER VLIET 1994). However, the major product of the Δ2 template (Table 1) results from read-through without jumping back, suggesting that there is enough space for pTP to reorientate without movement of the template strand. When a template was mutated at position G4 to A (3′-GTAATAG, Table 1), pTP-C formation started on G7, providing a long 3′-template overhang of six nucleotides. Remarkably, no further elongation was permitted in the presence of dATP, dCTP and dTTP, indicating a block in pTP movement possibly due to limited space available at the primer–template binding cleft. It is furthermore interesting to note that initiation of Ad DNA replication cannot start at an internal origin but needs to be linearized unless head-to-tail origins are provided (GRAHAM et al. 1989).

In conclusion, besides the recruitment of the pTP–Pol complex by the cellular replication proteins to the core origin, the preference for binding termini with a distinct length may provide an additional prerequisite to direct the exact initiation starting position.

Acknowledgements. We thank B. van Breukelen for Fig. 1 and for unpublished information. This work was supported in part by the Netherlands Organization for Scientific Research and by European Union Contract FMRX-CT97-0125.

References

VAN AMERONGEN H, VAN GRONDELLE R and VAN DER VLIET PC (1987) Interaction between adenovirus DNA-binding protein and single-stranded polynucleotides studied by circular dichroism and ultraviolet absorption. Biochemistry 26:4646–4652

BLASCO MA, MENDEZ J, LAZARO JM, BLANCO L and SALAS M (1995) Primer terminus stabilization at the phi 29 DNA polymerase active site. Mutational analysis of conserved motif KxY. J Biol Chem 270:2735–2740

BRENKMAN AB, HEIDEMAN MR, TRUNIGER V, SALAS M and DER VLIET PC (2001) The (I/Y)xGG motif of adenovirus dna polymerase affects template DNA binding and the transition from initiation to elongation. J Biol Chem 276:29846–29853

VAN BREUKELEN B, KANELLOPOULOS PN, TUCKER PA and DER VLIET PC (2000) The formation of a flexible DNA-binding protein chain is required for efficient DNA unwinding and adenovirus DNA chain elongation. J Biol Chem 275:40897–40903

CALDENTEY J, BLANCO L, BAMFORD DH and SALAS M (1993) In vitro replication of bacteriophage PRD1 DNA. Characterization of the protein-primed initiation site. Nucleic Acids Res 21:3725–3730

CHALLBERG MD and KELLY TJJ (1979) Adenovirus DNA replication in vitro. Proc Natl Acad Sci USA 76:655–659

CLEAT PH and HAY RT (1989) Co-operative interactions between NFI and the adenovirus DNA binding protein at the adenovirus origin of replication. EMBO J 8: 1841–1848

CLEGHON V and KLESSIG DF (1992) Characterization of the nucleic acid binding region of adenovirus DNA binding protein by partial proteolysis and photochemical cross-linking. J Biol Chem 267:17872–17881

DE JONG RN and VAN DER VLIET PC (1999) Mechanism of DNA replication in eukaryotic cells: cellular host factors stimulating adenovirus DNA replication. Gene 236:1–12

DE VEGA M, BLANCO L AND SALAS M (1999) Processive proofreading and the spatial relationship between polymerase and exonuclease active sites of bacteriophage phi29 DNA polymerase. J Mol Biol 292: 39–51

DEKKER J, KANELLOPOULOS PN, LOONSTRA AK, VAN OOSTERHOUT JA, LEONARD K, TUCKER PA and VAN DER VLIET PC (1997) Multimerization of the adenovirus DNA-binding protein is the driving force for ATP-independent DNA unwinding during strand displacement synthesis. EMBO J 16:1455–1463

DEKKER J, KANELLOPOULOS PN, VAN OOSTERHOUT JA, STIER G, TUCKER PA and VAN DER VLIET PC (1998) ATP-independent DNA unwinding by the adenovirus single-stranded DNA binding protein requires a flexible DNA binding loop. J Mol Biol 277: 825–838

DOUBLIE S, TABOR S, LONG AM, RICHARDSON CC and ELLENBERGER T (1998) Crystal structure of a bacteriophage T7 DNA replication complex at 2.2 A resolution. Nature 391: 251–258

ENOMOTO T, LICHY JH, IKEDA JE and HURWITZ J (1981) Adenovirus DNA replication in vitro: purification of the terminal protein in a functional form. Proc Natl Acad Sci USA 78: 6779–6783

ESTEBAN JA, SALAS M and BLANCO L (1993) Fidelity of phi 29 DNA polymerase. Comparison between protein-primed initiation and DNA polymerization. J Biol Chem 268:2719–2726

FIELD J, GRONOSTAJSKI RM and HURWITZ J (1984) Properties of the adenovirus DNA polymerase. J Biol Chem 259:9487–9495

FRANKLIN MC, WANG J and STEITZ TA (2001) Structure of the replicating complex of a pol alpha family DNA polymerase. Cell 105:657–667

GEORGAKI A, STRACK B, PODUST V and HUBSCHER U (1992) DNA unwinding activity of replication protein A. FEBS Lett 308:240–244

GRAHAM FL, RUDY J and BRINKLEY P (1989) Infectious circular DNA of human adenovirus type 5: regeneration of viral DNA termini from molecules lacking terminal sequences. EMBO J 8:2077–2085

HAY RT, FREEMAN A, LEITH I, MONAGHAN A and WEBSTER A (1995) Molecular interactions during adenovirus DNA replication. Curr Top Microbiol Immunol 199:31–48

HOPFNER KP, EICHINGER A, ENGH RA, LAUE F, ANKENBAUER W, HUBER R and ANGERER B (1999) Crystal structure of a thermostable type B DNA polymerase from Thermococcus gorgonarius. Proc Natl Acad Sci USA 96:3600–3605

ILLANA B, BLANCO L and SALAS M (1996) Functional characterization of the genes coding for the terminal protein and DNA polymerase from bacteriophage GA-1. Evidence for a sliding-back mechanism during protein-primed GA-1 DNA replication. J Mol Biol 264:453–464

ITO J and BRAITHWAITE DK (1991) Compilation and alignment of DNA polymerase sequences. Nucleic Acids Res 19:4045–4057

JOUNG I and ENGLER JA (1992) Mutations in two cysteine-histidine-rich clusters in adenovirus type 2 DNA polymerase affect DNA binding. J Virol 66:5788–5796

KANELLOPOULOS PN, TSERNOGLOU D, VAN DER VLIET PC and TUCKER PA (1996) Alternative arrangements of the protein chain are possible for the adenovirus single-stranded DNA binding protein. J Mol Biol 257:1–8

KANELLOPOULOS PN, VAN DER ZANDT H, TSERNOGLOU D, VAN DER VLIET PC and TUCKER PA (1995) Crystallization and preliminary X-ray crystallographic studies on the adenovirus ssDNA binding protein in complex with ssDNA. J Struct Biol 115:113–116

KENNY MK and HURWITZ J (1988) Initiation of adenovirus DNA replication. II. Structural requirements using synthetic oligonucleotide adenovirus templates. J Biol Chem 263:9809–9817

KING AJ and VAN DER VLIET PC (1994) A precursor terminal protein-trinucleotide intermediate during initiation of adenovirus DNA replication: regeneration of molecular ends in vitro by a jumping back mechanism. EMBO J 13:5786–5792

KING AJ, TEERTSTRA WR, BLANCO L, SALAS M and VAN DER VLIET PC (1997a) Processive proofreading by the adenovirus DNA polymerase. Association with the priming protein reduces exonucleolytic degradation. Nucleic Acids Res 25:1745–1752

KING AJ, TEERTSTRA WR and VAN DER VLIET PC (1997b) Dissociation of the protein primer and DNA polymerase after initiation of adenovirus DNA replication. J Biol Chem 272:24617–24623

KITAMURA N, ADLER CJ, ROTHBERG PG, MARTINKO J, NATHENSON SG and WIMMER E (1980) The genome-linked protein of picornaviruses. VII. Genetic mapping of poliovirus VPg by protein and RNA sequence studies. Cell 21:295–302

LICHY JH, HORWITZ MS and HURWITZ J (1981) Formation of a covalent complex between the 80,000-dalton adenovirus terminal protein and 5′-dCMP in vitro. Proc Natl Acad Sci USA 78:2678–2682

LINDENBAUM JO, FIELD J and HURWITZ J (1986) The adenovirus DNA binding protein and adenovirus DNA polymerase interact to catalyze elongation of primed DNA templates. J Biol Chem 261: 10218–10227

LIU H, NAISMITH JH and HAY RT (2000) Identification of conserved residues contributing to the activities of adenovirus DNA polymerase. J Virol 74:11681–11689

MARTIN AC, BLANCO L, GARCIA P, SALAS M and MENDEZ J (1996) In vitro protein-primed initiation of pneumococcal phage Cp-1 DNA replication occurs at the third 3′ nucleotide of the linear template: a stepwise sliding-back mechanism. J Mol Biol 260: 369–377

MENDEZ J, BLANCO L, ESTEBAN JA, BERNAD A and SALAS M (1992) Initiation of phi 29 DNA replication occurs at the second 3′ nucleotide of the linear template: a sliding-back mechanism for protein-primed DNA replication. Proc Natl Acad Sci USA 89: 9579–9583

MONAGHAN A, WEBSTER A and HAY RT (1994) Adenovirus DNA binding protein: helix destabilising properties. Nucleic Acids Res 22:742–748

MUL YM and VAN DER VLIET PC (1992) Nuclear factor I enhances adenovirus DNA replication by increasing the stability of a preinitiation complex. EMBO J 11:751–760

Mul YM and van der Vliet PC (1993) The adenovirus DNA binding protein effects the kinetics of DNA replication by a mechanism distinct from NFI or Oct-1. Nucleic Acids Res 21:641–647

Munn MM and Alberts BM (1991) DNA footprinting studies of the complex formed by the T4 DNA polymerase holoenzyme at a primer-template junction. J Biol Chem 266: 20034—20044

Nagata K, Guggenheimer RA, Enomoto T, Lichy JH and Hurwitz J (1982) Adenovirus DNA replication in vitro: identification of a host factor that stimulates synthesis of the preterminal protein-dCMP complex. Proc Natl Acad Sci USA 79:6438–6442

Nagata K, Guggenheimer RA and Hurwitz J (1983) Adenovirus DNA replication in vitro: synthesis of full-length DNA with purified proteins. Proc Natl Acad Sci USA 80: 4266–4270

Parker EJ, Botting CH, Webster A and Hay RT (1998) Adenovirus DNA polymerase: domain organisation and interaction with preterminal protein. Nucleic Acids Res 26:1240–1247

Paul AV, van Boom JH, Filippov D and Wimmer E (1998) Protein-primed RNA synthesis by purified poliovirus RNA polymerase. Nature 393:280–284

Paul AV, Rieder E, Kim DW, van Boom JH and Wimmer E (2000) Identification of an RNA hairpin in poliovirus RNA that serves as the primary template in the in vitro uridylylation of VPg. J Virol 74:10359–10370

Pronk R and van der Vliet PC (1993) The adenovirus terminal protein influences binding of replication proteins and changes the origin structure. Nucleic Acids Res 21:2293–2300

Pruijn GJ, Van Driel W and van der Vliet PC (1986) Nuclear factor III, a novel sequence-specific DNA-binding protein from HeLa cells stimulating adenovirus DNA replication. Nature 322:656–659

Rieder E, Paul AV, Kim DW, van Boom JH and Wimmer E (2000) Genetic and biochemical studies of poliovirus cis-acting replication element cre in relation to VPg uridylylation. J Virol 74:10371–10380

Rodriguez AC, Park HW, Mao C and Beese LS (2000) Crystal structure of a pol alpha family DNA polymerase from the hyperthermophilic archaeon Thermococcus sp. 9 degrees N-7. J Mol Biol 299:447–462

Salas M (1991) Protein-priming of DNA replication. Annu Rev Biochem 60:39–71

Salas M, Miller JT, Leis J and DePamphilis ML (1996) Mechanisms of priming DNA synthesis. In: DePamphilis ML (eds) DNA Replication in Eukaryotic Cells. Cold Spring Harbor Laboratory Press, 131–176

Shamoo Y and Steitz TA (1999) Building a replisome from interacting pieces: sliding clamp complexed to a peptide from DNA polymerase and a polymerase editing complex. Cell 99:155–166

Steitz TA (1999) DNA polymerases: structural diversity and common mechanisms. J Biol Chem 274:17395–17398

Stuiver MH and van der Vliet PC (1990) Adenovirus DNA-binding protein forms a multimeric protein complex with double-stranded DNA and enhances binding of nuclear factor I. J Virol 64:379–386

Stuiver MH, Bergsma WG, Arnberg AC, van Amerongen H, van Grondelle R and van der Vliet PC (1992) Structural alterations of double-stranded DNA in complex with the adenovirus DNA-binding protein. Implications for its function in DNA replication. J Mol Biol 225:999–1011

SWAMINATHAN S and THIMMAPAYA B (1995) Regulation of adenovirus E2 transcription unit. Curr Top Microbiol Immunol 199(Pt 3):177–194

TEMPERLEY SM and HAY RT (1992) Recognition of the adenovirus type 2 origin of DNA replication by the virally encoded DNA polymerase and preterminal proteins. EMBO J 11:761–768

TEMPERLEY SM, BURROW CR, KELLY TJ and HAY RT (1991) Identification of two distinct regions within the adenovirus minimal origin of replication that are required for adenovirus type 4 DNA replication in vitro. J Virol 65: 5037–5044

TUCKER PA, TSERNOGLOU D, TUCKER AD, COENJAERTS FE, LEENDERS H and VAN DER VLIET PC (1994) Crystal structure of the adenovirus DNA binding protein reveals a hook- on model for cooperative DNA binding. EMBO J 13:2994–3002

VAN DER VLIET PC (1995) Adenovirus DNA replication. Curr Top Microbiol Immunol 199:1–30

WARD P, DEAN FB, O'DONNELL ME and BERNS KI (1998) Role of the adenovirus DNA-binding protein in in vitro adeno-associated virus DNA replication. J Virol 72: 420–427

ZHAO Y, JERUZALMI D, MOAREFI I, LEIGHTON L, LASKEN R and KURIYAN J (1999) Crystal structure of an archaebacterial DNA polymerase. Structure Fold Des 7: 1189–1199

ZIJDERVELD DC, STUIVER MH and VAN DER VLIET PC (1993) The adenovirus DNA binding protein enhances intermolecular DNA renaturation but inhibits intramolecular DNA renaturation. Nucleic Acids Res 21:2591–2598

ZIJDERVELD DC and VAN DER VLIET PC (1994) Helix-destabilizing properties of the adenovirus DNA-binding protein. J Virol 68:1158–1164

Chapter 8

The Structure and Function of the Adenovirus Major Late Promoter

C. S. H. Young

Abstract. The adenovirus major late promoter (MLP) has played a preeminent role in the analysis of transcription initiation in mammalian cells, and is an outstanding example of the ways in which the study of adenovirus has led to fundamental insights into general cellular processes. The aim of this chapter is to give a comprehensive review of the structure and function of this model mammalian promoter. After a brief description of late transcription in the adenovirus replication cycle, the experimental evidence for the current consensus on the genetic structure of the MLP,

C. S. H. Young
Columbia University, College of Physicians and Surgeons, Department of Microbiology, 701W 168th Street, New York, NY 10032, USA
e-mail: csyl@columbia.edu

including a consideration of non-primate adenovirus MLPs, will be reviewed. Next, the functions of the MLP in the viral life cycle will be examined, and some of the problems that remain to be resolved will be addressed. The review ends with some ideas on how the knowledge of the structure and function of the MLP can be used in designing virus vectors for specific experimental purposes.

1 Introduction

The analysis of adenovirus-infected cells has led to fundamental insights into the mechanisms of RNA transcription at every step from the control of initiation, to transcript processing, to export from the nucleus. Among the most important discoveries in the late 1970s was that the adenovirus major late promoter (MLP) could be transcribed accurately and efficiently in cell-free extracts (Weil et al. 1979; Manley et al. 1980). This led the way not only to the definition of the DNA elements necessary for such accurate transcription, but more importantly to the fractionation and eventual cloning of the protein factors necessary for the formation of the pre-initiation complex, the so-called general transcription factors, and the definition at the atomic level of the binding of several of these factors to the DNA template. In addition, the MLP was among the first mammalian promoters in which the identification of upstream activating elements and the isolation of transcription factors that bind to them was achieved. This distinguished role in the development of our understanding of the mechanisms of transcription has made the MLP the model for mammalian promoters.

2 The Major Late Transcription Unit

Early results with KB cells infected by adenovirus serotype 2 (Ad2) showed that a new class of RNA was expressed after DNA replication had begun (Lucas and Ginsberg 1971). Some early RNAs continued to be synthesized, while others declined in abundance. Characterization of both early and late RNAs by an extensive array of hybridization techniques and by electron microscopy led to the amazing discovery of RNA splicing (Berget et al. 1977; Chow et al. 1977) and to the now familiar map of the transcriptional organization of the subgroup C adenovirus genome

(BROKER 1984). Reviews of the overall transcription program in adenovirus can be found in SHARP (1984), BERK (1986) and SHENK (2001), all of which have extensive references to the earlier literature. Subsequent analyses of other human subgroup adenoviruses and an ever-increasing number of viruses from a wide variety of non-primate mammalian and avian species have confirmed that the genome organization of early and late genes is largely conserved (see chapter by BENKÖ, this volume, and Sect. 4). Prominent in all adenoviruses is the major late transcription unit, controlled by the MLP. This transcription unit initiates at a site about 16.8 map units from the conventional left hand end of the genome in the human subgroup C viruses, and in equivalent positions in others. Although the MLP is active at low levels early in infection, yielding the mRNA for the L1 52/55-kDa polypeptide, after DNA replication begins, transcription continues almost to the right hand terminus. Processing of this large transcript at five polyadenylation sites and at several different splice sites leads to the production of five families of mRNAs (see SHAW and ZIFF 1980 and references cited therein) encoding all but one of the structural polypeptides of the virus, and two of the three late non-structural polypeptides. Thus proper functioning of the MLP is critical to the successful completion of the infectious cycle, and an understanding of the mechanisms and elements contributing to its function are of considerable interest to the virologist as well as to the student of transcription.

3
Defining the Elements That Control MLP Function

In this section, a brief historical summary will be given of how the current view of the organization of the MLP was obtained, and then the structure and function of each element will be considered in more detail. Early work using micro-injection and transient transfection assays, and a small set of deletion mutations, showed that the MLP had a relatively simple structure with a canonical TATA box and an upstream element(s) both of which were necessary for quantitative and accurate transcription (HEN et al. 1982). The results were broadly congruent with results obtained in vitro using whole cell extracts and plasmid templates (HU and MANLEY 1981). More detailed mutational analysis, again using model plasmid constructs and in vitro transcription assays (CONCINO et al. 1984a; YU and MANLEY 1984), confirmed and extended these observations so that by the mid-1980s a fairly clear picture of the MLP had emerged, namely that the TATA

CAAT box UPE MAZ TATA box MAZ/SP1
G**TGATTGGTTT**GTAGGTGTAG**GCCACGTGAC**CGGGTGTTCCT**GAAGGGGGGCTATAAAAGGGGGTGGGGG**CGCG
-80• -70• -60• -50• -40• -30• -20• -10•

INR
TTCGTCC**TCACTCT**CTTCCGCATCGCTGTCTGCGAGGGCCAGCTGTTGGGGTGAGTACTCCCTCTGAAAAGCGGG
>————————————————————
+1• +10• +20• +30• +40• +50• +60•

DE1 DE2b DE2a
CATGACTTCTGCGCTAAGA**TTGTCAGTTTC**CAAA**AACGAGGAGGATTTGA**TATTCACCTGGC
+70• +80• +90• +100• +110• +120•

Fig. 1. Transcription elements of the MLP of human subgroup C adenoviruses. The DNA sequence elements to which known transcription factors bind are shown in *bold*, and the names of the elements are given *above* the sequence. The *bold* sequence encompasses the consensus (CAAT box, TATA box, and Inr) or the footprints (MAZ binding sites, Sp1 binding site, UPE, DE1, and DE2b/a). The start site of transcription is shown as an *arrowhead* below the sequence and the *line* shows the extent of the first exon, which is 41 nucleotides long. The *numbers* below the sequence are relative to the start site at +1

box played a crucial role and that upstream sequences played an important enhancing role in transcription initiation. However, it soon became apparent that the MLP was more complex than this. Both in vitro transcriptional analysis and transfection assays showed that sequences surrounding the initiation site (Concino et al. 1984b; Lee et al. 1988) and within the first intron (Mansour et al. 1986; Jansen-Durr et al. 1988, 1989; Leong et al. 1990) were important to the function of the MLP, and subsequently, it was shown that the GC-rich regions prominent both immediately upstream and downstream of the TATA box played a role in ensuring high levels of transcription (Brunet et al. 1987; Parks and Shenk 1997). In addition, analysis of cellular factors binding to two different upstream elements showed that two ubiquitous proteins were important to transcription (Sawadogo and Roeder 1985; Gregor et al. 1990; Sirito et al. 1994; Carthew et al. 1985; Miyamoto et al. 1985; Chodosh et al. 1988a). Subsequent genetic analysis of mutations placed in the MLP in the correct location in the viral genome suggested that the two upstream elements acted in a functionally redundant manner (Reach et al. 1991; Sect. 3.2). Taken together, these results led to the current view of the organization of the MLP (Fig. 1). It is a text-book picture: a 'core' promoter consisting of a TATA box and an initiator element (Inr), upstream activating elements (the inverted CAAT box and upstream promoter element, UPE) and two adjacent elements downstream (DE1 and DE2b/a) that recognize the virus-specified IVa2 protein in a late phase-specific transcription-enhancing mechanism. The cellular proteins that bind to these various elements have been the subjects of intense investigation. Most have been identified, cloned and sequenced, and it seems probable that their roles in MLP function have been correctly deduced (although see later for more discussion on this point).

3.1 The 'Core' Promoter

The TATA Box. The MLP's TATA box sequence 5′-TATAAAAG-3′ conforms to the consensus sequence 5′-TATAWADR-3′ derived from 601 unrelated vertebrate promoter sequences (see http://www.epd.isb-sib.ch/promoter_elements). The first T is located 31 nucleotides upstream from one of the most prominent start sites in the major late transcription unit (MLTU) (ZIFF and EVANS 1978). Given this canonical sequence and location, it is not surprising that it plays a fundamental role in the function of the MLP. Indeed its primary importance has been established in a wide variety of assays including in vitro transcription (CONCINO et al. 1984a), transient transfection (LEE et al. 1988), and genetic tests in the correct location in the viral genome (REACH et al. 1991). The last mentioned is perhaps the most stringent test of function, as it measures the ability of the MLP to control transcription under true in vivo conditions in which the relevant transcription factors are present at the correct concentrations and stoichiometry, both with respect to each other and to the DNA template, and they are subjected to any potential modification by viral proteins. In the genetic analysis in the viral genome, the only change in a single promoter element that conferred a markedly reduced transcriptional ability was the double mutation T**G**TA**G**AA present in the virus 'TATA0' (REACH et al. 1991). However, single mutations T**G**TAAAA and TATA**G**AA, which gave rise to viruses 'TATA30' and 'TATA27' respectively, showed very little effect on transcription from the MLP (REACH et al. 1991). Note however that these single mutations, when coupled to mutations in other MLP transcription elements, revealed a deficiency in transcription (REACH et al. 1991), agreeing with previous results in vitro (CONCINO et al. 1984b).

Given its canonical sequence and primary importance in MLP function it is not surprising that the MLP TATA box has been used in determining the co-crystal structure of TATA box-binding protein (TBP) bound to DNA (KIM et al. 1993). The most marked aspect of the interaction between the saddle-shaped molecule of TBP and the DNA is an induced bend in the duplex, an observation confirmed by electron microscopy of MLP-containing templates (Griffith et al. 1995). Co-crystallographic studies with a set of naturally occurring variants in the TATA box sequence (Patikoglou et al. 1999) and subsequent computational analyses of the same set of sequences (Qian et al. 2001) have shown that quite wide divergences in sequence can be accommodated by the complex. One significant exception is the sequence TGTAAAA, which does not bind TBP and thus cannot be co-crystallized (Patikoglou et al. 1999). The computational analysis suggests that this may be because of a failure to undergo bending (Qian et al. 2001). However, as mentioned above, this mutation has no significant effect (by itself) when tested in vivo in the virus (Reach et al. 1991). On the other hand, in vitro assays, and in vivo tests in yeast using the MLP TATA box, showed that the TGTAAAA sequence had a marked transcriptional deficiency (Wobbe and Struhl 1990). These discrepancies between the biochemical and virological results again emphasize the relative robustness of the MLP in responding to mutational change in the natural context. Although the two single point mutations TGTAAAA and TATAGAA have no measurable effects by themselves, in combination with mutations in other elements they show significant transcriptional deficiency, leading to a marked reduction in viral yield or lethality (Reach et al. 1991; Lu et al. 1997). Those double mutations with severe transcriptional deficiencies have allowed the selection of revertants and the analysis of the sites of reversion. In the case of the double mutation-containing virus TATA27::CCCAT, individual revertants arising independently from 11 different viral populations were selected and analyzed. Of the 11, seven had reverted to the wild-type sequence TATAAAA, whereas the other four had gained the non-wild type (but still canonical) TATATAA sequence, and none had reverted within the mutated CAAT box (E. Minaya and H. Young, unpublished results). If we assume that the back mutation rates from G to A or from G to T are similar, this result would suggest that the two canonical sequences are close to being functionally equivalent in the MLP TATA box. In the comparison of the 601 vertebrate TATA boxes A and T at the fifth position (–27 in the MLP TATA box) are found in 68.2% and 30.4% of the cases respectively. A similar analysis of the double mutant virus TATA30::INR, which is transcriptionally deficient (Lu et al. 1997), yielded a single revertant containing a reversion from TGTAAAA to TATAAAA. Transversions to TTTAAAA and TCTAAAA cannot be recovered because they create stop codons in the DNA polymerase encoded on the opposite strand (see Fig. 2). Further use of point mutations in the

TATA box sequence will be discussed in the section concerning the role of inducible MLPs in the construction of adenovirus vectors.

The Initiator (Inr). Defining the Element. First indications that the sequence immediately surrounding the start site of transcription might play a role in MLP activity came from in vitro studies with templates containing mutations at the +1 position, although the effects were minor (CONCINO et al. 1984b). Later in vitro results showed a more marked effect and in addition were supported by transient transfection studies with recombinant plasmids (LEE et al. 1988). In the latter experiments mutations at the start site were as deleterious as those in the TATA box. Additional support for the idea that the sequence around the start site might constitute a transcription element came when it was recognized that the MLP sequence was functionally homologous to that of the newly defined element, termed Inr, present in the TATA-less promoter of the TdT gene (SMALE and BALTIMORE 1989). Both sequences were capable of directing transcription in vitro (SMALE and BALTIMORE 1989). Subsequent analysis of the transcriptional abilities of sequence variants and inspection of the sequences of many suspected Inr elements led to a proposed consensus sequence 5′-YYANWYY-3′ (JAVAHERY et al. 1994; LO and SMALE 1996) to which the Inr of the MLP conforms (5′-TCACTCT-3′). Genetic analysis of mutations in the MLP in its natural context in the virus also showed that the Inr played a role in determining the levels of transcription but, as with the single point mutations in the TATA box described above, this was detectable only when mutations were present in other elements (LU et al. 1997). This was true even in a virus containing multiple point mutations in the Inr so that the sequence diverged significantly from the consensus. In addition, it was possible to recover virus that contained mutations in both the Inr and in the UPE, in contrast to the observation that virus with mutations in the TATA box and in UPE were not recoverable (REACH et al. 1991). Taken together, these observations suggest that the Inr is not as crucial as the TATA box for the functioning of the MLP.

Proteins That Bind to the Inr. Several different proteins capable of binding to the Inr of the MLP have been identified, including CAP sequence DNA-binding protein (GARFINKEL et al. 1990) TFII-I (ROY et al. 1991) and components of TFIID (KAUFMANN et al. 1996). TFII-I was shown to be capable of binding not only to the Inr but also to the UPE, although binding is mediated by different domains of the protein (Roy et al. 1997). Furthermore, protein–protein interactions between TFII-I and the UPE-binding protein USF (see below) have been documented and show synergistic effects on transcription (Roy et al. 1991) suggesting that TFII-I might play a role in the activation of the MLP in addition to the establishment of the pre-initiation complex. Complicating the issue further, it has been shown that USF can also bind the Inr and stimulate transcription (Du et al. 1993). A suggestion has been made that there may be two distinct pathways of pre-initiation complex formation at those promoters containing a TATA box and those containing an Inr (Roy et al. 1993). However, although there is an impressive body of data showing that TFII-I is a transcription factor, there is no direct evidence that it plays a role in the activity of the MLP in vivo. This is a theme that will recur when reviewing the role of the upstream activating elements and the proteins that bind to them. Smale and his colleagues have examined the structure and function of Inr elements, including that from the MLP, in great detail (reviewed by Smale et al. 1998). They have shown that interactions between the Inr and one or more TBP–Associated Factors (TAFs) of the TFIID complex are important to its function (Kaufmann and Smale 1994; Kaufmann et al. 1996, 1998). In vitro, $TAF_{II}150$ is necessary for Inr function, but because it is not closely associated with TFIID and does not bind DNA in a sequence specific fashion, it is likely that it relies on protein–protein interactions with other TAF components such as $TAF_{II}250$ (Kaufmann et al. 1998) that are positioned close to the start site. Synergy between the Inr and the TATA box has been demonstrated in vitro (Concino et al. 1984b) and this synergy depends on the correct spacing between the two elements (Emami et al. 1997). In agreement with this observation, a mutant virus that contains mutations in both elements is transcriptionally deficient (Lu et al. 1997). However, placing a quarter or a half helical turn between the two elements by inserting three or six base pairs has no effect on transcriptional efficiency in vivo unless one or other element also contains a mutation (H. Lu and H. Young, unpublished results), although changes to the position of the start site are observed. Taking all of the in vivo results together, it seems that the MLP can function well without an intact Inr, unless there is some disruption to the function of another promoter element.

GC-Rich Regions Immediately Surrounding the TATA Box. The TATA box of the MLP is surrounded by a remarkably GC-rich sequence, with 7 consecutive G or C residues on the upper strand before the T at –31, and 13 G or C residues after the A at –25 interrupted by a single T residue at –19 (see Fig. 1). This striking asymmetry in base pair usage, first formally noted in a theoretical paper discussing the possible thermodynamic properties of eukaryotic promoters (BENSIMHON et al. 1983), has been investigated both in vitro and in vivo. Initial mutational analysis of the G-rich sequences using transcription assays showed little effect of multiple mutations from G to A in either region (YU and MANLEY 1984), but subsequent genetic analysis of similar mutations in the viral genome suggested that the upstream G-rich segment played a role in full activity of the MLP (BRUNET et al. 1987). Because of the close proximity of the elements to the TATA box, the role of the GC-rich sequences could be to create a specific DNA structure, recognized by the pre-initiation complex. Alternatively, they might simply be recognition sites for activating transcription factors. In support of the former view, the co-crystal structure of the general transcription factors TFIIB and TBP bound to the MLP showed that TFIIB made DNA contact in both GC-rich regions (NIKOLOV et al. 1995) a conclusion confirmed and extended by protein–DNA cross-linking studies (LEE and HAHN 1995; LAGRANGE et al. 1996). Selection of oligonucleotides having random sequences upstream of the TATA box revealed a sequence 5′-SSRCGCC-3′ to which the TBP/TFIIB complex bound with highest affinity (LAGRANGE et al. 1998). This selected sequence, which has a seven out of nine match to the MLP sequence 5′-GGGGGGC-3′, increased the ability of TFIIB to enter the pre-initiation complex and support transcription initiation (LAGRANGE et al. 1998). These observations led to the suggestion that the sequence constituted a core recognition element, the TFIIB recognition element (BRE), and the further speculation that promoter activity might be determined in part by how close the sequence immediately upstream of the TATA box conformed to the BRE consensus. However, later electrophoretic gel-shift analysis showed that the flanking GC-rich elements could bind the ubiquitous cellular transcription factors MAZ and Sp1 (PARKS and SHENK 1997). Furthermore, transient expression of either protein increased the activity of the MLP from both plasmid and viral templates. These data suggest that the GC-rich elements may play an activating role in the expression of the MLP in vivo. We are left with the question of whether these sequences are recognized by TFIIB, as part of the general pre-initiation complex recognition of the promoter DNA, or by transcriptional activators such as MAZ or Sp1. In passing, it should be noted that several non-primate MLPs do not have noticeably GC-rich sequences flanking the MLP TATA box (SONG et al. 1996) so the general importance of the BRE or of MAZ and Sp1 to MLP function is not clear.

3.2 Upstream Activating Elements

The UPE. Definition of the Element. Shortly after the demonstration that the MLP could be accurately and efficiently transcribed in vitro, the search began for the cis-acting sequence elements and protein factors required for this activity. The prominent role of the TATA box has been described above, but it was soon realized, both from comparisons with cellular promoters and from preliminary experiments using in vitro and transfection techniques, that the promoter was likely to require activating sequences located upstream (HEN et al. 1982; HU and MANLEY 1981). Two such sequences were revealed, the promoter-proximal upstream element UPE and the promoter-distal inverted CAAT box (see Fig. 1). In the former case, point mutations within an almost perfect palindromic sequence 5′-GGCCACGTGACC-3′ reduced transcription significantly (more than twofold), in two independent studies (YU and MANLEY 1984; LEE et al. 1988), an observation confirmed by transfection of recombinant plasmids into both 293 and HeLa cells (LEE et al. 1988). Subsequent mutational analysis of the MLP located in the viral genome showed that a multiple point mutation in the palindrome had only a minor effect on transcription unless mutations were present in other elements of the MLP (REACH et al. 1990; REACH et al. 1991). This demonstrated that the UPE played an important but not essential role in quantitative transcription from the MLP. The reasons behind this apparent functional redundancy will be addressed below.

Proteins That Bind to the UPE. The first protein to be identified that binds to the UPE was characterized by three groups and named upstream stimulatory factor (USF) (SAWADOGO and ROEDER 1985), major late transcription factor (MLTF) (CARTHEW et al. 1985), or upstream element factor (UEF) (MIYAMOTO et al. 1985). The former name has gained the widest acceptance and will be used here. Subsequent cloning and sequencing of the cDNA of the first member of the USF family of proteins, USF1, revealed that it is an E-box protein with helix-loop-helix/leucine zipper motifs, and that it binds to the UPE as a dimer (GREGOR et al. 1990). USF DNA-binding activity is ubiquitous, although its ability to activate transcription is diminished in some tumor cell lines including the Saos-2 line derived from an osteosarcoma (Qyang et al. 1999) and several breast tumor lines (Ismail et al. 1999). The data further suggest that the lack of activity may result either from alterations to USF itself or to the absence of a putative co-factor necessary for its transcriptional activity. However, although the data are unequivocal that USF is capable of activating transcription from UPE-containing promoters such as the MLP, the issue that arises is whether or not USF is the protein that is used to activate transcription at the MLP in the correct location in the viral genome. Several other proteins are now known to bind to the UPE. As mentioned above, these include TFII-I with which USF interacts (Roy et al. 1991, 1997), and COUP-TF (P. Hearing, personal communication). The latter protein, which is a member of the family of steroid/thyroid receptor proteins, binds to a sequence overlapping the UPE. Over-expression of COUP-TF leads to a marked reduction in transcription from the wild-type MLP, and a severely depressed viral yield, but the mutant USF0, in which the COUP-TF recognition site is disrupted, replicates efficiently (P. Hearing, personal communication). Whether this inhibition is important to the virus in its natural host is unknown, but the observations serve as a reminder that negative regulators might play a role in MLP function. Experimental tests to see if USF is the sole transcription factor acting on the MLP in virus have not been attempted, although in principle this is now possible using the cell lines deficient in USF transactivating activity, and the range of viruses with different mutations in the MLP.

The Inverted CAAT Box. Identification of the Element. The first suggestion that a CAAT box consensus might play a role in MLP transcriptional activity came from the deletional analysis of Hen et al. (1982). They noted that deletions leading to a reduction in MLP activity encompassed a sequence showing some similarity to a CAAT box found in the promoter of the rabbit β-globin gene. Although the sequence identified was not that of the true inverted CAAT box located a number of nucleotides upstream, the gathering consensus that many promoters did indeed contain such an element (as in the herpes TK gene promoter; Coen et al. 1986), led to the search for factors that might bind to the sequence 5′-CCAAT-3′ (see below). Genetic analysis confirmed that this sequence does play a role in the activity of the MLP in vivo, although it seems to be less important than the UPE. This conclusion is based on several observations. First, a transition mutation in the first C had minimal effects in an in vitro assay system, where changes to the UPE or to the core elements significantly reduced transcription (LEE et al. 1988). Second, it was possible to recover virus containing mutations in both the CAAT box and the TATA box (although it had a significant transcriptional deficit) whereas double mutations in the UPE and the TATA box were lethal (REACH et al. 1991). Third, double mutations in the CAAT box and Inr were without discernible viral phenotype (SONG and YOUNG 1998) whereas double mutations in the UPE and Inr were transcriptionally deficient (LU et al. 1997). Nevertheless, the fact that the virus that contains mutations in the CAAT and TATA boxes is much more transcriptionally deficient than the viruses containing either single mutation alone shows that the CAAT box does play a role in transcription (REACH et al. 1991).

Proteins That Bind to the CAAT Box. Fractionation of uninfected HeLa cells revealed a protein, referred to as CP1, which binds with high affinity to the CAAT box (CHODOSH et al. 1988a). CP1 is a heterotrimer, whose subunits are homologous to a yeast protein capable of binding to a similar CAAT sequence (CHODOSH et al. 1988b). Other groups isolated the protein, using other cell types and promoters, and referred to it as NF-Y or CBF (reviewed by MAITY and DE CROMBRUGGHE 1998). Analysis of sequence databases (BUCHER 1990) and in vitro selection techniques on a set of duplex oligonucleotides containing a central CCAAT consensus and randomized flanking sequences (BI et al. 1997), showed that the consensus binding site is 5′-YRRCCAATCA-3′ to which the MLP sequence conforms except for the most 5′ site which is A. As with the UPE, and its binding protein USF, the assumption that CP1 plays an important role in MLP transcription is probably correct, but direct evidence for its role in vivo is lacking. Genetic experiments with dominant-negative versions of one of the three subunits, and using the set of available MLP mutant viruses, might confirm the importance of CP1 to MLP function.

3.3 Downstream Activating Elements

Defining the Elements. The first evidence that elements lying within the first intron played a role in MLP expression came from construction of hybrid adenovirus vector genomes (MANSOUR et al. 1986). Further analysis, using a copy of the MLP placed in the E3 region of the genome, showed that there were two or perhaps three such elements with some functional redundancy (LEONG et al. 1990). In addition, evidence was obtained that there is functional interaction between the downstream elements and the UPE since mutations in both caused a marked reduction in MLP activity in transcription studies in vitro (MONDÉSERT and KEDINGER 1991). Although these conclusions were not supported by initial genetic analysis of the MLP in the correct viral genomic context, in which a double mutation in the UPE and DE1 had a wild-type phenotype (REACH et al. 1991), more recent evidence has shown that a virus genome containing extensive point mutations in both DE1 and DE2b/a is transcriptionally deficient. Furthermore, it has proved impossible to create a compound mutation with USF0 and this more extensively mutated DE sequence (A. Pardo Mateos and H. Young, unpublished results), as predicted from the earlier results (MONDÉSERT and KEDINGER 1991).

The Proteins That Bind DE1 and DE2. Kedinger and his colleagues partially purified from late phase-infected cells two protein complexes termed DEF-A and DEF-B that bound to the elements DE1 and DE2a, and DE2b, respectively (JANSEN-DURR et al. 1989; MONDÉSERT et al. 1992) (see Fig. 1). Partial purification of the proteins allowed a detailed analysis of the DNA sequence necessary for efficient binding to be conducted and, most importantly, the identification of one of the proteins present in the complexes. It was found to be the adenovirus non-structural protein IVa2 (TRIBOULEY et al. 1994; LUTZ and KEDINGER 1996). When bound to adenovirus DNA, DEF-B is a homodimer of pIVa2, while DEF-A is a heterodimer of pIVa2 and an as yet unidentified protein. The identification of pIVa2 was surprising as until then, it was generally thought that pIVa2 was involved in some step in assembly (see PETTERSSON 1984 for an early review).

The Role of IVa2 in Transcription from the MLP. The evidence that pIVa2 is a transcription factor is compelling in the infection–transfection experiments (TRIBOULEY et al. 1994). Furthermore, the idea that pIVa2 is the missing component in the early to late switch in transcription from the MLP is very attractive because IVa2 transcription is detected before the MLP becomes maximally active (BINGER and FLINT 1984). Moreover, all adenoviruses from a wide variety of species, and belonging to the four genera (see chapter by BENKÖ, this volume), contain a IVa2 homolog, which is located in the same position relative to the MLP and is transcribed in a divergent direction. This strongly suggests that it plays a critical role in the life cycle. However, the mechanism by which pIVa2 exerts its transcriptional role is far from clear. There are no obvious transcription factor motifs and the DNA sequence of the DE1 and DE2 binding sites do not give any clues as to the relatedness of pIVa2 to other cellular transcription factors. Furthermore, it is not active as a transactivator unless it is expressed in the milieu of the late phase infected cell (Tribouley et al. 1994). While this is to be expected for the role it is thought to play in the infectious cycle, it does not lead to any obvious parallels with better understood cellular transactivators. Published structure–function analysis of the IVa2 protein has been limited to a broad deletion analysis of the regions necessary for DNA binding (Lutz and Kedinger 1996), and a more detailed mutagenic analysis is lacking. In addition, although the primary structure of all IVa2 proteins contains two motifs found in proteins capable of ATP-binding and hydrolysis – so-called Walker A and B boxes (Walker et al. 1982) – no experimental evidence has been presented to show if these putative motifs are important to function, nor indeed if IVa2 is an ATPase. Recently, strong evidence has indicated that pIVa2 plays a role in encapsidation of viral genomes. Several observations support this idea. First, pIVa2 is capable of binding to the packaging signals at the left hand end of the genome (Zhang and Imperiale 2000). These share considerable homology to the sequence of DE1 and DE2. Second, yeast two-hybrid screens with the L1 52/55-kDa protein gene, a protein with a known function in viral assembly (Hasson et al. 1989; Gustin and Imperiale 1998), identified pIVa2 as a binding partner (Gustin et al. 1996). Third, there seems to be serotype-specificity in encapsidation in that the packaging signal and pIVa2 must be from the same serotype for effective packaging (Zhang et al. 2001). Thus pIVa2 may play two roles in the adenovirus life-cycle, one in stimulating transcription from the MLP after DNA replication commences and also a later role in encapsidation. Genetic analysis of the role of pIVa2 in vivo has not been reported yet. However, an amino terminal deletion mutation creates a hypomorphic mutant phenotype in which packaging efficiency is reduced but transcription activation activity is intact (A. Pardo Mateos and H. Young, unpublished results). Mutations creating large C-terminal mutations appear to be lethal (A. Pardo Mateos and H. Young, unpublished results), but these of course do not distinguish between encapsidation, transactivation, or both as being critical for virus replication, nor does it rule out a dominant-negative phenotype for the specific mutations. Similarly, although the serotype specificity of pIVa2 for its cognate packaging signal is consistent with the idea that it is essential for packaging, a dominant-negative effect of any heterotypic interaction cannot be ruled out. It will be very interesting to dissect more precisely the regions of the protein that play roles in these seemingly very dissimilar functions. The role of pIVa2 in the early to late transcriptional switch will be discussed in a later section.

3.4
Do Other Elements Play a Role in MLP Function?

The Effects of Elements Upstream of the MLP. The MLP is downstream of three other adenovirus promoters, the early E1A and E1B promoters, which are transcribed in the same direction as the MLP, and the immediately adjacent but divergent IVa2 promoter. The question that arises is whether elements in these promoters affect the expression of the MLP by some cis-acting mechanism either to enhance or to repress transcription. The principle that enhancer elements near E1A can operate over the entire length of the genome was established by HEARING and SHENK (1986). The analysis was restricted to expression from other early promoters and no evidence was gathered concerning either early or late expression from the MLP. The mechanism underlying this enhancement is unknown but if either of the authors' speculations that it promoted recruitment of transcription factors or served to locate the viral genome in a transcriptionally competent compartment of the nucleus is correct, it might be anticipated that this would apply to the MLP too. The idea that transcription from the E1A and/or E1B promoters might 'read-through' the MLP and thereby occlude transcription from the latter was given some support by the observations of CONNOLLY and MANLEY (1989a, b) that the inverted CAAT box can act as a barrier to transcription between it and the normal start of transcription of the MLTU. These observations have not been confirmed in wild-type virus infections, nor have there been any obvious effects of mutation in the CAAT box (REACH et al. 1991; SONG and YOUNG 1998). Thus there is currently no compelling evidence that far upstream elements or transcription from upstream promoters plays a role in regulating MLP transcription, but it must be emphasized that such effects may be subtle and not easily recognized.

The Divergent IVa2 Promoter. One of the striking genomic arrangements conserved in all adenoviruses examined to date is the adjacent but transcriptionally divergent position of the IVa2 promoter and the MLP (Fig. 2). In subgroup C viruses, the two initiation sites are just ~210 nucleotides apart and the upstream elements are of course even closer. There has been some controversy over whether there is competition between the two promoters for shared transcription factors and whether they are co-regulated (Natarajan et al. 1985; Adami and Babiss 1992). Although this issue has not been fully resolved, the evidence from mutational analysis of the MLP elements in the viral genome does not suggest that either competition or co-regulation is a major feature in vivo. No changes in IVa2 expression could be detected in viruses containing mutations conferring severe defects in MLP expression (Reach et al. 1991). As yet there are no reports of the phenotypes of mutations in which IVa2 expression is altered by mutating the known activating promoter elements (Kasai et al. 1992; Li et al. 1992; Chen and Flint 1992; Chen et al. 1994) in the context of the virus genome. As discussed above and in a later section, because IVa2 protein is thought to have a role in activating the MLP, any cis-acting effects would have to be examined under conditions where the protein is supplied in trans, such as in a cell line expressing IVa2 (Winter and D'Halluin 1991; Zhang et al. 2001).

The 'Mitigator'. Although human adenovirus can bind to and enter a wide variety of mammalian cells, gene expression in non-native hosts is usually restricted to a subset of viral genes. The reasons for this restriction are not immediately obvious because the cellular gene products involved in adenovirus transcription are highly conserved between species. Furthermore, the interactions between the viral E1A proteins, known transcriptional activators, and cellular targets are conserved, suggesting that at least the early stages of the transcription cycle might be similar in different species. Restriction points occur at very late stages in infection, as in the abortive infection of monkey cells by human viruses (reviewed by Klessig 1984), or at the level of MLP transcription, as in the abortive infection of human Ad12 (subgroup A) in BHK21 cells derived from Syrian hamsters. These cells are permissive for Ad2 or Ad5 (subgroup C) and this difference allowed Doerfler and his colleagues to identify a 33-base pair sequence downstream of the initiation site that conferred inhibition to Ad12 MLP function (Zock and Doerfler 1990). Removal of this sequence, termed the mitigator, allowed expression of an Ad12 MLP in the otherwise non-permissive hamster cells, but transfer of this negative regulator to the equivalent position in the Ad2 MLP did not confer inhibition (Zock et al. 1993). Further analysis of the Ad12 MLP suggested that the negative effects may require upstream elements from the same serotype. Gel-shift analyses showed that the mitigator can bind the transcription factor YY1, but so does the sequence at the equivalent position in the Ad2 MLP, suggesting that inhibition may depend on the concerted action of several elements and protein factors. As yet, it is not clear if the mitigator plays a role in the regulation of the MLP in the natural host, but negative regulation may be important to prevent expression from the MLP at early times. Precedent for this comes not only from other viruses such as the SV40 late promoter (WILEY et al. 1993; ZUO and MERTZ 1995) but also from recent work with the IVa2 promoter, in which a cellular protein binds to an upstream element to prevent premature expression (LIN and FLINT 2000). In both of these cases, it is thought that as the viral genome number increases with DNA replication, the inhibitory effects decline as the proportion of genomes bound by the factor decreases. This observation will be considered again in the section dealing with the early to late switch.

4
Comparisons of the MLP Among Adenoviruses from Diverse Species

The majority of experimental work on MLP structure and function has been conducted with human subgroup C viruses Ad2 and Ad5. It was of some interest to see if the general promoter structure described above is conserved among other members of the *Mastadenoviridae*, both to gain insight into the roles of specific elements and to contribute to the investigation of the evolution of adenoviruses. Conservation of specific elements not only would suggest the importance and universality of their function but also would lend support to other data indicating evolution of all extant adenoviruses from a common ancestor. The first set of data obtained from several members of the *Mastadenoviridae* showed that the overall promoter structure is highly conserved and is located at the same position in the genome (SONG et al. 1996). Subsequently, sequence data from several other members of the *Mastadenoviridae* have been determined and in every case the MLP has the same overall structure. The co-ordinates are most easily pinpointed by locating the highly conserved motifs I and III of the DNA polymerase (WONG et al. 1988) encoded on the other strand. These motifs bracket the MLP (Fig. 2). Not surprisingly, every MLP, except canine adenovirus types 1 and 2, has a canonical TATA box, and a recognizable Inr; all except mouse adenovirus type 1 have a recognizable promoter proximal UPE (the canonical CACGTG E box motif) and all have a promoter-distal consensus inverted CAAT box. Thus all of the true *Mastadenoviridae* have the same promoter structure with the same promoter

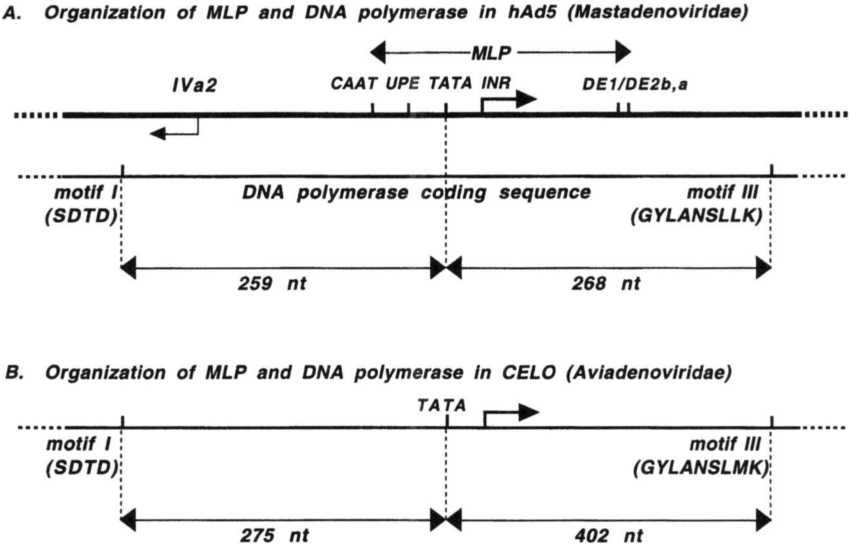

Fig. 2A–D. Organization of the MLP and overlapping DNA polymerase coding region in viruses representing the four proposed genera of the *Adenoviridae*. **A** Human Ad5. The *thick continuous line with dashed extensions* represents the adenovirus DNA sequence that includes the MLP, some of whose transcription elements are shown above the line. The MLTU and the IVa2 transcription unit are represented by the *rightward and leftward bent arrows*, respectively. The DNA polymerase ORF is shown by the *thinner continuous line with dashed extensions*, and amino acids within motifs I and III (SDTD, and GYLANSLLK, respectively) are indicated. Because the DNA polymerase ORF extends from right to left, these motifs are given in the opposite direction to those normally presented (Wong et al. 1988). The 259 nucleotides (nt) shown *between the arrows* include the first nucleotide of the proximal D codon in motif I and the first T of the TATA box. The 268 nt include the first T of the TATA box and the first A of the K codon. Other *Mastadenoviridae* have very similar MLP and DNA polymerase organizations including the relative spacing of the motifs and TATA boxes. **B** CELO (FAV-1). Experimental evidence has identified the MLP of CELO and the initiation site of the MLTU (Payet et al. 1998). The *continuous line* represents the DNA, with the TATA box of the MLP and the DNA polymerase motifs indicated. Other conventions are as for A. MLP elements, other than the TATA box are not easily identified from the sequence (Chiocca et al. 1996) and the IVa2 transcription initiation site has not been determined experimentally. Note that the distance between the TATA box and motif III is considerably larger than in the *Mastadenoviridae*. This increase is also seen in the only other completely sequenced member of the *Aviadenoviridae* FAV-8 (Ojkic and Nagy 2000)

C. Organization of MLP and DNA polymerase in OAV287 (Atadenoviridae)

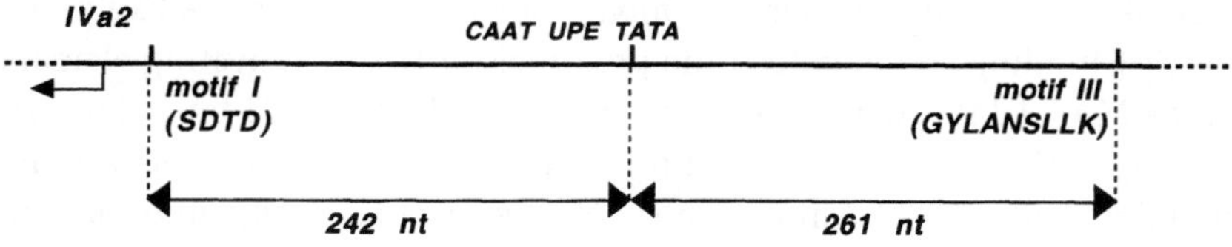

D. Organization of DNA polymerase in FrAdV-1 (proposed new genus)

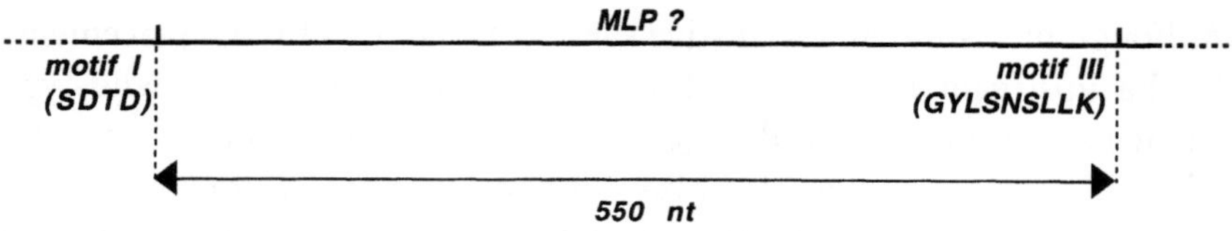

Fig. 2C–D. C OAV287. The conventions are as in **B**. Experimental evidence has identified the initiation sites of both the MLTU and IVa2 transcription units (VRATI et al. 1996; KHATRI and BOTH 1998). Sequence inspection shows a consensus TATA box, E box, and a modified CAAT box (see text). Unlike the *Mastadenoviridae*, the IVa2 transcription unit initiates to the left of motif I and there is no evidence for a spliced mRNA, the first exon of which would encode primary amino acid sequence in the same frame as the DNA polymerase (KHATRI and BOTH 1998). Other members of the *Atadenoviridae* have a very similar organization with respect to both the MLP and the DNA polymerase and IVa2 ORFs (HESS et al. 1997; YAGUBI et al. 1998; SONG et al. 1996). **D** Region of the Frog adenovirus-1 expected to contain the MLP. The conventions are as in **B** and **C**. The presence of the DNA polymerase motifs is readily apparent from the sequence, but there are no obvious MLP elements. No experimental evidence is available to test the prediction that the MLP does indeed lie between the two motifs. The only other member of this proposed genus to be sequenced, turkey adenovirus-3 (HEV) also lacks obvious MLP promoter elements (PITCOVSKI et al. 1998)

elements and although the spacing of the upstream elements does vary, the binding sites are predicted to be an integral number of helical turns apart on the same face of the DNA. These data suggest that the MLP sequence has been conserved throughout the evolution of the *Mastadenoviridae* and that the various promoter elements function optimally in a wide variety of host species. The one exception to strict conservation of the elements, the presence of a predicted Sp1 binding site as the promoter proximal upstream element in Mouse adenovirus type 1 (MAV-1), suggests that other ubiquitous factors can be used instead. The predicted MAV-1 MLP was demonstrated to be active in infected cells, the initiation site was determined, and the predicted Sp1 site shown to be capable of binding purified Sp1 in vitro (Song and Young 1997). Taking the experimental data and the sequence comparisons together, it seems likely that the Sp1 binding site is indeed the proximal upstream activating element in the MAV-1 MLP. Replacement of the UPE in human Ad5 by an Sp1 site is compatible with full activity, although it is not as active in a background containing the TATA27 mutation (E. Minaya and H. Young, unpublished results). Similarly, the CAAT box can be replaced in hAd5 by ATF1 and OCT1 sites, but they are not as effective as the original element when examined in a mutated UPE background (Song and Young 1998). Although this is not an exhaustive survey, the failure of the replacements in the upstream elements to function as efficiently as the wild-type elements they have replaced, suggests that the canonical E box and inverted CAAT box are optimal for MLP function in most *Mastadenoviridae*. It is less easy to determine from sequence inspection if the non-primate *Mastadenoviridae* contain equivalents to the downstream elements DE1 and DE2. Despite the conservation of a recognizable IVa2 ORF in all adenoviruses, the DNA binding site and protein domain of pIVa2 may have co-evolved much faster than the up-stream sites which bind highly conserved cellular transcription factors. Indeed one of the DNA binding domains mapped by Lutz and Kedinger (1996) lie in the highly diverged N terminus of pIVa2.

Complete sequences are now available for 19 adenoviruses from the four proposed groups, *Mastadenoviridae*, *Aviadenoviridae*, *Atadenoviridae*, and the newly proposed fourth genus, which contains frog adenovirus type 1 and turkey adenovirus type 3 (hemorrhagic enteritis virus) (Davison et al. 2000). Avian adenovirus type 1 (CELO), the type virus for the *Aviadenoviridae*, has been the subject of considerable experimental analysis, and the site of late transcription from the MLP has been determined (Payet et al. 1998). As with the *Mastadenoviridae*, the MLP is located between motifs I and III on the putative DNA polymerase (Fig. 2B), and it has a canonical TATA box 5′-TATATAAG-3′ located at a position consistent with the initiation site determined empirically. Inspection of the region upstream of the TATA box reveals a possible variant of an inverted CAAT box but no obvious proximal promoter E box, although a possible GC-rich motif is present in approximately the correct position. One drawback to this kind of sequence inspection is that there is more information concerning mammalian than avian upstream elements, so the consensus used is heavily weighted towards the former. More empirical information on the *Aviadenoviridae* promoters is clearly necessary. The *Atadenoviridae* comprise four species in either mammalian or avian hosts, with empirical data available for ovine virus OAV287 (VRATI et al. 1996; KHATRI and BOTH 1998). Again the MLP is located between the consensus DNA polymerase motifs, and there is a canonical TATA box, an E-box motif, and a close variant of the inverted CAAT box (Fig. 2C). This variant 5′-CCTAT-3′ is found in three members of the *Atadenoviridae*, while the fourth from duck adenovirus type 1 (EDS-76) has the canonical 5′-CCAAT-3′. Whether the variant can function as a true CAAT box has not been determined. Finally, the two members of the fourth proposed genus have a genomic organization very similar to those of all other adenoviruses, and the DNA polymerase motifs that bracket the MLP in the three other virus genera can be identified (Fig. 2D), but no canonical TATA box is apparent (DAVISON et al. 2000). Again, empirical data will be invaluable in further characterizing the MLP of this novel genus.

5
Temporal Control of MLP Expression. Early Observations

The early to late switch is a hallmark of the transcription program of all DNA-containing viruses from bacteriophages to the cytoplasmic pox viruses. The temporal defining point is the onset of DNA replication, but the mechanism of switching from an early mode to a late mode is not clearly understood in any of the animal virus families and is almost certainly different in each family and even within families. Nevertheless, it seems self-evident that the temporal defining point of DNA replication has evolved so that large quantities of late structural proteins are not made until new genomes have accumulated in order that assembly is maximally productive.

As mentioned in the Introduction, it was clear early on that there was a major switch in the regions of the genome undergoing transcription as infection progressed. Once mapping data had identified the MLTU and its initiation site and the MLP had been defined, the question that arose is whether or not the MLP is indeed silent prior to DNA replication. It became apparent that it is not; transcription proceeds at least through the L1 polyA signal (KITCHINGMAN and WESTPHAL 1980; CHOW et al. 1979; SHAW and ZIFF 1980), and this transcription is quantitatively dependent on prior E1A expression (NEVINS 1981), i.e. the MLP is behaving like an early promoter, but transcription is arrested in some manner. In addition, as discussed by Sharp (1984), although the level of mRNAs arising from the MLTU is prodigious at late times (Flint and Sharp 1976), template numbers are orders of magnitude higher at this stage of the replication cycle, so that the apparent 'strength' of the promoter may be less than it appears. Conversely, the number of templates actively undergoing transcription at late times is considerably less than the total (Wolgemuth and Hsu 1981; Beyer et al. 1981), which would raise estimates of the apparent 'strength'. A conservative view of the early to late switch therefore would be that the MLP is active but transcription is prematurely terminated at early times, and that this termination is lost and the activity is increased, by some unspecified amount, at late times. The mechanistic issues then become one of trying to understand the change in polyA site usage (see chapter by Akusjärvi, this volume) and the reason for any increased activity of the MLP itself. In passing, it should pointed out that in addition to this temporal change, there is evidence for a phenomenon akin to bacterial attenuation in the MLTU, in that a majority of initiated transcripts do not elongate more than a few hundred nucleotides (Evans et al. 1979). The biological significance of this observation is not clear, although it is found in cellular transcription units such as that for c-*myc* (Krumm et al. 1992, 1993).

Genetic evidence that there is a change in transcriptional activity of the MLP (as opposed to a change in post-transcriptional processing) came from the analysis in virus of a second copy of the MLP substituted for the E1A promoter and controlling expression of the E1A gene (Logan and Shenk 1986). A segment of the MLP from –55 to +33 was transcriptionally active at early times but showed no relative increase as DNA replication commenced. This fragment contains the TATA box and Inr but no upstream or downstream elements. In contrast, a segment from –122 to +33, which contains both upstream elements but still lacks the downstream elements, was active early and was enhanced in activity at late times. This suggested that the early to late switch may involve the upstream elements, and in support of this idea, it was shown by in vivo foot-print analysis that occupancy of the UPE is markedly enhanced at late times despite a constant level of UPE-binding protein (Toth et al. 1992). Although not demonstrated unequivocally, the pattern of the footprint was consistent with enhanced occupancy by USF at late times. This enhancement depended on prior DNA replication. However, these observations must be reconciled with later data showing that pIVa2, which binds to the downstream elements, is involved in this switch. This will be discussed further below.

5.1
Possible Changes in Template Structure and/or Nuclear Location

Early data showed that inhibitors of viral DNA replication led to a block in late transcription and late protein accumulation. This was confirmed using temperature sensitive mutations in genes involved in DNA replication (Carter and Ginsberg 1976). Infected cells maintained at the non-permissive temperature failed to enter the late phase, although once replication had begun, ongoing replication was not required for continuing transcription (Carter and Ginsberg 1976). Thus late transcription depends on prior DNA replication. One possibility for this observation is that there is a cis-acting change in template structure accompanying DNA replication that allows the MLP to be fully active. Evidence both for (Tate and Philipson 1979; Dery et al. 1985) and against (Chatterjee et al. 1986) such changes has been gathered. Alternatively, the intranuclear milieu in cells undergoing active DNA replication might be different from that at early times and more 'permissive' for MLP activity. Thomas and Mathews (1980), in a technically daunting set of super-infection experiments using distinguishable virus serotypes and genotypes and appropriate inhibitors of DNA replication, showed conclusively that even if cells had entered the late phase of infection following the initial infection, the super-infecting genomes could not undergo late transcription unless they too had been replicated. This strongly argues that a cis-acting change in the template is necessary for quantitative levels of MLP driven transcription, rather than a trans-acting change in the intranuclear milieu. What this template change might be is not clear, although it might reflect changes in the nucleoprotein structure of the template, and/or changes in intranuclear location of viral genomes, attendant on DNA replication. Despite the fact that these observations are more than 20 years old, virtually no progress has been made in understanding the mechanisms underlying them. One of the problems is that no genetic approach can be envisaged that easily addresses such general concepts as 'template structure' (but see Chatterjee et al. 1986 in which a mutant in pVII was used) or 'intranuclear genome localization'. Certainly none of the existing mutants with defects in MLP function have altered the temporal program.

5.2 A Role for IVa2 Protein in Regulating the Early to Late Switch

Although the experiments outlined above show that cis-acting changes in template structure are necessary for late activity from the MLP, they do not indicate if they are sufficient. As discussed by Thomas and Mathews, the early to late switch might require a virus-encoded or -induced protein acting in trans, as well as changes in template structure. This idea has received strong support from the observations with the IVa2 protein. Received opinion for many years was that pIVa2 had a morphogenetic role (see review by Pettersson 1984). This idea, supported by the observation that small amounts of pIVa2 are found in immature virions (Persson et al. 1979), did not explain why IVa2 is transcribed as an 'intermediate' RNA, first detected immediately after the onset of DNA replication and accumulating before large quantities of late RNAs are found (Binger and Flint 1984). As described in an earlier section however, this temporal pattern is explicable if pIVa2 plays a role in up-regulating the MLP (Tribouley et al. 1994). Furthermore, recent evidence from in vitro transcription studies using the IVa2 promoter shows that it is probably regulated by the binding of a cellular inhibitor (Chen et al. 1994; Lin and Flint 2000), whose effects would be predicted to diminish as templates increase in number following the onset of DNA replication. Replication is required because the activity of the inhibitory protein is not reduced during the infectious cycle (Lin and Flint 2000). Taking all these facts together, leads to a model first suggested by Lin and Flint (2000). At early times, the cellular repressor would inhibit expression from the IVa2 promoter. As template number increases, repression of the IVa2 promoter would diminish, leading to the expression of pIVa2. In turn, pIVa2 would bind to its partner protein(s) and to the DE1 and DE2 elements of the MLP leading to stimulation of MLP activity. This model is consistent with all of the currently available information on the temporal regulation of both the MLP and IVa2 promoters. Nor is it incompatible with the template modification results suggested by Thomas and Mathews (1980) provided that pIVa2 activity is necessary but not sufficient for the early to late switch. Furthermore, the model can be tested in virus by using mutations in the inhibitory factor-binding site. To reconcile the model with the earlier genetic data of Logan and Shenk (1986), which showed that the upstream region was sufficient for temporal control (at least for an MLP in an ectopic location), it would be necessary to invoke recruitment of pIVa2 to the MLP by protein–protein interactions, for example with USF. This idea has not been tested. Finally, it should be pointed out that just as the cis-acting DNA replication, template-modification mechanism may be necessary but not sufficient for activation of the MLP, the trans-acting IVa2 protein may also be necessary but not sufficient. For example, the other protein in the DEF-A heterodimer has not been identified (despite the expectation from the yeast two-hybrid results that it might be the L1 52/55-kDa protein) and it might be required for IVa2 function. Alternatively, another viral protein might substitute for pIVa2 function. In this regard, a report suggested that the single-stranded DNA binding protein (DBP) can stimulate the MLP (CHANG and SHENK 1990). Although this protein, which is essential for DNA replication, is obviously expressed prior to the onset of replication, it may act at the MLP only after a sufficient pool has been built up in the infected cell, thus activating the MLP only after DNA replication has begun.

6
Changes in Early Gene Expression Contingent on MLP Activation

Ever since the first description of the temporal transcription program in adenovirus-infected cells, it has been recognized that there are changes to early gene transcription after the entry into the late phase (reviewed by SHARP 1984). On a per template basis, some of the early transcription units show a relative or absolute decline in activity, as the MLP becomes the dominant promoter. The mechanism underlying this change is not understood, but the creation of viruses with mutations affecting the activity of the MLP gave some insight into this phenomenon. It was discovered that the steady state level of E1B mRNA was significantly increased in virus containing the MLP mutations TATA0 or TATA27::CCCAT, both of which have pronounced deficiencies in MLP transcription (REACH et al. 1991). This was later confirmed and extended to include the E2A transcription unit, in which it was shown that the expression of the DNA binding protein was markedly enhanced in TATA0-infected cells (FESSLER and YOUNG 1998). The effects on E2A expression are mediated in trans because they are abolished by co-infection with a wild-type virus. Other early transcription units were examined and they showed a variety of changes, ranging from increases in the E1A unit to a decrease in the E3 unit and alterations to the set of mRNAs detected in the E4 unit (FESSLER and YOUNG 1998). This survey suggested that the activity of the MLP and/or the products of the MLTU affect early gene expression in complex ways. Because of the ability to reverse the effects on early gene expression in trans, it was suggested that these effects arise either from the repressive activity of a product of the MLTU or from promoter competition between the very active MLP and the early promoters. Efforts to identify possible repressive proteins included an examination of the L4 33kDa protein. A mutant virus containing a stop codon at position 20 does not have a detectable transcriptional defect although it shows a viral replication deficiency probably at the level of packaging (Fessler and Young 1999). Promoter competition as a model for down-regulation of early promoters by the 'stronger' MLP has not been tested so far in any methodical manner. As a first attempt to see if this model has any validity, the non-canonical TATA box of the E2L promoter was altered to the canonical sequence. This change is known to have a significant positive effect in transcription assays in vitro (Huang et al. 1988). However, no effects on E2 expression could be detected when this mutation was placed in virus either in an MLP wild-type or TATA0 genetic background (S. P. Fessler and H. Young, unpublished results). The likelihood is that the control circuits operating at late times in infection are multi-component, and that down-regulation of early gene functions may occur at the transcriptional and post-transcriptional levels, depending on the particular transcription unit. Furthermore, because several early gene products are transcriptional activators, changing the expression of early genes will also have significant contingent effects.

7
Modifying the MLP to Improve Adenovirus' Function as a Vector

Although the potential use of adenovirus as a vector has been recognized for many years (see Gluzman et al. 1982 and Berkner 1988 for early reviews), the last decade has seen an upsurge of interest in the creation and testing of such vectors for purposes ranging from protein expression to tumor therapy to gene therapy (as reviewed in *Gene Therapy*, vol. 273). Can the increasing understanding of the structure and function of the MLP contribute to this effort? From an experimental standpoint, the results discussed in Sect. 6 would be easier to gather and to interpret if an inducible MLP could be created. Entering the late phase of the infection, as measured by the initiation of DNA replication, could be uncoupled from activation of the MLP and the magnitude of the contingent effects on early gene expression would be expected to be greater than with the partially MLP-defective viruses described above. As a practical matter, a vector in which the MLP is inducible would be invaluable both as a 'destructive' targeting vector in which the late proteins were turned on as needed, and as a gene therapy vector in which an inactive MLP would be an added layer of protection against unwanted expression of highly immunogenic late gene products. The difficulty in designing an inducible MLP stem in large part from the fact that the promoter is embedded in the sequence of the DNA polymerase encoded on the opposite strand (see Fig. 2). Although the DNA polymerase primary sequence in the region of the MLP can tolerate many point mutations (Brunet et al. 1987; Song and Young 1998), to replace the MLP with a complete inducible promoter may not be possible. Fortunately, the results of the genetic analysis of the MLP described in Sect. 3 suggest a way round this dilemma. Genomes containing four point mutations in the UPE (USF0) and a single point mutation in the TATA box (TATA30) are not recoverable as virus, suggesting that this combination of mutations, neither of which by itself is lethal, leads to a severely deficient MLP. Fortunately, the mutation in the TATA box 5′-TGTAAAA-3′ is recognized by an extended specificity variant of the TATA box-binding protein referred to as TBP-m3e (Strubin and Struhl 1992). This altered specificity variant has been used in a number of investigations of transcriptional mechanisms (see, for example, Tansey and Herr 1997). To test the ability of m3e to suppress the lethality of TATA30::USF0, the ORF for TBP-m3e was placed under the control of the E3 promoter and levels of protein accumulation were measured and found to be about 10-fold higher than that of the endogenous TBP (H. Lu and H. Young, unpublished results). The E3 vector cassette was then placed in the TATA30::USF0 background to see if this level of expression was sufficient to suppress the lethality of the double MLP mutation (Fig. 3A). Virus was obtained and the presence of the m3e and MLP mutations confirmed. Control constructions employing the wild-type version of TBP were unproductive. Together, these results show that the altered specificity variant of TBP is capable of suppressing the lethal phenotype of the TATA30::USF0 double mutation. This proof of principle then encouraged us to design the vector depicted in Fig. 3B. Taking advantage of the tet-inducible cytomegalovirus (CMV) promoter placed in E1 to control expression of m3e, we have shown that m3e can be expressed to various levels, dependent on the concentration of the inducer doxycycline in the medium (A. Pardo Mateos and H. Young, unpublished results). Preliminary results suggest that the levels of m3e

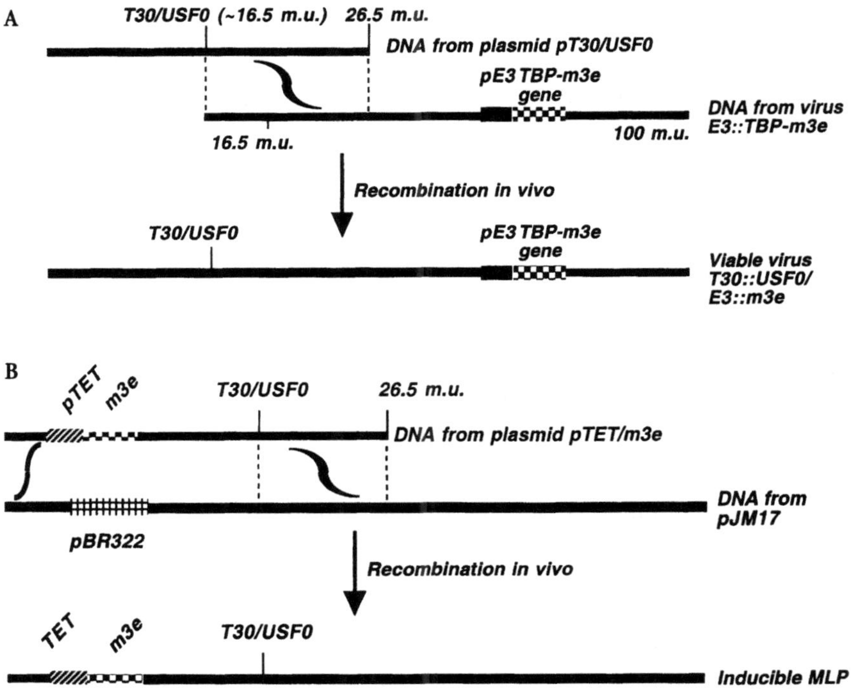

Fig. 3 A, B. The development of an inducible MLP. **A** The lethality of the MLP double mutation TATA30::USF0 can be suppressed by expression of TBP-m3e, a form of TBP with an extended specificity that can recognize the mutant TATA box sequence TGTAAAA. A plasmid containing the double mutation (pT30/USF0) was co-transfected into human A549 cells with cleaved DNA from a viral genome containing the ORF for TBP-m3e under the control of the early region 3 promoter (virus E3::TBP-m3e). Overlap recombination in vivo yielded a virus, T30::USF0/E3::m3e, whose replicative yield is about an order of magnitude lower than that of the E3::TBP-m3e parent virus. The E3 promoter is shown as a *thicker black box* and the m3e ORF by the *checkered box*. B Creation of a virus with an inducible MLP. A plasmid with the *tet* operator and CMV promoter controlling the expression of the TBP-m3e ORF was constructed and the TATA30 and USF0 MLP mutations were substituted for the wild-type MLP sequences. This plasmid was co-transfected with pJM17 (McGrory et al. 1988) into 293 cells expressing the 'tet-on' version of the *tet* repressor (Gossen and Bujard 1992), and yields were obtained under inducing conditions. Recombination to the right of the MLP mutations will yield a virus containing all the elements necessary for the experimental induction of the MLP. Recombination to the left of the pTET/CMV promoter is necessary to generate a unit length linear viral genome from the two circular plasmids. The *tet* operator and CMV promoter are shown together as a *striped box* and the m3e ORF as a *checkered box*. The pBR322 plasmid vector sequence present in pJM17 is indicated

expression following induction are sufficient to suppress the lethal phenotype of the TATA30::USF0 mutation. If this proves true, then the goal of a truly inducible MLP will have been achieved. This should allow a series of vectors to be created in which late gene expression is under the control of the investigator, giving a much clearer insight into the regulatory circuits late in infection. In addition, this inducible MLP could be incorporated into vectors expressing specific transgenes for therapeutic purposes, with the MLP under the control of a class of antibiotics approved for human use.

8
Concluding Remarks

The adenovirus MLP has played a pivotal role in elucidating the mechanism of transcription initiation, as it was the promoter of choice in the reductive biochemical approach. It continues to be used in detailed investigations of the mechanism of transcription initiation and elongation. Its biological role is no less fascinating as it lies at the heart of our conception of the early to late switch in viral transcription. Here, it must be said, we do not fully understand the mechanisms underlying the dependence of full activation of the MLP upon prior DNA replication, nor the reasons behind the contingent down-regulation of early gene expression. The demonstration that pIVa2 up-regulates the promoter in a DNA replication-dependent manner shows the way forward and it will be instructive to know more about the way in which this protein acts, and the identity of its protein partners. Detailed structure–function analysis of pIVa2 is just beginning, and the results of genetic and biochemical analyses are likely to be of considerable interest. More understanding of how the MLP functions in vivo will come as the structure of the nucleoprotein complex, of which the genome is a part, is determined, and also as our understanding of potentially important sites of nuclear localization improves. Currently, these are both poorly defined concepts. Finally, it should be emphasized again that although many cellular proteins that can bind to, and activate, the MLP have been identified, the proteins that carry out these tasks in vivo are not known with certainty. The set of mutations now available in the MLP, and the ability to create vectors capable of expressing wild-type or potentially dominant-negative forms of specific transactivating proteins may allow a definitive answer to these questions.

Acknowledgments. The initial stimulus for my laboratory to work on the MLP came from D. Mills and S. Silverstein, who recognized the value of studying the promoter in its correct genomic location. The continued support of the National Institute of General Medical Sciences is gratefully acknowledged.

References

Adami G, Babiss LE (1992) Evidence that USF can interact with only a single general transcription complex at one time. Mol Cell Biol 12:1630–1638
Bensimhon M, Gabarro-Arpa J, Ehrlich R, Reiss C (1983) Physical characteristics in eucaryotic promoters. Nucleic Acids Res 11:4521–4540
Berget SM, Moore C, Sharp PA (1977) Spliced segments at the 5′ terminus of adenovirus 2 late mRNA. Proc Natl Acad Sci USA 74:3171–3175
Berk AJ (1986) Adenovirus promoters and E1A transactivation. Ann Rev Genetics 20: 45–79
Berkner KL (1988) Development of adenovirus vectors for the expression of heterologous genes. BioTechniques 6:616–629
Beyer AL, Bouton AH, Hodge LD, Miller OL, Jr (1981) Visualization of the major late R strand transcription unit of adenovirus serotype 2. J Mol Biol 147:269–285
Bi WM, Wu L, Coustry F, de Crombrugghe B, Maity SN (1997) DNA binding specificity of the CCAAT-binding factor CBF/NF-Y. J Biol Chem 272:26562–26572
Binger MH, Flint SJ (1984) Accumulation of early and intermediate mRNA species during subgroup C adenovirus productive infections. Virology 136:387–403
Broker TR (1984) Animal virus RNA processing. In: Apirion D (ed) Processing of RNA. CRC Press, Boca Raton, pp 181–212
Brunet LJ, Babiss LE, Young CSH, Mills DR (1987) Mutations in the adenovirus major late promoter: effects upon viability and transcription during infection. Mol Cell Biol 7:1091–1100
Bucher P (1990) Weight matrix descriptions of four eukaryotic RNA polymerase II promoter elements derived from 502 unrelated promoter sequences. J Mol Biol 212:563–578
Carter TH, Ginsberg HS (1976) Viral transcription in KB cells infected by temperature-sensitive "early" mutants of adenovirus type 5. J Virol 18:156–166
Carthew RW, Chodosh LA, Sharp PA (1985) An RNA polymerase II transcription factor binds to an upstream element in the adenovirus major late promoter. Cell 43:439–448
Chang L-S, Shenk T (1990) The adenovirus DNA-binding protein stimulates the rate of transcription directed by adenovirus and adeno-associated virus promoters. J Virol 64:2103–2109
Chatterjee PK, Vayda ME, Flint SJ (1986) Adenoviral protein VII packages intracellular viral DNA throughout the early phase of infection. EMBO J 5:1633–1644
Chen H, Flint SJ (1992) Mutational analysis of the adenovirus 2 IVa$_2$ initiator and downstream elements. J Biol Chem 267: 25457–25465
Chen H, Vinnakota R, Flint SJ (1994) Intragenic activating and repressing elements control transcription from the adenovirus IVa$_2$ initiator. Mol Cell Biol 14:676–685

Chiocca S, Kurzbauer R, Schaffner G, Baker A, Mautner V, Cotten M (1996) The complete DNA sequence and genomic organization of the avian adenovirus CELO. J Virol 70:2939–2949

Chodosh LA, Baldwin AS, Carthew RW, Sharp PA (1988a) Human CCAAT-binding proteins have heterologous subunits. Cell 53:11–24

Chodosh LA, Olesen J, Hahn S, Baldwin AS, Guarente L, Sharp PA (1988b) A yeast and a human CCAAT-binding protein have heterologous subunits that are functionally interchangeable. Cell 53:25–35

Chow L, Gelinas R, Broker T, Roberts R (1977) An amazing sequence arrangement at the 5′ ends of adenovirus 2 messenger RNA. Cell 12:1–8

Chow LT, Broker TR, Lewis JB (1979) Complex splicing patterns of RNAs from the early regions of adenovirus-2. J Mol Biol 134:265–303

Coen DM, Weinheimer SP, McKnight SL (1986) A genetic approach to promoter recognition during trans induction of viral gene expression. Science 234:53–59

Concino M, Goldman RA, Caruthers MH, Weinmann R (1984a) Point mutations of the adenovirus major late promoter with different transcriptional efficiencies in vitro. J Biol Chem 258:8493–8496

Concino MF, Lee RF, Merryweather JP, Weinmann R (1984b) The adenovirus major late promoter TATA box and initiation site are both necessary for transcription in vitro. Nucleic Acids Res 12:7423–7433

Connelly S, Manley JL (1989a) RNA polymerase II transcription termination is mediated specifically by protein binding to a CCAAT box sequence. Mol Cell Biol 9:5254–5259

Connelly S, Manley JL (1989b) A CCAAT box sequence in the adenovirus major late promoter functions as part of an RNA polymerase II termination signal. Cell 57:561–571

Davison AJ, Wright KM, Harrach B (2000) DNA sequence of frog adenovirus. J Gen Virol 81:2431–2439

Dery CV, Toth M, Brown M, Horvath J, Allaire S, Weber JM (1985) The structure of adenovirus chromatin in infected cells. J Gen Virol 66:2671–2684

Du H, Roy AL, Roeder RG (1993) Human transcription factor USF stimulates transcription through the initiator elements of the HIV-1 and the Ad-ML promoters. EMBO J 12:501–511

Emami KH, Jain A, Smale ST (1997) Mechanism of synergy between TATA and initiator: synergistic binding of TFIID following a putative TFIIA-induced isomerization. Genes Dev 11:3007–3019

Evans R, Weber J, Ziff E, Darnell JE (1979) Premature termination during adenovirus transcription. Nature 278:367–370

Fessler SP, Young CSH (1998) Control of adenovirus early gene expression during the late phase of infection. J Virol 72:4049–4056

Fessler SP, Young CSH (1999) The role of the L4 33 K gene in adenovirus infection. Virology 263:507–516

Flint SJ, Sharp PA (1976) Adenovirus transcription V. Quantitation of viral RNA sequences in adenovirus 2-infected and transformed cells. J Mol Biol 106:749–771

Garfinkel S, Thompson JA, Jacob WF, Cohen R, Safer B (1990) Identification and characterization of an adenovirus 2 major late promoter CAP sequence DNA-binding protein. J Biol Chem 265:10309–10319

Gluzman Y, Reichl H, Solnick D (1982) Helper-free adenovirus type 5 vectors. In: Gluzman Y (ed) Eukaryotic viral vectors. Cold Spring Harbor Laboratory, Cold Spring Harbor, pp 187–192

Gossen M, Bujard H (1992) Tight control of gene expression in mammalian cells by tetracycline-responsive promoters. Proc Natl Acad Sci USA 89:5547–5551

Gregor PD, Sawadogo M, Roeder RG (1990) The adenovirus major late transcription factor USF is a member of the helix-loop-helix group of regulatory proteins and binds to DNA as a dimer. Genes Dev 4:1730–1740

Griffith JD, Makhov A, Zawel L, Reinberg D (1995) Visualization of TBP oligomers binding and bending the HIV-1 and adeno promoters. J Mol Biol 246: 576–584

Gustin KE, Lutz P, Imperiale MJ (1996) Interaction of the adenovirus L1 52/55-kilodalton protein with the IVa2 gene product during infection. J Virol 70: 6463–6467

Gustin KE, Imperiale MJ (1998) Encapsidation of viral DNA requires the adenovirus L1 52/55-kilodalton protein. J Virol 72:7860–7870

Hasson TB, Soloway PD, Ornelles DA, Doerfler W, Shenk T (1989) Adenovirus L1 52- and 55-kilodalton proteins are required for assembly of virions. J Virol 63: 3612–3621

Hearing P, Shenk T (1986) The adenovirus type 5 E1A enhancer contains two functionally distinct domains: one is specific for E1A and the other modulates all early units in *cis*. Cell 45: 229–236

Hen R, Sassone-Corsi P, Corden J, Gaub MP, Chambon P (1982) Sequences upstream from the TATA box are required in vivo and in vitro for efficient transcription from the adenovirus serotype 2 major late promoter. Proc Natl Acad Sci USA 79: 7132–7136

Hess M, Blöcker H, Brandt P (1997) The complete nucleotide sequence of the egg drop syndrome virus: An intermediate between mastadenoviruses and aviadenoviruses. Virology 238:145–156

Hu S-L, Manley JL (1981) DNA sequence required for initiation of transcription *in vitro* from the major late promoter of adenovirus 2. Proc Natl Acad Sci USA 78: 820–824

Huang D-H, Horikoshi M, Roeder RG (1988) Activation of the adenovirus EIIa late promoter by a single-point mutation which enhances binding of transcription factor IID. J Biol Chem 263:12596–12601

Ismail PM, Lu T, Sawadogo M (1999) Loss of USF transcriptional activity in breast cancer cell lines. Oncogene 18:5582–5591

Jansen-Durr P, Boeuf P, Kedinger C (1988) Replication-induced stimulation of the major late promoter of adenovirus is correlated to the binding of a factor to sequences in the first intron. Nucleic Acids Res 16:3771–3786

Jansen-Durr P, Mondésert G, Kedinger C (1989) Replication-dependent activation of the adenovirus major late promoter is mediated by the increased binding of a transcription factor to sequences in the first intron. J Virol 63:5124–5132

Javahery R, Khachi A, Lo K, Zenzie-Gregory B, Smale ST (1994) DNA sequence requirements for transcriptional initiator activity in mammalian cells. Mol Cell Biol 14:116–127

Kasai Y, Chen H, Flint SJ (1992) Anatomy of an unusual RNA polymerase II promoter containing a downstream TATA element. Mol Cell Biol 12:2884–2897

Kaufmann J, Smale ST (1994) Direct recognition of initiator elements by a component of the transcription factor IID complex. Genes Dev 8:821–829

Kaufmann J, Verrijzer CP, Shao J, Smale ST (1996) CIF, an essential cofactor for TFIID-dependent initiator function. Genes Dev 10:873–886

Kaufmann J, Ahrens K, Koop R, Smale ST, Müller R (1998) CIF150, a human cofactor for transcription factor IID-dependent initiator function. Mol Cell Biol 18: 233–239

Khatri A, Both GW (1998) Identification of transcripts and promoter regions of ovine adenovirus OAV287. Virology 245:128–141

Kim JL, Nikolov DB, Burley SK (1993) Co-crystal structure of TBP recognizing the minor groove of a TATA element. Nature 365:520–527

Kitchingman GR, Westphal H (1980) The structure of adenovirus 2 early nuclear and cytoplasmic RNAs. J Mol Biol 137:23–48

Klessig DF (1984) Adenovirus-simian virus 40 interactions. In: Ginsberg HS (ed) The adenoviruses. Plenum Press, New York, pp 399–449

Krumm A, Meulia T, Brunvand M, Groudine M (1992) The block to transcriptional elongation within the human c-*myc*gene is determined in the promoter-proximal region. Genes Dev 6:2201–2213

Krumm A, Meulia T, Groudine M (1993) Common mechanisms for the control of eukaryotic transcriptional elongation. BioEssays 15:659–665

Lagrange T, Kim TK, Orphanides G, Ebright YW, Ebright RH, Reinberg D (1996) High-resolution mapping of nucleoprotein complexes by site-specific protein-DNA photocrosslinking: Organization of the human TBP-TFIIA-TFIIB-DNA quaternary complex. Proc Natl Acad Sci USA 93:10620–10625

Lagrange T, Kapanidis AN, Tang H, Reinberg D, Ebright RH (1998) New core promoter element in RNA polymerase II-dependent transcription: sequence-specific DNA binding by transcription factor IIB. Genes Dev 12:34–44

Lee RF, Concino MF, Weinmann R (1988) Genetic profile of the transcriptional signals from the adenovirus major late promoter. Virology 165:51–56

Lee S, Hahn S (1995) Model for binding of transcription factor TFIIB to the TBP-DNA complex. Nature 376:609–612

Leong K, Lee W, Berk AJ (1990) High-level transcription from the adenovirus major late promoter requires downstream binding sites for late-phase-specific factors. J Virol 64:51–60

Li X-C, Huang WL, Flint SJ (1992) The downstream regulatory sequence of the adenovirus type 2 major late promoter is functionally redundant. J Virol 66:5685–5690

Lin HJL, Flint SJ (2000) Identification of a cellular repressor of transcription of the adenoviral late IVa_2 gene that is unaltered in activity in infected cells. Virology 277: 397–410

Lo K, Smale ST (1996) Generality of a functional initiator consensus sequence. Gene 182:13–22

Logan J, Shenk T (1986) In vivo identification of sequence elements required for normal function of the adenovirus major late transcriptional control region. Nucleic Acids Res 14:6327–6335

Lu H, Reach MD, Minaya E, Young CSH (1997) The initiator element of the adenovirus major late promoter has an important role in transcription initiation in vivo. J Virol 71:102–109

Lucas JJ, Ginsberg HS (1971) Synthesis of virus-specific RNA in KB cells infected with type 2 adenovirus. J Virol 8:203–213
Lutz P, Kedinger C (1996) Properties of the adenovirus IVa2 gene product, an effector of late-phase-dependent activation of the major late promoter. J Virol 70:1396–1405
Maity SN, de Crombrugghe B (1998) Role of the CCAAT-binding protein CBF/NF-Y in transcription. Trends Biochem Sci 23:174–178
Manley JL, Fire A, Cano A, Sharp PA, Gefter ML (1980) DNA-dependent transcription of adenovirus genes in a soluble whole cell extract. Proc Natl Acad Sci USA 77:3855–3859
Mansour JL, Grodzicker T, Tjian R (1986) Downstream sequences affect transcription initiation from the adenovirus major late promoter. Mol Cell Biol 6:2684–2694
McGrory WJ, Bautista DS, Graham FL (1988) A simple technique for the rescue of early region 1 mutations into infectious human adenovirus type 5. Virology 163: 614–617
Miyamoto NG, Moncollin V, Egly JM, Chambon P (1985) Specific interaction between a transcription factor and the upstream element of the adenovirus-2 major late promoter. EMBO J 4:3563–3570
Mondésert G, Tribouley C, Kedinger C (1992) Identification of a novel downstream binding protein implicated in late-phase-specific activation of the adenovirus major late promoter. Nucleic Acids Res 20:3881–3889
Mondésert G, Kedinger C (1991) Cooperation between upstream and downstream elements of the adenovirus major late promoter for maximal late phase-specific transcription. Nucleic Acids Res 19:3221–3228
Natarajan V, Madden MJ, Salzman NP (1985) Positive and negative control sequences within the distal domain of the adenovirus IVa2 promoter overlap with the major late promoter. J Virol 55:10–15
Nevins JR (1981) Mechanism of activation of early viral transcription by the adenovirus E1a gene product. Cell 26:213–220
Nikolov DB, Chen H, Halay ED, Usheva AA, Hisatake K, Lee DK, Roeder RG, Burley SK (1995) Crystal structure of a TFIIB-TBP-TATA-element ternary complex. Nature 377:119–128
Ojkic D, Nagy É (2000) The complete nucleotide sequence of fowl adenovirus type 8. J Gen Virol 81:1833–1837
Parks CL, Shenk T (1997) Activation of the adenovirus major late promoter by transcription factors MAZ and Sp1. J Virol 71:9600–9607
Patikoglou GA, Kim JL, Sun LP, Yang SH, Kodadek T, Burley SK (1999) TATA element recognition by the TATA box-binding protein has been conserved throughout evolution. Genes Dev 13:3217–3230
Payet V, Arnauld C, Picault JP, Jestin A, Langlois P (1998) Transcriptional organization of the avian adenovirus CELO. J Virol 72:9278–9285
Persson H, Mathisen B, Philipson L, Pettersson U (1979) A maturation protein in adenovirus morphogenesis. Virology 93:198–208
Pettersson U (1984) Structural and nonstructural adenovirus proteins. In: Ginsberg HS (ed) The adenoviruses. Plenum Press, New York and London, pp 205–270

Pitcovski J, Mualem M, Rei-Koren Z, Krispel S, Shmueli E, Peretz Y, Gutter B, Gallili GE, Michael A, Goldberg D (1998) The complete DNA sequence and genome organization of the avian adenovirus, hemorrhagic enteritis virus. Virology 249:307–315

Qian XL, Strahs D, Schlick T (2001) Dynamic simulations of 13 TATA variants refine kinetic hypotheses of sequence/activity relationships. J Mol Biol 308:681–703

Qyang YB, Luo X, Lu T, Ismail PM, Krylov D, Vinson C, Sawadogo M (1999) Cell-type-dependent activity of the ubiquitous transcription factor USF in cellular proliferation and transcriptional activation. Mol Cell Biol 19:1508–1517

Reach M, Babiss LE, Young CSH (1990) The upstream factor binding site is not essential for activation of transcription from the adenovirus major late promoter. J Virol 64:5851–5860

Reach M, Xu L-X, Young CSH (1991) Transcription from the adenovirus major late promoter uses redundant activating elements. EMBO J 10:3439–3446

Roy AL, Meisterernst M, Pognonec P, Roeder RG (1991) Cooperative interaction of an initiator-binding transcription initiation factor and the helix-loop-helix activator USF. Nature 354:245–248

Roy AL, Malik S, Meisterernst M, Roeder RG (1993) An alternative pathway for transcription initiation involving TFII-I. Nature 365:355–359

Roy AL, Du H, Gregor PD, Novina CD, Martinez E, Roeder RG (1997) Cloning of an Inr- and E-box-binding protein, TFII-I, that interacts physically and functionally with USF1. EMBO J 16:7091–7104

Sawadogo M, Roeder RG (1985) Interaction of a gene-specific transcription factor with the adenovirus major late promoter upstream of the TATA box. Cell 43:165–175

Sharp PA (1984) Adenovirus transcription. In: Ginsberg HS (ed) The adenoviruses. Plenum Press, New York, pp 173–204

Shaw AR, Ziff EB (1980) Transcripts from the adenovirus-2 major late promoter yield a single early family of 3′ coterminal mRNAs and five late families. Cell 20:905–916

Shenk TE (2001) Adenoviridae: the viruses and their replication. In: Knipe DM, Howley PM (eds) Fields Virology. 4th edn Lippincott Williams and Wilkins, Philadelphia, pp 2265–2300

Sirito M, Lin Q, Maity T, Sawadogo M (1994) Ubiquitous expression of the 43- and 44-kDa forms of transcription factor USF in mammalian cells. Nucleic Acids Res 22:427–433

Smale ST, Baltimore D (1989) The "initiator" as a transcription control element. Cell 57:103–113

Smale ST, Jain A, Kaufmann J, Emami KH, Lo K, Garraway IP (1998) The initiator element: A paradigm for core promoter heterogeneity within metazoan protein-coding genes. Cold Spring Harbor Symp Quant Biol 63:21–31

Song B, Hu S-L, Darai G, Spindler KR, Young CSH (1996) Conservation of DNA sequence in the predicted major late promoter regions of selected mastadenoviruses. Virology 220:390–401

Song B, Young CSH (1997) Functional characterization of the major late promoter of mouse adenovirus type 1. Virology 235:109–117

Song B, Young CSH (1998) Functional analysis of the CAAT box in the major late promoter of the subgroup C human adenoviruses. J Virol 72:3213–3220

Strubin M, Struhl K (1992) Yeast and human TFIID with altered DNA-binding specificity for TATA elements. Cell 68:721–730

Tansey WP, Herr W (1997) Selective use of TBP and TFIIB revealed by a TATA-TBP-TFIIB array with altered specificity. Science 275:829–831

Tate VE, Philipson L (1979) Parental adenovirus DNA accumulates in nucleosome-like structures in infected cells. Nucleic Acids Res 6:2769–2785

Thomas GP, Mathews MB (1980) DNA replication and the early to late transition in adenovirus infection. Cell 22:523–533

Toth M, Doerfler W, Shenk T (1992) Adenovirus DNA replication facilitates binding of the MLTF/USF transcription factor to the viral major late promoter within infected cells. Nucleic Acids Res 20:5143–5148

Tribouley C, Lutz P, Staub A, Kedinger C (1994) The product of the adenovirus intermediate gene IVa2 is a transcriptional activator of the major late promoter. J Virol 68:4450–4457

Vrati S, Brookes DE, Boyle DB, Both GW (1996) Nucleotide sequence of ovine adenovirus tripartite leader sequence and homologues of the IVa2, DNA polymerase and terminal proteins. Gene 177:35–41

Walker JE, Saraste M, Runswick MJ, Gay NJ (1982) Distantly related sequences in the alpha- and beta-subunits of ATP synthase, myosin, kinases, and other ATP-requiring enzymes and a common nucleotide binding fold. EMBO J 1:945–951

Weil PA, Luse DS, Segall J, Roeder RG (1979) Selective and accurate initiation of transcription at the adenovirus major late promoter in a soluble system dependent on purified RNA polymerase II and DNA. Cell 18:469–484

Wiley SR, Kraus RJ, Zuo F, Murray EE, Loritz K, Mertz JE (1993) SV40 early-to-late switch involves titration of cellular transcriptional repressors. Genes Dev 7: 2206–2219

Winter N, D'Halluin J-C (1991) Regulation of the biosynthesis of subgroup C adenovirus protein IVa2. J Virol 65:5250–5259

Wobbe CR, Struhl K (1990) Yeast and human TATA-binding proteins have nearly identical DNA sequence requirements for transcription in vitro. Mol Cell Biol 10: 3859–3867

Wolgemuth DJ, Hsu M-T (1981) Visualization of nascent RNA transcripts and simultaneous transcription and replication in viral nucleoprotein complexes from adenovirus 2-infected HeLa cells. J Mol Biol 147:247–268

Wong SW, Wahl AF, Yuan P-M, Arai N, Pearson BE, Arai K-i, Korn D, Hunkapiller MW, Wang TS-F (1988) Human DNA polymerase α gene expression is cell proliferation dependent and its primary structure is similar to both prokaryotic and eukaryotic replicative DNA polymerases. EMBO J 7:37–47

Yagubi A, Ojkic D, Bautista DS, Haj-Ahmad Y (1998) Sequencing analysis of the region encoding the DNA polymerase of bovine adenovirus serotypes 2 and 3. Intervirology 41: 69–79

Yu Y-T, Manley JL (1984) Generation and functional analysis for base-substitution mutants of the adenovirus major late promoter. Nucleic Acids Res 12:9309–9321

Zhang W, Imperiale MJ (2000) Interaction of the adenovirus IVa2 protein with viral packaging sequences. J Virol 74:2687–2693

Zhang W, Low JA, Christensen JB, Imperiale MJ (2001) Role for the adenovirus IVa2 protein in packaging of viral DNA. J Virol 75:10466–10454

Ziff EB, Evans RM (1978) Coincidence of the promoter and capped 5′ terminus of RNA from the adenovirus major late promoter. Cell 15:1463–1475
Zock C, Doerfler W (1990) A mitigator sequence in the downstream region of the major late promoter of adenovirus type 12 DNA. EMBO J 9:1615–1623
Zock C, Iselt A, Doerfler W (1993) A unique mitigator sequence determines the species specificity of the major late promoter in adenovirus type 12 DNA. J Virol 67:682–693
Zuo FR, Mertz JE (1995) Simian virus 40 late gene expression is regulated by members of the steroid/thyroid hormone receptor superfamily. Proc Natl Acad Sci USA 92:8586–8590

3
Host Cell Interactions

Chapter 9

Remodelling of the Host Cell RNA Splicing Machinery During an Adenovirus Infection

G. Akusjärvi, J. Stévenin

Abstract. Adenovirus makes extensive use of RNA splicing to produce a complex set of spliced mRNAs during virus replication. All transcription units, except pIX and IVa2, encode multiple alternatively spliced mRNAs. The accumulation of viral mRNAs is subjected to a temporal regulation,

G. Akusjärvi
Department of Medical Biochemistry and Microbiology, BMC, Uppsala University, 751 23 Uppsala, Sweden

J. Stévenin
Institut de Génétique et de Biologie Moléculaire et Cellulaire, CNRS/INSERM/ULP, 1 Rue Laurent Fries, BP 163, 67404 Illkirch Cedex, France

a mechanism that ensures that proteins that are needed at certain stages of the viral life cycle are produced. The complex interaction between host cell RNA splicing factors and viral regulatory elements has been studied intensely during the last decade. Such studies have begun to produce a picture of how adenovirus remodels the host cell RNA splicing machinery to orchestrate the shift from the early to the late profile of viral mRNA accumulation. Recent progress has to a large extent focused on the mechanisms regulating E1A and L1 alternative splicing. Here we will review the current knowledge of cis-acting sequence element, trans-acting factors and mechanisms controlling E1A and L1 alternative splicing.

1 Exons, Introns and RNA Splicing: General Considerations

During a hectic few months during spring 1977, scientists working at the Cold Spring Harbor Laboratory and Massachusetts Institute of Technology discovered that adenovirus (Ad) mRNAs were encoded by discontinuous segments on the viral genome (BERGET et al. 1977; CHOW et al. 1977; DUNN and HASSELL 1977; KLESSIG 1977). The results were first presented to a larger scientific audience at the Cold Spring Harbor Symposium on 'Chromatin' in May 1977. The description of the split gene concept represented a major turning point in scientific thinking. As described by PIERRE CHAMBON in his summary of the symposium (CHAMBON 1977), scientists working in other systems were able to rapidly explain inconsistencies in their own data. Thus, within half a year after the public presentation of split genes and RNA splicing in the adenovirus system, cellular genes like β-globin, immunoglobulin light chain genes, ovalbumin and the 28S ribosomal RNA genes were similarly shown to be encoded by discontinuous gene segments.

The work in the adenovirus system took another leap forward when it was demonstrated that not only are adenovirus genes spliced, but the majority of the transcription units produces multiple alternatively spliced mRNAs, in a temporally regulated manner (CHOW and BROKER 1978; CHOW et al. 1979) (Fig. 1); many that are translated into proteins with unique biological functions (reviewed in AKUSJÄRVI et al. 1986). A further advance came when it was demonstrated that the accumulation of alternatively spliced adenovirus mRNAs is a regulated process, requiring late viral protein synthesis (AKUSJÄRVI and PERSSON 1981; NEVINS and WILSON 1981).

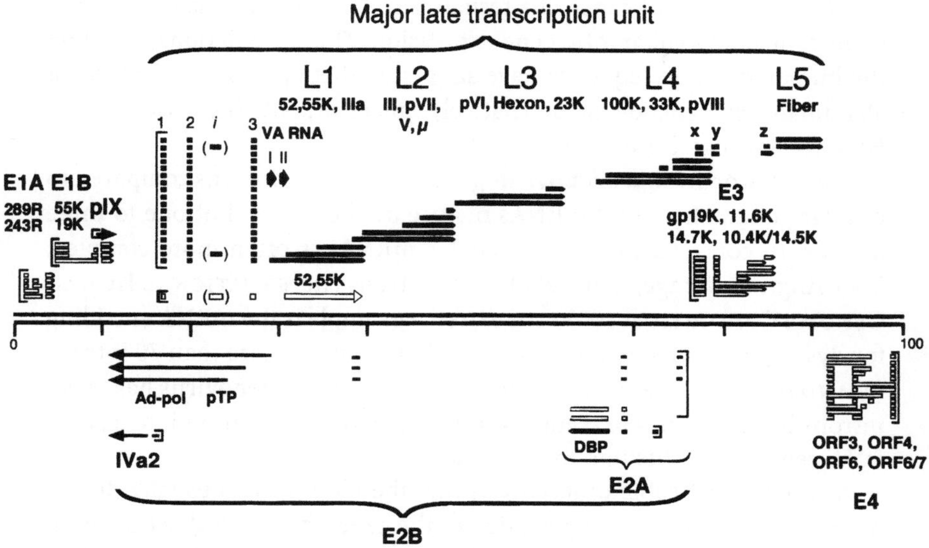

Fig. 1. The adenovirus type 2 transcription map. *Arrows* illustrate the structure of mRNAs expressed at different times of infection. *Thick arrows*, mRNAs expressed early; *thin arrows*, mRNAs expressed at intermediate times; *open arrows*, mRNAs expressed late during infection. Some of the more abundant polypeptides that have been assigned to the different transcription units are indicated. (Adapted from Chow et al. 1979; Akusjärvi et al. 1986)

The adenovirus system provides an excellent opportunity to study regulatory events that are manifested at the level of RNA splicing, as most of the viral transcription units encode for two or more alternatively spliced mRNAs (Fig. 1). The potential to produce alternatively spliced mRNAs from a pre-mRNA is of great importance as it allows for a novel mechanism to regulate adenovirus gene expression. It is clear that the temporal control of alternative splicing is of vital importance for lytic virus growth by controlling the synthesis of proteins that are needed at certain stages of the life cycle; for example, maturation of the multiple mRNAs encoding the structural proteins of the viral capsid late during infection (Fig. 1). The general tendency is that shorter mRNAs, produced by splicing out larger introns, accumulate at later time points of infection (Chow et al. 1979). For example, the E1A 13S and the E1B 22S mRNAs are the most abundant E1 mRNAs expressed at early times of infection whereas the shorter E1A 9S mRNA and E1B 13S mRNAs predominate at late times of infection. The recent completion of the DNA sequence of multiple eukaryotes has further highlighted the significance of alternative splicing as a regulatory mechanism controlling eukaryotic gene expression. Thus, predictions based on the human genome sequence have suggested that approximately 60% of all human genes encode for alternatively spliced mRNAs (International Human Genome Consortium 2001).

It is noteworthy that adenovirus genes contain few introns compared to cellular genes. Most viral mRNAs mature by the removal of one to three introns. In contrast, splicing of cellular mRNAs is often more complex, involving, on average, removal of seven introns (International Human Genome Consortium 2001). An extreme example is the mRNA encoding for dystrophin which matures by the splicing out of more than 70 introns (Mandel 1989). A valid argument is, of course, that adenovirus has fewer introns because it has to compress much genetic information into a small DNA genome that fits into the viral capsid.

A second noticeable difference is that the viral introns usually do not interrupt the protein coding portion of the gene (the noticeable exception is E1A). Most often introns are positioned in the 5′ or the 3′ non-coding portion of the precursor-RNA (pre-mRNA). This is in striking contrast to cellular genes where introns typically interrupt the coding portion of the gene.

1.1 The SR Family of Splicing Factors and Spliceosome Assembly

As will become evident later, the SR family of splicing factors is of particular importance for this review. We will therefore give a brief description of the function of SR proteins in spliceosome assembly. For a more comprehensive review on spliceosome assembly and splicing we refer to other reviews (Burge et al. 1998; Graveley 2000).

Pre-mRNA splicing takes place in the nucleus and occur either co- or post-transcriptionally. The exonic sequences are joined to generate the mature mRNA that is exported to the cytoplasm and translated into protein. Assembly of the spliceosome is a dynamic process and occurs via a sequential recruitment of multiple small nuclear ribonucleoproteins (snRNPs) and many non-snRNP protein factors, like U2AF and SR proteins, to the pre-mRNA (reviewed in Burge et al. 1998). SR proteins are a family of essential splicing factors that is required for constitutive splicing and which function as regulators of alternative splicing (reviewed in Fu 1995; Manley and Tacke 1996; Graveley 2000). The function of SR pro-

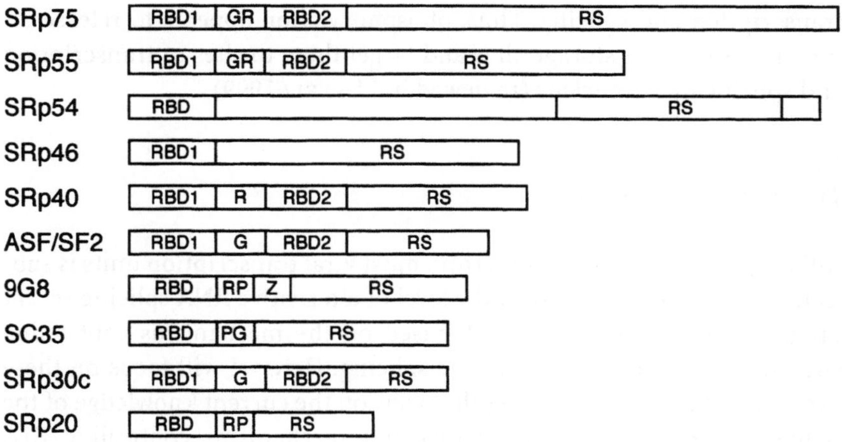

Fig. 2. Schematic diagram of the domain structure of human SR proteins. *RBD*, RNA binding domain; *Z*, zinc knuckle; *RS*, arginine/serine-rich domain. The regions between the RBDs and RS domains are rich in glycine (*G*), arginine (*R*), and proline (*P*) residues. (Modified from GRAVELEY 2000)

teins in spliceosome formation resembles that of enhancer binding transcription factors in pre-initiation complex formation at a promoter. Thus, SR proteins bind to RNA sequences called splicing enhancer elements and help to initiate spliceosome assembly, by recruiting U1 snRNP and U2AF to the pre-mRNA. SR proteins also function as exon bridging factors, and make simultaneous contact with factors bound to the 5′ and 3′ splice sites (WU and MANIATIS 1993; STARK et al. 1998). SR proteins range in size between 20 and 75 kDa (Fig. 2). Like transcription factors, SR proteins can be subdivided into a nucleic acid binding domain and an effector domain. Thus, SR proteins have one or two amino-terminal RNA binding domains and a carboxy-terminal effector domain rich in arginine and serine dipeptide repeats (RS-domain), which mediates protein interaction with other components of the splicing machinery (reviewed in FU 1995; MANLEY and TACKE 1996; GRAVELEY 2000).

SR proteins are highly phosphorylated in vivo, mainly in the RS domain, a modification that is required for their function in spliceosome assembly (reviewed in MISTELI 1999). Thus, phosphorylated SR proteins are needed for initiation of spliceosome assembly, and SR protein dephosphorylation appears to be required for splicing catalysis. The phosphorylated status of SR proteins has also been shown to control their localization and transport in the nucleus between speckles and sites of active transcription and splicing. Thus, phosphorylation appears to release SR proteins from their storage sites and target them to sites of transcription and splicing in the nucleus (reviewed in Misteli 1999).

1.2
The Scope of This Article

Although mRNA accumulation from most viral transcription units is subjected to a temporal control at the level of alternative RNA splicing recent progress has to a large extent focused on the mechanisms controlling adenovirus E1A and L1 alternative splicing. Here we will focus on these two units and provide an in depth review on the current knowledge of the sequence elements, splicing factors and mechanisms controlling E1A and L1 alternative splicing. For a more general review on adenovirus splicing we refer to two previous reviews (Akusjärvi et al. 1986; Imperiale et al. 1995).

2
Regulation of E1A Alternative Splicing

During the early period of infection, six major transcription units; E1A, E1B, E2A, E2B, E3 and E4 are expressed from the adenovirus genome. Among the resulting pre-mRNAs, those of the E1A, E1B, E3 and E4 units are submitted to alternative splicing. The E1A transcription unit is the shortest of these units [1,130 base pairs (bp) long], a fact that has greatly facilitated its in vivo, as well as in vitro study. Soon after the discovery of the split gene concept in the adenovirus major late unit, it was shown that the E1A pre-mRNA gives rise to three major mRNAs, the 13S, 12S, and 9S mRNAs, by alternative splicing using three 5″ splice sites and a common 3′ splice site (Fig. 3A, Berk and Sharp 1978; Spector et al. 1978; Perricaudet et al. 1979). In addition, a strong temporal shift in E1A alternative splicing occurs during the infection cycle, such that the 9S mRNA, which is almost undetectable at early times, becomes the predominant E1A mRNA at late times of infection. Although early studies suggested that the increase in 9S mRNA accumulation, late in infection, might result from an increased cytoplasmic stability (Wilson and Darnell 1981), it became rapidly clear that the shift from 13S to 9S mRNA production is regulated primarily at the level of RNA splicing (Svensson et al. 1983).

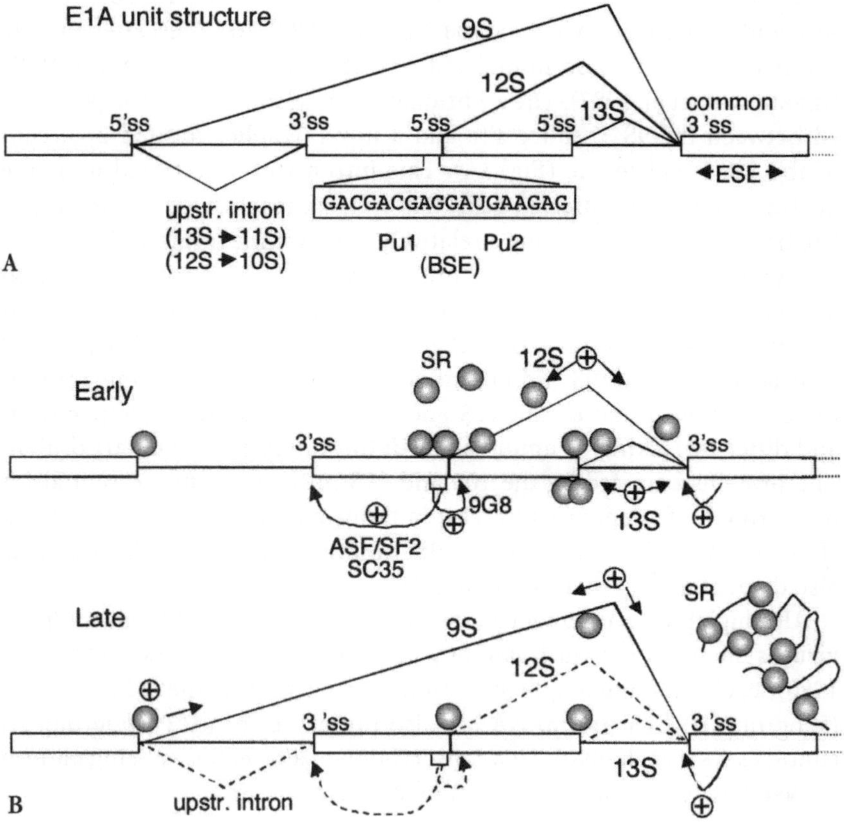

Fig. 3 A, B. The adenovirus E1A unit; structure and regulation. **A** Schematic drawing showing the various alternative splicing reactions that occur in the E1A pre-mRNA. The positions of the identified splicing enhancers (BSE and ESE) are indicated, and the nucleotide sequence of the BSE is shown expanded. **B** A model for the temporal shift from 13S to 9S mRNA splicing, late in infection. At early times of infection, SR proteins (shown as *beads*) bind to the three E1A 5′ splice sites, based on their intrinsic strength, leading to the almost exclusive selection of the proximal 13S and 12S 5′ splice sites. After transition to the late phase of infection, SR proteins are sequestered by the massive amount of transcript synthesized from the MLTU. This results in a decrease in 13S and 12S 5′ splice site selection and a concomitant activation of the 9S splicing reaction

A second minor alternative splicing reaction has been discovered in Ad2 and Ad5, first in vitro (Schmitt et al. 1987) and thereafter rapidly confirmed to occur also in infected cells (Stephens and Harlow 1987; Ulfendahl et al. 1987). The additional 216 nucleotide intron is positioned between the 9S 5′ splice site and a novel 3′ splice located upstream of the 12S 5′ splice site (Fig. 3A). This intron will be referred to as the upstream intron. In infected cells, this splicing reaction occurs mainly at late times of infection, during a relatively narrow time frame and results in the synthesis of the 10S and 11S mRNAs (Fig. 3A). This splicing reaction occurs also moderately in culture cells transfected with the E1A unit (Svensson et al. 1983; Caceres et al. 1994; Yang et al. 1994). The proteins translated from the 10S and 11S mRNAs are encoded in the same translational reading frame as the 12S and 13S mRNA products respectively, and differ by lacking 72 amino acids in the amino-terminal part of E1A. The proteins encoded by the 10S and 11S mRNAs are non-essential for lytic growth of adenovirus, at least in tissue culture cells (Stephens and Harlow 1987; Ulfendahl et al. 1987). Similarly, the 9S mRNA protein is also dispensable for virus growth in HeLa cells.

Although the E1A sequences of subgroups A, B and C of human adenoviruses show an overall homology of only 50% (van Ormondt et al. 1980), each individual splicing reaction occurring in the Ad2 and Ad5 (subgroup C) E1A pre-mRNA are also preserved in Ad12 (subgroup A) (Brockmann et al. 1990). This conservation of alternative splicing pattern probably reflects a specific feature of all adenoviruses.

2.1 The Cis-Acting Elements Involved in the Temporal Control of 13S to 9S mRNA Splicing

The early to late shift from 13S to 9S splicing is probably regulated by the intrinsic properties of the 5′ splice site regions. From the primary sequence, the expected strength of the three 5′ splice sites is 13S > 12S > 9S. Therefore the selection of the most proximal 13S and 12S 5′ splice sites at early times, as well as in in vitro in nuclear extracts prepared from uninfected HeLa cells (HeLa-NE) or early infected HeLa cells (Schmitt et al. 1987; Gattoni et al. 1991) suggests that the 13S and 12S splicing reactions represent the default splicing pathway of the E1A pre-mRNA. Interestingly, varying the ionic conditions in HeLa-NE, or inactivation of the 13S, or both the 13S and the 12S 5′ splice sites, results in a significant increase in

9S mRNA splicing. This finding indicates that all three 5′ splice sites are indeed in direct competition and that the 9S mRNA splicing reaction is not absolutely dependent on late viral protein expression (SCHMITT et al. 1987; GATTONI et al. 1991).

In addition, a novel cis-acting bidirectional splicing enhancer (BSE), located just upstream of the 12S 5′ splice site (Fig. 3A), has been shown to play a critical role in E1A pre-mRNA splicing (BOURGEOIS et al. 1999). The BSE consists of two contiguous 9 nucleotide purine motifs, called Pu1 and Pu2. The Pu1 motif activates specifically the downstream 12S 5′ splice site by binding the SR protein 9G8, whereas the Pu1 and Pu2 motifs together activate the minor upstream 3′ splice site used in the upstream intron splicing reaction (see Sect. 2.3). Mutations that destroy the activity of the BSE inhibits both 12S mRNA and upstream intron splicing in vitro and in transfected cells. These results suggest that this splicing enhancer is of crucial importance for E1A alternative splicing in infected cells (BOURGEOIS et al. 1999).

Besides the three alternative 5′ splice sites, the common 3′ splice site in the E1A unit also exhibits some specific features. Based on an inspection of its primary sequence (a short U rich sequence flanked on each side by purine rich motifs), this 3′ splice site would be considered as a suboptimal 3′ splice site; however, it is used very efficiently in its natural context, both in vivo and in vitro. Interestingly, it resembles the 3′ splice site of the IIIa mRNA. The suboptimal character of the E1A 3′ splice site was confirmed by the demonstration that it requires a second exon splicing enhancer for activity (ESE; Fig. 3A). Thus, mutating the exonic sequences downstream of it resulted in a strong reduction of E1A 3′ splice site usage. Interestingly, insertion of an SR dependent splicing enhancer in place of the mutated sequence restored an efficient splicing (CAVALOC et al. 1999), a result demonstrating that the common E1A 3′ splice site requires a second exon splicing enhancer. However, it should be noted that this finding does not necessarily demonstrate that the natural second exon splicing enhancer functions by recruiting SR proteins. Replacing the common E1A 3′ splice site with the 3′ splice site from the constitutively active β-globin pre-mRNA resulted in a high 9S mRNA splicing in transiently transfected HeLa cells (ULFENDAHL et al. 1989), suggesting that the common E1A 3′ splice site plays an active role in the default and regulated splicing of the E1A pre-mRNA during virus infection.

2.2 Sequestering/Titration of SR Proteins Triggers the Shift from 13S to 9S mRNA Splicing

The fact that Ad2 and Ad5 grow lytically in HeLa cells, which were initially the only cell type available for the preparation of highly active nuclear extracts for in vitro splicing experiments, has greatly facilitated the biochemical analysis of the mechanisms controlling E1A alternative splicing.

The shift in E1A pre-mRNA splicing has been reproduced in vitro using nuclear extracts prepared at early [5 h post infection (pi)] and late times of infection (12–14 hpi) (GATTONI et al. 1991). Importantly, such late nuclear extract also showed an enhanced capacity to splice the first intron (between leader 1 and 2; Fig. 1) in the major late transcription unit (MLTU). This finding suggests that the late extracts exhibit the expected property of late infected cells. Two complementary approaches have been used to identify the molecular mechanisms controlling E1A alternative splicing. Both approaches were aimed at determining what kind of treatments were required to convert an uninfected, or an early nuclear extract, to an extract with the same characteristics as a late nuclear extract, or conversely convert a late extract to an extract with the characteristics of an uninfected nuclear extract (GATTONI et al. 1991; HIMMELSPACH et al. 1995).

Interestingly, nuclease treatment of a late nuclear extract resulted in a loss of its late splicing phenotype. This was scored as an improvement of 13S mRNA splicing in a complementation assay (GATTONI et al. 1991). In contrast, the addition of adenoviral RNA purified from late nuclear extracts, or an excess of the E1A pre-mRNA, to an uninfected or an early infected nuclear extract resulted in a specific gain of 9S relative to 13S mRNA splicing, similar to that observed in a late nuclear extract (GATTONI et al. 1991; HIMMELSPACH et al. 1995). Collectively, these results strongly suggest that the dramatic accumulation of adenoviral transcripts, expressed from the MLTU at late times of infection (accounts for 50%–60% of the non-ribosomal RNA at late times of infection, MANLEY et al. 1979), triggers a sequestration and a titration of RNA binding splicing regulators which become limiting for 13S mRNA splicing, but not for the 9S mRNA splicing (Fig. 3B) (GATTONI et al. 1991).

U1 snRNP and SR proteins, which both participate in the early recognition of 5′ splice sites, represented the candidate splicing factors potentially involved in the modulation of E1A alternative splicing. Subsequent studies have shown that U1 snRNP is not the primary target involved in this modulation (HIMMELSPACH et al. 1995). In contrast, partial inactivation of the SR family of splicing factors present in uninfected nuclear extracts by the addition of a 9G8-specific antibody or an antibody directed against the whole family of SR proteins resulted in a shift from 13S to 9S mRNA splicing (HIMMELSPACH et al. 1995). Conversely, addition of purified SR proteins to a late nuclear extract restored 13S mRNA splicing while blocking 9S mRNA splicing. Significantly, direct analysis of SR proteins in extracts prepared at early or late times of infection demonstrated that the total steady-state amount of SR proteins, and their activity as splicing regulators were not changed at 12 hpi (HIMMELSPACH et al. 1995). Importantly, at this time of infection the phosphorylated status of SR proteins was not significantly altered, a modification that appears to occur later in infection, at 20–24 hpi when IIIa mRNA splicing becomes highly active (Sect. 3.2; KANOPKA et al. 1998; ESTMER NILSSON et al. 2001).

Collectively, these data strongly support the idea that SR proteins are recruited and sequestered late in infection (around 12 hpi) by the excess of transcripts synthesized from the MLTU, resulting in a reduction in the functional concentration of some SR proteins needed for 13S mRNA splicing, but not for 9S mRNA splicing (Fig. 3B). Interestingly, several studies, using various approaches, are fully consistent with the proposed mechanism. First, it has been shown that the spatial distribution of splicing factors, including U snRNP, SR proteins and U2AF is temporally regulated during an adenovirus infection, and that these factors are preferentially detected at or near sites of viral transcription (PUVION-DUTILLEUL et al. 1994; BRIDGE et al. 1995, 1996; GAMA-CARVALHO et al. 1997; ASPEGREN et al. 1998). Thus, a colocalization of splicing factors and viral transcripts, which accumulate dramatically in limited regions of the nucleus may facilitate the recruiting and sequestering of specific RNA binding proteins of the splicing machinery. Second, a rough survey of the MLTU sequence (~28,000 bp) shows that purine rich motifs of 8–10 bp, containing triplets of GAA, GGA, GAG or GAC are frequently present in MLTU transcripts (between 8 and 10 motifs per 1,000 bp). Such motifs resemble the cis-acting elements identified in the E1A unit (the BSE) or in the L1 unit (the 3RE, see Sect. 3). They are likely to be high affinity binding sites for SR proteins like ASF/SF2, SC35 or 9G8 (TACKE and MANLEY 1995; CAVALOC et al. 1999), and potentially may serve as strong elements recruiting SR proteins. Third, analysis of the splicing of adenoviral transcripts has been performed in the absence of late viral protein synthesis, under conditions which preserve efficient viral DNA replication and transcription (LARSSON et al. 1991). Under such conditions, a substantial shift from 13S to 9S mRNA splicing occurs, demonstrating that the 13S to 9S modulation does not require late viral protein expression. In contrast, the shift in L1 pre-mRNA splicing was significantly distorted under these conditions (see also Sect. 3), indicating that the temporal regulation of E1A and L1 alternative splicing differ mechanistically. It is noteworthy that the 13S to 9S modulation persists during the late phase of infection (Ulfendahl et al. 1987). However, this is not completely unexpected as the partial SR protein dephosphorylation, which takes place at the very late stage of infection, is probably functionally equivalent to the initial SR protein sequestration that occurs at the early to late transition.

Another important question to be answered is whether the sequestration mechanism used in E1A alternative splicing is encountered in other biological situations. Although it appears that such a sequestration mechanism may be rare in non-pathological conditions, several examples of titration of protein factors by RNA have been proposed in pathological situations. For example, viral transcription in cells infected by vesicular stomatitis virus (VSV) is necessary and sufficient to trigger the inhibition of the maturation of U snRNP, most probably because the 5′ part of the early RNA of VSV sequesters the Sm proteins necessary for U snRNP assembly (Crone and Keene 1989). Another unexpected example concerns myotonic dystrophy, a neuromuscular disorder associated with CTG expansion in the 3′ untranslated region of the myotonin protein kinase gene (Philips et al. 1998). A proposed model for this disease is that the expanded CUG repeats in the 3′ untranslated region of the mRNA sequester specific RNA-binding proteins interacting with CUG repeats, namely the CUG-binding protein, and other related proteins (Lu et al. 1999).

2.3 Specific Features of the Upstream Intron Splicing Reaction

As described earlier, the upstream intron splicing reaction occurs at the late stage of infection, but only during a relatively narrow window of time, and with a moderate efficiency (Stephens and Harlow 1987; Ulfendahl et al. 1987). Thus, splicing of the upstream intron is also an example of an alternative splicing reaction that is subjected to viral regulation. From a functional point of view, it is also important that this splicing reaction is inefficient early during infection as removal of the upstream intron excises a portion of the 13S or 12S mRNA that is essential for establishment of an efficient lytic virus replication (STEPHENS and HARLOW 1987; ULFENDAHL et al. 1987).

The cis-acting elements, which are responsible for the weak splicing phenotype of the upstream reaction, have been studied in vitro. First, the multiple branch sites used in splicing of this intron are located unusually far away from the 3′ splice site (at 51, 55 and 59 nucleotides, respectively) and should normally render these branch sites non-functional (SCHMITT et al. 1987; GATTONI et al. 1988). However, a detailed mutational analysis indicated that a hairpin structure formed just upstream of the 3′ splice site induces a shortening of the operational distance between the branch sites and the minor 3′ splice site such that these branch sites retain a moderate activity (CHEBLI et al. 1989). Second, the 3′ splice site itself appears to be weak, and splicing enhancers capable of activating the upstream reaction have been searched downstream of the 216 nucleotide intron. Evidence has been presented that two cis-acting elements act coordinately to make the 216 nucleotide intron functional in vitro and in vivo (POPIELARZ et al. 1993; BOURGEOIS et al. 1999). The first element consists of the 13S 5′ splice site, located 259 nucleotides downstream of the 216 nucleotide intron, and its absence results in a significant inhibition of upstream intron splicing (POPIELARZ et al. 1993). Surprisingly, the 12S 5′ splice site, although positioned at an ideal distance (121 nucleotides) does not exhibit such a cis-acting splicing enhancer activity. The second cis-acting splicing enhancer element is the BSE, located immediately upstream of the 12S 5′ splice site, which activates both the downstream 12S 5′ splice site as well as the upstream minor 3′ splice site (BOURGEOIS et al. 1999). Whereas the 12S mRNA splicing reaction is activated by the SR protein 9G8, the upstream reaction is activated by ASF/SF2 and SC35, indicating that each splicing reaction depends on the presence of a specific SR protein.

Thus, regulation of the upstream splicing reaction, during the infectious cycle may be explained as follows (Fig. 3B). At early times of infection, the inefficient recognition of the 9S 5′ splice site precludes efficient upstream intron splicing, although the minor 3′ splice site may be optimally active. Thus, at this time point the BSE stimulates the minor 3′ splice site by interacting with ASF/SF2 and/or SC35. In contrast, at late times of infection the excess of MLTU transcripts produced sequesters the SR family of splicing factors resulting in an increase in 9S mRNA splicing, and also a moderate production of the 10S and 11S mRNAs. However, the production of the 10S and 11S mRNAs is not optimal at late times of infection, most probably because the activity of the BSE is reduced because SR proteins become severely limiting and/or become less active due to a decrease of their phosphorylated status (see Sect. 3.2). Thus, splicing of the upstream intron is a minor splicing reaction because efficient recognition of the 5′ splice site and the minor 3′ splice site are regulated by conflicting mechanisms. When the minor 3′ splice site is highly active the 9S 5′ splice site is inefficiently used, and when the 9S 5′ splice site is optimally active the minor 3′ splice site is downregulated.

2.4 The E1A pre-mRNA Is a Preferred Model System for Studies of Splicing Factor Activity

The E1A pre-mRNA has became a standard model system to assess the function of various splicing factors with a putative function in alternative RNA splicing (Ge and Manley 1990; Krainer et al. 1990). Thus, the E1A unit represents a natural pre-mRNA substrate where alternative 5′ splice site usage can be studied. The alternative method, which is sometimes used, is to study alternative splicing in chimeric substrates containing duplicated 5′ splice sites. Additionally, the E1A pre-mRNA has been shown to be very sensitive to changes in various parameters, both in vitro (Schmitt et al. 1987; Gattoni et al. 1991; Harper and Manley 1992) and in vivo (Larsson et al. 1991).

These initial studies have paved the way for E1A as a general model system to characterize the function of SR proteins as regulators of alternative RNA splicing. Comparison of the activity of SR proteins on E1A alternative splicing, either in vitro or in transient cotransfection assays has generated data of general interest. Thus, the activity of the whole family of SR proteins (Fig. 2) has been tested using the E1A unit as model substrate (Zahler et al. 1993; Screaton et al. 1995; Wang and Manley 1995; Zhang and Wu 1996). The results show that many of the tested SR proteins have unique abilities to activate one of the three 5′ splice sites in the E1A pre-mRNA. Thus, whereas ASF/SF2 or SC35 stimulates 13S mRNA splicing, SRp20 enhances 12S mRNA splicing and SRp54 9S mRNA splicing. Most probably the activation by SRp20 and SRp54 occurs via an interaction with specific cis-acting splicing enhancer elements. However, these enhancer elements have not yet been identified.

The function of SR protein kinases has been studied also by using E1A unit as a substrate. Thus, Clk/Sty, which phosphorylates SR proteins, has been shown to modulate the activity of SR protein in E1A pre-mRNA splicing (Duncan et al. 1997; Prasad et al. 1999). The results show that over-phosphorylation of SR proteins leads to an inhibition of 13S and 12S mRNA splicing and in a concomitant activation of 9S mRNA splicing, both in vivo and in vitro. These results together with other results (Kanopka et al. 1998; Estmer Nilsson et al. 2001) strongly suggest that both hyperphosphorylation and hypophosphorylation of SR proteins could potentially lead to a similar shift in E1A alternative splicing. This apparent contradiction is, however, easily explained as both an over-phosphorylation and a dephosphorylation of SR proteins result in a decrease in activity (Gui et al. 1994; Mermoud et al. 1994).

E1A has also served as an important model system to study the function of splicing factors that antagonize the activity of SR proteins. For instance, hnRNP A1 which binds with a high affinity to sequences containing UAGGGA/U motifs (Burd and Dreyfuss 1994) has been shown to enhance 9S 5′ splice site selection and counteract the stimulatory activity of ASF/SF2 on 13S mRNA splicing (Mayeda and Krainer 1992; Caceres et al. 1994; Yang et al. 1994). A similar effect on E1A 5′ splice site selection has also been observed with another RNA binding protein, TLS, which does not, however, recognize the same RNA motif as hnRNP A1 (Lerga et al. 2001).

Finally, it is worth pointing out that although the E1A unit has been well studied, several important questions remain to be answered. For example, the role of the common E1A 3′ splice site in alternative splice site selection remains to be elucidated. Current data clearly indicate that the SR family of splicing factors plays a significant role in the regulation of 13S to 9S mRNA splicing. However, the functional status, and therefore the contribution, of each SR protein in E1A alternative splicing during a virus infection remain to be established.

3 Temporal Control of L1 Alternative Splicing

The adenovirus L1 unit is part of the so-called major late transcription unit (Fig. 1). The MLTU generates a primary transcript of approximately 28,000 nucleotides, that is processed into a minimum of 20 cytoplasmic mRNAs, which are grouped into five families (L1 to L5, Fig. 1), where each family consists of multiple alternatively spliced mRNAs with co-terminal 3′ ends. An important consequence of the processing pathway is that all mRNAs expressed from the MLTU have a common 201 nucleotide long non-coding tripartite leader sequence at their 5′ end. This leader allows the MLTU mRNAs to initiate translation by a cap independent mechanism (see Yueh and Schneider 2000). In fact, the 5′ leader comes in two variants, either with or without the extra 440-nucleotide long i-leader exon (Fig. 1). The biological significance of the i-leader exon inclusion/skipping reaction is far from clear. However, it should be noted that the i-leader encodes a stable and abundant 16-kDa protein that is expressed at intermediate and late times of infection (Symington et al. 1986). Although the i-leader protein is dispensable for virus growth in tissue culture cells (Soloway and Shenk 1990) it may have an important function connected with virus growth in humans. Previous studies have shown that the virus-encoded E4-ORF3 and E4-ORF6 proteins control i-leader inclusion/-skipping (reviewed in Imperiale et al. 1995).

The major late promoter is active both early and late after infection. However, at early times transcription terminates near the middle of the genome resulting in an exclusive expression of the L1 unit (Fig. 1). L1 represents an example of an alternatively spliced gene where the last intron is spliced using a common 5′ splice site and two alternative 3′ splice sites, generating two predominant cytoplasmic mRNAs; the 52/55K (proximal 3′ splice site) and the IIIa (distal 3′ splice site) mRNAs, respectively (Fig. 4). The pattern of splicing of the L1 pre-mRNA is regulated during the infectious cycle (Chow et al. 1979; Akusjärvi and Persson 1981; Nevins and Wilson 1981). Thus, the 52/55K mRNA is produced both early and late after infection, whereas the IIIa mRNA is produced exclusively at late times of infection (Fig. 4).

Although, L1 gene expression is subjected to a tight control at the level of alternative splicing, there is no obvious functional reason for this temporal expression of the 52/55K and IIIa proteins. Thus, the absence of 52/55K expression at early times (Gustin and Imperiale 1998) or expression of the IIIa protein from the start of infection (Molin et al. 2002) does not have a detrimental effect on virus growth in tissue culture cells. Despite the fact that the 52/55K mRNA is detected very early after infection the 52/55-kDa protein appears to serve its main function late during infection, as a protein stabilizing association of viral DNA with empty capsids – a result that suggests a function in viral DNA encapsidation (Gustin and Imperiale 1998). This study also showed that the 52/55-kDa protein does not provide an essential function at early times but appears to be required for the proper activation of late gene expression at the transition from the early to the late stage of infection. IIIa is a structural

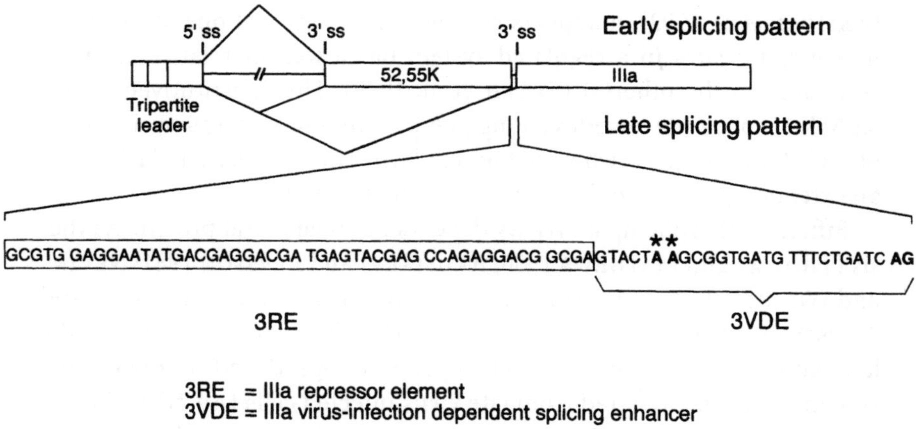

Fig. 4. Schematic drawing showing the splicing pattern of the L1 pre-mRNA at early and late times of infection. *Below*, the nucleotide sequence of the IIIa 3′ splice site, with the position of the 3RE and 3VDE indicated. *Asterisks* indicate the two A residues used as branch sites in IIIa splicing. (Kreivi et al. 1991)

protein, found interspersed between the peripentonal hexons in the virus capsid (reviewed in Stewart and Burnett 1995). Overexpression of the IIIa protein, from the start of infection, results in a slight (two- to fivefold) decrease in late protein expression (Molin et al. 2002). This study also showed that unscheduled expression of the IIIa protein results in a change in the relative expression of L1 mRNAs. Thus, the IIIa protein may partake in an auto-regulatory loop that amplifies IIIa mRNA production by inhibiting 52/55K mRNA splicing, late after infection. It is conceivable that the small alterations in late gene expression observed in tissue culture cells with viruses lacking 52/55K expression, or overexpressing the IIIa protein may exert a more profound effect on the capacity of adenovirus to grow in an infected human.

Although IIIa mRNA splicing is a late specific event, the IIIa 3′ splice site is recognized also in uninfected cells, both in vivo (Delsert et al. 1989) and in vitro (Kreivi et al. 1991). However, under such conditions, the efficiency of IIIa splicing is low, only a few percent compared to that of 52/55K splicing (Kreivi et al. 1991). This difference in splice site activity correlates with a reduced affinity of the IIIa 3′ splice site for cellular polypyrimidine tract binding proteins, such as the general splicing factor U2AF (Mühlemann et al. 1995, 2000); this protein binds with a strong preference for long pyrimidine rich sequences, the longer the better

(Zamore et al. 1992). It appears common in nature that one of the alternative splice sites in a regulated system has a weaker relative strength compared to the other(s) (Smith et al. 1989). The constitutively active 52/55K 3′ splice site encodes a long polypyrimidine tract (18 pyrimidines out of 19 nucleotides) whereas the regulated IIIa 3′ splice site lacks such an extended polypyrimidine tract (Kreivi et al. 1991).

Efficient IIIa splicing is strictly dependent on late viral protein synthesis (Thomas and Mathews 1980; Akusjärvi and Persson 1981; Nevins and Wilson 1981; Johnston et al. 1985; Pilder et al. 1986; Svensson and Akusjärvi 1986; Larsson et al. 1991). The shift from the early to the late pattern of L1 alternative splicing can be reproduced in vitro, using splicing extracts prepared from late virus infected cells (Ad-NE) (Kreivi and Akusjärvi 1994; Mühlemann et al. 2000). However, it should be acknowledged that the shift towards an enhanced IIIa mRNA splicing is only partially restored in Ad-NE. This failure to accurately reproduce the in vivo shift in 52/55K/IIIa mRNA splicing may result from experimental limitations. Thus, to be able to study splicing in vitro a substantial portion of the authentic L1 sequences has been removed to create mini-gene construct suitable for in vitro experimentation. Potentially important cis-acting regulatory elements are also deleted in such mini-gene constructs.

The activation of IIIa splicing in the context of mini-gene constructs have shown that the shift in splice site usage results from an enhanced efficiency of IIIa 3′ splice site recognition combined with a virus-induced repression of 52/55K splicing (Kreivi and Akusjärvi 1994). An important finding was that the shift in L1 3′ splice site usage does not require cis competition between the 52/55K and IIIa 3′ splice sites. Thus, the enhancement of IIIa splicing and inhibition of 52/55K splicing in Ad-NE was observed on transcripts encoding the individual units (Kreivi and Akusjärvi 1994).

More recent work has demonstrated that IIIa splicing is tightly regulated by two cis-acting viral elements: the IIIa repressor element (3RE), and the IIIa virus-infection dependent splicing enhancer (3VDE; Fig. 4). In the following sections we will summarize the current views on how the 3RE and the 3VDE contribute to the regulated expression of the IIIa mRNA in adenovirus-infected cells.

3.1 Function of the 3RE

The 3RE functions as a IIIa splicing repressor element, by binding highly phosphorylated SR proteins present in uninfected, and probably also in early virus-infected cells (Kanopka et al. 1996). The 3RE is a 49 nucleotide long RNA segment that bind all tested members of the SR family of splicing factors. It is located immediately upstream of the IIIa branch site (Fig. 4), and SR protein binding to the 3RE inhibits IIIa splicing by preventing U2 snRNP recruitment to the IIIa branch site, in part by steric interference (Kanopka et al. 1996). All of the classical SR proteins [SRp20, the SRp30 fraction (includes SC35, 9G8, SRp30c, ASF/SF2), SRp40, SRp55 and SRp75] can suppress IIIa splicing by binding to the 3RE, with the SRp30 fraction and SRp55 being most effective. The absence of IIIa splicing in early virus-infected cells can therefore, in part (see Sect. 3.4), be explained by a direct inhibitory effect of HeLa SR proteins on IIIa 3′ splice site usage.

The inhibitory effect of the 3RE is simply due to its high affinity for SR proteins. Thus, replacing the 3RE with consensus binding sites for ASF/SF2 results in splicing repression (Kanopka et al. 1996). Also, moving the 3RE to the second exon in the IIIa pre-mRNA converts the 3RE from a repressor element to a classical splicing enhancer element (Kanopka et al. 1996). Collectively, our results have suggested that, dependent on where in the pre-mRNA SR proteins bind, they either activate or repress splicing. More recent experiments have shown that the splicing enhancer and splicing repressor functions of ASF/SF2 can be separated and attributed to distinct domains in the protein. Thus the RS domain functions as the splicing enhancer domain, whereas the second RNA binding domain (RBD2) encodes the splicing repressor domain (Dauksaite and Akusjärvi 2002). Interestingly, the functions of RBD2 and the RS domains in splicing regulation are opposite and position dependent. Thus, the ASF/SF2-RBD2 domain inhibits IIIa splicing when tethered to the IIIa intron, but has no effect when bound to the second exon. In contrast, the ASF/SF2-RS domain activates IIIa splicing when tethered to the IIIa second exon, but has no effect on splicing when bound at an intronic position.

In the context of mini-gene constructs the activity of SR proteins as regulators of splicing appears to be directional: resulting in an activation of upstream 3′ splice site usage while causing a repression of downstream 3′ splice site usage (Kanopka et al. 1996). Thus, in the mini-52/55K–IIIa transcript the 3RE functions as a splicing repressor element for IIIa splicing while simultaneously functioning as a classical second exon splicing enhancer for 52/55K splicing (Kanopka et al. 1996). However, in the authentic L1 unit the 3RE is positioned around 1,200 nucleotides downstream of the 5255K 3′ splice site. At this distance the 3RE apparently does not function as a 52/55K splicing enhancer element. Thus, in an interesting study Gustin and Imperiale (1998) constructed a mutant virus containing five point mutations in a region located 50–61 nucleotides downstream of the 52/55K 3′ splice site. These mutations were introduced in order to eliminate translation of the 52/55-kDa protein. Surprisingly, these mutations also abolished 52/55K mRNA expression, suggesting that the mutations introduced inactivated an essential splicing enhancer required for 52/55K mRNA splicing. The existence of such a 52/55K specific splicing enhancer was not unexpected. Although the 52/55K 3′ splice sites contain consensus types of splicing signals (Kreivi et al. 1991), we have shown that 52/55K mRNA splicing is strictly dependent on a second exon splicing enhancer for activity (Yue and Akusjärvi 1999). Apparently, Gustin and Imperiale (1998) identified this enhancer. Despite a lack of 52/55K mRNA splicing, the IIIa mRNA was not expressed earlier or at higher levels in mutant infected cells, corroborating the previous conclusion, derived from in vitro experimentation, that the temporal shift in L1 alternative splicing does not require cis competition between the 52/55K and IIIa 3′ splice sites (Kreivi and Akusjärvi 1994).

3.2
Activation of IIIa pre-mRNA by SR Protein Dephosphorylation

Activation of IIIa splicing in Ad-NE is accompanied by a virus-induced partial dephosphorylation of SR proteins (Kanopka et al. 1998). This SR protein dephosphorylation renders them non-functional as splicing enhancer, or splicing repressor proteins. They no longer bind RNA efficiently, and therefore lose much of their splicing regulatory properties. Importantly, re-phosphorylation of SR proteins isolated from late adenovirus-infected cells restores their activity as splicing repressor proteins (Huang et al. 2002). As SR proteins are essential for generic pre-mRNA splicing (Fu 1995; Manley and Tacke 1996), this virus-induced dephosphorylation results in an inhibition of splicing of pre-mRNAs with consensus-type of splicing signals (Kanopka et al. 1998). As such this mechanism may contribute to the virus induced inhibition of host cell gene expression seen late after infection.

The adenovirus E4-ORF4 protein is multifunctional viral regulator that has been shown to block E1A induced transcription activation (MULLER et al. 1992; KLEINBERGER and SHENK 1993; BONDESSON et al. 1996), induce hypophosphorylation of various viral and cellular proteins (MULLER et al. 1992; KANOPKA et al. 1998), and regulate adenovirus alternative RNA splicing (KANOPKA et al. 1998). It also induces p53-independent apoptosis selectively in transformed cells (MARCELLUS et al. 1998; SHTRICHMAN and KLEINBERGER 1998), and G_2/M arrest in yeast and mammalian cells (KORNITZER et al. 2001). Current data is in agreement with the hypothesis that E4-ORF4 exerts these multiple activities by binding to the cellular serine/threonine specific protein phosphatase 2A (PP2A) (KLEINBERGER and SHENK 1993). The E4-ORF4 protein activates IIIa splicing by inducing SR protein dephosphorylation (KANOPKA et al. 1998). E4-ORF4 activation of splicing requires the 3RE. Thus, deletion of the 3RE from the IIIa pre-mRNA eliminates the stimulatory effect of E4-ORF4 on splicing (KANOPKA et al. 1998; ESTMER NILSSON et al. 2001). Conversely, transfer of the 3RE to the β-globin pre-mRNA reduces, as expected, basal β-globin splicing but makes this transcript sensitive to E4-ORF4 activation of splicing (ESTMER NILSSON et al. 2001).

Surprisingly, E4-ORF4 was shown to interact specifically with only a subset of the SR family of splicing factors. Thus, E4-ORF4 interacts efficiently with ASF/SF2 and SRp30c, but not with SRp20, SC35, SRp40, SRp55 or SRp75 (ESTMER NILSSON et al. 2001). As all SR proteins, except SRp20, appear to be hypophosphorylated in Ad-NE, these results suggest that there must be at least one more viral protein yet to be discovered that regulates SR protein phosphorylation. Furthermore, E4-ORF4 mutant proteins that failed to bind strongly to ASF/SF2 or PP2A also did not relieve the repressive effect of SR proteins on IIIa splicing (ESTMER NILSSON et al. 2001). Although E4-ORF4 also interacts with SRp30c, this interaction did not correlate with E4-ORF4 induced activation of IIIa splicing, suggesting that ASF/SF2 is the primary target for E4-ORF4. Current data is compatible with the hypothesis that E4-ORF4 functions as a bridging factor that recruits ASF/SF2 for dephosphorylation, by making simultaneous contact with ASF/SF2 and PP2A. In agreement with this hypothesis E4-ORF4 selectively interacts with the highly phosphorylated SR proteins present in uninfected HeLa cells. Also, re-phosphorylation of SR proteins prepared from adenovirus infected cells restored E4-ORF4 interaction (ESTMER NILSSON et al. 2001).

3.3 Control of SR Protein Function in Other Viral Systems

Recent studies have shown that SR proteins purified from *Vaccinia virus* infected cells also are dephosphorylated and functionally inactivated as splicing enhancer or splicing repressor proteins (HUANG et al. 2002). In fact, SR proteins purified from *Vaccinia virus* infected cells are further impaired as splicing regulatory proteins compared to those purified from an adenovirus infection. This could be explained by the fact that *Vaccinia virus* is an unusual DNA virus in that it replicates in the cytoplasm and encodes for genes lacking introns, and hence does not require the cellular splicing machinery for its gene expression. In contrast, adenovirus replicates in the cell nucleus and depends on a functional splicing machinery for expression of viral genes. Thus adenovirus, instead of abolishing RNA splicing, appears to redirect the specificity of the host cell splicing machinery such that splicing of the late viral mRNAs is enhanced.

A lesson that we have learnt from the last two decades of research is that mammalian viruses typically target the same key cellular factors controlling cellular growth (reviewed in FLINT et al. 2000). The best examples are the multiple viruses that encode for proteins that interact with the tumour suppressor proteins pRb and p53, and the interferon induced protein kinase PKR. Potentially future studies will show that the SR family of splicing factors also are major targets of viral interference. It is also worth pointing out that other viruses encode for proteins that bind PP2A. Thus, SV40 small-t and polyoma virus middle-t and small-t, and HIV Vpr also interact with PP2A (PALLAS et al. 1990; WALTER et al. 1990; HRIMECH et al. 2000). Potentially, these viral proteins may also function as regulators of RNA splicing.

The cellular splicing factor-associated protein, p32, has also been shown to block ASF/SF2 as a IIIa splicing repressor protein, by preventing stable ASF/SF2 interaction with the 3RE (PETERSEN-MAHRT et al. 1999). Mechanistically, p32 blocks ASF/SF2 as a IIIa splicing repressor protein by inhibiting ASF/SF2 phosphorylation. Therefore, the viral E4-ORF4 protein and the cellular p32 protein both inactivate ASF/SF2 as a splicing regulatory protein by inducing accumulation of hypophosphorylated ASF/SF2. However, they do so by opposite mechanisms. Current data are compatible with the hypothesis that E4-ORF4 induces ASF/SF2 dephosphorylation while p32 accomplishes the same by inhibiting ASF/SF2 phosphorylation.

A surprisingly large number of viruses encodes for proteins that interact with p32. Examples include HIV Rev (Luo et al. 1994), HIV Tat (Yu et al. 1995), HSV-1 Orf-P (Bruni and Roizman 1996), *Epstein-Barr virus* EBNA I (Wang et al. 1997; Chen et al. 1998), *Herpes simplex virus* ICP27 (Bryant et al. 2000) and the adenovirus polypeptide V (Matthews and Russell 1998). The interaction between p32 and the HIV Rev protein is particularly interesting as it has been proposed that p32 de-commits the HIV pre-mRNA from the spliceosome assembly pathway, thereby promoting transport of unspliced HIV RNA required for new virus formation (Tange et al. 1996). Preliminary experiments suggest that adenovirus polypeptide V interaction with p32 regulates the activity of p32 as an ASF/SF2 inhibitory protein (C. Öhrmalm, S. Petersen-Mahrt and G. Akusjärvi, unpublished results).

3.4 Function of the 3VDE

The finding that SR proteins are functionally inactivated in Ad-NE (Kanopka et al. 1998) was, and to some extent still is, an enigma. Why are adenovirus pre-mRNAs efficiently spliced in the absence of fully functional SR proteins? After all SR proteins have been shown to be essential for generic pre-mRNA splicing (reviewed in Fu 1995; Manley and Tacke 1996). With this perspective in mind it is important to note that IIIa splicing is not SR protein independent. Thus, the IIIa pre-mRNA is not spliced in cytoplasmic S100 extracts (contains all splicing factors except the SR family of splicing factors; Krainer et al. 1990). SR proteins appear to perform two independent functions; one that requires binding to RNA, and one function independent of RNA binding (reviewed in Graveley 2000). Most likely, IIIa splicing does not require the RNA binding activity of SR proteins, but depends on the non-RNA binding activity of SR proteins. Also, the identification of the virus-infection dependent splicing enhancer (the 3VDE, Mühlemann et al. 2000) appears to provide, at least, part of the answer to this puzzle. In summary, current experiments have shown that both the 3RE and the 3VDE contribute to the enhanced IIIa splicing phenotype seen late during infection (Figs. 4, 5), with the 3VDE making the most significant contribution (Mühlemann et al. 2000).

The first indication that the 3RE was not the primary element controlling IIIa pre-mRNA splicing came from an inspection of the sequence composition of the IIIa 3′ splice site (Fig. 4). This showed that the IIIa 3′

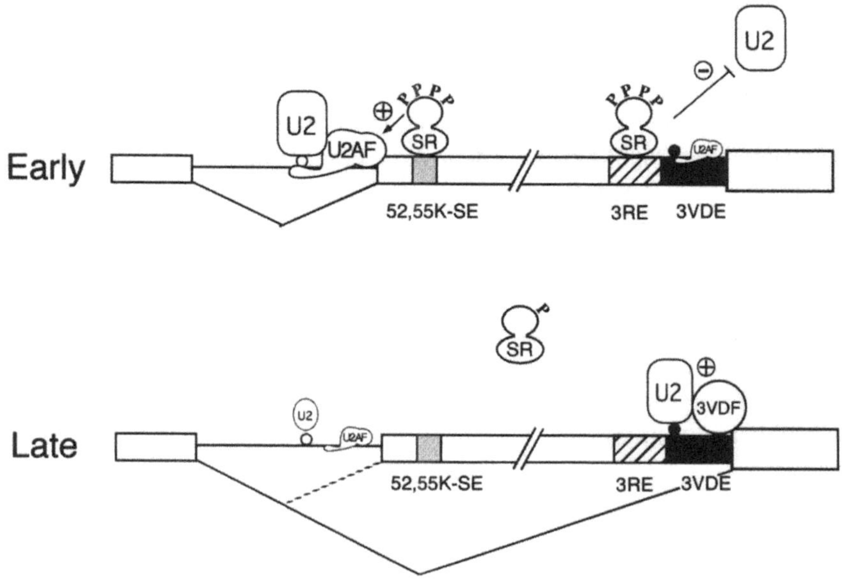

Fig. 5. Temporal regulation of L1 alternative splicing: a model. Early during infection IIIa splicing is low, or absent, because highly phosphorylated SR proteins bind to the 3RE and block U2 snRNP recruitment (*U2*). Also, the general splicing factor U2AF binds inefficiently to the IIIa pyrimidine tract. At the same time highly phosphorylated SR proteins bind to the 52/55K specific second exon splicing enhancer (*52,55K-SE*), stimulating 52/55K splicing. At late times of infection, the repressive effect of SR proteins on IIIa splicing is relieved by a virus-induced dephosphorylation. Also, this hypophosphorylation results in a reduction in 52/55K splicing (*dotted line*). In addition a hypothetical viral, or virus-induced/activated pyrimidine binding factor (*3VDF*), binds to the IIIa pyrimidine tract and stimulates IIIa splicing by recruiting U2 snRNP to the IIIa branch site. *Open circles* denote the branch sites for the 52/55K (*open*) and IIIa (*grey*) 3′ splice sites. See text for further details

splice site is weak and has a short pyrimidine tract (Kreivi et al. 1991) that does not bind the general splicing factor U2AF efficiently (Mühlemann et al. 1995, 2000). U2AF binds with a strong preference to long pyrimidine rich sequences, the longer the better (Zamore et al. 1992; Singh et al. 1995). Interestingly, splicing of pre-mRNAs, like 52/55K or β-globin, that contain long prototypical polypyrimidine tracts, and bind U2AF efficiently, is repressed in Ad-NE, while splicing of the IIIa and pV 3′ splice sites, which are weak, are enhanced in Ad-NE (Mühlemann et al. 1995). Thus, the shift in specificity of RNA splice site choice from pyrimidine to more purine rich 3′ splice sites may be a general feature in the control of RNA splicing in late adenovirus-infected cells. These observations led to the hypothesis that inactivation of the repressive activity of SR proteins on IIIa splicing might not be sufficient to explain the high IIIa splicing activity observed in Ad-NE.

In vitro, IIIa pre-mRNA splicing is activated more than 200-fold in Ad-NE, whereas β-globin splicing is slightly repressed (MÜHLEMANN et al. 2000). To identify the minimal IIIa sequence element conferring an enhanced splicing phenotype to a pre-mRNA in Ad-NE, chimeric pre-mRNAs between IIIa and β-globin were constructed. Using this approach it was shown that the IIIa branch site/pyrimidine tract (Fig. 4) was the key element controlling IIIa splicing. The 3VDE functions as a 'Janus element'; it blocks splicing in HeLa-NE and functions as a splicing enhancer selectively in Ad-NE. Because of these properties this sequence element was named the IIIa virus-infection dependent splicing enhancer (3VDE). The 3VDE is essential for regulated IIIa pre-mRNA splicing and sufficient to convert β-globin from a pre-mRNA that is repressed to a pre-mRNA that is activated in Ad-NE (MÜHLEMANN et al. 2000). Replacing the 3VDE, in the IIIa pre-mRNA with the corresponding sequence from β-globin, resulted as expected, in a constitutively high splicing activity of the IIIa pre-mRNA, and also eliminated the enhancement of IIIa splicing in Ad-NE.

The mechanism by which the 3VDE activates splicing is currently not known. A virus-encoded splicing factor(s) that show a specificity for the viral type of non-consensus 3′ splice sites may specifically be required for 3VDE function. Alternatively, virus infection may lead to an increased synthesis, or post-translational modification, of a cellular 3′ splice site-binding factor. It appears unlikely that such a hypothetical virus-induced modulation would involve the general splicing factor U2AF. First, the steady-state amounts of U2AF does not change in virus-infected cells (KREIVI and AKUSJÄRVI 1994; MÜHLEMANN et al. 2000). Second, U2AF interaction with pyrimidine tracts is reduced in Ad-NE compared to HeLa-NE (MÜHLEMANN et al. 2000). This finding provides an explanation for our observation that splicing of pre-mRNAs with prototypical 3′ splice sites is repressed in Ad-NE, but does not explain how the 3VDE functions in the absence of efficient U2AF recruitment. In constitutive splicing U2AF binds to the polypyrimidine tract at the 3′ splice site, and helps to recruit U2 snRNP to the branch site (reviewed in BURGE et al. 1998). Current experimental evidence is compatible with the model that splicing activation by the 3VDE operates through a novel mechanism that may be

U2AF independent, or alternatively does not require stable binding of U2AF to the IIIa pyrimidine tract. However, U2 snRNP is required for 3VDE function (Mühlemann et al. 2000). In our working hypothesis we predict that an unknown factor (the 3VDE interacting factor, 3VDF in Fig. 5) replaces U2AF as the pyrimidine tract binding factor, and aids in the recruitment of U2 snRNP to the IIIa branch site.

The existence of an alternative HeLa cell 3′ splice site factor that stimulates IIIa splicing is supported by the observation that, using alternative extraction protocols to prepare HeLa-NE, a trans-acting factor that activates IIIa splicing, but does not have a significant impact on 52/55K or β-globin splicing was retrieved (Zerivitz et al. 1992).

3.5
Temporal Regulation of L1 Alternative Splicing: A Model

In Fig. 5 we present a model for how L1 alternative splicing might be regulated during an adenovirus infection. Early during infection IIIa splicing is low or absent because highly phosphorylated SR proteins binds to the 3RE and block U2 snRNP recruitment to the IIIa branch site. Also, the highly phosphorylated SR proteins binds to the 52/55K splicing enhancer and stimulates U2AF recruitment to the 52/55K 3′ splice site, resulting in an efficient 52/55K mRNA formation. Late during infection, a virus-induced partial SR protein dephosphorylation reduces the RNA binding capacity of the SR family of splicing factors. This results in a reduced binding of SR proteins to the 52/55K splicing enhancer and the 3RE. As a result 52/55K splicing will decrease and IIIa splicing increase. At late times of infection IIIa splicing is further augmented by the 3VDF factor which becomes induced and/or activated, and via a binding to the 3VDE further enhances IIIa splicing. This model is compatible with available data but contains several speculative elements that need to be experimentally addressed.

4
Concluding Remarks

Adenovirus has served, and most likely will continue to serve, as a productive model system to study basic mechanisms controlling alternative RNA splicing. Particularly the E1A unit has become 'the' standard substrate to study the function of splicing regulators controlling alternative 5′

splice site selection. Although not reviewed in this chapter, it is worth pointing out that the major late leader 1–leader 2 splicing reaction is one of the preferred model systems to study basic mechanisms in spliceosome assembly and splicing catalysis (Burge et al. 1998). With the increased awareness of the significance of alternative splicing as a regulatory mechanism differentiating higher eukaryotes from lower eukaryotes and bacteria it seems likely that adenovirus will continue to serve as a central model system to study these effects.

The results summarized here highlight the central role that the SR family of splicing factors plays in the regulation of alternative splicing. Thus, viral inactivation of SR proteins appears to partake in the control of the temporal shift in both E1A and L1 alternative splicing. Interestingly, from a mechanistic point of view, the shift in E1A and L1 splicing appears to differ slightly. Thus, the enhanced E1A–9S mRNA splicing, which can be detected in vitro in extracts prepared at 12 hpi, appears to result from a titration/sequestration of SR proteins by the massive amounts of MLTU transcripts synthesized after initiation of viral DNA replication. In contrast, enhanced L1–IIIa mRNA splicing, which is best seen in extracts prepared at 20–24 hpi, appears to result from a virus-induced dephosphorylation of SR proteins. However, the net result in both cases is the same; a functional inactivation of the SR family of splicing factors.

Viruses typically inhibit host cell gene expression to gain full access to the biosynthetic machinery of the cell. Thus, many viruses inhibit host cell RNA processing and RNA transport. Since RNA splicing is a pre-requisite for nuclear to cytoplasmic export of most cellular mRNAs, a virus-induced suppression of RNA splicing may be an important regulatory mechanism by which viruses inhibit host cell gene expression. Some viruses, like *Vaccinia virus* and most RNA viruses, encode for genes lacking introns. Such viruses could theoretically completely shut-off host cell RNA splicing without a significant impact on virus-specific gene expression. However, most DNA viruses replicate in the cell nucleus and depend on a functional splicing machinery for viral gene expression. Thus, all adenovirus genes, with the exception of pIX, contain introns (Fig. 1). Therefore adenovirus, instead of abolishing RNA splicing, appears to redirect the specificity of the splicing machinery such that splicing of pre-mRNAs with consensus-type splicing signals are reduced, and splicing of viral mRNAs is enhanced.

Work on the L1 unit has demonstrated that the authentic IIIa 3′ splice site plays a crucial role for the regulated expression of the IIIa mRNA.

Thus, the IIIa branch site/pyrimidine tract functions as a virus-infection dependent splicing enhancer (the 3VDE; Fig. 4). We note that the common E1A 3′ splice site, which also has a suboptimal sequence content, shows primary sequence similarity to the 3VDE. Potentially E1A and many more viral 3′ splice sites may be regulated by virus infection-dependent splicing enhancer elements.

Acknowledgements. We are grateful to all previous and present members of our laboratories who have contributed to the elucidation of the mechanistic details of E1A and L1 splicing. The work done in our laboratories was supported mainly by the Swedish Cancer Society and the Association pour la Recherche contre le Cancer.

References

Akusjärvi G and Persson H (1981) Controls of RNA splicing and termination in the major late adenovirus transcription unit. Nature 292:420–426

Akusjärvi G, Pettersson U and Roberts RJ (1986) Structure and function of the adenovirus-2 genome. In: Doerfler W (ed) Adenovirus DNA: the viral genome and its expression. Martin Nijhoff Publishing, Oxford 8:53–95

Aspegren A, Rabino C and Bridge E (1998) Organization of splicing factors in adenovirus-infected cells reflects changes in gene expression during the early to late phase transition. Exp Cell Res 245:203–213

Berget SM, Moore C and Sharp PA (1977) Spliced segments at the 5′ terminus of adenovirus 2 late mRNA. Proc Natl Acad Sci USA 74:3171–3175

Berk AJ and Sharp PA (1978) Structure of the adenovirus 2 early mRNAs. Cell 14: 695–711

Bondesson M, Öhman K, Mannervik M, Fan S and Akusjärvi G (1996) Adenovirus E4 open reading frame 4 protein autoregulates E4 transcription by inhibiting E1A transactivation of the E4 promoter. J Virol 70:3844–3851

Bourgeois CF, Popielarz M, Hildwein G and Stévenin J (1999) Identification of a bidirectional splicing enhancer: differential involvement of SR proteins in 5′ or 3′ splice site activation. Mol Cell Biol 19 7347–7356

Bridge E, Riedel KU, Johansson BM and Pettersson U (1996) Spliced exons of adenovirus late RNAs colocalize with snRNP in a specific nuclear domain. J Cell Biol 135:303–314

Bridge E, Xia DX, Carmo-Fonseca M, Cardinali B, Lamond AI and Pettersson U (1995) Dynamic organization of splicing factors in adenovirus-infected cells. J Virol 69:281–290

Brockmann D, Tries B and Esche H (1990) Isolation and characterization of novel adenovirus type 12 E1A mRNAs by cDNA PCR technique. Virology 179:585–590

Bruni R and Roizman B (1996) Open reading frame P – A herpes simplex virus gene repressed during productive infection encodes a protein that binds a splicing factor and reduces synthesis of viral proteins made from spliced mRNA. Proc Natl Acad Sci USA 93:10423–10427

Bryant HE, Matthews DA, Wadd S, Scott JE, Kean J, Graham S, Russell WC and Clements JB (2000) Interaction between herpes simplex virus type 1 IE63 protein and cellular protein p32. J Virol 74:11322–11328

Burd CG and Dreyfuss G (1994) RNA binding specificity of hnRNP A1: significance of hnRNP A1 high- affinity binding sites in pre-mRNA splicing. EMBO J 13:1197–1204

Burge CB, Tuschl T and Sharp P (1998) Splicing of precursors to mRNAs by the spliceosome. In: Gesteland RF, Cech TR, Atkins JF (eds) The RNA world, Second edition. Cold Spring Harbor Press, Cold Spring Harbor NY, pp 525–560: 525–560

Caceres JF, Stamm S, Helfman DM and Krainer AR (1994) Regulation of alternative splicing in vivo by overexpression of antagonistic splicing factors. Science 265: 1706–1709

Cavaloc Y, Bourgeois CF, Kister L and Stevenin J (1999) The splicing factors 9G8 and SRp20 transactivate splicing through different and specific enhancers. RNA 5: 468–483

Chambon P (1977) Summary: The molecular biology of the eukaryotic genome is coming of age. Cold Spring Harbor Symposium Series, Vol. XLII, Cold Spring Harbor Press, Cold Spring Harbor NY, Vol. XLII: pp 1209–1234

Chebli K, Gattoni R, Schmitt P, Hildwein G and Stevenin J (1989) The 216-nucleotide intron of the E1A pre-mRNA contains a hairpin structure that permits utilization of unusually distant branch acceptors. Mol Cell Biol 9:4852–4861

Chen MR, Yang JF, Wu CW, Middeldorp JM and Chen JY (1998) Physical association between the EBV protein EBNA-1 and P32/TAP/hyaluronectin. J Biomed Sci 5:173–179

Chow LT and Broker TR (1978) The spliced structures of adenovirus 2 fiber message and the other late mRNAs. Cell 15:497–510

Chow LT, Broker TR and Lewis JB (1979) Complex splicing patterns of RNAs from the early regions of adenovirus-2. J Mol Biol 134:265–303

Chow LT, Gelinas RE, Broker TR and Roberts RJ (1977) An amazing sequence arrangement at the 5′ ends of adenovirus 2 messenger RNA. Cell 12:1–8

Crone DE and Keene JD (1989) Viral transcription is necessary and sufficient for vesicular stomatitis virus to inhibit maturation of small nuclear ribonucleoproteins. J Virol 63:4172–4180.

Dauksaite V and Akusjärvi G (2002) Human splicing factor ASF/SF2 encodes for a repressor domain required for its inhibitory activity on pre-mRNA splicing. J Biol Chem 277:12579–12586

Delsert C, Morin N and Klessig DF (1989) cis-acting elements and a trans-acting factor affecting alternative splicing of adenovirus L1 transcripts. Mol Cell Biol 9:4364–4371

Duncan PI, Stojdl DF, Marius RM and Bell JC (1997) In vivo regulation of alternative pre-mRNA splicing by the Clk1 protein kinase. Mol Cell Biol 17:5996–6001

Dunn AR and Hassell JA (1977) A novel method to map transcripts: evidence for homology between an adenovirus mRNA and discrete multiple regions of the viral genome. Cell 12:23–36

Estmer Nilsson C, Petersen-Mahrt S, Durot C, Shtrichman R, Krainer AR, Kleinberger T and Akusjärvi G (2001) The adenovirus E4-ORF4 splicing enhancer protein interacts with a subset of phosphorylated SR proteins. EMBO J 20: 864–871

Flint SJ, Enquist LW, Krug RM, Racaniello VR and Skalka AM (2000) The principles of Virology; Molecular Biology, Pathogenesis and Control. ASM Press, Washington, DC

Fu XD (1995) The superfamily of arginine/serine-rich splicing factors. RNA 1:663–680

Gama-Carvalho M, Krauss RD, Chiang L, Valcarcel J, Green MR and Carmo-Fonseca M (1997) Targeting of U2AF65 to sites of active splicing in the nucleus. J Cell Biol 137: 975–987

Gattoni R, Chebli K, Himmelspach M and Stévenin J (1991) Modulation of alternative splicing of adenoviral E1A transcripts: factors involved in the early-to-late transition. Genes Dev 5:1847–1858

Gattoni R, Schmitt P and Stévenin J (1988) In vitro splicing of adenovirus E1A transcripts: characterization of novel reactions and of multiple branch points far from the 3′ splice site. Nucleic Acids Res 16:2389–2409

Ge H and Manley JL (1990) A protein factor, ASF, controls cell-specific alternative splicing of SV40 early pre-mRNA in vitro. Cell 62:25–34

Graveley BB (2000) Sorting out the complexity of SR protein functions. RNA 6:1197–1211

Gui JF, Lane WS and Fu XD (1994) A serine kinase regulates intracellular localization of splicing factors in the cell cycle. Nature 369:678–682

Gustin KE and Imperiale MJ (1998) Encapsidation of viral DNA requires the adenovirus L1 52/55-kilodalton protein. J Virol 72:7860–7870

Harper JE and Manley JL (1992) Multiple activities of the human splicing factor ASF. Gene Expr 2:19–29

Himmelspach M, Cavaloc Y, Chebli K, Stévenin J and Gattoni R (1995) Titration of serine/arginine (SR) splicing factors during adenoviral infection modulates E1A pre-mRNA alternative splicing. RNA 1:794–806

Hrimech M, Yao XJ, Branton PE and Cohen EA (2000) Human immunodeficiency virus type 1 Vpr-mediated G(2) cell cycle arrest: Vpr interferes with cell cycle signaling cascades by interacting with the B subunit of serine/threonine protein phosphatase 2 A. EMBO J 19:3956–3967

Huang T-S, Estmer-Nilsson C, Punga T and Akusjärvi G (2002) Functional inactivation of the SR family of splicing factors during a vaccinia virus infection. EMBO reports 3:1088–1093

Imperiale M, Akusjärvi G and Leppard K (1995) Post-transcriptional control of adenovirus gene expression. In: Doerfler W and Böhm P (eds) Current Topics in Microbiology and Immunology. Springer Verlag, Vol. 199/II, pp 139–171

Johnston JM, Anderson KP and Klessig DF (1985) Partial block to transcription of human adenovirus type 2 late genes in abortively infected monkey cells. J Virol 56:378–385

Kanopka A, Mühlemann O and Akusjärvi G (1996) Inhibition by SR proteins of splicing of a regulated adenovirus pre-mRNA. Nature, 381:535–538

Kanopka A, Mühlemann O, Petersen-Mahrt S, Estmer C, Öhrmalm C and Akusjärvi G (1998) Regulation of adenovirus alternative RNA splicing by dephosphorylation of SR proteins. Nature 393:185–187

Kleinberger T and Shenk T (1993) Adenovirus E4orf4 protein binds to protein phosphatase-2 A, and the complex down regulates E1A-enhanced junB transcription. J Virol 67:7556–7560

Klessig DF (1977) Two adenovirus mRNAs have a common 5′ terminal leader sequence encoded at least 10 kb upstream from their main coding regions. Cell 12:9–21

Kornitzer D, Sharf R and Kleinberger T (2001) Adenovirus E4orf4 protein induces PP2A-dependent growth arrest in Saccharomyces cerevisiae and interacts with the anaphase-promoting complex/cyclosome. J Cell Biol 154:331–344

Krainer AR, Conway GC and Kozak D (1990) The essential pre-mRNA splicing factor SF2 influences 5′ splice site selection by activating proximal sites. Cell 62: 35–42

Kreivi J-P and Akusjärvi G (1994) Regulation of adenovirus alternative RNA splicing at the level of commitment complex formation. Nucleic Acids Res 22: 332–337

Kreivi J-P, Zerivitz K and Akusjärvi G (1991) Sequences involved in the control of adenovirus L1 alternative RNA splicing. Nucleic Acids Res 19:2379–2386

Larsson S, Kreivi J-P and Akusjärvi G (1991) Control of adenovirus alternative RNA splicing: effect of viral DNA replication on splice site choice. Gene 107:219–227

Lerga A, Hallier M, Delva L, Orvain C, Gallais I, Marie J and Moreau-Gachelin F (2001) Identification of an RNA binding specificity for the potential splicing factor TLS. J Biol Chem 276:6807–6816

Lu X, Timchenko NA and Timchenko LT (1999) Cardiac elav-type RNA-binding protein (ETR-3) binds to RNA CUG repeats expanded in myotonic dystrophy. Hum Mol Genet 8:53–60

Luo Y, Yu H and Peterlin BM (1994) Cellular protein modulates effects of human immunodeficiency virus type 1 Rev. J Virol 68:3850–3856

Mandel JL (1989) Dystrophin: The gene and its product. Nature 339:584–586

Manley JL, Sharp PA and Gefter ML (1979) RNA synthesis in isolated nuclei: identification and comparison of adenovirus 2 encoded transcripts synthesized in vitro and vivo. J Mol Biol 135:171–197

Manley JL and Tacke R (1996) SR proteins and splicing control. Genes & Dev 10: 1569–1579

Marcellus RC, Lavoie JN, Boivin D, Shore GC, Ketner G and Branton PE (1998) The early region 4 orf4 protein of human adenovirus type 5 induces p53- independent cell death by apoptosis. J Virol 72:7144–7153

Matthews DA and Russell WC (1998) Adenovirus core protein V interacts with p32–a protein which is associated with both the mitochondria and the nucleus. J Gen Virol 79:1677–1685

Mayeda A and Krainer AR (1992) Regulation of alternative pre-mRNA splicing by hnRNP A1 and splicing factor SF2. Cell 68:365–375

Mermoud JE, Cohen PT and Lamond AI (1994) Regulation of mammalian spliceosome assembly by a protein phosphorylation mechanism. EMBO J 13:5679–5688

Misteli T (1999) RNA splicing: What has phosphorylation got to do with it? Curr Biol 9:198–200

Molin M, Bouakas L, Berenjian S and Akusjärvi G (2002) Unscheduled expression of capsid protein IIIa results in defects in adenovirus major late mRNA and protein expression. Virus Res 83 197–206

Muller U, Kleinberger T and Shenk T (1992) Adenovirus E4orf4 protein reduces phosphorylation of c-Fos and E1A proteins while simultaneously reducing the level of AP-1. J Virol 66:5867–5878

Mühlemann O, Kreivi J-P and Akusjärvi G (1995) Enhanced splicing of nonconsensus 3′ splice sites late during adenovirus infection. J Virol 69:7324–7327

Mühlemann O, Yue BG, Petersen-Mahrt S and Akusjärvi G (2000) A novel type of splicing enhancer regulating adenovirus pre-mRNA splicing. Mol Cell Biol 20: 2317–2325

Nevins JR and Wilson MC (1981) Regulation of adenovirus-2 gene expression at the level of transcriptional termination and RNA processing. Nature 290:113–118

Pallas DC, Shahrik LK, Martin BL, Jaspers S, Miller TB, Brautigan DL and Roberts TM (1990) Polyoma small and middle T antigens and SV40 small t antigen form stable complexes with protein phosphatase 2 A. Cell 60:167–176

Perricaudet M, Akusjärvi G, Virtanen A and Pettersson U (1979) Structure of two spliced mRNAs from the transforming region of hmna subgroup C adenoviruses. Nature 281:694–696

Petersen-Mahrt SK, Estmer C, Öhrmalm C, Matthews DA, Russell WC and Akusjärvi G (1999) The splicing factor-associated protein, p32, regulates RNA splicing by inhibiting ASF/SF2 RNA binding and phosphorylation. EMBO J 18: 1014–1024

Philips AV, Timchenko LT and Cooper TA (1998) Disruption of splicing regulated by a CUG-binding protein in myotonic dystrophy. Science 280:737–741

Pilder S, Moore M, Logan J and Shenk T (1986) The adenovirus E1B-55 K transforming polypeptide modulates transport or cytoplasmic stabilization of viral and host cell mRNAs. Mol Cell Biol 6:470–476

Popielarz M, Gattoni R and Stevenin J (1993) Contrasted cis-acting effects of downstream 5′ splice sites on the splicing of a retained intron: the adenoviral E1A pre-mRNA model. Nucleic Acids Res 21:5144–5151

Prasad J, Colwill K, Pawson T and Manley JL (1999) The protein kinase Clk/Sty directly modulates SR protein activity: both hyper- and hypophosphorylation inhibit splicing. Mol Cell Biol 19:6991–7000

Puvion-Dutilleul F, Bachellerie JP, Visa N and Puvion E (1994) Rearrangements of intranuclear structures involved in RNA processing in response to adenovirus infection. J Cell Sci 107:1457–1468

Schmitt P, Gattoni R, Keohavong P and Stévenin J (1987) Alternative splicing of E1A transcripts of adenovirus requires appropriate ionic conditions in vitro. Cell 50:31–39

Screaton GR, Caceres JF, Mayeda A, Bell MV, Plebanski M, Jackson DG, Bell JI and Krainer AR (1995) Identification and characterization of three members of the human SR family of pre-mRNA splicing factors. EMBO J 14:4336–4349

Shtrichman R and Kleinberger T (1998) Adenovirus type 5 E4 open reading frame 4 protein induces apoptosis in transformed cells. J Virol 72:2975–2982

Singh R, Valcarcel J and Green MR (1995) Distinct binding specificities and functions of higher eukaryotic polypyrimidine tract-binding proteins. Science 268: 1173–1176

Smith CWJ, Patton JG and Nadal-Ginard B (1989) Alternative splicing in the control of gene expression. Ann Rev Genet 23:527–577

Soloway PD and Shenk T (1990) The adenovirus type 5 i-leader open reading frame functions in cis to reduce the half-life of L1 mRNAs. J Virol 64:551–558

Spector DJ, McGrogan M and Raskas HJ (1978) Regulation of the appearance of cytoplasmic RNAs from region 1 of the adenovirus 2 genome. J Mol Biol 126: 395–414
Stark JM, Bazett-Jones DP, Herfort M and Roth MB (1998) SR proteins are sufficient for exon bridging across an intron. Proc Natl Acad Sci USA 95:2163–2168
Stephens C and Harlow E (1987) Differential splicing yields novel adenovirus 5 E1A mRNAs that encode 30 kd and 35 kd proteins. EMBO J 6:2027–2035
Stewart PL and Burnett RM (1995) Adenovirus structure by X-ray crystallography and electron microscopy. In: Doerfler W and Böhm P (eds) Current Topics in Microbiology and Immunology. Springer Verlag, Berlin, Vol. 199/I, pp 25–38
Svensson C and Akusjärvi G (1986) Defective RNA splicing in the absence of adenovirus-associated RNAI. Proc Natl Acad Sci USA 83:4690–4694
Svensson C, Pettersson U and Akusjärvi G (1983) Splicing of adenovirus 2 early region 1 A mRNAs is non-sequential. J Mol Biol 165:475–495
Symington JS, Lucher LA, Brackmann KH, Virtanen A, Pettersson U and Green M (1986) Biosynthesis of adenovirus type 2 i-leader protein. J Virol 57:848–856
Tacke R and Manley JL (1995) The human splicing factors ASF/SF2 and SC35 possess distinct, functionally significant RNA binding specificities. EMBO J 14:3540–3551
Tange TO, Jensen TH and Kjems J (1996) In vitro interaction between human immunodeficiency virus type 1 rev protein and splicing factor ASF/SF2-associated protein, p32. J Biol Chem 271:10066–10072
Thomas GP and Mathews MB (1980) DNA replication and the early to late transition in adenovirus infection. Cell 22:523–532
Ulfendahl PJ, Kreivi JP and Akusjärvi G (1989) Role of the branch site/3′-splice site region in adenovirus-2 E1A pre- mRNA alternative splicing: evidence for 5′- and 3′-splice site co- operation. Nucleic Acids Res 17:925–938
Ulfendahl PJ, Linder S, Kreivi JP, Nordqvist K, Sevensson C, Hultberg H and Akusjärvi G (1987) A novel adenovirus-2 E1A mRNA encoding a protein with transcription activation properties. EMBO J 6:2037–2044
van Ormondt H, Maat J and Dijkema R (1980) Comparison of nucleotide sequences of the early E1a regions for subgroups A, B and C of human adenoviruses. Gene 12:63–76
Walter G, Ruediger R, Slaughter C and Mumby M (1990) Association of protein phosphatase 2 A with polyoma virus medium tumor antigen. Proc Natl Acad Sci USA 87:2521–2525
Wang J and Manley JL (1995) Overexpression of the SR proteins ASF/SF2 and SC35 influences alternative splicing in vivo in diverse ways. RNA 1:335–346
Wang YL, Finan JE, Middeldorp JM and Hayward SD (1997) P32/TAP, a cellular protein that interacts with EBNA-1 of Epstein-Barr virus. Virology 236:18–29
Wilson MC and Darnell JE, Jr (1981) Control of messenger RNA concentration by differential cytoplasmic half- life. Adenovirus messenger RNAs from transcription units 1 A and 1B. J Mol Biol 148:231–251
Wu JY and Maniatis T (1993) Specific interactions between proteins implicated in splice site selection and regulated alternative splicing. Cell 75:1061–1070
Yang X, Bani MR, Lu SJ, Rowan S, Ben-David Y and Chabot B (1994) The A1 and A1B proteins of heterogeneous nuclear ribonucleoparticles modulate 5′ splice site selection in vivo. Proc Natl Acad Sci USA 91:6924–6928

Yu L, Loewenstein PM, Zhang Z and Green M (1995) In vitro interaction of the human immunodeficiency virus type 1 Tat transactivator and the general transcription factor TFIIB with the cellular protein TAP. J Virol 69:3017–3023

Yue BG and Akusjärvi G (1999) A downstream splicing enhancer is essential for in vitro pre-mRNA splicing. FEBS Lett 451:10–14

Yueh A and Schneider RJ (2000) Translation by ribosome shunting on adenovirus and hsp70 mRNAs facilitated by complementarity to 18S rRNA. Genes Dev 14: 414–421

Zahler AM, Neugebauer KM, Lane WS and Roth MB (1993) Distinct functions of SR proteins in alternative pre-mRNA splicing. Science 260:219–222

Zamore PD, Patton JG and Green MR (1992) Cloning and domain structure of the mammalian splicing factor U2AF. Nature 355:609–614

Zerivitz K, Kreivi J-P and Akusjärvi G (1992) Evidence for a HeLa cell splicing activity that is necessary for activation of a regualted adenovirus 3′ splice site. Nucleic Acids Res 20:3955–3961

Zhang WJ and Wu JY (1996) Functional properties of p54, a novel SR protein active in constitutive and alternative splicing. Mol Cell Biol 16:5400–5408

CHAPTER 10

Regulation of mRNA Production by the Adenoviral E1B 55-kDa and E4 Orf6 Proteins

S. J. FLINT, R. A. GONZALEZ

Abstract. The E1B 55-kDa and E4 Orf6 proteins of human subgroup C adenoviruses both counter host cell defenses mediated by the cellular p53 protein and regulate viral late gene expression. A complex containing the two proteins has been implicated in induction of selective export of viral late mRNAs from the nucleus to the cytoplasm, with concomitant inhibi-

S. J. FLINT, R. A. GONZALEZ
Department of Molecular Biology, Princeton University, Princeton, New Jersey 08844, USA
e-mail: sjflint@molbio.princeton.edu

tion of export of the majority of newly synthesized cellular mRNAs. The molecular mechanisms by which these viral proteins subvert cellular pathways of nuclear export are not yet clear. Here, we review recent efforts to identify molecular and biochemical functions of the E1B 55-kDa and E4 Orf6 proteins required for regulation of mRNA export, the several difficulties and discrepancies that have been encountered in studies of these viral proteins, and evidence indicating that the reorganization of the infected cell nucleus and production of viral late mRNA at specific intra-nuclear sites are important determinants of selective mRNA export in infected cells. In our view, it is not yet possible to propose a coherent molecular model for regulation of mRNA export by the E1B 55-kDa and E4 Orf6 proteins. However, it should now be possible to address specific questions about the roles of potentially relevant properties of these viral proteins.

1 Introduction

Human adenoviruses rely on the RNA polymerase II transcriptional system and the RNA processing machinery of the host cell for production of all viral mRNAs (see Shenk 1996). Nevertheless, a characteristic feature of the late phase of infection of transformed human cell lines with the subgroup C serotypes 2 or 5 (Ad2 or Ad5) is inhibition of cellular gene expression such that only viral proteins are synthesized by 18–24 h after infection (Anderson et al. 1973; Beltz and Flint 1979). Such preferential expression of viral genetic information is the result of post-transcriptional regulatory mechanisms. Several viral gene products, as well as features of viral late mRNA species, allow selective translation of these mRNAs during the late phase of infection, when the activity of at least two cellular translation initiation proteins is compromised (see Schneider 1995). In addition, the E1B 55-kDa and E4 Orf6 early proteins induce preferential export of newly synthesized, viral late mRNAs from the nucleus to the cytoplasm (Babiss et al. 1985; Bridge and Ketner 1990; Cutt et al. 1987; Halbert et al. 1985; Pilder et al. 1986; Sandler and Ketner 1989; Williams et al. 1986). In addition to facilitating efficient viral late gene expression during productive infection, both the E1B 55-kDa (see Jones 1990; Shenk 1996) and the E4 Orf6 (Moore et al. 1996; Nevels et al. 1997) proteins possess oncogenic activity. Each can bind to the cellular p53 protein and inhibit its transcriptional activity (Dobner et al. 1996; Sarnow et al. 1982; Teodoro and Branton 1997; Yew et al. 1994) and the two proteins cooperate to induce accelerated degradation of this cellular tumor suppressor (Querido et al. 1997; Roth et al. 1998; Steegenga et al. 1998). The adenoviral 243R E1A protein induces p53-dependent apoptosis by increasing p53 protein concentration (Braithwaite et al. 1990; Debbas and White 1993; Lowe and Ruley 1993). Thus, down-regulation of p53 concentration and activity by the E1B 55-kDa and E4 Orf6 proteins is believed to facilitate both transformation and viral replication by preventing premature apoptosis upon synthesis of the essential E1A proteins (see Dobner and Kzhyshkowska 2001; Querido et al. 2001).

Efforts to elucidate the molecular mechanisms by which the E1B 55-kDa and E4 Orf6 proteins fulfill their important regulatory functions have steadily gained momentum during the past decade. The potential of an Ad5 mutant that cannot direct synthesis of the E1B 55-kDa protein dl1520 (Barker and Berk 1987) also known as ONYX-015, for anti-cancer therapy (see Kirn 2000) has provided an additional impetus for elucidation of the molecular functions of this protein. A substantial body of new information about the E1B 55-kDa and E4 Orf6 proteins has therefore been collected in the past few years. Nevertheless, the mechanism(s) by which these proteins alter mRNA export specificity in infected cells are not understood in detail: several long-standing questions remain to be addressed, while new uncertainties or controversies have arisen.

2 Regulation of Viral Late Gene Expression by the E1B 55-kDa and E4 Orf6 Proteins

Although they both are products of adenoviral early transcription units, the E1B 55-kDa and E4 Orf6 proteins are required for maximally efficient expression of viral late genes. Genetic analyses originally established that these proteins facilitate accumulation of high concentrations of viral late mRNAs in the cytoplasm of infected HeLa cells, although their functions are not identical. However, adenovirus reproduction does not invariably require these proteins, because the phenotypes exhibited by mutant viruses defective for their production are influenced by the type of host cell, the phase of the cell cycle at the time of infection, and the temperature or multiplicity of infection.

The absence of functional E1B 55-kDa protein from infected HeLa cells leads to reduced cytoplasmic concentrations of viral late mRNAs (Babiss et al. 1985; Leppard and Shenk 1989; Leppard 1993; Pilder et al. 1986;

WILLIAMS et al. 1986), primarily because the selective export of viral mRNAs and concomitant inhibition of cellular mRNA export characteristic of the late phase of infection (BABICH et al. 1983; BELTZ and FLINT 1979; CASTIGLIA and FLINT 1983; FLINT et al. 1983) are not induced. The magnitude of this defect is greater at lower temperature (LEPPARD and SHENK 1989; WILLIAMS et al. 1986) and such mutant viruses exhibit cold-sensitive replication (HO et al. 1982). This property was exploited to demonstrate that the E1B 55-kDa protein facilitates an intranuclear reaction in viral late mRNA production (LEPPARD and SHENK 1989). Analysis of the flow of newly synthesized RNA through a series of biochemically defined, nuclear fractions indicated that major late (ML) RNA is made in a nuclear matrix fraction of infected cells and subsequently accumulates in a nuclear soluble fraction, prior to transit through a nuclear membrane fraction and entry into the cytoplasm (LEPPARD and SHENK 1989). The major defect observed in the absence of the E1B protein was greatly reduced accumulation of major late ML RNA in the nuclear soluble fraction, with degradation of the RNA within the nucleus (LEPPARD and SHENK 1989). As the majority of ML RNAs reaching this fraction were correctly processed, it was concluded that the E1B 55-kDa protein mediates a rate-limiting, intranuclear reaction necessary for efficient export of late mRNAs to the cytoplasm (LEPPARD and SHENK 1989). It has also been reported that in Ad2-infected cells, cellular mRNAs are no longer recovered in a nuclear soluble fraction, as they are in uninfected HeLa cells, but are replaced by viral late mRNAs (DENOME et al. 1989). Although the requirement for the E1B protein was not examined, this observation suggests that a single, E1B 55-kDa protein-dependent process both stimulates viral mRNA export and inhibits transport of cellular mRNAs from the nucleus.

As the late phase continues, the rate of ML transcription becomes progressively lower in E1B null virus than in wild-type virus infected cells (LEPPARD 1993; PILDER et al. 1986) and the stability of viral late mRNAs is modestly reduced (PILDER et al. 1986; WILLIAMS et al. 1986). As these two parameters are unperturbed during the initial period of the late phase of infection in the absence of the E1B protein, such phenotypes are likely to be secondary consequences of the defect in viral late mRNA production, and thus in synthesis of late proteins that regulate viral gene expression. One obvious candidate would be the IVa_2 protein, which has been implicated (TRIBOULEY et al. 1994) in the 20- to 30-fold stimulation of ML transcription that accompanies entry into the late phase of infection (SHAW and ZIFF 1980), for efficient export of IVa_2 mRNA from the nucleus requires the E1B 55-kDa protein (LEPPARD 1993). The absence of functional E1B 55-kDa protein also results in greater inhibition of viral late protein synthesis than can be accounted for by the reduced cytoplasmic viral mRNA concentrations, and failure to inhibit cellular protein synthesis (BABICH et al. 1983; BABISS et al. 1985; PILDER et al. 1986; WILLIAMS et al. 1986). Such aberrations in the normal pattern of protein synthesis during the late phase of infection might also be a secondary effect of lowered concentrations of viral late proteins: the L4 100-kDa protein is required for efficient translation of all late mRNAs (HAYES et al. 1990) and an as yet unidentified viral late protein appears to induce the dephosphorylation of the cellular translation initiation protein eIF-4E that contributes to inhibition of cellular protein synthesis (HUANG and SCHNEIDER 1991; ZHANG et al. 1994). On the other hand, the particularly large disparity between the degrees of inhibition of viral late protein synthesis and reductions in concentrations of the corresponding cytoplasmic mRNAs observed in H1299 cells (HARADA and BERK 1999) and the observation that eIF-4E remains largely phosphorylated in cells infected by E1B deletion mutants (ZHANG et al. 1994) have led to the suggestion that the E1B 55-kDa protein plays a more direct role in translation (HARADA and BERK 1999).

The phenotypes associated with mutations of the E1B 55-kDa protein coding sequence described above were identified using HeLa cells, a typical laboratory host for subgroup C human adenovirus, in which replication of the mutants is significantly restricted, particularly at low temperatures (HO et al. 1982). It was subsequently reported that replication of the mutant dl1520/ONXY-015, which cannot direct synthesis of the E1B protein (BARKER and BERK 1987), is cell-type dependent (BISCHOFF et al. 1996). This property was originally attributed to the p53 status of the host cells, as the mutant replicated, and/or induced cytopathic effect, almost as efficiently as wild-type Ad5 in cell lines lacking wild-type p53 protein (BISCHOFF et al. 1996). The cell-type dependence of the requirement for the E1B 55-kDa protein during adenovirus replication has been confirmed using a wide variety of established lines of transformed human cells, as well as normal cell strains, but does not, in fact, correlate with absence of wild-type p53 (GOODRUM and ORNELLES 1998; HARADA and BERK 1999; ROTHMANN et al. 1998; TURNELL et al. 1999). The basis of this phenomenon remains unclear, and a matter of much debate (see DIX et al. 2001). However, comparisons of the replication of Ad5 and E1B deletion mutants in the p53-negative cell line H1299, and a derivative of this line in which a temperature-sensitive p53 protein is synthesized, have demonstrated that both p53-dependent and p53-independent functions of the E1B 55-kDa protein are required for maximally efficient replication (Goodrum and Ornelles 1998; Harada and Berk 1999). The mutant viruses replicated to near wild-type yields at 39°C, but exhibited 20- to 50-fold reductions in virus yield at 32°C in p53-null H1299 cells. Mutant virus replication was further impaired by production of functional p53 protein at 32°C in Hl299 cells synthesizing the temperature-sensitive protein (Goodrum and Ornelles 1998; Harada and Berk 1999), because viral DNA synthesis was delayed and reduced (Harada and Berk 1999). It is clear that the p53-independent function of the E1B 55-kDa protein required for efficient virus replication at low temperature facilitates viral late gene expression (Goodrum and Ornelles 1998; Harada and Berk 1999; Leppard and Shenk 1989; Williams et al. 1986). However, whether the absence of the E1B protein results primarily in a defect in late mRNA production or in its translation also appears to be cell-type dependent. In E1B null virus-infected HeLa cells at 32°C, export of newly synthesized viral late mRNA and accumulation of viral late mRNAs in the cytoplasm are inhibited by factors of 5–20 (Leppard and Shenk 1989; Williams et al. 1986). In contrast, the cytoplasmic concentration of L5 fiber mRNA was reported to be decreased only twofold in the absence of the E1B protein in Hl299 cells at 32°C, but synthesis of the fiber protein was decreased by greater than a factor of 10 (Harada and Berk 1999).

Replication of mutant viruses that cannot direct synthesis of the E1B 55-kDa protein is regulated by a third parameter, the phase of the cell cycle at the time of infection. In contrast to the wild-type, such mutant viruses replicate more efficiently at 37°C in HeLa cells infected during S phase than in cells infected when in G_1 phase (Goodrum and Ornelles 1997, 1999). Thus, the E1B protein relieves cell-cycle restrictions imposed on adenovirus replication. Both synthesis of viral late proteins and export of late mRNAs are more efficient in HeLa cells infected by E1B null mutant viruses during the S compared to during the G_1 phase (Goodrum and Ornelles 1997, 1999), as they are in HeLa and other cells maintained at high temperature. It has therefore been suggested that the temperature- and cell cycle phase-dependent replication exhibited by such mutant viruses may be mechanistically related (Goodrum and Ornelles 1998). However, the cellular components that facilitate adenovirus replication in the absence of the E1B 55-kDa protein at elevated temperatures or when cells are infected in S phase have not yet been identified.

It has been generally accepted for some time that regulation of mRNA export in adenovirus infected cells is mediated by the E1B 55-kDa protein in association with the E4 Orf6 protein (see DOBNER and KZHYSHKOWSKA 2001; IMPERIALE et al. 1995). These two proteins form a complex (subsequently designated the E1B–E4 complex), originally identified by co-immunoprecipitation of the E4 Orf6 with the E1B 55-kDa protein from infected cell extracts (SARNOW et al. 1984). Like its components, this complex is predominantly nuclear in infected cells (CUTT et al. 1987; ORNELLES and SHENK 1991; SMILEY et al. 1990), and increases in concentration until well into the late phase of infection (BOIVIN et al. 1999; ROWE et al. 1984; RUBENWOLF et al. 1997; SMILEY et al. 1990; YEW et al. 1994). Both immunoprecipitation and immunofluorescence indicate that infected cell nuclei also contain significant quantities of E1B 55-kDa and E4 Orf6 proteins that are not bound to one another (CUTT et al. 1987; ORNELLES and SHENK 1991; SMILEY et al. 1990). These populations might represent E1B and E4 protein molecules engaged in different activities than those performed by the complex (see Sect. 1 and below). Another possibility, not necessarily exclusive, is that these viral proteins associate transiently, with the complex continuously forming and dissociating.

The similar phenotypes exhibited by mutant viruses defective for production of either the E1B or the E4 protein, or both of these early proteins, initially implicated the E1B–E4 complex in regulation of viral late mRNA production (BABISS et al. 1985; BRIDGE and KETNER 1989; BRIDGE and KETNER 1990; CUTT et al. 1987; HALBERT et al. 1985; PILDER et al. 1986; WEINBERG and KETNER 1986; WILLIAMS et al. 1986). The E4 Orf6 protein-dependent localization of the E1B 55-kDa protein to intranuclear sites of viral mRNA production (ORNELLES and SHENK 1991; see Sect. 4) and the ability of each protein, as well as of the complex, to shuttle between nucleus and cytoplasm under various experimental circumstances (DOBBELSTEIN et al. 1997; DOSCH et al. 2001; GOODRUM and ORNELLES 1997; KRATZER et al. 2000; RABINO et al. 2000; see Sect. 3.3), provided additional support for this conclusion. The E4 Orf6 deletion mutant dl355 also exhibits a cell-cycle restriction to virus growth and viral late mRNA transport similar to, although more modest than, those observed during infection in the absence of the E1B 55-kDa protein (GOODRUM and ORNELLES 1999). On the other hand, it has been suggested recently that the binding of the E1B to the E4 protein is not required for regulation of viral late mRNA production. This conclusion was based on the failure of a Thr255Ala substitution in the E1B 55-kDa protein that strongly inhibited interaction with the E4 Orf6 protein to impair viral protein synthesis or inhibition of cellular protein synthesis (SHEN et al. 2001). At face value, this observation appears in contradiction to previous genetic studies, in particular the finding that double mutants carrying deletions in the E1B 55-kDa and E4 Orf6 protein coding sequences exhibit no greater defects in viral late mRNA metabolism than do single mutants (BRIDGE and KETNER 1990; CUTT et al. 1987). However, the apparent discrepancy can be explained by the cell type-dependence of the phenotypes exhibited by viruses with E1B or E4 mutations.

In the recent experiments, viral and cellular protein synthesis were examined in infected A549 cells (SHEN et al. 2001), in which neither the E1B 55-kDa (GOODRUM and ORNELLES 1998; HARADA and BERK 1999; ROTHMANN et al. 1998; TURNELL et al. 1999) nor the E4 (HALBERT et al. 1985; SMILEY et al. 1990) proteins are required for efficient virus reproduction. Indeed, in these experiments, dl1520/ONYX-015 did not exhibit significant defects in late protein synthesis in infected A549 cells (SHEN et al. 2001), despite the complete absence of the E1B 55-kDa protein. The effect of the Thr255Ala substitution on mRNA export regulation has not been examined in HeLa, or other, cells in which the E1B 55-kDa protein is needed for selective viral mRNA export. However, the properties of a different E1B mutation that also specifically inhibits association of the two early proteins directly implicate the E1B-E4 complex in mRNA export regulation in such cells. Insertion of four amino acids at residue 143 in a hybrid Ad2–Ad5 E1B 55-kDa protein (YEW et al. 1990) strongly inhibits interaction with the E4 Orf6 protein (RUBENWOLF et al. 1997), but not binding to either the p53 protein (KAO et al. 1990) or a second cellular protein termed E1B-AP5 (GABLER et al. 1998; see Sect. 3.2.1). Measurement of cytoplasmic:nuclear ratios of processed mRNAs has shown that this mutation results in impaired export of viral late mRNAs in infected HeLa cells (GONZALEZ and FLINT 2002).

As discussed above, there is no evidence for direct participation of the E1B 55-kDa protein in reactions by which viral late mRNAs are produced within the nucleus, such as transcription or processing. In contrast, genetic analyses identified an additional role for the E4 Orf6 protein: its absence results in decreased intranuclear concentrations of viral ML mRNA precursors greater than can be accounted for by reduced rates of ML transcription and which are therefore attributed to decreased stability of these RNA molecules (BRIDGE and KETNER 1989; BRIDGE and KETNER 1990; SANDLER and KETNER 1989). The severity of this defect is magnified over

20-fold when the E4 Orf3 protein is also absent, but is not altered by absence of the E1B 55-kDa protein (BRIDGE and KETNER 1990). These two E4 proteins are independently capable of allowing near wild-type levels of replication and viral late protein synthesis in mutant virus-infected HeLa cells, although the E4 Orf6 protein is somewhat more effective (BRIDGE and KETNER 1989; HUANG and HEARING 1989). Thus, the E4 Orf3 and Orf6 proteins appear to function in parallel, and independently of the E1B–E4 Orf6 protein complex, in modulation of the stability of viral RNA within the nucleus. This effect is probably the result of regulation of splicing. In transient expression assays, each of these E4 proteins alone can stimulate nuclear and cytoplasmic accumulation of processed ML mRNAs by a splice site-dependent mechanism (NORDQVIST and AKUSJÄRVI 1990; OHMAN et al. 1993) and each regulates alternative splicing of ML and other mRNAs (NORDQVIST et al. 1994). In infected cells the E4 Orf6 protein facilitates exon skipping and thus production of ML mRNAs lacking the i leader between the l2 and l3 segments of the tripartite leader sequence (NORDQVIST et al. 1994). The observation that the E4 Orf3 and Orf6 proteins are required for nuclear stabilization of only ML mRNAs (BRIDGE et al. 1991) is consistent with the view that they regulate the complex splicing to which these large primary transcripts are subjected (see SHENK 1996).

3 Properties of the E1B 55-kDa and E4 Orf6 Proteins

3.1 Sequence and Structural Features

3.1.1 *The E1B 55-kDa Protein*

The E1B 55-kDa proteins of Ad2 and Ad5 (495 and 496 amino acids, respectively) are phosphoproteins (MALETTE et al. 1983; SARNOW et al. 1982). Three C-terminal phosphorylation sites, Ser490 and Ser491, which lie within consensus sequences for casein kinase II, and Thr495, within a casein kinase I consensus, account for phosphorylation of the protein made in infected cells (TEODORO and BRANTON 1997; TEODORO et al. 1994). These modifications are necessary for the transforming activity of the E1B protein, suppression of E1A protein-induced, p53 protein-dependent apoptosis and repression of transcription by the E1B protein fused to a GAL4 DNA-binding domain (TEODORO and BRANTON 1997; TEODORO et al. 1994). A second modification of the E1B 55-kDa protein, addition of the small ubiquitin-related modifier protein 1 (SUMO-1) has also been implicated recently in transformation of baby rat kidney cells by the E1B 55-kDa protein in cooperation with E1A gene products (ENDTER et al. 2001). The modifier protein is attached to Lys104 in the Ad5 E1B protein, a residue that lies within a match to the proposed consensus sequence for SUMO-1 conjugation, and replacement of this amino acid by arginine severely inhibited both transformation activity and repression of p53-dependent transcription (ENDTER et al. 2001). When the SUMO-1 protein was over-produced in stably transformed rat cells, the E1B 55-kDa protein was re-localized from concentration in a large cytoplasmic body to nuclear track or dots that contained SUMO-1 (ENDTER et al. 2001). Substitutions with the sumoylation site of the E1B protein prevented such relocalization, establishing that this modification can dramatically influence the site of accumulation of the viral protein (ENDTER et al. 2001). In Ad5-infected cells, only a small fraction of E1B 55-kDa protein (<5%) is stably conjugated to SUMO-1 (ENDTER et al. 2001), and it is not known whether this modification contributes to the partitioning of the protein among multiple intracellular sites during the infectious cycle (see Sect. 4.2). Indeed, the requirements for either sumoylation or phosphorylation of the E1B 55-kDa protein during productive infection have not yet been investigated.

In addition to consensus sequences for modification by cellular enzymes, the E1B 55-kDa protein contains sequences matching motifs characteristic of a C_2H_2 zinc finger, an ribonucleoprotein (RNP) RNA-binding domain and a leucine-rich nuclear export signal (NES; Fig. 1). The NES is discussed in Sect. 3.3. The RNP motif was identified following the demonstration that a glutathione-S-transferase (GST)–E1B fusion protein synthesized in *Escherichia coli* bound to sequences of viral late mRNAs in both electrophoretic mobility shift and filter-binding assays (HORRIDGE and LEPPARD 1998). Either of two substitutions within the RNP1 portion of RNP motif (Fig. 1) eliminated the in vitro RNA-binding activity, suggesting the viral protein contains a bona fide RNA-binding domain (HORRIDGE and LEPPARD 1998). In these assays, no preference for viral mRNA over antisense RNA sequences was observed. However, as noted by HORRIDGE and LEPPARD (1998), such non-specific RNA-binding might reflect the ability of the E1B 55-kDa protein to bind specifically to viral mRNA sequences or structural features not tested in these experiments, or the several respects in which such in vitro assays differ from the intranuclear milieu. Such features include the packaging of viral mRNAs to be exported by cellular hnRNP proteins (GATTONI et al. 1980; VAN EEKELEN et al. 1981) and association of the E1B 55-kDa with the E4 Orf6 protein. Whether the RNP motif is necessary for regulation of mRNA export in infected cells is not yet clear. Two insertions of four amino acids within the RNP motif were found to reduce virus yield and impair viral late protein synthesis in infected HeLa cells (YEW et al. 1990). These mutations also result in large reductions in the efficiency with which viral late mRNAs are exported from the nucleus, but in both cases this phenotype can be attributed to the failure of the altered E1B 55-kDa proteins to enter the nucleus (GONZALEZ and FLINT 2002).

Although it has not been established that the E1B protein sequence matching the zinc finger motif located between amino acids 350 and 376 (Fig. 1) binds zinc, and is thus a true structural motif, insertion of four amino acids at residue 354 was reported to inhibit virus reproduction and viral late protein synthesis (YEW et al. 1990), as well as binding of the E1B to the p53 protein (KAO et al. 1990). This alteration was not observed to lead to severe mislocalization of the protein, but did reduce E1B protein concentration to a degree that appeared sufficient to account for the modest defect in viral mRNA export observed (GONZALEZ and FLINT 2002). It is therefore possible that this sequence is indeed required for proper folding and stability of the E1B protein.

As summarized in Fig. 1, many of the insertions introduced into the E1B 55-kDa protein coding sequence by (YEW et al. 1990) either induce accelerated protein turnover or prevent its entry into the nucleus (GONZALEZ and FLINT 2002; RUBENWOLF et al. 1997; YEW et al. 1990). These properties, particularly the unexpected finding that several of the altered E1B proteins that accumulate to close to wild-type concentrations are mislocalized, emphasize the need for structural information to provide a rational guide for the introduction of informative mutations. None is yet available, and the E1B 55-kDa protein made in bacteria becomes largely fragmented (HORRIDGE and LEPPARD 1998). However, the intact protein can be synthesized in insect cells from a baculovirus vector, when it retains such activities as binding to the p53 protein and repression of transcription (MARTIN and BERK 1998).

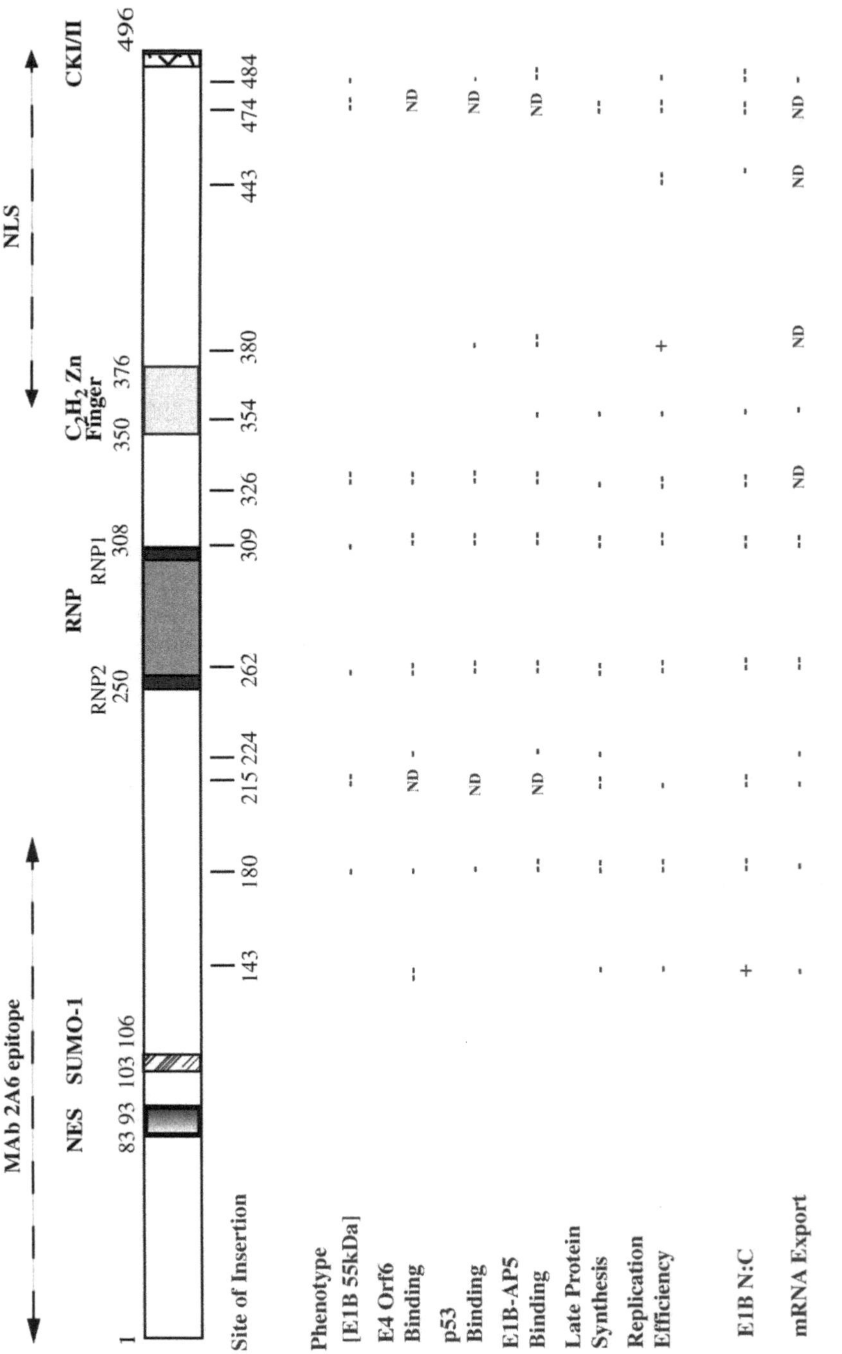
MAb 2A6 epitope
NLS
NES
SUMO-1
RNP
RNP2
RNP1
C2H2 Zn Finger
CKI/II
1
83 93 103 106
250
308
350 376
496
Site of Insertion
143
180
215 224
262
309
326
354
380
443
474 484
Phenotype [E1B 55kDa]
E4 Orf6 Binding
p53 Binding
E1B-AP5 Binding
Late Protein Synthesis
Replication Efficiency
E1B N:C
mRNA Export

3.1.2
The E4 Orf6 Protein

The 294 amino acid Ad5 E4 Orf6 protein contains a sequence resembling the leucine-rich NES consensus, discussed in Sect. 3.3, and the sequence HCHC conserved among many adenoviral E4 proteins (Fig. 2). The latter sequence has been suggested to contribute to an unusual zinc-binding motif (BOYER and KETNER 2000; BROWN et al. 2001). The E4 Orf6 protein solubilized from insect cells infected with a baculovirus vector was shown to bind zinc in a blotting assay (BOYER and KETNER 2000). Substitution of any of seven histidine and cysteine residues conserved among most adenoviral E4 Orf6 proteins impaired several activities, including complementation of an E4 mutant virus by transiently synthesized Orf6 protein, binding to and re-localization of the E1B 55-kDa protein (see Sect. 3.2) and destabilization of the p53 protein in the presence of the E1B protein. It was therefore proposed that the HCHC sequence is part of a zinc-binding domain that is required for several functions of the protein (BOYER and KETNER 2000). While the effects of these substitutions on binding of the metal has not been reported, a structural model of the E4 Orf6 protein

Fig. 1. Sequence and functional features of the Ad5 E1B 55-kDa protein. The 496 residue protein is represented to scale at the *top* by the *open box*, within which are shown the positions of a leucine-rich NES necessary for export of an E1B–GFP fusion protein, and sufficient for export of a GST-E1B NES-GFP protein (KRATZER et al. 2000); a sequence that matches the ribonucleoprotein (RNP) motif of many RNA-binding proteins in which specific substitutions impair non-specific RNA-binding activity of the E1B protein in vitro (HORRIDGE and LEPPARD 1998); a sequence matching the consensus for C_2H_2 zinc fingers, a C-terminal segment containing CKI and CKII consensus sequences in which the protein is phosphorylated at Thr 495, and Ser 490 and Ser491, respectively (TEODORO and BRANTON 1997); and a sequence required for addition of SUMO-1 (ENDTER et al. 2001). The region containing the epitope recognized by the monoclonal antibody 2A6 (cited in KAO et al. 1990) and a sequence required for nuclear localization (NLS) of an E1B-GFP fusion protein in human cells (KRATZER et al. 2000) are indicated above. The *short, vertical lines* drawn below the protein list the sites of insertions introduced into an Ad2/5 E1B 55-kDa protein coding sequence in the viral genome. With the exception of the 18 amino acid addition at residue 215, these insertions comprise four amino acids (YEW et al. 1990). The effects of these alterations on the concentration of the protein and other properties listed at the *left* are summarized below, where "--" and "-" indicate severe and moderate defects, respectively, and "+" indicates an increase in the parameter listed compared to the wild-type. (GABLER et al. 1998; GONZALEZ and FLINT 2002; KAO et al. 1990; YEW et al. 1990)

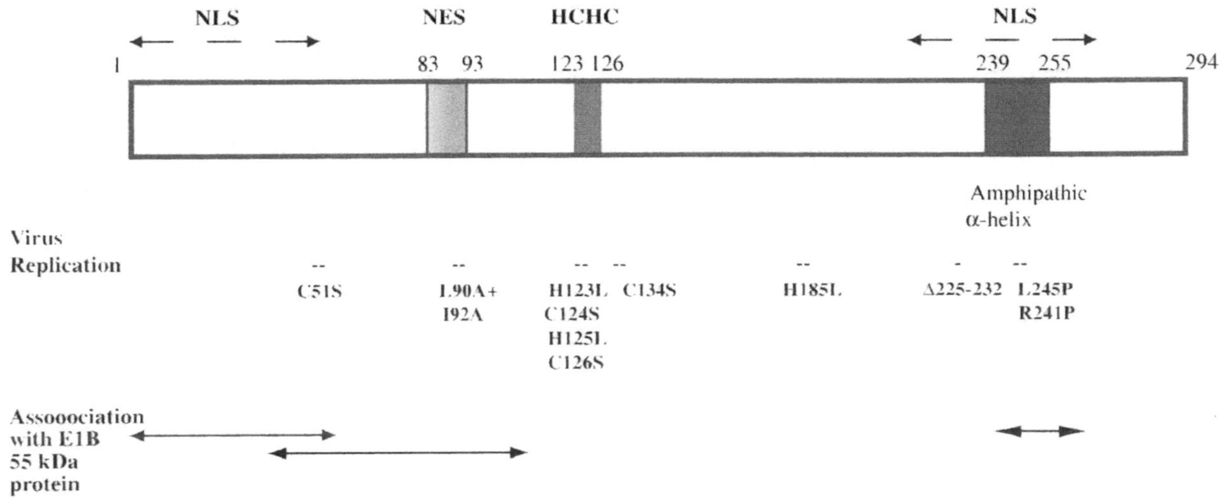

Fig. 2. Sequence and functional features of the Ad2/5 E4 Orf6 protein. The 294 residues protein is represented to scale at the *top* by the *box*, within which are shown the positions of possible nuclear localization signals (*NLS*; cited in Orlando and Ornelles 1999; Weigel and Dobbelstein 2000); a nuclear export signal (*NES*) reported to mediate the shuttling of the E1B 55-kDa–E4 Orf6 protein complex (Dobbelstein et al. 1997), a His- and Cys-containing motif (HCHC) conserved among adenoviral E4 Orf6 proteins (Boyer and Ketner 2000; Nevels et al. 2000), and a C-terminal sequence that forms an amphipathic α-helix as a synthetic peptide (Orlando and Ornelles 1999). The effects of the alterations listed on the ability of transiently expressed protein to complement the growth of mutant viruses lacking E4 sequences are summarized below, where "--" and "-" indicate severe and moderate defects, respectively (Boyer and Ketner 2000; Orlando and Ornelles 1999; Weigel and Dobbelstein 2000). Three regions reported to mediate binding to the E1B protein (Querido et al. 2000; Rubenwolf et al. 1997; Weigel and Dobbelstein 2000) are also indicated

derived by application of fold-recognition and comparative protein modeling methods (threading) is consistent with this hypothesis (Brown et al. 2001). As illustrated in Fig. 3A, the E4 protein was best modeled with a core, TIM-barrel-like structure formed by residues 69 to 294, with the N-terminal 68 amino acids in an α non-bundle topology lying within the open end of the TIM barrel. In the predicted structure, the conserved HCHC sequence (Fig. 2) forms a β-strand in the interior of the TIM barrel (Fig. 3B). The positions of Cys51 and Cys134, which are also internal, relative to that of the HCHC sequence (Fig. 3B) suggested that this β-strand forms the core of a binuclear zinc co-ordination motif (Fig. 3C). This motif resembles the zinc-binding element of the yeast transcriptional activator Gal4, in which two cysteine residues co-ordinate with each of two zinc atoms (Pan and Coleman 1990a,b). The structural model can account for the effects of various alterations introduced into the protein in mutational analyses. It also makes a number of specific predictions, for example, that the sequence linking the C-terminal TIM barrel and N-terminal domains (amino acids 64–69) must be structurally flexible to allow sufficiently close proximity of Cys51 to the proposed b-strand core of the zinc co-ordination motif (Brown et al. 2001). However, the predicted structure has yet to be verified by biochemical, structural or genetic methods.

This model also included the only structural feature of the E4 Orf6 protein for which there is experimental evidence, a C-terminal, amphipathic α-helix formed by residues 239–255: a peptide analogue of this arginine-rich sequence was shown to adopt helical conformation in solution (Orlando and Ornelles 1999). Deletion of residues 241–250 or proline substitution of Arg239, Arg241 or Leu245 eliminated the ability of the E4 protein to direct nuclear localization of the E1B 55-kDa protein when the two were synthesized in HeLa cells (Orlando and Ornelles 1999). The latter two substitutions also impaired complementation of replication of the mutant virus dl1014, which carries deletions of E4 Orf 1, 2, 3 and 6 (Bridge and Ketner 1989), by transiently synthesized E4 Orf6 protein. It was therefore concluded that the α-helix is important for the function of the E4 protein because it is required to direct nuclear localization of the E1B 55-kDa protein (Orlando and Ornelles 1999). As discussed in Sect. 3.3, the demonstration that the E1B protein actively shuttles between nuclear and cytoplasmic compartments in the absence of other viral proteins (Kratzer et al. 2000) suggests that the E4 Orf6 protein alters the intracellular distribution of the E1B protein rather than inducing its nuclear entry.

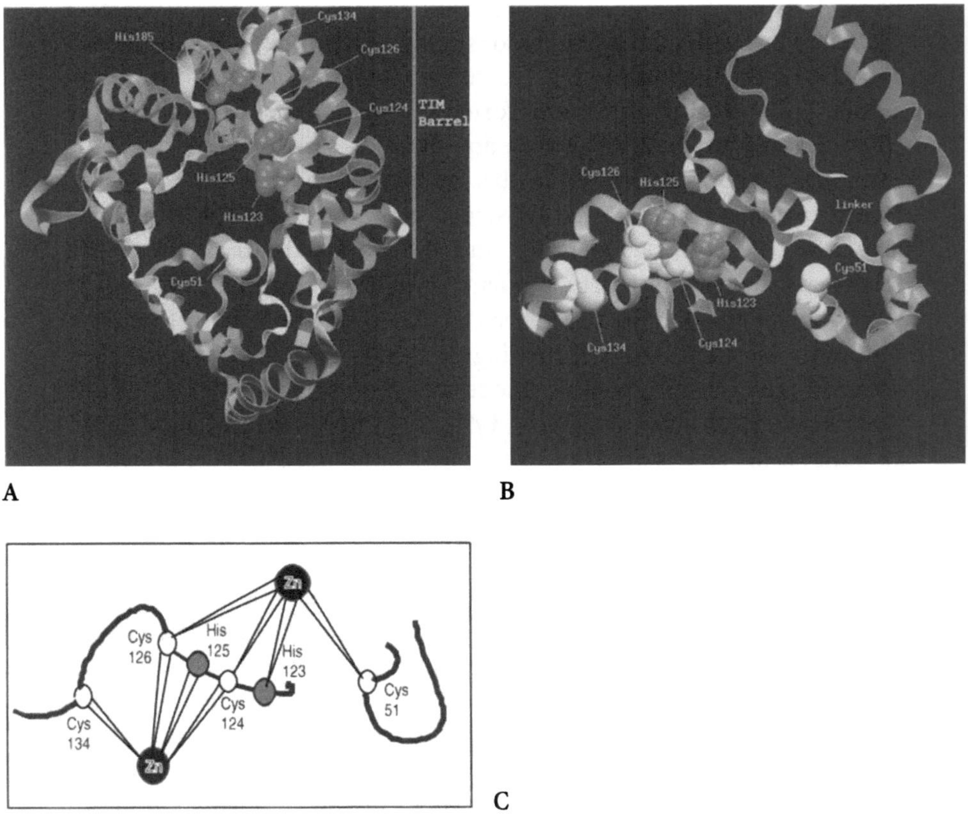

Fig. 3 A–C. Conserved histidines and cysteines required for several functions of the E4 Orf6 protein. **A** The chimeric model of the entire E4 Orf6 protein is shown, with the TIM barrel indicated, and conserved histidine and cysteine residues depicted as *gray* and *white*, respectively, van der Waals surfaces. **B** The N-terminal 161 residues of the protein are separated on the *right* for clearer visualization of a potential binuclear zinc-coordination element. **C** In this cartoon rendering of the proposed binuclear zinc-coordination element, the peptide backbone and metal coordination orbitals are depicted in *black*. The zinc ions coordinated with distal Cys51 and Cys134 are projected *into* and *out of* the page, respectively, to indicate proximity in three-dimensional space

The C-terminal amphipathic α-helix includes the sequence RRL, which matches the RXL motif identified in cellular proteins that interact with cyclin A or that are substrates of cyclin A–Cdk-2 phosphorylation (GRIFMAN et al. 1999). This sequence was identified following the finding that the Orf6 protein greatly increases the efficiency of transduction of adeno-associated virus vectors when it is stably synthesized in 293 cells (GRIFMAN et al. 1999). This effect appears to be the result of induction of accumulation of cells in S phase via degradation of cyclin A and inhibition of Cdk-2 (GRIFMAN et al. 1999).

3.2 Interactions with Other Proteins

In addition to their interactions with each other (Sect. 2) and with the cellular p53 protein (Sect. 1), both the E1B 55-kDa and E4 Orf6 proteins associate with several other viral or cellular proteins. These additional interactions have been identified relatively recently, and for the most part their functional significance has yet to be established.

3.2.1 *The E1B 55-kDa Protein*

The E1B protein co-immunoprecipitates with a second E4 protein, Orf3, from nuclear matrix fractions of infected human cells and co-localizes with it in infected cell nuclei (LEPPARD and EVERETT 1999). This interaction can re-localize the E1B protein from the cytoplasm to the nucleus, when the two proteins are transiently synthesized in the absence of other viral proteins (KONIG et al. 1999). No evidence has been obtained for formation of a ternary complex containing the E1B and both the E4 Orf6 and Orf3 proteins. Rather, the fraction of the Orf3 protein co-immunoprecipitating with the E1B protein increased in the absence of the Orf6 protein, as did staining of the E1B protein in E4 Orf3-containing nuclear foci, indicating that the two E4 proteins compete for binding to the E1B 55-kDa protein (KONIG et al. 1999; LEPPARD and EVERETT 1999). During Ad5 infection, the Orf3 protein is synthesized before the Orf6 protein and thus the association of the E1B 55-kDa protein with the Orf3 protein in nuclear tracks precedes relocalization to Orf6 protein-containing structures (KONIG et al. 1999). When the E1B 55-kDa and p53 proteins are transiently synthesized in HeLa cells, the viral protein relocalizes p53 from the nucleus to cytoplasmic clusters (KONIG et al. 1999) like those observed in adenovirus-transformed cells (BLAIR-ZAJDEL and BLAIR 1988; ZANTEMA et al. 1985). However, when the Orf3 protein is also made, both the restriction to the cytoplasm and the inhibition of the transcriptional activity of p53 by the E1B protein are suppressed (KONIG et al. 1999). It has therefore been suggested (KONIG et al. 1999) that during infection the Orf3 protein releases the p53 protein from its interaction with the E1B 55-kDa protein for degradation by this and the E4 Orf6 protein (see Sect. 1).

The track-like structures containing the E4 Orf3 and E1B 55-kDa proteins are formed by remodeling of the nucleus during adenovirus infection. When viral genomes first enter the nucleus, they are not randomly distributed, but rather preferentially associated with the normal nuclear structures termed PML bodies, PML oncogenic domains (PODs), Kr bodies or nuclear domain 10s (ND10s) (ISHOV and MAUL 1996). These structures appear in the electron microscope as dense, spherical particles of 0.3–0.5 μm diameter. They are found at an average frequency of 10 per nucleus, and contain, among other proteins, the PML protein first identified as part of a fusion with the retinoic acid receptor created by a chromosomal translocation associated with acute promyelocytic leukemia (see GRIMWADE and SOLOMON 1997; MAUL 1998). This specific nuclear microenvironment appears to confer some general advantage for virus reproduction, as it is also initially colonized by other DNA viruses that replicate their genomes within eukaryotic cell nuclei (see MAUL 1998; STERNSDORF et al. 1997). However, adenovirus infection rapidly induces a drastic reorganization of the PML protein form the variable number of brightly staining nuclear bodies characteristic of uninfected cells to a larger number of track-like or filamentous structures (CARVALHO et al. 1995; DOUCAS et al. 1996; ISHOV and MAUL 1996). Other proteins normally present in PML bodies undergo a similar reorganization. This change is detected during the early phase of infection before viral replication centers form (see Sect. 4.2), and is induced by the E4 Orf3 protein: PML body reorganization is not observed in infected cells infected by mutant viruses unable to direct synthesis of the Orf3 protein and transient synthesis of this protein in the absence of other viral proteins is sufficient for formation of PML protein-containing tracks (CARVALHO et al. 1995; DOUCAS et al. 1996; ISHOV and MAUL 1996; LEPPARD and EVERETT 1999). As discussed previously, both the Orf3 and the E1B 55-kDa proteins co-localize with the PML protein in these infected cell-specific structures (CARVALHO et al. 1995; DOUCAS et al. 1996; KONIG et al. 1999; LEPPARD and EVERETT 1999).

The significance of this reorganization of nuclear components is not clear. On the one hand, transient or stable over-production of the PML protein in infected cells in which the Orf3 protein is made has been reported to inhibit both track formation and viral DNA synthesis (Doucas et al. 1996). Nevertheless, mutant viruses lacking E4 Orf3 protein coding sequences grow nearly as well as wild-type virus in standard laboratory host cells (Bridge and Ketner 1989; Halbert et al. 1985; Huang and Hearing 1989). It is possible that the activities of the Orf3 protein that are required in normal cells are more representative of those in which the virus reproduces in nature, or at low multiplicities of infection (see Leppard and Everett 1999).

The E1B 55-kDa protein also interacts with a second cellular protein, termed E1B-AP5, which was identified by screening a HeLa cell expression library for proteins that bound to labeled GST-E1B 55-kDa fusion protein (Gabler et al. 1998). This cellular protein also bound to the E1B protein when the two proteins were made in vitro or in 293 cells, and co-immunoprecipitated with the E1B protein from Ad5-infected MCF-7 cells (Gabler et al. 1998). The E1B-AP5 protein is predominantly nuclear in location and contains several sequences related to those of cellular RNA-binding proteins. These include residues 195 to 595 that are 56% identical with isoform I of the human hnRNP-U protein, and a C-terminal sequence of some 280 amino acids related to the N-terminus of the RNA-binding protein hnRNP-G. It was therefore proposed that E1B-AP5 is a member of the hnRNP protein family (Gabler et al. 1998). Consistent with this hypothesis, the protein was shown to bind to homopolymeric RNAs and single-stranded DNA in vitro, with a preference for poly(G) similar to that of the hnRNP-G protein (Gabler et al. 1998). Like other hnRNP proteins, E1B-AP5 is also methylated within a sequence related to RGG domains, arginine-rich motifs found in many proteins that bind to RNA (Kzhyshkowska et al. 2001). The most intriguing property of the E1B-AP5 protein is its ability, when present in extracts of uninfected HeLa cells or synthesized in vitro, to bind to the Tap protein (Bachi et al. 2000). The latter protein was identified by virtue of its specific binding to the sequence (the constitutive transport element) required for export of unspliced RNAs of some simple retroviruses that do not encode proteins analogous to the HIV-1 Rev protein (Braun et al. 1999; Grüter et al. 1998). Both vertebrate Tap and its *Saccaromyces cerevisae* homologue, Mex67p, have been implicated in export of cellular mRNAs from the nucleus (Grüter et al. 1998; Kang and Cullen 1999; Katahira et al.

1999; Pritchard et al. 1999). The binding of the E1B 55-kDa protein to E1B-AP5 therefore seems likely to modulate this cellular export pathway. However, whether this interaction facilitates export of viral mRNAs and/or inhibits transport of cellular mRNAs from the nucleus is not yet clear. In one approach to address this question, viral and cellular gene expression were compared in cell lines (derived from H1299 cells) that do and do not over express an exogenous E1B-AP5 coding sequence (Gabler et al. 1998). A threefold increase in the total concentration of E1B-AP5 increased somewhat the accumulation of viral L3 and L5 mRNAs in the cytoplasm and, more strikingly, reversed the large decreases in cytoplasm concentrations of cellular β-actin and E1B-AP5 mRNAs observed during the late phase of infection in the control cell line (Gabler et al. 1998). Such reductions in concentrations of stable cellular mRNAs like β-actin mRNA have not been observed by other investigators in infected HeLa cells (Babich et al. 1983; Pilder et al. 1986; Williams et al. 1986) and may therefore represent a cell type-specific response. Be that as it may, this property indicates that the E1B-AP5 protein could regulate nucleocytoplasmic transport indirectly, by stabilization of nuclear or cytoplasmic mRNAs, or directly.

The mutant viruses carrying insertions within the E1B 55-kDa protein coding sequence (Yew et al. 1990) mentioned previously have been used to identify and compare sequences of this protein required for association with the E1B-AP5, p53 and E4 Orf6 proteins. Several of the insertions, notably those at residues 180, 262, 309 and 326, impaired immunoprecipitation of each of these three proteins with the E1B protein (Gabler et al. 1998; Kao et al. 1990; Rubenwolf et al. 1997), and also reduced virus yield and viral late protein synthesis (Yew et al. 1990; see Fig. 1). Under some experimental conditions, these mutations also resulted in reduced accumulation of the E1B protein, although in most cases the decreases were not sufficient to account for observed reductions in binding activity (Gabler et al. 1998; Kao et al. 1990; Rubenwolf et al. 1997). Each of the insertions listed above, as well as that at residue 484, was subsequently shown to lead to severe mislocaliztion of the altered E1B 55-kDa proteins, which failed to enter infected cell nuclei (Gonzalez and Flint 2002). This property, the pleiotropic phenotypes exhibited by the viruses carrying these mutations (Fig. 1), and the fact that the majority lie in or close to the sequence encoding the only portion of the E1B protein that is strongly predicted to be both hydrophobic and buried, indicate that these altered E1B proteins are likely to be misfolded (Gonzalez and Flint 2002).

Insertions at residues 354 and 380 of the E1B protein were also observed to inhibit co-immunoprecipitation of the E1B-AP5 with the E1B protein (GABLER et al. 1998). In the former case, this property can probably be attributed to the decreases in E1B protein concentration induced by the mutation (GABLER et al. 1998; GONZALEZ and FLINT 2002; YEW et al. 1990). In contrast, the insertion at position 380 appears to disrupt specifically the E1B 55kDa–E1B-AP5 interaction (GABLER et al. 1998; see Fig. 1). However, this mutation was shown to increase virus yield in HeLa cells some twofold, with no effect on either viral late protein synthesis or inhibition of cellular protein synthesis (YEW et al. 1990). Thus, additional studies are required to delineate the E1B protein binding site for the E1B–AP5 protein and thus to define unambiguously the role of this interaction in regulation of RNA metabolism in infected cells.

The insertion at residue 380 also reduced, but did not eliminate, binding of the E1B protein synthesized in infected cells to the p53 protein made in vitro (KAO et al. 1990). Using the same assay, it has been reported more recently that substitution of alanine for either Arg240 or His260 severely impairs binding to p53, without reducing the steady-state concentration of the E1B protein or, in the case of the former substitution, binding to the E4 Orf6 protein (SHEN et al. 2001). Comparison of the interactions with p53 of a series of chimeric Ad2/Ad12 E1B proteins synthesized in transformed rat cells mapped a major determinant of the strong binding characteristic of subgroup C adenoviral E1B 55-kDa proteins to amino acids 216–235 (GRAND et al. 1999). Residues 240–280 also increased somewhat the binding of an otherwise Ad12 viral protein to p53 (GRAND et al. 1999) The data currently available therefore indicate that the sequence from residues 216 to 260 of the subgroup C E1B protein is required for p53 binding, but additional segments may also contribute to this binding site. The substitution mutations mentioned above resulted in increased steady-state concentrations of the p53 protein in infected A549 cells (SHEN et al. 2001), consistent with several reports that interaction of the E1B and p53 proteins is necessary for destabilization of the latter (CATHOMEN and WEITZMAN 2000; QUERIDO et al. 2001; ROTH et al. 1998). Whether this interaction also contributes to other functions of the E1B 55-kDa protein is not clear (but see DIX et al. 2001).

Introduction of four amino acids at residue 143 of the E1B 55-kDa protein was found to inhibit severely co-immunoprecipitation of the E4 Orf6 protein from infected A549 cell extracts (RUBENWOLF et al. 1997). This mutation did not impair any other interactions of the E1B protein that have been examined (GABLER et al. 1998; KAO et al. 1990), nor has it been observed to destabilize the protein or prevent its entry into the nucleus (GONZALEZ and FLINT 2002; RUBENWOLF et al. 1997; YEW et al. 1990). Thus, it appears to define a sequence specifically participating in interaction with the E4 Orf6 protein. It has been reported recently that alanine substitution of either Thr255 or His260 also inhibits binding to the E4 Orf6 protein in infected cell extracts (SHEN et al. 2001). These observations suggest that the binding site for the E4 Orf6 protein may be formed by portions of the E1B protein that are widely spaced in the primary sequence.

3.2.2
The E4 Orf6 Protein

Current information about the E1B 55-kDa protein sequences that mediate interactions with other proteins is clearly incomplete. However, efforts to locate the E4 Orf6 protein-binding site for the E1B protein have yielded inconsistent, sometimes contradictory results. Using a protein-blotting assay with full length E4 Orf6 or N- and C-terminal segments synthesized as GST-fusion proteins, RUBBENWOLF et al. (1997) observed that the N-terminal 55 amino acids were necessary and sufficient for binding to the E1B protein. Consistent with this result, the E4 Orf6/7 protein, which shares its N-terminal 55 amino acids with Orf6, also bound to the E1B protein (RUBENWOLF et al. 1997). The same requirement for the Orf6 N-terminus was seen when wild-type and truncated E4 proteins synthesized in 293 cells were tested by co-immunoprecipitation with the E1B protein, while removal of the C-terminal 152 residues did not impair the interaction (RUBENWOLF et al. 1997). However, other reports have contradicted these observations. In the first place, no binding of the E1B to the E4 Orf6/7 protein could be detected by other investigators in adenovirus-infected or transfected cells (CUTT et al. 1987; QUERIDO et al. 2001), and the Orf6/7 protein cannot alter the cytoplasmic accumulation of the E1B protein when the two are transiently synthesized in HeLa cells (WEIGEL and DOBBELSTEIN 2000). In addition, WEIGEL and DOBBELSTEIN (2000) observed that removal of residues 225–232, a segment shown to be required for relocalization of the E1B 55-kDa protein to the nucleus in HeLa cells synthesizing just these two viral proteins, prevented co-immunoprecipitation of the E4 protein made by in vitro transcription and translation with the E1B protein present in 293 cell extracts. The C-terminal, amphipathic α-helix (Fig. 2), which begins some seven amino acids downstream, is also necessary for Orf6-dependent re-localization of the E1B protein to the nucleus (ORLANDO and ORNELLES 1999) and substitution of alanine for Arg243 and Leu245 within the a-helix has been shown to impair both this activity and co-immunoprecipitation of the Orf6 with the E1B protein (CATHOMEN and WEITZMAN 2000). In conjunction, these observations suggest that C-terminal sequences located between amino acids 225 and 245 are necessary for physical association of the E1B and E4 proteins. On the other hand, a similar co-immunoprecipitation approach applied to H1299 cells transiently expressing E1B 55-kDa and E4 Orf6 protein coding sequences did not detect any decrease in association of the E1B protein with altered Orf6 proteins lacking various C-terminal segments, including residues 228–248 (QUERIDO et al. 2001). Rather, in this study, removal of either amino acids 1–55 or 1–108 eliminated interaction of Orf6 with the E1B protein, but deletion of the first 38 residues did not (QUERIDO et al. 2001) As several smaller deletions within the required segment, amino acids 39–107, did not alter association of the two proteins, it was concluded that several parts of this Orf6 sequence mediate, or establish the appropriate conformation for, the interaction (QUERIDO et al. 2001).

Why N- or C-terminal regions of the E4 Orf6 proteins are necessary for association with the E1B protein in different experiments is not at all clear, especially as all the studies summarized above comprised or included co-immunoprecipitation of transiently synthesized viral proteins. Nevertheless, it must be that differences in experimental conditions account for the discrepancies reported. Such differences might include the absolute and relative concentrations attained by the E1B and/or E4 proteins, variations in post-translational modifications, and consequences of variation in cell type, or of aberrant splicing of E4 mRNA made from E4 Orf6 cDNA, which results in synthesis of a major, abnormal protein of some 8 kDa (QUERIDO et al. 2001). It is obviously of considerable importance to extend mutational analyses of the E1B–E4 protein interaction from simplified experimental systems to virus-infected cells.

The binding site for p53 has not yet been defined, but several regions of the E4 Orf6 protein are required for accelerated degradation of this cellular protein in co-operation with the E1B 55-kDa protein. These include the N terminus, the conserved HCHC motif and a C-terminal segment spanning the amphipathic α-helix (BOYER and KETNER 2000; NEVELS et al. 2000; QUERIDO et al. 2001). The first and last of these regions contain sequences that can mediate nuclear import and export of the protein (see Fig. 2), signals that may be required for localization of p53 to specific nuclear sites of infected or transformed cells (see Nevels et al. 2000; Querido et al. 2001). An interaction between the p53-related protein p73 and Orf6 that inhibits the activities of the former has been observed by some investigators (Higashino et al. 1998; Steegenga et al. 1998) but not by others (Querido et al. 2001; Roth et al. 1998).

In addition to binding to the E1B and p53 proteins, the E4 Orf6 protein co-immunoprecipitates with several other cellular and infected cell-specific proteins from Ad5-infected H1299 cell extracts (Boivin et al. 1999). With the exception of the viral L4 100-kDa protein, which appeared to be recovered non-specifically in the E4 protein-containing immunoprecipitates (Boivin et al. 1999), the identities of these proteins have not been reported. However, the interaction of Orf6 with two of the cellular proteins, named p14 and p19, has been correlated with induction of p53 degradation (Querido et al. 2001).

The E4 Orf6 and Orf3 proteins have also been reported to bind to the catalytic subunit of the cellular DNA-dependent protein kinase and inhibit double-strand break repair activity dependent on this enzyme (Boyer et al. 1999). Both these E4 proteins can also inhibit repair of double-stranded breaks in DNA made by the *S. cerevisiae* HO endonuclease (Nicolas et al. 2000). Such activity may prevent viral DNA concatenation, by end-to-end joining of the products of viral DNA synthesis by the double-stranded break repair system: viral DNA concatomers accumulate in cells infected by mutant viruses lacking all E4 sequences, but their production is suppressed by either the Orf3 or the Orf6 protein (Weiden and Ginsberg 1994). As the DNA-dependent protein kinase also contributes to activation of p53 in response to DNA damage, the inhibition of this enzyme by the E4 proteins may also help prevent apoptosis of adenovirus infected cells (Boyer et al. 1999).

3.3
Nucleocytoplasmic Transport: Who Shuttles and How?

A characteristic feature of eukaryotic proteins that convey macromolecules from the nucleus to the cytoplasm or vice versa is the ability to shuttle between these cellular compartments (see Nakielny and Dreyfuss 1999; Weis 1998). This property is also exhibited by the mRNA export regulator encoded by human immunodeficiency virus type 1 (HIV-1), the Rev protein, which promotes export of viral mRNAs that retain introns and splice sites (see EMERMAN and MALIM 1998). The Rev protein binds to an intronic sequence in these RNAs, the Rev-response element. This protein contains a short, leucine-rich sequence necessary and sufficient for its shuttling between the nucleus and the cytoplasm. This nuclear export signal directs Rev, and the RNAs to which it is bound, to the cellular exportin-1 (Crm-1) export receptor (see MATTAJ and ENGLMEIER 1998; STUTZ and ROSBASH 1998; WEIS 1998). This was the first export receptor for RNA to be identified unambiguously, in part because its ability to form an export-competent complex with a NES-containing substrate and the essential export co-factor Ran-GTP is specifically inhibited by the drug leptomycin B (FORNEROD et al. 1997; FUKUDA et al. 1997). This reagent, or peptides containing the Rev NES, inhibit transit of various small cellular RNAs, such as 5S rRNA, and NES-containing proteins from the nucleus to the cytoplasm (FISCHER et al. 1995; FORNEROD et al. 1997). Thus, the HIV-1 Rev protein circumvents the host cell restriction on export of intron-containing mRNAs by directing such viral mRNAs to a cellular pathway that normally handles small RNAs (see EMERMAN and MALIM 1998). These properties of Rev and Rev-dependent RNA export have guided investigations of shuttling of the adenoviral E1B 55-kDa and E4 Orf6 proteins, investigations that have produced more questions than answers about the shuttling activities of these proteins, and their relevance to regulation of mRNA export.

It was initially reported that the E1B–E4 protein complex possessed shuttling activity (DOBBELSTEIN et al. 1997). In these experiments, HeLa cells in which the E1B and E4 proteins are nuclear when both are made (GOODRUM et al. 1996) were fused to mouse cells in the presence of a protein synthesis inhibitor, and partition of the viral proteins between human and murine nuclei was monitored by immunofluorescence. The proteins were detected in the murine nuclei, indicating that they can shuttle between nucleus and cytoplasm, but only when both proteins were present (DOBBELSTEIN et al. 1997). The short, leucine-rich sequence occupying positions 83–93 in the Orf6 protein, which resembles the NES of the HIV-1 Rev protein, was shown to be necessary for shuttling of the complex, and sufficient to mediate export when fused to the heterologous GST protein (DOBBELSTEIN et al. 1997). It was therefore concluded that this sequence directs export of the E1B–E4 protein complex. Furthermore, competition experiments suggested that export dependent on the E4 protein NES was by means of a pathway that includes components responsible for export of the Rev protein (Dobbelstein et al. 1997). In these experiments, the E4 Orf6 protein was retained within the nucleus unless its C-terminal α-helix was altered, or the E1B 55-kDa protein was also synthesized. It was therefore proposed that the C-terminal α-helix includes a nuclear retention signal that is blocked upon binding of the E1B 55-kDa protein (Dobbelstein et al. 1997). However, Ornelles and colleagues have observed shuttling of the wild-type E4 Orf6 protein when synthesized in the absence of the E1B protein in HeLa cell, rat cell and HeLa–rat cell heterkaryons (Goodrum et al. 1996; cited in Orlando and Ornelles 1999). Other discrepancies emerged upon examination of E1B 55-kDa protein localization and trafficking in human and rodent cell lines, using wild-type and altered viral proteins fused to the green fluorescent protein (GFP) (Kratzer et al. 2000).

As observed in previous studies (Goodrum et al. 1996), the E1B–GFP protein was predominantly cytoplasmic when synthesized in the absence of other viral proteins. However, either treatment with leptomycin B, or over-expression of a protein that inhibits the exportin-1 export pathway induced nuclear accumulation of the fusion protein (Kratzer et al. 2000). Consistent with the view that the E1B shuttles between nucleus and cytoplasm, both a nuclear localization sequence (NLS) and an NES were identified (Kratzer et al. 2000). The latter, the leucine-rich sequence comprising amino acids 82–93, was sufficient to direct transport to the cytoplasm of a GST–E1BNES–GFP chimera microinjected into uninfected Vero cell nuclei, or Ad5-infected cell nuclei. It was also necessary for shuttling activity of the E1B–GFP fusion protein in both uninfected and Ad5-infected cells (Dosch et al. 2001; Kratzer et al. 2000). Active shuttling of the native E1B protein made in Ad5-infected HeLa cells was also detected using the heterokaryon assay, with export via a leptomycin B-sensitive mechanism (Dosch et al. 2001). In contrast, in parallel assays, the Orf6 sequence identified as an NES by Dobbelstein et al. (1997) failed to direct export of an analogous fusion protein, even when altered to match more closely the leucine-rich NES consensus sequence (Kratzer et al. 2000). These investigators did detect shuttling of the Orf6 protein made in Ad5-infected cells, but its export was less efficient than that of the E1B protein and was insensitive to leptomycin B (Dosch et al. 2001). The partition of an E4Orf6–GFP fusion protein between nucleus and cytoplasm was also unaffected by this drug (Dosch et al. 2001). On the other hand, leptomycin B has been reported by Rabino et al. (2000) to reduce shuttling of the Orf6 protein from infected cell nuclei by some 75%. As this drug is an irreversible inhibitor of exportin-1, it has been suggested that transport of the Orf6 protein from the nucleus is non-specific (by diffusion) and that the inhibition observed by RABINO et al (2000) was the result of cytoxicity induced by prolonged treatment with this reagent (DOSCH et al. 2001). The differences in treatment times used in the two studies were not markedly different, 1 h versus 3.5 h, and DOSCH et al. (2001) observed no cytotoxicity in experiments in which leptomycin B was applied at a twofold higher concentration for 2 h.

It is difficult to provide a rationalization that can account satisfactorily for these contradictory observations, beyond the traditional catch-all of differences in experimental conditions. Thus, the questions of how the Orf6 protein leaves the nucleus and whether it contains a leucine-rich NES cannot be settled. And the mechanism by which the E1B 55-kDa protein is exported has yet to be confirmed. However, the most crucial issue to be addressed concerns the contributions of shuttling and export of either of these early proteins to regulation of mRNA export, or other functions of the proteins, during the infectious cycle.

The ability of transiently synthesized E4 Orf6 protein to compensate for defects in viral late gene expression exhibited by mutants lacking the E4 transcription unit (KETNER et al. 1989) has been exploited in an attempt to assess the role of the putative Orf6 NES. In one series of experiments, double substitution of alanine for Leu90 and Ile92 within this Orf6 sequence was reported to impair complementation of the mutant virus dl366, which lacks the entire E4 transcription unit (WEIGEL and DOBBELSTEIN 2000). These substitutions, as well as removal of the C-terminal segment, residues 225–232, which is required for Orf6-dependent nuclear accumulation of the E1B protein in the absence of other viral proteins, inhibited the ability of the Orf6 protein to stimulate virus production, and synthesis of viral DNA and viral late proteins (WEIGEL and DOBBELSTEIN 2000). The Orf6 protein containing the Leu90Ala/Ile92Ala substitutions was able to support some viral DNA synthesis, but accumulation of viral late mRNAs in the cytoplasm was severely impaired, and the cytoplasmic: nuclear ratio of the late fiber mRNA was reduced at very late times in the infectious cycle (WEIGEL and DOBBELSTEIN 2000). Furthermore, the inhibitor of leucine-rich, NES-dependent export that was previously found to inhibit Orf6 protein-mediated shuttling of the E1B–E4 protein complex (DOBBELSTEIN et al. 1997) suppressed virus production when the wild type E4 Orf6 was transiently synthesized in dl366-infected cells (WEIGEL and DOBBELSTEIN 2000). It was therefore concluded that NES-dependent export of the Orf6 protein is required for virus reproduction and accumulation of viral late mRNAs (Weigel and Dobbelstein 2000).

In a similar protocol with the E4 mutant virus dl1011, which lacks Orf6 (Bridge et al. 1993), Bridge and colleagues observed that the same Orf6 NES substitutions did not significantly reduce complementation of viral late protein synthesis (Rabino et al. 2000. Furthermore, treatment of Ad5-infected cells with leptomycin B did not inhibit viral late gene expression, as judged by measurement of the percentage of E4-producing cells synthesizing viral late proteins and comparison of production of the viral penton and fiber proteins in untreated and drug-treated cells. As Orf6 export was significantly reduced, these authors could therefore find no evidence for a requirement for either the Orf6 NES or an NES-dependent export pathway for viral late gene expression during the adenovirus infectious cycle (Rabino et al. 2000). It has been suggested that synthesis of much larger quantities of the Orf6 protein in transfected cells than in virus-infected cells overcame the requirement for the E4 NES (Dobbelstein 2000). However, this explanation does not appear consistent with the observations that transfected cells containing the higher concentrations of the Orf6 protein did not generally synthesize viral late proteins and that the level of the E4 protein detected in cells in which the E4 mutant virus was complemented was similar to that observed during the late phase of Ad5 infection (Bridge 2000). Some of the differences may be attributable to multiplicity dependent phenotypes exhibited by mutant viruses lacking E4 sequences (Rabino et al. 2000): at low mutiplicities of infection, such mutations result in delayed onset of viral DNA synthesis, often with substantial decreases in the quantities of viral DNA made, while replication defects are often not seen following high multiplicity infection (Bridge and Ketner 1989; Halbert et al. 1985; Huang and Hearing 1989; Weinberg and Ketner 1986; Yoder and Berget 1986). Indeed, Weigel and Dobbelstein (2000) observed delayed and reduced viral DNA synthesis when the complementing Orf6 protein carried the NES substitutions, defects that could account for impaired production of viral late mRNAs and proteins. Furthermore, when such inhibition of viral late gene expression was clearly evident, the NES substitutions did not alter the partition of late L5 mRNA between the nucleus and cytoplasm (Weigel and Dobbelstein 2000). The very late (72 h post infection) decrease in the cytoplasmic : nuclear ratio of this late mRNA induced by the NES substitutions was attributed to inhibition of viral mRNA export

(WEIGEL and DOBBELSTEIN 2000). However, the nuclear and cytoplasmic concentrations of the mRNA were still increasing at this time in the presence of the altered E4 protein, but constant, or even decreasing (nuclear) when complementation was by the wild-type protein. Thus, this late difference may well be the result of delayed entry into, and progression through, the late phase of infection when the Orf6 protein carried the NES substitutions. Be that as it may, the complementation protocol has yielded no clearcut evidence that the E4 Orf6 NES contributes to regulation of viral late mRNA production, although it is required for the ability of the protein to support viral DNA synthesis and virus reproduction under some conditions. This conclusion and the other discrepancies summarized in this section underscore the urgent need for analysis of sequences that may mediate export and import of the E4 Orf6 and E1B proteins, and their contributions to the functions fulfilled by these proteins, in the normal context of adenovirus-infected cells.

4 Discrimination of mRNAs for Selective Export: A Function of Specialized Nuclear Microenvironments?

4.1 Parameters That Govern Export Efficiency

The majority of adenoviral late mRNAs are processed from ML primary transcripts and therefore contain the common, 5′-terminal tripartite leader sequence (see SHENK 1996). When linked to a heterologous, non-viral mRNA, this sequence can increase the efficiency of export during the late phase of adenovirus infection some three- to fivefold (HUANG and FLINT 1998). However, the tripartite leader sequence it is not necessary for such export, and is absent from other viral mRNA that are selectively transported from the nucleus, including the IVa_2 and pIX mRNAs (DIX and LEPPARD 1993; LEPPARD 1993). Although the viral mRNAs that are selectively exported during the late phase of infection lack a common sequence, they might be distinguished by means of a structural feature present in all. This mechanism can be excluded by the observation that export of the L1 52/55-kDa protein mRNA synthesized during the early phase is unaffected by absence of the E1B 55-kDa protein, but becomes E1B protein-dependent when the mRNA is made during the late phase (LEPPARD 1993). Similarly, export of the cellular heat shock protein (Hsp)70 mRNA is strongly inhibited in cells infected at 37 °C, but activation of transcription of the Hsp70 gene by heat shock during the late phase of infection induces export of the newly synthesized Hsp70 mRNA (MOORE et al. 1987; YANG et al. 1996), by an E1B protein-dependent mechanism (YANG et al. 1996). Thus, rather than features of the mRNA export substrates, production of an mRNA during the late phase of infection is important for selective export (LEPPARD 1993).

In fact, activation of transcription during this period of the infectious cycle correlates with efficient export of the mRNA product: all viral late promoters are either first activated or substantially stimulated following the onset of viral DNA synthesis and the mRNAs they encode are selectively exported (LEPPARD 1993; PILDER et al. 1986; WILLIAMS et al. 1986); viral mRNA products of early transcription units that are not specifically activated during the late phase, are subject to export inhibition; and all of several additional cellular mRNAs that escape export inhibition are products of transcription units activated during the late phase of infection (YANG et al. 1996). Furthermore, efficient export of these cellular mRNAs, like that of viral late mRNAs, requires the E1B 55-kDa protein (YANG et al. 1996). A mechanistic link between transcription and export may represent an adenovirus-specific adaptation of a normal cellular process: it was reported some time ago that promoter sequences can influence mRNA export efficiency (DE LA PEÑA and ZASLOFF 1987; ENSSLE et al. 1993), and recently that mRNAs are marked for export during transcription in *S. cerevisiae* (LEI et al. 2001). The conclusion that activation of transcription during the late phase of infection is an important, or primary, determinant of E1B protein mediated export implies that activated transcription units and/or their RNA products (the majority of which are viral) occupy distinct nuclear microenvironments functionally specialized to facilitate transport of mature mRNAs from the nucleus, analogous to the gating mechanisms originally proposed by BLOBEL (1985). As discussed in the next subsection, this inference is consistent with previous and subsequent cytological studies of the locations of the viral E1B 55-kDa and E4 Orf6 proteins, of viral late RNAs and of cellular splicing components in adenovirus infected cells.

4.2
Late Phase-Specific Nuclear Sites of Production of Viral mRNA

As discussed in Sect. 3.2, the early phase of adenovirus infection is characterized by E4 Orf3 protein-mediated reorganization of cellular PML body proteins into track-like structures and localization of the E1B 55-

kDa protein to these sites. The onset of viral DNA synthesis and entry into the late phase of infection is accompanied by further substantial reorganization of the nucleus with redistribution of both viral and cellular macromolecules. The most obvious feature is the formation of viral nuclear inclusion bodies, which appear as phase dense structures. The organization of these large inclusion bodies has been examined in some detail by both optical and electron microscopy. Their definitive feature is the presence of high concentrations of the viral E2 DNA-binding protein (DBP) (REICH et al. 1983; SUGAWARA et al. 1977; VOELKERDING and KLESSIG 1986) with which single-stranded viral DNA accumulates (POMBO et al. 1994; PUVION-DUTILLEUL and PUVION 1990; PUVION-DUTILLEUL et al. 1992). Both viral genomes and viral RNAs are found around the periphery of the DBP-containing inclusion bodies, regions termed peripheral replicative zones because replication takes place within them, as does transcription (POMBO et al. 1994; PUVION-DUTILLEUL et al. 1992; see BRIDGE and PETTERSSON 1996). Simultaneous examination of DNA and RNA synthesis has established that these reactions occur largely at separate sites, replication in foci at the edges of the single-stranded DNA accumulation sites, and transcription adjacent to such replicative foci and extending further from the centers of the inclusion bodies (POMBO et al. 1994). Both newly synthesized ML transcripts and spliced tripartite leader sequences (ASPEGREN et al. 1998) are present at the sites of transcription, to which cellular small nuclear (sn)RNPs and splicing proteins are reorganized (ASPEGREN et al. 1998; BRIDGE et al. 1995; POMBO et al. 1994; PUVION-DUTILLEUL et al. 1994; REBELO et al. 1996). Components of PML bodies are also further redistributed, and become associated with peripheral replicative zones (DOUCAS et al. 1996; PUVION-DUTILLEUL et al. 1995), as does the E1B 55-kDa protein.

Indirect immunofluorescence identified three distinct patterns of nuclear staining of the E1B protein during the late phase of Ad5 infection of HeLa cells, in addition to both a reticular distribution and localization in discrete perinuclear bodies in the cytoplasm (ORNELLES and SHENK 1991). Within the nucleus, the E1B protein exhibited diffuse staining, upon which were evident filamentous structures and brightly stained, round bodies organized around less intensely stained centers (ORNELLES and SHENK 1991). The former structures appear to correspond to the E4 Orf3-containing tracks described previously. Comparison of the locations of the E1B protein with that of the DBP indicated that the round bodies are present in peripheral replicative zones. This localization of the E1B

55-kDa protein requires the E4 Orf6 protein, for it is not observed in cells infected by a mutant virus that cannot direct production of the E4 protein (Ornelles and Shenk 1991). The presence of the E1B protein in the peripheral replicative zones is the direct result of its interaction with E4 Orf6, rather than an indirect consequence of other activities of the E4 protein, for example in facilitating late mRNA splicing (Sect. 2): the insertion at position 143 in the E1B 55-kDa protein that specifically inhibits binding to the Orf6 protein (Rubenwolf et al. 1997) eliminates such localization (Gonzalez and Flint 2002). That of the Orf6 protein was not changed, indicating that this viral protein recognizes one or more molecules present in the peripheral replicative zones. This E1B mutation also reduced the efficiency with which viral late mRNAs are exported (Gonzalez and Flint 2002), directly implicating E4 Orf6 protein-dependent association of the E1B 55-kDa protein with peripheral replicative zones in regulation of mRNA export.

During the initial period of the late phase of infection, both reorganized splicing components and spliced ML exons appear as ring-like structures around the sites of single-stranded DNA accumulation (Aspegren et al. 1998; Bridge and Pettersson 1996). As the infectious cycle progresses and ML transcription attains its maximal rate, cellular splicing snRNPs and proteins form large clusters that correspond to the enlarged interchromatin granules (IGs) that are a characteristic feature of late phase nuclei (Bridge et al. 1993; Bridge and Pettersson 1996; Bridge et al. 1995; Puvion-Dutilleul et al. 1994). Transcription does not take place in these structures, as they are not labeled with RNA precursors, but rather continues in the peripheral replicative zones (Bridge et al. 1993; Puvion-Dutilleul and Puvion 1991; Puvion-Dutilleul et al. 1992). Under the electron microscope, sites of transcription appear as a fibrillogranular network, with which splicing components are associated and in which the enlarged IGs, as well as the DBP-containing, single-stranded DNA accumulation sites, are embedded (Puvion-Dutilleul et al. 1992, 1994; Puvion-Dutilleul and Puvion 1991). The enlarged IGs contain polyadenylated RNA and spliced ML tripartite leader sequences, suggesting that they correspond to sites at which one or more post-transcriptional reactions take place (Aspegren et al. 1998; Bridge et al. 1996; Puvion-Dutilleul et al. 1994). Although it is not yet clear whether the viral RNAs that accumulate in these IGs are fully spliced, it has been proposed that these structures may participate in reactions subsequent to splicing, such as sorting of mRNAs or of shuttling and non-shuttling splicing components, or transport of mRNA through the nucleus to the cytoplasm (ASPEGREN et al. 1998; BRIDGE and PETTERSSON 1996; PUVION-DUTILLEUL et al. 1994). This hypothesis is based in part on the view that association of splicing components with the sites of viral RNA synthesis is indicative of splicing where the RNA is made (PUVION-DUTILLEUL et al. 1994), consistent with co-transcriptional splicing of ML RNAs (BEYER and OSHEIM 1988). In addition, accumulation of viral ML mRNAs in late phase IGs correlates with their efficient export to the cytoplasm (ASPEGREN et al. 1998; BRIDGE and PETTERSSON 1996). Notably, examination of individual, infected cells established that those in which snRNP clusters had formed contained much higher concentrations of processed ML late mRNAs and viral late proteins in the cytoplasm than did those in which snRNPs were still organized in ring structures (ASPEGREN et al. 1998).

The observations summarized above are consistent with the hypothesis that E4 Orf6 protein-dependent recruitment of the E1B 55-kDa protein to late phase-specific structures in which viral late RNA (partially or fully processed) accumulates promotes selective export. It should be noted, however, that this view rests upon the untested assumption that the E1B–E4 protein containing structures observed around single-stranded DNA accumulation zones (ORNELLES and SHENK 1991) correspond to enlarged IGs containing cellular splicing components and spliced ML exons (ASPEGREN et al. 1998; BRIDGE and PETTERSSON 1996). As the organization of processed viral mRNA sequences and splicing components has not been examined in the absence of either early protein, it is also possible that association of the E1B–E4 protein complex with the peripheral replicative zones during the initial period of the late phase, when this localization was most prevalent (ORNELLES and SHENK 1991), is necessary for the subsequent formation of enlarged IGs.

5 Molecular Actions of the E1B 55-kDa and E4 Orf6 Proteins in mRNA Export Regulation

As discussed in the previous section, recognition of peripheral replicative zones by the E4 Orf6 protein and recruitment of the E1B 55-kDa protein to these sites appears to be a prerequisite for mRNA export regulation in adenovirus-infected cells. However, the molecular mechanisms by which these viral proteins then subvert one or more cellular pathways that mediate or regulate nuclear export remain largely a matter of speculation.

A priori, it is clear that the E1B 55-kDa and/or E4 Orf6 proteins must direct viral late mRNAs matured within the infected cell nucleus to a cellular export pathway. To date, attention has focused on leucine-rich, NES-dependent nuclear export by means of the exportin-1 export receptor (Sect. 3.3). The ability of the HIV-1 Rev protein to detour viral RNAs that retain introns to the exportin-1 pathway established that cellular export receptors can be diverted from their normal substrates, in this case small RNAs and cellular proteins. As many of the adenoviral ML mRNAs retain 3′ splice sites (although not complete introns defined by both 5′ and 3′ splice sites), it is not unreasonable to suppose that they might be transported from the nucleus by a mechanism not normally used for mRNA. However, at this juncture there is no clear evidence implicating the exportin-1 pathway in export of viral late mRNAs (Sect. 3.3). Indeed at face value, the failure of leptomycin B to inhibit viral late gene expression reported by Rabino et al. 2000 suggests that this pathway is not responsible for late mRNA export, even though it may be required for nucleocytoplasmic shuttling of the E1B 55-kDa and/or E4 Orf6 proteins (Sect. 3.3). Such movement of these viral proteins between nucleus and cytoplasm could be required for other regulatory functions they fulfill, for example in controlling the concentration of the cellular p53 protein (Sect. 1). It is to be hoped that future experiments will establish a clear consensus on the contribution of the exportin-1 pathway of nuclear export to the adenoviral infectious cycle.

Regardless of how viral late mRNAs leave the nucleus, the E1B 55-kDa and/or E4 Orf6 proteins must affect the cellular mRNA export pathway, at a minimum to block transport of host mRNAs and perhaps to usurp this route for viral late mRNAs. So far, only one candidate cellular protein by which the viral proteins might regulate an mRNA export pathway has been identified, and this is largely a case of 'guilt by association': the E1B 55-kDa protein binds to the cellular hnRNP protein E1B-AP5, which in turn can bind to the mRNA export protein Tap (Sect. 3.2.1). It will therefore be of considerable interest to establish whether, and how, the interaction of the E1B protein with E1B-AP5 controls export of cellular and/or viral mRNAs during the infectious cycle.

An important characteristic of cellular mRNA export in mammalian cells is the requirement for removal of introns and retention of RNAs carrying splice sites within the nucleus (see Stutz and Rosbash 1998; Zenklusen and Stutz 2001). It has been reported that, in the absence of the E1B 55-kDa protein, the cytoplasmic concentrations of the longest mRNAs of the ML L1, L2 and L3 families, which retain 1 (L1) to 3 (L2) 3′ splice sites, become progressively lower than those observed in wild-type virus-infected cells, while the concentrations of the shortest mRNAs in each family are less severely reduced (LEPPARD 1993). As this difference could not be attributed to altered splice site usage (LEPPARD 1993), it was suggested that the E1B 55-kDa protein is required to overcome such nuclear retention of splice site-containing RNAs. Both the E1B 55-kDa and E4 Orf6 protein-dependent, cytoplasmic accumulation of an E4 mRNA that is partially spliced (DIX and LEPPARD 1993) and the delayed release of newly synthesized ML RNA sequences from association with a nuclear matrix fraction in the absence of the E1B protein (LEPPARD and SHENK 1989) are consistent with this hypothesis. Some effect of one or both of these proteins on cellular proteins that mark spliced, cellular mRNAs by binding stably to them, such as the Aly/REF-1 protein (ZHOU et al. 2000), might therefore be anticipated.

6 Perspectives

It would be gratifying to close this review with at least the bare bones of a model of the mechanism(s) by which the E1B 55-kDa and E4 Orf6 proteins ensure efficient production of viral late mRNAs in adenovirus-infected cells. In our opinion, the combination of current controversies about some crucial issues and the dearth of definitive evidence to address others render such an endeavor so speculative, or biased, as to be unhelpful. Although no coherent hypothesis describing mRNA export regulation by these viral proteins can be proposed, a considerable body of information about potentially relevant properties of the E1B and E4 proteins has now been collected. The challenge that remains is to establish unambiguously the contributions to mRNA export regulation of such activities, which include NES-dependent export of either or both proteins, the binding of the E1B 55-kDa protein to RNA and the cellular hnRNP protein E1B-AP5, and regulation of viral late mRNA splicing by the E4 Orf6 protein. In addition to binding to one another, both the E1B 55-kDa in the E4 Orf6 protein can engage in multiple interactions with other proteins (Sect. 3.2) and at least the former is found at several locations within infected cells. These properties, and the apparently intimate connection between efficient export of viral, late mRNAs and formation of specific, nuclear micro-environments (Sect. 4.2), indicate that such issues can be addresses only using virus-infected cells to reproduce the complex milieu in which mRNA export regulation takes place.

In addition to addressing the fundamental virological issue of how human adenoviruses achieve selective expression of their genetic information, elucidation of the molecular basis of regulation of mRNA export by the E1B 55-kDa and E4 Orf6 proteins may well provide new insights into cellular RNA export pathways and coordination among the many reactions needed to produce functional mRNA in mammalian cells. Identification of the cellular export pathway(s) that are modified should also help solve the mystery of why the requirement for these proteins during viral replication depends on the type of host cell, temperature, and phase of the cell cycle (Sect. 3). Such information, as well as clarification of any mechanistic relationship between regulation of mRNA metabolism and modulation of p53 concentration, or activity, by the E1B 55-kDa and E4 Orf6 proteins, is also likely to be of practical value, facilitating the design of adenoviral vectors for gene or anti-cancer therapies. Despite the complexities and difficulties encountered in studies of these proteins, it therefore seems certain that the molecular basis of their important regulatory functions will eventually be elucidated.

References

Anderson CW, Baum PR, Gesteland RF (1973) Processing of adenovirus 2-induced proteins. J Virol 12:241–252

Aspegren A, Rabino C, Bridge E (1998) Organization of splicing factors in adenovirus-infected cells reflects changes in gene expression during the early to late phase transition. Exp Cell Res 245:203–213

Babich A, Feldman CT, Nevins JR, Darnell JE, Weinberger C (1983) Effect of adenovirus on metabolism of specific host mRNAs: transport and specific translational discrimination. Mol Cell Biol 3:1212–1221

Babiss LE, Ginsberg HS, Darnell JE (1985) Adenovirus E1B proteins are required for accumulation of late viral mRNA and for effects on cellular mRNA translation and transport. Mol Cell Biol 5:2552–2558

Bachi A, Braun IC, Rodrigues JP, Pante N, Ribbeck K, von Kobbe C, Kutay U, Wilm M, Gorlich D, Carmo-Fonseca M, Izaurralde E (2000) The C-terminal domain of TAP interacts with the nuclear pore complex and promotes export of specific CTE-bearing RNA substrates. RNA 6:136–158

Barker DD, Berk AJ (1987) Adenovirus proteins from both E1B reading frames are required for transformation of rodent cells by viral infection and DNA transfection. Virology 156:107–121

BELTZ GA, FLINT SJ (1979) Inhibition of HeLa cell protein synthesis during adenovirus infection: restriction of cellular messenger RNA sequences to the nucleus. J Mol Biol 131:353–373

BEYER AL, OSHEIM YN (1988) Splice site selection, rate of splicing and alternative splicing on nascent transcripts. Genes Dev 2:754–765

BISCHOFF JR, KIRN DH, WILLIAMS A, HEISE C, HORN S, MUNA M, NG L, NYE JA, SAMPSON-JOHANNES A, FATTAEY A, MCCORMICK F (1996) An adenovirus mutant that replicates selectively in p53-deficient human tumor cells. Science 274:373–376

BLAIR-ZAJDEL ME, BLAIR GE (1988) The intracellular distribution of the transformation-associated protein p53 in adenovirus-transformed rodent cells. Oncogene 2:579–584

BOIVIN D, MORRISON MR, MARCELLUS RC, QUERIDO E, BRANTON PE (1999) Analysis of synthesis, stability, phosphorylation, and interacting polypeptides of the 34-kilodalton product of open reading frame 6 of the early region 4 protein of human adenovirus type 5. J Virol 73:1245–1253

BOYER JL, KETNER G (2000) Genetic analysis of a potential zinc-binding domain of the adenovirus E4 34 k protein. J Biol Chem 275:14969–14978

BOYER TG, MARTIN ME, LEES E, RICCIARDI RP, BERK AJ (1999) Mammalian Srb/Mediator complex is targeted by adenovirus E1 A protein. Nature 399:276–279

BRAITHWAITE A, NELSON C, SKULIMOWSKI A, MCGOVERN J, PIGOTT D, JENKINS J (1990) Transactivation of the p53 oncogene by E1a gene products. Virology 177:595–605

BRAUN IC, ROHRBACH E, SCHMITT C, IZAURRALDE E (1999) TAP binds to the constitutive transport element (CTE) through a novel RNA-binding motif that is sufficient to promote CTE-dependent RNA export from the nucleus. EMBO J 18:1953–1965

BRIDGE E (2000) Letter to the Editor. J Virol 74:1200–1201

BRIDGE E, HEMSTROM C, PETTERSSON U (1991) Differential regulation of adenovirus late transcriptional units by the products of early region. Virology 183:260–266.

BRIDGE E, KETNER G (1989) Redundant control of adenovirus late gene expression by early region 4. J Virol 63:631–638

BRIDGE E, KETNER G (1990) Interaction of adenoviral E4 and E1b products in late gene expression. Virology 174:345–343

BRIDGE E, MEDGHALCHI S, UBOL S, LEESONG M, KETNER G (1993) Adenovirus early region 4 and viral DNA-synthesis. Virology 193:794–801

BRIDGE E, PETTERSSON U (1996) Nuclear organization of adenovirus RNA biogenesis. Exp Cell Res 229:233–239

BRIDGE E, RIEDEL KU, JOHANSSON BM, PETTERSSON U (1996) Spliced exons of adenovirus late RNAs colocalize with snRNP in a specific nuclear domain. J Cell Biol 135:303–314

BRIDGE E, XIA DX, CARMO-FONESCA M, CARDINALI B, LAMOND AI, PETTERSSON U (1995) Dynamic organization of splicing factors in adenovirus-infected cells. J Virol 69:281–290

BROWN LM, GONZALEZ RA, NOVOTNY J, FLINT SJ (2001) The structure of the adenovirus E4 Orf 6 protein predicted by fold recognition and comparative protein modeling. Proteins 44:97–109

CARVALHO T, SEELER JS, OHMAN K, JORDAN P, PETTERSSON U, AKUSJÄRVI G, CARMO-FONSECA M, DEJEAN A (1995) Targeting of adenovirus E1 A and E4-ORF3 proteins to nuclear matrix-associated PML bodies. J Cell Biol 131:45–56

Castiglia CL, Flint SJ (1983) Effects of adenovirus infection on rRNA synthesis and maturation in HeLa cells. Mol Cell Biol 3:662–671

Cathomen T, Weitzman MD (2000) A functional complex of adenovirus proteins E1B-55 kDa and E4orf6 is necessary to modulate the expression level of p53 but not its transcriptional activity. J Virol 74:11407–11412.

Cutt JR, Shenk T, Hearing P (1987) Analysis of adenovirus early region 4-encoded polypeptides synthesized in productively infected cells. J Virol 61:543–552

de la Peña P, Zasloff M (1987) Enhancement of mRNA transport by promoter elements. Cell 50:613–619

Debbas M, White E (1993) Wild-type p53 mediates apoptosis by E1A, which is inhibited by E1B. Genes Dev 7:546–54

Denome RM, Werner EA, Patterson RJ (1989) RNA metabolism in nuclei: adenovirus and heat shock alter intranuclear RNA compartmentalization. Nucl Acids Res 17:2081–2098

Dix BR, Edwards SJ, Braithwaite AW (2001) Does the antitumor adenovirus ONYX-015/dl1520 selectively target cells defective in the p53 pathway? J Virol 75:5443–5447

Dix I, Leppard KN (1993) Regulated splicing of adenovirus type 5 E4 transcripts and regulated cytoplasmic accumulation of E4 mRNA. J Virol 67:3226–31

Dobbelstein M (2000) The nuclear export signal within the adenovirus E4orf6 protein contributes to several steps in the viral life cycle. J Virol 74:1200–1201

Dobbelstein M, Roth J, Kimberly WT, Levine AJ, Shenk T (1997) Nuclear export of the E1B 55-kDa and E4 34-kDa adenoviral oncoproteins mediated by a rev-like signal sequence. EMBO J 16:4276–4284

Dobner T, Horikoshi N, Rubenwolf S, Shenk T (1996) Blockage by adenovirus E4orf6 of transcriptional activation by the p53 tumor suppressor. Science 272:1470–1473

Dobner T, Kzhyshkowska J (2001) Nuclear export of adenovirus RNA. Curr Top Microbiol Immunol 259:25–34

Dosch T, Horn F, Schneider G, Kratzer F, Dobner T, Hauber J, Stauber RH (2001) The adenovirus type 5 E1B-55 k oncoprotein actively shuttles in virus- infected cells, whereas transport of E4Orf6 is mediated by a CRM1-independent mechanism. J Virol 75:5677–5683.

Doucas V, Ishov AM, Romo A, Juguilon H, Weitzman MD, Evans RM, Maul GG (1996) Adenovirus replication is coupled with the dynamic properties of the PML nuclear structure. Genes Dev 10:196–207

Emerman M, Malim MH (1998) HIV-1 regulatory/accessory genes: keys to unraveling viral and host cell biology. Science 280:1880–1884

Endter C, Kzhyshkowska J, Stauber R, Dobner T (2001) SUMO-1 modification required for transformation by adenovirus type 5 early region 1B 55-kDa oncoprotein. Proc Natl Acad Sci USA 98:11312–11317

Enssle J, Kugler W, Hentze MW, Kulozik AE (1993) Determination of mRNA fate by different RNA polymerase II promoters. Proc Natl Acad Sci USA 90:10091–10095

Fischer A, Huber J, Boulens WC, Mattaj IW, Luhrmann R (1995) The HIV-1 Rev activation domain is a nuclear export signal that accesses an export pathway used by specific cellular RNAs. Cell 82:475–483

Flint SJ, Beltz GA, Linzer D (1983) Synthesis and processing of SV40-specific RNA in adenovirus-infected, SV40-transformed human cells. J Mol Biol 167:335–359

FORNEROD M, OHNO M, YOSHIDA M, MATTAJ IW (1997) CRM1 is an export receptor for leucine-rich nuclear export signals. Cell 90:1051–1060

FUKUDA M, ASANO S, NAKAMURA T, ADACHI M, YOSHIDA M, YANAGIDA M, NISHIDA E (1997) CRM1 is responsible for intracellular transport mediated by the nuclear export signal. Nature 390:308–311.

GABLER S, SCHUTT H, GROITL P, WOLF H, SHENK T, DOBNER T (1998) E1B 55-kilodalton-associated protein: a cellular protein with RNA- binding activity implicated in nucleocytoplasmic transport of adenovirus and cellular mRNAs. J Virol 72:7960–7971

GATTONI R, STÉVININ J, JACOB M (1980) Comparison of the nuclear ribonucleoproteins containing transcripts of adenovirus 2 and HeLa cell DNA. Eur J Biochem 108:203–210

GONZALEZ RA, FLINT SJ (2002) Effects of mutations in the adenoviral E1B 55 kDa protein coding sequence on viral late mRNA metabolism. J Virol submitted

GOODRUM FD, ORNELLES DA (1997) The early region 1B 55-kilodalton oncoprotein of adenovirus relieves growth restrictions imposed on viral replication by the cell cycle. J Virol 71:548–561

GOODRUM FD, ORNELLES DA (1998) p53 status does not determine outcome of E1B 55-kilodalton mutant adenovirus lytic infection. J Virol 72:9479–9490.

GOODRUM FD, ORNELLES DA (1999) Roles for the E4 orf6, orf3, and E1B 55-kilodalton proteins in cell cycle-independent adenovirus replication. J Virol 73:7474–7488

GOODRUM FD, SHENK T, ORNELLES DA (1996) Adenovirus early region 4 34-kilodalton protein directs the nuclear localization of the early region 1B 55-kilodalton protein in primate cells. J Virol 70:6323–6335

GRAND RJ, PARKHILL J, SZESTAK T, ROOKES SM, ROBERTS S, GALLIMORE PH (1999) Definition of a major p53 binding site on Ad2E1B58 K protein and a possible nuclear localization signal on the Ad12E1B54 K protein. Oncogene 18:955–965

GRIFMAN M, CHEN NN, GAO GP, CATHOMEN T, WILSON JM, WEITZMAN MD (1999) Overexpression of cyclin A inhibits augmentation of recombinant adeno-associated virus transduction by the adenovirus E4orf6 protein. J Virol 73:10010–9

GRIMWADE D, SOLOMON E (1997) Characterisation of the PML/RAR alpha rearrangement associated with t(15;17) acute promyelocytic leukaemia. Curr Top Microbiol Immunol 220:81–112

GRÜTER P, TABERNERO C, VON KOBBE C, SCHMITT C, SAAVEDRA C, BACHI A, WILM M, FELBER BK, IZAURRALDE E (1998) TAP, the human homolog of Mex67p, mediates CTE-dependent RNA export from the nucleus. Mol Cell 1:649–659

HALBERT DN, CUTT JR, SHENK T (1985) Adenovirus early region 4 encodes functions required for efficient DNA replication, late gene expression, and host cell shutoff. J Virol 56:250–257

HARADA JN, BERK AJ (1999) p53-Independent and -dependent requirements for E1B-55 K in adenovirus type 5 replication. J Virol 73:5333–5344

HAYES BW, TELLING GC, MYAT MM, WILLIAMS JF, FLINT SJ (1990) The adenovirus L4 100-kilodalton protein is necessary for efficient translation of viral late mRNA species. J Virol 64:2732–2742

HIGASHINO F, PIPAS JM, SHENK T (1998) Adenovirus E4orf6 oncoprotein modulates the function of the p53-related protein, p73. Proc Natl Acad Sci USA 95:15683–15687

Ho YS, Galos R, Williams JF (1982) Isolation of type 5 adenovirus mutants with a cold-sensitive phenotype: Genetic evidence of an adenovirus transformation maintainence function. Virology 122:109–110

Horridge JJ, Leppard KN (1998) RNA-binding activity of the E1B 55-kilodalton protein from human adenovirus type 5. J Virol: 9374–9379

Huang J, Schneider RJ (1991) Adenovirus inhibition of cellular protein synthesis involves inactivation of cap-binding protein. Cell 65:271–280

Huang M-M, Hearing P (1989) The adenovirus early region 4 open reading frame 6/7 protein regulates the DNA binding activity of the cellular transcription factor, E2F, through a direct complex. Genes Dev 3:1699–1710

Huang W, Flint SJ (1998) The tripartite leader sequence of subgroup C adenovirus major late mRNAs can increase the efficiency of mRNA export. J Virol 72:225–235

Imperiale MJ, Akusjärvi G, Leppard KN (1995) Post-transcriptional control of adenovirus gene expression. Curr Top Microbiol Immunol 199:139–171

Ishov AM, Maul GG (1996) The periphery of nuclear domain 10 (ND10) as site of DNA virus deposition. J Cell Biol 134:815–826.

Jones NC (1990) Transformation by the human adenoviruses. Semin Cancer Biol 1:425–435.

Kang Y, Cullen BR (1999) The human Tap protein is a nuclear mRNA export factor that contains novel RNA-binding and nucleocytoplasmic transport sequences. Genes Dev 13:1126–1139

Kao CC, Yew PR, Berk AJ (1990) Domains required for *in vitro* association between the cellular p53 and the adenovirus 2 E1B 55 K proteins. Virology 179:806–814

Katahira J, Strasser K, Podtelejnikov A, Mann M, Jung JU, Hurt E (1999) The Mex67p-mediated nuclear mRNA export pathway is conserved from yeast to human. Embo J 18:2593–2609

Ketner G, Bridge E, Virtanen A, Hemstrom C, Pettersson U (1989) Complementation of adenovirus E4 mutants by transient expression of E4 cDNA and deletion plasmids. Nucleic Acids Res 17:3037–48

Kirn D (2000) Replication-selective oncolytic adenoviruses: virotherapy aimed at genetic targets in cancer. Oncogene 19:6660–6669

Konig C, Roth J, Dobbelstein M (1999) Adenovirus type 5 E4orf3 protein relieves p53 inhibition by E1B-55- kilodalton protein. J Virol 73:2253–2262

Kratzer F, Rosorius O, Heger P, Hirschmann N, Dobner T, Hauber J, Stauber RH (2000) The adenovirus type 5 E1B-55 K oncoprotein is a highly active shuttle protein and shuttling is independent of E4orf6, p53 and Mdm2. Oncogene 19: 850–857

Kzhyshkowska J, Schutt H, Liss M, Kremmer E, Stauber R, Wolf H, Dobner T (2001) Heterogeneous nuclear ribonucleoprotein E1B-AP5 is methylated in its Arg-Gly-Gly (RGG) box and interacts with human arginine methyltransferase HRMT1L1. Biochem J 358:305–314

Lei EP, Krebber H, Silver PA (2001) Messenger RNAs are recruited for nuclear export during transcription. Genes Dev 15:1771–1782

Leppard KN, Everett RD (1999) The adenovirus type 5 E1b 55 K and E4 Orf3 proteins associate in infected cells and affect ND10 components. J Gen Virol 80:997–1008

Leppard KN, Shenk T (1989) The adenovirus E1B 55 kd protein influences mRNA transport via an intranuclear effect on RNA metabolism. EMBO J 8:2329–2336
Leppard N (1993) Selective effects on adenovirus late gene expression of deleting the E1B 55K protein. J Gen Virol 74:575–582
Lowe SW, Ruley HE (1993) Stabilization of the p53 tumor suppressor is induced by adenovirus 5 E1A and accompanies apoptosis. Genes Dev 7:535–545
Malette P, Yee SP, Branton PE (1983) Studies on the phosphorylation of the 58000 dalton early region 1B protein of human adenovirus type 5. J Gen Virol 64:1069–1078
Martin ME, Berk AJ (1998) Adenovirus E1B 55K represses p53 activation in vitro. J Virol 72:3146–3154
Mattaj IW, Englmeier L (1998) Nucleocytoplasmic transport: the soluble phase. Annu Rev Biochem 67:265–306
Maul GG (1998) Nuclear domain 10, the site of DNA virus transcription and replication. Bioessays 20:660–667
Moore M, Horikoshi N, Shenk T (1996) Oncogenic potential of the adenovirus E4orf6 protein. Proc Natl Acad Sci USA 93:11295–11301
Moore M, Schaack J, Baim SR, Morimoto RI, Shenk T (1987) Induced heat shock mRNAs escape the nucleocytoplasmic transport block in adenovirus-infected HeLa cells. Mol Cell Biol 7:4505–4512
Nakielny S, Dreyfuss G (1999) Transport of proteins and RNAs in and out of the nucleus. Cell 99:677–690
Nevels M, Rubenwolf S, Spruss T, Wolf H, Dobner T (1997) The adenovirus E4orf6 protein can promote E1A/E1B-induced focus formation by interfering with p53 tumor suppressor function. Proc Natl Acad Sci USA 94:1206–1211
Nevels M, Rubenwolf S, Spruss T, Wolf H, Dobner T (2000) Two distinct activities contribute to the oncogenic potential of the adenovirus type 5 E4orf6 protein. J Virol 74:5168–5181
Nicolas AL, Munz PL, Falck-Pedersen E, Young CS (2000) Creation and repair of specific DNA double-strand breaks in vivo following infection with adenovirus vectors expressing Saccharomyces cerevisiae HO endonuclease. Virology 266:211–224
Nordqvist K, Akusjärvi G (1990) Adenovirus early region 4 stimulates mRNA accumulation via 5′ introns. Proc Natl Acad Sci USA 87:9543–9547
Nordqvist K, Ohman K, Akusjärvi G (1994) Human adenovirus encodes two proteins which have opposite effects on accumulation of alternatively spliced mRNAs. Mol Cell Biol 14:437–445
Ohman K, Nordqvist K, Akusjärvi G (1993) Two adenovirus proteins with redundant activities in virus growth facilitates tripartite leader mRNA accumulation. Virology 194:50–58
Orlando JS, Ornelles DA (1999) An arginine-faced amphipathic alpha helix is required for adenovirus type 5 e4orf6 protein function. J Virol 73:4600–4610
Ornelles D, Shenk T (1991) Location of the adenovirus early region 1B 55 kilodalton protein during lytic infection: association with nuclear viral inclusions requires the early region 4 34 kilodalton protein. J Virol 65:424–439
Pan T, Coleman JE (1990a) The DNA binding domain of GAL4 forms a binuclear metal ion complex. Biochemistry 29:2023–2029
Pan T, Coleman JE (1990b) GAL4 transcription factor is not a "zinc finger" but forms a Zn(II)2Cys6 binuclear cluster. Proc Natl Acad Sci USA 87:2077–2081

Pilder S, Moore M, Logan J, Shenk T (1986) The adenovirus E1B-55kd transforming polypeptide modulates transport or cytoplasmic stablization of viral and host cell mRNAs. Mol Cell Biol 6:470–476

Pombo A, Ferreira J, Bridge E, Carmo-Fonseca M (1994) Adenovirus replication and transcription sites are spatially separated in the nucleus of infected cells. EMBO J 13:5075–5085

Pritchard CE, Fornerod M, Kasper LH, van Deursen JM (1999) RAE1 is a shuttling mRNA export factor that binds to a GLEBS-like NUP98 motif at the nuclear pore complex through multiple domains. J Cell Biol 145:237–254

Puvion-Dutilleul F, Bachellerie JP, Visa N, Puvion E (1994) Rearrangements of intranuclear structures involved in RNA processing in response to adenovirus infection. J Cell Sci 107:1457–1468

Puvion-Dutilleul F, Chelbi-Alix MK, Koken M, Quignon F, Puvion E, de The H (1995) Adenovirus infection induces rearrangements in the intranuclear distribution of the nuclear body-associated PML protein. Exp Cell Res 218:9–16

Puvion-Dutilleul F, Puvion E (1990) Analysis by in situ hybridization and autoradiography of sites of replication and storage of single- and double-stranded adenovirus type 5 DNA in lytically infected HeLa cells. J Struct Biol 103:280–289

Puvion-Dutilleul F, Puvion E (1991) Sites of transcription of adenovirus type 5 genomes in relation to early viral DNA replication in infected HeLa cells. A high resolution in situ hybridization and autoradiographical study. Biol Cell 71:135–147

Puvion-Dutilleul F, Roussev R, Puvion E (1992) Distribution of viral RNA molecules during the adenovirus type 5 infectious cycle in HeLa cells. J Struct Biol 108: 209–220

Querido E, Chu-Pham-Dang H, Branton PE (2000) Identification and elimination of an aberrant splice product from cDNAs encoding the human adenovirus type 5 e4orf6 protein. Virology 275:263–266

Querido E, Marcellus RC, Lai A, Charbonneau R, Teodoro JG, Ketner G, Branton PE (1997) Regulation of p53 levels by the E1B 55-kilodalton protein and E4orf6 in adenovirus-infected cells. J Virol 71:3788–3798

Querido E, Morisson MR, Chu-Pham-Dang H, Thirlwell SW, Boivin D, Branton PE (2001) Identification of three functions of the adenovirus e4orf6 protein that mediate p53 degradation by the E4orf6-E1B55 K complex. J Virol 75:699–709

Rabino C, Aspegren A, Corbin-Lickfett K, Bridge E (2000) Adenovirus late gene expression does not require a Rev-like nuclear RNA export pathway. J Virol 74: 6684–6688

Rebelo L, Almeida F, Ramos C, Bohmann K, Lamond AI, Carmo-Fonesca M (1996) The dynamics of coiled bodies in the nucleus of adenovirus-infected cells. Mol Biol Cell 7:1137–1151

Reich NC, Sarnow P, Duprey E, Levine AJ (1983) Monoclonal antibodies which recognise native and denatured forms of the adenovirus DNA-binding protein. Virology 128:480–484

Roth J, Konig C, Wienzek S, Weigel S, Ristea S, Dobbelstein M (1998) Inactivation of p53 but not p73 by adenovirus type 5 E1B 55-kilodalton and E4 34-kilodalton oncoproteins. J Virol 72:8510–8516

Rothmann T, Hengstermann A, Whitaker NJ, Scheffner M, zur Hausen H (1998) Replication of ONYX-015, a potential anticancer adenovirus, is independent of p53 status in tumor cells. J Virol 72:9470–9478

Rowe DT, Branton PE, Graham FL (1984) The kinetics of synthesis of early viral proteins in KB cells infected with wild-type and transformation-defective host-range mutants of human adenovirus type 5. J Gen Virol 65:585–597

Rubenwolf S, Schutt H, Nevels M, Wolf H, Dobner T (1997) Structural analysis of the adenovirus type 5 E1B 55-kilodalton-E4orf6 protein complex. J Virol 71:1115–1123

Sandler AB, Ketner G (1989) Adenovirus early region 4 is essential for normal stability of late nuclear RNAs. J Virol 63:624–630

Sarnow P, Hearing P, Anderson CW, DN Halbert, Shenk T, Levine AJ (1984) Adenovirus early region 1B 58,000 dalton tumor antigen is physically associated with an early region 4 25,000-dalton protein in productively infected cells. J Virol 49:692–700

Sarnow P, Sullivan CA, Levine AJ (1982) A monoclonal antibody detecting the Ad5 E1B-58 K tumor antigen in adenovirus-infected and transformed cells. Virology 120:387–394

Shaw AR, Ziff EB (1980) Transcripts from the adenovirus-2 major late promoter yield a single early family of 3′ co-terminal mRNAs and five late families. Cell 22:905–916

Shen Y, Kitzes G, Nye JA, Fattaey A, Hermiston T (2001) Analyses of single-amino-acid substitution mutants of adenovirus type 5 E1B-55 K protein. J Virol 75:4297–4307

Shenk T (1996) Adenoviridae and their replication. In: Fields Virology B Fields, P Howley and D Knipe (ed) Raven Press, New York, NY pp 2111–2148

Smiley JK, Young MA, Flint SJ (1990) Intranuclear location of the adenovirus type 5 E1B 55-kilodalton protein. J Virol 64:4558–4564

Steegenga WT, Riteco N, Jochemsen AG, Fallaux FJ, Bos JL (1998) The large E1B protein together with the E4orf6 protein target p53 for active degradation in adenovirus infected cells. Oncogene 16:349–357

Sternsdorf T, Grotzinger T, Jensen K, Will H (1997) Nuclear dots: actors on many stages. Immunobiology 198:307–331

Stutz F, Rosbash M (1998) Nuclear RNA export. Genes Dev 12:3303–3319

Sugawara K, Gilead Z, Wold WSM, Green M (1977) Immunofluorescence study of the adenovirus type 2 single-stranded DNA binding protein in infected and transformed cells. J Virol 22:527–539

Teodoro JG, Branton PE (1997) Regulation of p53-dependent apoptosis, transcriptional repression, and cell transformation by phosphorylation of the 55-kilodalton E1B protein of human adenovirus type 5. J Virol 71:3620–3627

Teodoro JG, Halliday T, Whalen SG, Takayesu D, Graham FL, Branton PE (1994) Phosphorylation at the carboxy terminus of the 55-kilodalton adenovirus type 5 E1B protein regulates transforming activity. J Virol 68:776–786

Tribouley C, Lutz P, Staub A, Kedinger C (1994) The product of the adenovirus intermediate gene IVa_2 is a transcription activator of the major late promoter. J Virol 68:4450–4457

Turnell AS, Grand RJ, Gallimore PH (1999) The replicative capacities of large E1B-null group A and group C adenoviruses are independent of host cell p53 status. J Virol 73:2074–2083

Van Eekelen CA, Mariman EC, Reinders RJ, Van Venrooij WJ (1981) Adenoviral heterogeneous nuclear RNA is associated with host cell proteins. Eur J Biochem 119:461–467

Voelkerding K, Klessig DF (1986) Identification of two nuclear subclasses of the adenovirus type 5-encoded DNA-binding protein. J Virol 60:353–362

Weiden MD, Ginsberg HS (1994) Deletion of the E4 region of the genome produces adenovirus DNA concatemers. Proc Natl Acad Sci USA 91:153–157

Weigel S, Dobbelstein M (2000) The nuclear export signal within the E4orf6 protein of adenovirus type 5 supports virus replication and cytoplasmic accumulation of viral mRNA. J Virol 74:764–772

Weinberg DH, Ketner G (1986) Adenoviral early region 4 is required for efficient viral DNA replication and for late gene expression. J Virol 57:833–838

Weis K (1998) Importins and exportins: how to get in and out of the nucleus. Trends Biochem Sci 23:185–189

Williams J, Karger BD, Ho YS, Castiglia CL, Mann T, Flint SJ (1986) The adenovirus E1B 495R protein plays a role in regulating the transport and stability of the viral late messages. Cancer Cells 4:275–284

Yang U-C, Huang W, Flint SJ (1996) mRNA export correlates with activation of transcription in human subgroup C adenovirus-infected cells. J Virol 70:4071–4080

Yew PR, Kao CC, Berk AJ (1990) Dissection of functional domains in the adenovirus 2 early 1B 55 k polypeptide by suppressor-linker-insertional mutagenesis. Virology 179:795–805

Yew PR, Liu X, Berk AJ (1994) Adenovirus E1B oncoprotein tethers a transcriptional repression domain to p53. Genes Dev 8:190–202

Yoder SS, Berget SM (1986) Role of adenovirus type 2 early region 4 in the early-to-late switch during productive infection. J Virol 60:779–781

Zantema A, Fransen JA, Davis-Olivier A, Ramaekers FC, Vooijs GP, DeLeys B, Van der Eb AJ (1985) Localization of the E1B proteins of adenovirus 5 in transformed cells, as revealed by interaction with monoclonal antibodies. Virology 142:44–58

Zenklusen D, Stutz F (2001) Nuclear export of mRNA. FEBS Lett 498:150–156

Zhang Y, Feigenblum D, Schneider RJ (1994) A late adenovirus factor induces eIF-4B dephosphorylation and inhibition of cellular protein synthesis. J Virol 68:7040–7050

Zhou Z, Luo MJ, Straesser K, Katahira J, Hurt E, Reed R (2000) The protein Aly links pre-messenger-RNA splicing to nuclear export in metazoans. Nature 407:401–405

Chapter 11

Adenovirus Interaction with Its Cellular Receptor CAR

J. Howitt, C.W. Anderson, P. Freimuth

Abstract. Representative adenoviruses from four of the five major virus subgroups have been shown to interact with the 46-kDa coxsackievirus and adenovirus receptor (CAR) that is widely expressed on many human cell types, suggesting that the ability to bind CAR may be a conserved feature of many of the ~50 known adenovirus serotypes. Receptor binding is a function of the distal 'knob' domain of the trimeric viral fiber protein. Here we review recent structural characterizations of knob, CAR and knob–CAR complexes, and we discuss how knob architecture may have evolved to accommodate opposing selective pressures to vary anti-

J. Howitt, C.W. Anderson, P. Freimuth
Biology Department, Brookhaven National Laboratory, Upton, NY 11973, USA
e-mail: freimuth@bnl.gov

genic structure while conserving receptor binding specificity. In contrast to the hypervariability of the solvent-exposed surface of knob, the CAR receptor was found to be non-polymorphic.

1 Introduction

The mechanism of adenovirus (Ad) attachment to the host cell plasma membrane has been revealed in detail by research over the past 10 years. It has long been known that receptor binding activity is associated with the viral fibers, trimeric spike proteins that protrude radially from the vertices of the icosahedral capsid (Philipson et al. 1968). In some adenovirus serotypes, fiber and other virus structural proteins are synthesized in excess and accumulate in the cell nucleus during late stages of infection. Fiber protein can be readily purified from lysates of cells infected with subgroup C viruses, for example Ad2 and Ad5 (Boulanger and Puvion 1973). Addition of purified fiber protein to virus suspensions during adsorption strongly inhibits infection, indicating that fiber and intact virus particles compete for binding sites on host cells (Philipson et al. 1968; Hautala et al. 1998). Cell binding studies using purified radiolabeled fiber demonstrated that fiber binds specifically and with high affinity to the cell plasma membrane, and that cell lines typically used for laboratory propagation of adenovirus have approximately 10^4 high-affinity receptor sites per cell (Persson et al. 1985; Freimuth 1996). Similar numbers of high-affinity binding sites for radiolabeled intact virus particles also were observed (Seth et al. 1994).

Adenovirus fibers have two distinct structural domains, a rod-like shaft which is attached to the capsid vertex by an amino-terminal anchor, and a distal globular domain referred to as the knob or head domain. The shaft domain consists of repetitive sequence motifs that vary in number between different serotypes and fold into a novel coiled-coil type structure (Stouten et al. 1992; van Raaij et al. 1999). The domain boundary between the shaft and knob is, therefore, easy to recognize in alignments of fiber protein sequences from different serotypes. Knob domains from several different serotypes have been produced as recombinant protein fragments in insect cell and bacterial expression systems, and this led to the key observation that receptor binding activity is associated with the knob domain (Henry et al. 1994; Louis et al. 1994). Recombinant knobs from some serotypes assemble into trimers in heterologous expression systems. The trimeric form of knob is necessary for its receptor binding activity. Recombinant full-length fiber also has been produced, using a vaccinia virus vector for expression in mammalian cells (HONG and ENGLER 1996). Mutant analysis demonstrated that the knob domain plays a critical role in trimerization of the full-length fiber protein. Trimeric fiber protein is resistant to denaturation by sodium dodecylsulfate at room temperature (HONG and ENGLER 1996), indicating that the knob and shaft domains have highly stable protein folds.

CAR, the cellular receptor recognized by fibers of the adenoviruses that are commonly used in the laboratory (Ad2, Ad5), was recently identified and molecularly cloned (BERGELSON et al. 1997; TOMKO et al. 1997). As its name indicates, CAR serves as a receptor for both subgroup B coxsackieviruses and adenovirus. Many adenovirus serotypes recognize CAR with the notable exception of subgroup B viruses (Ads 3 and 7) (DEFER et al. 1990; ROELVINK et al. 1998). CAR has a mass of 46 kDa and is a member of the immunoglobulin superfamily. The amino-terminal extracellular region of CAR consists of two Ig-like domains, a distal Ig variable-type domain (CAR D1) and a proximal C2-type Ig domain (CAR D2). CAR has a single hydrophobic membrane-spanning region and a ~110 residue carboxy-terminal cytoplasmic domain. CAR D1 and the entire CAR ectodomain have been expressed as recombinant protein fragments in both bacterial and mammalian cell expression systems. Ad12 knob and CAR D1 that were produced in *Escherichia coli* formed specific complexes in vitro, and both bacterially expressed proteins inhibited infection of HeLa cells by Ad2 and Ad12 virus (FREIMUTH et al. 1999). Thus the CAR D1 domain alone is sufficient for interaction with fiber knob. A glycosylphosphatidylinositol (GPI)-linked form of CAR also supported adenovirus infection (WANG and BERGELSON 1999), indicating that the membrane-spanning and cytoplasmic domains of CAR are not required for infection of cultured cells. From these results it can be surmised that the knob–CAR interaction serves mainly to attach virus particles to the host cell plasma membrane. Subsequent endocytosis of bound particles can be accelerated in certain cell types by interaction of virus–CAR complexes with integrin coreceptors (BAI et al. 1993; WICKHAM et al. 1993). Integrin-independent pathways for infection do exist, however, as mutant viruses that have lost the ability to bind to integrins do infect cells with wild-type efficiency, albeit at a slower rate on tightly adherent cells (BAI et al. 1993). The role of integrin coreceptors in adenovirus infection is beyond the scope of this discussion but has been reviewed in detail elsewhere (NEMEROW 2000).

The objective of this chapter is to review the structural biology and biochemistry of the knob–CAR interaction. To date, knob–CAR and HIV gp120–CD4 are the only examples where the interface between virus particles and their cellular protein receptors have been examined by X-ray crystallography at high resolution (Kwong et al. 1998; Bewley et al. 1999). These two systems have certain striking similarities and unusual characteristics which may reflect the struggle that all animal viruses have in conserving specificity for a particular receptor while simultaneously varying the antigenic structure of the receptor-binding protein itself. The relatively large number of adenovirus serotypes that bind CAR and the ease with which recombinant knob–CAR systems can be derived from these serotypes, therefore provide excellent model systems to investigate how receptor specificity is maintained during virus evolution, and conversely, how viruses manage to change receptor specificity.

2 Crystal Structures

2.1 Ad5 Knob

The Ad5 fiber protein knob domain was the first knob to have its atomic resolution structure solved by X-ray crystallography (Xia et al. 1994). For this analysis, the Ad5 knob was expressed as a soluble protein fragment in *E. coli*. Knob was found to have an eight-stranded β-sandwich structure containing two distinct antiparallel β-sheets that accommodate a high number of reverse turns with nearly all hydrogen-bond donors and acceptors found to be properly paired. The majority of the protein (65%) is made up of surface loops with approximately 35% of the structure containing β-strands. The knob fold was found to be similar to other structures solved at that time, the closest being superoxide dismutase (Tainer et al. 1982); however, the folding topologies of knob differed from these other structures. Knob, therefore, represents a new class of antiparallel β-structure.

The nomenclature adopted by Xia et al. labeled the two β-sheets R and V, for receptor and virion, respectively. The R sheet is made up of β-strands D, G, H and I, which mainly are exposed to solvent. The V sheet is made up of the β-strands A, B, C and J and is partially buried in the trimer structure. Two very small strands, labeled E and F, also occur within the V sheet (see Fig. 1A for details). Loop regions are labeled according to the strands

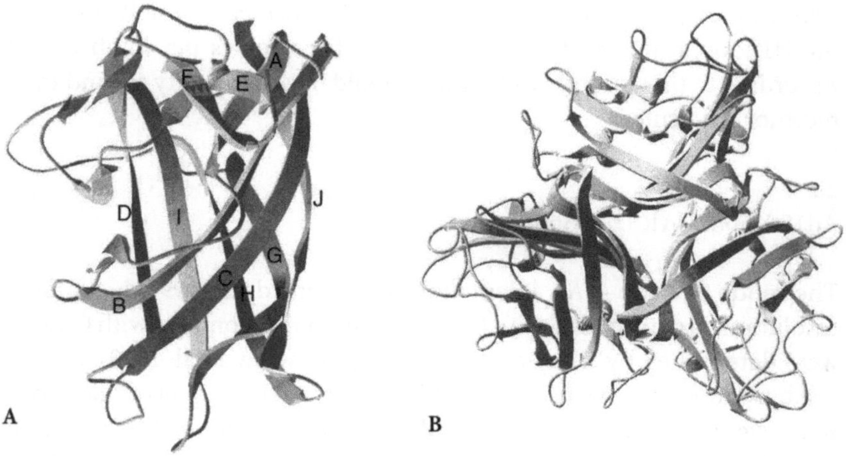

Fig. 1 A, B. Ribbon diagrams of Ad5 knob monomer and trimer. **A** Knob monomer, with *β* strands labeled according to the nomenclature adopted by XIA et al. (1994). The R-sheet is composed of *β*-strands D, G, H and I; the V sheet is composed of *β*-strands A, B, C and J. **B** Knob trimer, as viewed from the cellular receptor

that they connect. An angle of ~30 degrees separates the two *β*-sheets creating a hydrophobic core within the molecule. Previously, it had been postulated that the fiber protein might be a dimer; however, stoichiometric analysis of radiolabeled protein indicated that the fiber knob was trimeric (VAN OOSTRUM and BURNETT 1985). In agreement with this geometry, Ad5 knob monomers were found to form homotrimers in the crystal lattice (Fig. 1B) and in solution (XIA et al. 1994). The trimer is approximately 62 Å in diameter by 40 Å high and contains extensive contacts between monomers which involve hydrogen bonding, salt bridges, van der Waals' contacts and water-mediated charge interactions. Strands G and J of each monomer play a critical role in the formation of the trimeric protein with a loss of accessible surface area upon trimerization of ~1950 Å^2.

The cellular receptor for adenovirus had not been identified at the time the Ad5 knob structure was solved, and it was difficult to predict the location of the receptor-binding site(s) based on sequence conservation given the extensive sequence variation between knobs from different adenovirus serotypes. It was postulated, however, that a large cavity located along the central threefold axis of symmetry might accommodate binding of a single receptor molecule. Also suggested as a possible receptor binding site was the floor of the valley formed by the R-sheet and the HI loop, features that are present in three copies per knob trimer. According to this second model, knob would have a capacity to bind three receptor molecules.

2.2 Ad12 Knob–CAR D1 Complexes

The mode of knob–CAR interaction was clarified in 1999 when atomic resolution structures of the Ad12 knob alone and in complex with CAR D1 were determined by X-ray crystallography (BEWLEY et al. 1999). Ad12 is a subgroup A adenovirus, and by this time it had been shown that representatives of all subgroups except subgroup B bound to CAR (DEFER et al. 1990; BERGELSON et al. 1997; TOMKO et al. 1997; ROELVINK et al. 1998). Both Ad12 knob and CAR D1 were expressed as soluble protein fragments in *E. coli*. The structures of Ad12 and Ad5 knob trimers are very similar, with a root mean square (rmsd) deviation of 1.2 Å when equivalent $C\alpha$ atoms are superimposed. The only major difference between the structures centers around the HI loop, which is well-ordered in Ad12 knob but extended and disordered in Ad5 knob. CAR D1 has a β-sandwich fold that is characteristic of immunoglobulin variable (IgV) domains, with an insertion of two short β strands, C′ and C″, between strands C and D (Fig. 2 A). Surprisingly, the binding site for CAR D1 was not located in the floor of the R sheets as predicted from structural analysis of the Ad5 knob, but rather were located at the interface between adjacent knob monomers (Fig. 2 B). All three receptor binding sites were fully saturated with CAR D1 molecules in the Ad12 knob–CAR D1 crystal structure, indicating that individual fibers on the virus capsid potentially can make multivalent contact with two or three CAR molecules on the host cell surface. The receptor binding site is made up of four loop regions from the knob trimer interacting with a single surface of CAR D1. The AB loop in particular contributes over 50% of the interfacial protein–protein interactions (discussed in Sect. 3). Mutational analysis of fiber knob and CAR D1 confirmed the structure of the bound complex (BEWLEY et al. 1999; ROELVINK et al. 1999). No large conformational rearrangements of knob are required for receptor binding, but small backbone movements are observed within contact regions of CAR D1.

CAR D1 also was crystallized independently of fiber knob, and its structure was solved at high resolution (VAN RAAIJ et al. 2000). In this study

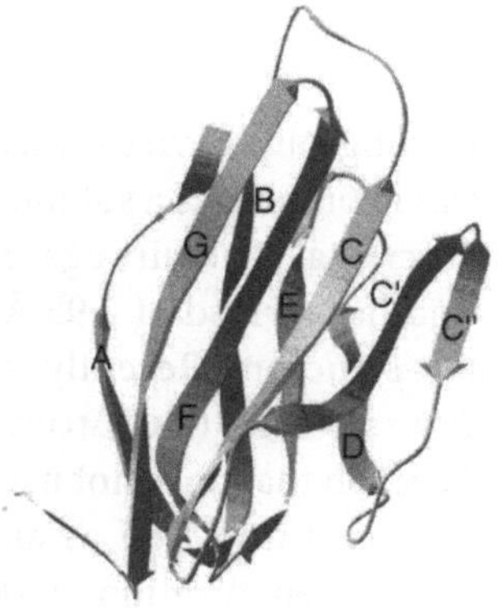

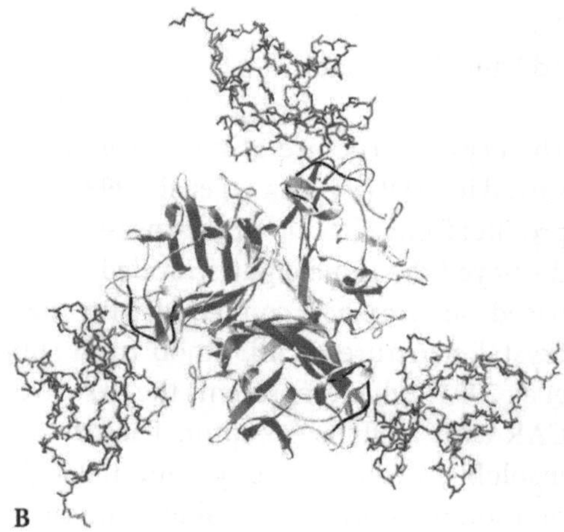

A B

Fig. 2A, B. Diagrams of CAR D1 and the Ad12 knob–CAR D1 complex. **A** Ribbon diagram of the CAR D1 domain with β strands labeled. Strands C, C′ and C″ contain residues important for interfacial contacts with knob. **B** The Ad12 knob–CAR D1 complex viewed down the fiber shaft. Ad12 knob trimer is shown as a *ribbon structure* and CAR D1 monomers as a *backbone structure*. The AB loop, which contains ~50% of the interfacial residues from knob, is shown in *black*

CAR D1 was expressed as a soluble protein within the periplasm of *E. coli*. The structure shows CAR D1 to be a homodimer in the crystal lattice; solution based ultracentrifugation experiments also indicated a dimeric form of the protein with a dissociation constant of 16 μM. These findings are consistent with results suggesting that CAR may be a homophilic adhesion molecule likely to mediate tissue organization as its native function (Honda et al. 2000). The high resolution structure of the dimeric CAR D1 is almost identical to the structure of CAR D1 in complex with knob; the BC and FG loops, which are involved in homodimerization and knob binding, show minor rearrangements in conformation. However, a larger difference is observed in the C″D loop, which adopts a different backbone conformation resulting in up to 4 Å movement of this loop relative to its position in the knob–CAR complex. Interestingly, the CAR D1 surface involved in the formation of the homodimer is very similar to that used for fiber knob binding, suggesting that this surface may need to be conserved for proper CAR function in normal, uninfected cells. This perhaps accounts for the low frequency of single nucleotide polymorphisms that map to this region of the CAR gene (discussed below).

2.3 Ad2 and Ad3 Knob

The crystal structure of knob from Ad2, another subgroup C virus, was solved in 1999 (van Raaij et al. 1999). Ad2 knob was expressed as a soluble product from a baculovirus expression system. The crystal structure again displayed the same eight-stranded β-sheet fold and has a rmsd of 1.95 Å based on Cα when compared with the Ad5 knob structure. Recently, a crystal structure of Ad3 knob from subgenus B was solved (Durmort et al. 2001), which represents the first structure of a knob that does not use CAR as its cellular receptor. The protein was expressed in *E. coli* as an insoluble inclusion body, and through subsequent denaturation and dilution renaturation, a stable, trimeric protein was obtained. The overall topology of the Ad3 structure is again very similar to that of the other solved adenovirus knobs with each monomer consisting of two anti-parallel β-sheets. Observed differences are a shorter β-strand F and an insertion of two short helical folds in the CD and EG loop regions. A comparison of Cα atoms between the Ad3 and Ad5 knobs revealed a rmsd of 1.62 Å, indicating the close relationship between the two subgroups. In keeping with the notion that major antigenic determinants of knob map to variable loop regions of the subgenera, larger differences in structure were observed between the surface loops of Ad3 knob compared to Ad5 knob. In particular, the CD and EG loops, and part of the AB loop of Ad3 knob were shown to vary considerably from that of the other solved structures. A number of key residues in these loops are substituted in Ad3 knob, accounting for its inability to bind CAR D1.

3 Knob–CAR D1 Interface

The interface between Ad12 knob and CAR D1 is made up of a number of contacts that contribute to the formation of a high affinity complex. Four loop regions of knob have been found to interact with a single face of CAR D1; there is no structural or mutational evidence for an interaction with the second domain of CAR (D2). The complex forms between two adjacent knob monomers with residues from each subunit required for binding (Fig. 3 A–C). Specifically, this interaction involves the AB, CD and DG loops of one monomer, and the FG loop of the second monomer. The majority of contacts occur within the AB loop region of knob where hydrogen

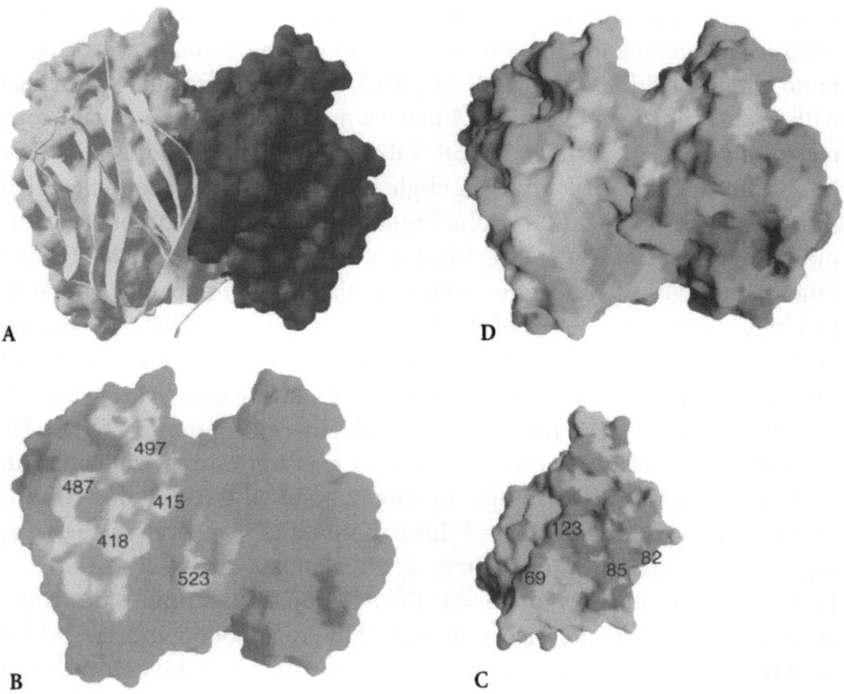

Fig. 3 A–D. Molecular surface representations of the interface in the Ad12 knob–CAR D1 complex. **A** Surface diagram of two knob monomers in complex with a monomer of CAR D1 (CAR D1 shown as a *ribbon structure*). The *light gray* monomer of knob contains the majority of interfacial contacts with the AB, CD and DG loops all interacting with CAR D1, the *dark gray* monomer contains the FG loop which interacts with CAR D1. **B** Surface diagram of two adjacent Ad12 knob monomers (same view as A) *colored* on a sliding scale from *white* (contact) to *gray* (no contact). Residues 415 and 418 are contained within the AB loop, and residues 487 and 497 are contained within the DG loop. The *gray* depression shown between residues 415, 418, 487 and 497 forms part of the large interfacial cavity that is observed in the Ad12 knob–CAR D1 complex. **C** Surface diagram of the CAR D1 surface that interacts with Ad12 knob. The surface is *colored* on a sliding scale: *dark gray* (contact) to *light gray* (no contact). **D** Sequence conservation diagram of two knob monomers viewed at the CAR D1 interface (same view as **A**). Molecules are highlighted on a sliding scale from *white* (highest relative level of amino acid sequence conservation) to *dark gray* (lowest relative level of conservation)

bonds, salt bridges and van der Waals' contacts all take place and appear to form a key anchor for the complex. Residue Asp415 forms a hydrogen bond and salt bridge with Lys125 of CAR D1 and also a second salt bridge with Lys123; prolines 417 and 418 make a number of van der Waals' contacts with CAR D1 residues Glu58, Val72, Leu75, Tyr85 and Lys125. The importance of the AB loop is highlighted by mutational analyses which showed that amino acid (aa) substitutions inhibit formation of the complex (Bewley et al. 1999; Roelvink et al. 1999; Kirby et al. 2000). Two other loops within the same monomer also show important contacts; residues Val450 and Lys451 of the CD loop and residues Gln487, Gln494, Ser497 and Val498 of the DG loop (which contains the E and F strands) all interact with CAR D1. The DE loop residue Gln487 forms a hydrogen bond with Ser77 of CAR D1. Residues from the adjacent monomer of the FG loop, Pro517, Pro519, Asn520 and Glu523 also have direct contact with CAR D1, with Asn520 forming a hydrogen bond with Asp70 of CAR D1. An important residue conserved throughout all human adenovirus serotypes, apart from the short fibers of subgenus F (Ad40 and Ad41), is Lys429. This residue lies just below the AB loop on β-strand B down the central y-axis of the interface and forms a hydrogen bond and salt bridge with Glu58 of CAR D1. The strict conservation of this residue throughout the subgenera and the important contacts made with CAR makes this one of the few residues that are crucial for cellular attachment.

The protein-protein interface formed by knob and CAR D1 is atypical due to the low topological complementarity between the two molecules. Although cavities within protein interfaces are common (Hubbard and Argos 1994), the size of the pocket formed within the complex is larger than usually observed. This cavity is most likely filled with solvent and allows for water-mediated bonding between the two molecules. The cavity sits between the AB and DE loops of knob and is approximately 240 $Å^3$ in size (Fig. 3 B). The cavity increases the footprint at the interface by approximately 15% to a value of ~1880 $Å^2$ (820 $Å^2$ from one knob monomer, 140 $Å^2$ from the adjacent monomer and 920 $Å^2$ by CAR D1) compared to what would be predicted from the values of the buried surface area. A mixture of hydrophobic and polar groups line the walls of the cavity, with over 60% being either backbone atoms or side chains that are conserved in other adenoviruses. This cavity theoretically can accommodate four water molecules, although in the structural model, only one water molecule was observed (Bewley et al. 1999). The use of water as a bonding agent to stabilize the knob-CAR interface lowers the constraints on the virus to maintain specific amino acid side chain contacts with CAR. As a result, there is not strict conservation of residues in the AB and DE loops which form the walls of the water-trapping cavity (Fig. 3D). This indirect, water-mediated mode of binding allows a greater degree of antigenic drift of residues which form the receptor binding site without loss of binding specificity.

Interestingly, the structurally unrelated HIV gp120–CD4 complex shows a similar unusual mismatch in surface topology which results in the formation of large cavities (Kwong et al. 1998). The largest of these is 273 Å^3 and is formed by an equal number of gp120 and CD4 residues. The authors note that the cavity acts as a water buffer which allows for variation on the gp120 surface and the formation of an 'anti-hotspot', which may help the virus escape from antibodies directed against the CD4 binding site. Thus, the gp120–CD4 and knob–CAR complexes, although structurally different, may both employ a similar mechanism for evasion of the host immune response. More recent studies of the gp120–CD4 complex suggest that conformational rearrangements of the gp120 protein also provide a mechanism for evasion of the immune system (Myszka et al. 2000). Such rearrangements are not observed with the adenovirus knob protein, where little conformational change accompanies formation of the knob–CAR complex.

The subgroup B adenoviruses are suggested to use a still unknown cellular receptor for viral attachment (Defer et al. 1990; Roelvink et al. 1998). Sequence alignments of this subgroup with other CAR-binding serotypes suggests that differences within the AB loop of the proteins may affect receptor binding (Fig. 4). Although it is clear that the AB loop is a major determinant of binding specificity, the contribution of individual residues within the AB loop or other loops is difficult to assess from

```
Ad2     LTLWTTPDPSPNCRIH
Ad9     RTLWTTPDTSPNCKID
Ad12    LTLWTTPDPPPNCSLI
Ad3     NTLWTGPKPEANCIIE
         **** * . .** :
```

Fig. 4. Amino acid sequence alignment of residues within the AB loop. Ad2, Ad9 and Ad12 interact with CAR, while Ad3 is suggested to recognize an unknown cellular receptor. *Asterisks* indicate conserved residues, and *dots* indicate similar residues. Highlighted in *light gray* are two residues of Ad3 knob that are thought to interfere with CAR binding, Lys138 and Glu140. *Underlined* are Ad12 knob residues in contact with CAR D1 in the Ad12 knob–CAR D1 complex

sequence alignments alone, due to underlying local differences in the structures or surface properties of knobs from different serotypes. Thus, mutational analysis, binding studies and structural characterization together are required for a complete understanding of the contributions that individual residues make to the interaction of knob with the cellular receptor and for the precise localization of the receptor binding sites on these knobs. Models based on structural data of Ad3 knob predict that AB loop residue Glu140 would form charge clashes with two acidic residues of CAR, Asp56 and Glu58, in a hypothetical Ad3–CAR D1 complex (Durmort et al. 2001). Other differences between the Ad3 and Ad12 knobs that probably account for the inability of Ad3 knob to interact with CAR D1 are substitution of critical Ad12 knob AB loop residue Asp415 for lysine at the equivalent position in Ad3 knob, and also differences in conformation of the DE and EG loops between these two serotypes. This change of conformation within the DE and EG loops of Ad3 knob may allow for interaction with receptors larger than CAR D1.

3.1 Binding Studies and Mutational Analysis of Knob–CAR Complexes

The adenovirus knob domain has been extensively studied through site-directed mutational analysis with a view to understanding the interactions important for receptor binding. A number of methods have been used to quantify the binding of mutant knob proteins to CAR D1. Techniques, such as cell binding assays, have been employed by a number of investigators, who used different indirect detection mechanisms (Roelvink et al. 1999; Santis et al. 1999; Kirby et al. 2000). These assays present the native receptor in an orientation that mimics that of a normal cellular infection and gives rise to affinity values in the subnanomolar to nanomolar range for knob–CAR binding. More recently, surface plasmon resonance (SPR) has been used to investigate binding of recombinant knob and CAR molecules (Kirby et al. 2000, 2001; Lortat-Jacob et al. 2001). This technique more readily lends itself to high throughput analysis of mutant proteins. Results of Lortat-Jacob et al. show the importance of the orientation of knob and its receptor in such assays. Ad2 knob binding to immobilized CAR, as found in the native state on cells, has an affinity of ~1 nM; however, if knob is immobilized and exposed to soluble CAR D1 then the affinity decreases to ~24 nM. Analysis of these data indicated that in the native orientation, the first binding event is 1:1 for Ad2 knob and

CAR D1; subsequently, the complex is further stabilized by the association of two further CAR D1 molecules with the trivalent knob. This avidity mechanism of trivalent binding results in the high overall affinity that is observed between fiber knob and CAR and is analogous to the common bivalent antibody–antigen binding.

Other SPR studies using CAR D1, CAR D2 and CAR D1D2 indicate that the knob–CAR interaction requires only the first domain of CAR for efficient binding (Kirby et al. 2000) as reported previously through direct and indirect methods (Freimuth et al. 1999). Different serotypes of knob have been shown, through SPR, to have varying affinities for CAR D1. Ad5 and Ad12 knobs and the long fiber knob from Ad41 have affinities between 8 and 15 nM for CAR D1, whereas Ad9 knob has an affinity of 6,400 nM (Kirby et al. 2001). Thus, the subgroup D Ad9 and the subgroup C Ad5 knobs differ by about 800-fold in binding affinity for CAR D1. Substituting Ad9 knob residue Asp222 with lysine, the structurally homologous residue in Ad5 knob, increased the affinity 20-fold to 310 nM. The converse mutation in Ad5 knob (Lys442Asp) decreased its affinity for CAR D1 by 350-fold to 2800 nM. Other mutations substituting Ad5 and Ad12 knob residues into Ad9 knob (e.g., Lys260Pro and Lys260Gln, respectively) did not have any measurable effect on affinity. Earlier studies indicated that Ad2 and Ad9 knob utilize the same primary receptor (later found to be CAR); however, the shorter fiber length of Ad9 was suggested to allow fiber-independent binding of the penton base to integrins on the cell surface (Roelvink et al. 1996). This situation suggests a limited role for knob in the attachment of Ad9 to cells, which, therefore, would not require such high affinity interactions as those seen for the longer fibers. Importantly, these results indicate that the binding sites on adenovirus knob, although similar between serotypes, often do not involve identical residues or contacts.

In an effort to further define the differences between serotypes within the knob domain, we have developed a fluorescence polarization method of measuring knob binding to CAR D1. This technique is entirely solution-based and thus removes the effects of immobilization of the cellular receptor. Therefore, results pertain solely to the affinity of individual binding sites on knob for CAR D1 without the avidity effects observed in other techniques. These studies show that Ad2 has a fivefold greater affinity for CAR D1 than Ad12 knob (J. Howitt and P. Freimuth, unpublished results).

It was inferred from modeling studies overlaying the Ad2 knob structure with the Ad12 knob–CAR D1 structure that two residues within Ad2 knob could form hydrogen bonds that were not observed in the Ad12 knob–CAR D1 complex. Using this structural model as a guide, it has been possible to change the affinity of both Ad2 and Ad12 knobs for CAR D1. Substitution of Ad12 knob residue Pro417 with serine, the structurally equivalent residue in Ad2 knob, increased the binding affinity of the resulting Ad12 knob mutant Pro417Ser for CAR D1 by threefold. A crystal structure of the Ad12 Pro417Ser mutant–CAR D1 complex confirmed the mutation and the formation of the predicted extra hydrogen bond (J. Howitt and P. Freimuth, unpublished results). Interestingly, Glu58 of CAR D1, the residue which forms the new hydrogen bond with Ser417 of the Ad12 knob mutant Pro417Ser, also is hydrogen bonded to Lys429 of Ad12 knob, a conserved residue in the majority of fiber knobs.

The second key residue is Tyr477 of Ad2 knob, which corresponds to Ser489 in Ad12 knob. Mutation of Ad12 knob Ser489 to tyrosine increased the affinity of Ad12 knob for CAR D1 by tenfold, to a value greater than that observed for Ad2 knob (J. Howitt and P. Freimuth, unpublished results). This mutation also is thought to add a hydrogen bond at the interface; however, the bonding pair in CAR D1 is not currently known as no crystal structure has been solved. Tyrosine residues are commonly found in the interfaces of complexes (Hubbard and Argos 1994) and have considerably lower hydrophobic transfer free energies compared with aliphatic side chains (Stites 1997). This configuration may provide a way to cover large, normally solvent-filled surfaces without destabilizing the subunit but while allowing for the contribution of an important hydrophobic functional group within the receptor–ligand interface. The addition of this residue to the binding interface may, therefore, have a twofold effect and lead to the higher affinity binding that we observed.

Why do the knob domains from different adenovirus serotypes vary in binding affinity for CAR D1? It is clear that viral receptor-binding proteins are under strong selective pressure both to evade neutralization by varying their antigenic structure and to simultaneously maintain the ability to interact with the cellular receptor. Interestingly, although residues which form the receptor-binding sites of CAR-binding knobs are relatively more highly conserved than other regions of knob, these residues in fact are not strictly conserved (Bewley et al. 1999). The receptor-binding sites on knob, therefore, may be designed to tolerate a certain degree of sequence variation without complete loss of receptor binding affinity and specificity. One feature that may serve to buffer the binding sites against sequence variations is that they are constructed primarily from loops, which have a greater degree of flexibility than other regions of regular secondary structure. Another aspect that may accommodate sequence variations is the avidity effect of having a trivalent structure; although certain residue changes may weaken the affinity of individual binding sites for CAR D1, the overall binding affinity increases when one or two additional receptors are recruited into the initial knob–CAR complex.

Clearly not all adenovirus serotypes have evolved to bind CAR at maximal affinity. This finding may imply either that low affinity knobs are younger in evolutionary terms and have not matured to form high affinity contacts, or that the high affinity binding knobs represent a dead-end in evolutionary terms as they are irreversibly constrained to a single receptor. Lowering the affinity of interaction between the receptor and ligand would probably have the same effect on efficiency of virus infection as decreasing the concentration of available receptors. Indeed, the efficiency of Ad2 infection was found to vary in direct proportion to the concentration of receptors in the cell membrane (Freimuth 1996); therefore, it appears that the transmission efficiency of viruses with low binding affinities would be poor. In viral evolution, this arrangement may allow for emergence of viruses that have novel receptor-binding specificities. Supporting this view is the notion that viruses cannot survive a complete loss of receptor binding at any time during their evolution. Thus, it is likely that changes in receptor specificity would occur through a stepwise reduction in affinity for an initial receptor, with an accompanying decrease in transmission efficiency, followed by an increase in affinity for a new receptor. Alternatively, a second site for receptor recognition may be used by the virus at certain times, which would allow for large mutations within the former receptor binding site in a single or multi-step process. Although actual pathways for evolution of new specificities are unknown, the current structural and functional data determined for knob suggests that viral tropism is mediated through a stepwise process and results in similar structures which are able to recognize different receptors.

3.2
Modification of Binding Specificity – Implications for Adenovirus-Based Vectors for Gene Therapy

Adenoviruses have a number of advantages as vehicles for gene therapy, including broad host range, high efficiency of gene transfer in vivo and failure to integrate the viral genome into host cell chromosomal DNA

(Wilson 1996; Robbins et al. 1998; Benihoud et al. 1999). However, if adenoviruses are to be used in more sophisticated applications, retargeting of the viral vector towards specific cell types will be required. Such retargeting may be achieved through two major methods of altering the cellular tropism. The first is the removal of the innate receptor-binding specificity through the in vitro addition of bispecific molecular bridges that contain other ligands known to bind to other cellular markers (Douglas et al. 1996; Dmitriev et al. 1998, 2000; Haisma et al. 2000; Trepel et al. 2000). The second technique is genetic engineering of the knob domain to recognize other receptors and inhibit the natural tropism for CAR. The first method of retargeting through the fusion of separate ligands is outside the scope of this review. The genetic method of vector targeting requires knowledge of the residues in functional contact with the receptor and an understanding of the binding process. The use of computational techniques combined with known structural data may facilitate redirecting the ligand towards specific cells. Also, the use of in vitro evolution methods may accelerate the selection of variant knob domains that recognize novel receptors. The recent development of a phage display system for knob (Pereboev et al. 2001) may represent the first step towards high-throughput techniques for altering knob binding specificity. The architecture of the knob protein appears to be well suited for the development of novel binding specificities, as the exposed loops used for binding receptors have a high tolerance for changes without impacting the structural integrity of the protein. Also, the trivalent nature of the domain may allow for low receptor affinity to be enhanced through avidity effects to levels where cellular recognition may occur. Therefore, future studies aimed at redirecting knob to bind to novel receptors may prove highly beneficial for the development of gene delivery vectors for treatment of specific cell types.

4
Structure of Full-Length CAR Protein

CAR serves as a cellular receptor for both adenovirus and group B coxsackieviruses. In collaboration with Michael Rossmann's group at Purdue University, we investigated the interaction of coxsackievirus B3 (CVB3) with human CAR (He et al. 2001). Initial attempts to visualize complexes of CVB3 bound to the recombinant CAR D1 domain by cryoelectron microscopy were disappointing in that electron density corresponding to

CAR D1 was very weak, an indication that binding sites on the virus capsid were not completely saturated with CAR D1. We therefore focused on production of the complete CAR ectodomain (CAR D1D2) and the full-length intact CAR protein, in the hope that these forms of CAR might interact more strongly with the CVB3 capsid.

To produce the complete CAR ectodomain, we turned to mammalian cell expression systems as the CAR D1D2 fragment was not soluble when expressed in bacteria, and its solubility was not increased by carboxy-terminal peptide extensions that enhanced the solubility and proper folding of CAR D1. Just prior to embarking on this project, Rose and colleagues had described a system in which cDNAs were expressed in mammalian cells using vectors based on vesicular stomatitis virus (VSV) (Schnell et al. 1996). Due to the aggressive nature of VSV infection, this system was attractive for rapid, high-level protein expression. Furthermore, because VSV overtakes the host cell biosynthetic pathways very rapidly, cell cultures can be maintained in serum-free media during the course of infection. These features, in the case of CAR D1D2 and also for other secreted proteins, greatly facilitate protein purification as the recombinant protein accounts for a large fraction of the total protein contained in the spent culture medium.

In collaboration with Michael Whitt and colleagues at UT-Memphis, we constructed a VSV-based expression vector for production of the CAR D1D2 ectodomain fragment. Earlier work from Whitt and others had shown that the VSV glycoprotein (G) gene could be deleted and that the resulting ΔG virus could be propagated on host cells that were transfected with plasmid DNA encoding the *G* gene (Robison and Whitt 2000). In our initial construct, the VSV G gene was substituted by a cDNA fragment encoding the CAR signal peptide, domains D1 and D2, and a stop codon immediately after D2. Although detectable CAR D1D2 was present in the culture fluid of BHK cells infected with this initial VSV–CAR virus, the levels of secreted protein were substantially lower than seen for other secreted proteins produced from comparable vectors. We noticed that the CAR signal peptide is shorter and less hydrophobic than typical mammalian signal peptides; therefore, the cDNA sequence encoding the CAR signal peptide was deleted from the initial construct and replaced by a cDNA fragment encoding the VSV-G protein signal peptide. This substitution improved secretion levels, and the chimeric CAR D1D2 protein accumulated in the culture fluid to approximately 0.01 mg/ml. The secreted CAR D1D2 protein was further purified and concentrated by affinity

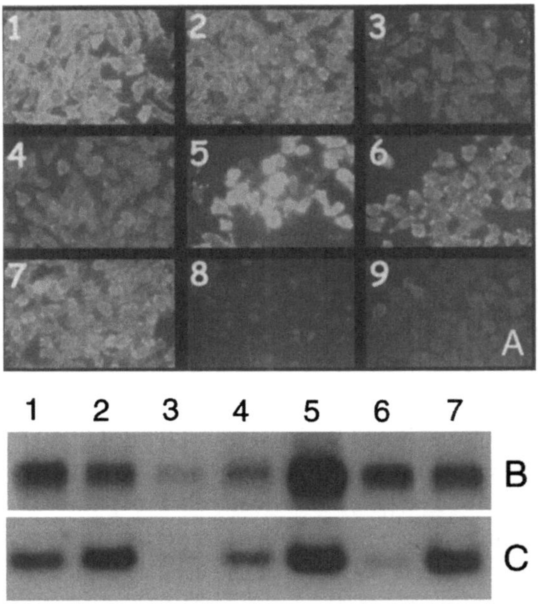

Fig. 5A–C. Expression levels of human CAR protein and mRNA in mouse A9 cells transfected with BAC-CXADR. **A** Monolayers of mouse A9 cell subclones stably transfected with BAC-CXADR (*panels 1–7*), parental A9 cells (*panel 8*) and HeLa cells (*panel 9*) grown on glass coverslides were incubated sequentially with biotinylated Ad12 knob then fluorescein-conjugated anti-biotin antibodies. Bound antibody, an indirect indicator of CAR protein expression, was then visualized by fluorescence microscopy. Northern blot of total RNA (**B**) and Southern blot (**C**) of *Eco*RI-digested genomic DNA from transfected A9 cell subclones 1–7, using [^{32}P]-labeled human CAR cDNA as hybridization probe (only a single representative band from each analysis is shown)

chromatography on a matrix of Ad12 knob protein covalently coupled to Sepharose beads.

A mammalian cell expression system also was developed for production of full-length CAR protein. As described below, we isolated a bacterial artificial chromosome (BAC) containing the intact, functional human CAR gene and the 3′ end of an adjacent gene, ANA. When transfected into mouse A9 cells along with a selectable neomycin phosphotransferase gene, this BAC conferred a receptor-positive phenotype to a large percentage of neomycin-resistant colonies. A number of A9 subclones that were stably transfected with this BAC were characterized for cell surface expression of CAR protein. Cell monolayers grown on coverslides were incubated with biotinylated Ad12 knob and then stained with fluorescein-labeled anti-biotin antibody. As shown in Fig. 5, CAR protein levels varied over a wide range in individual subclones. Furthermore, Southern and Northern blot analyses showed that, in most cases, CAR protein and mRNA levels were proportional to the number of CAR genes integrated in these cell subclones. Such dosage-dependent and position-independent transgene expression has been observed in other systems and often results from the activity of DNA elements that maintain the integrated transgene in an open chromatin conformation (Bell et al. 2001). The close proximity of the CAR and ANA genes at this chromosomal locus suggests a possible need for regulatory elements to insulate these two genes from each other's regulatory signals, possibly accounting for the dosage-dependent expression we observed in transfected cell lines.

One A9 subclone, A9.5, expressed very high levels of CAR protein on the cell surface and was used as a source of full-length CAR protein. CAR protein was purified from Triton X100 extracts of A9.5 cells in a single step, by chromatography on a knob–Sepharose column as described above. The eluate from this column contained a major protein species of ~46 kDa, which subsequently was identified as CAR by Western blotting and amino-terminal sequencing.

Gradient-purified CVB3 virions were incubated with full-length CAR or the VSV-produced CAR D1D2 fragment and then examined by cryo-electron microscopy (cryoEM). This analysis generated ~20 Å resolution electron density maps for both complexes, which revealed the relative positions of the CAR D1 and D2 domains (Fig. 6). In addition, the cryoEM structure provided sufficient detail to allow unambiguous fitting of the high resolution structure of the CAR D1 domain, which had previously been determined by X-ray crystallography. The resulting model demonstrated that CVB3 also interacts solely with the CAR D1 domain, and it revealed that the surface of CAR D1 in contact with the CVB3 capsid is different from the surface which contacts the adenovirus fiber knob domain.

More important for this discussion, however, was the striking observation that full-length CAR forms dimers when bound to the CVB3 capsid. The cryoEM structure shows that the cytoplasmic domains of adjacent CAR subunits form a region of ordered electron density, suggesting that the cytoplasmic domain mediates CAR dimerization. Although no supporting biochemical evidence for this CAR dimer has yet been reported, comparison of the cryoEM structures of the CVB3–CAR D1D2 and CVB3–full length CAR structures revealed that the orientation of the CAR

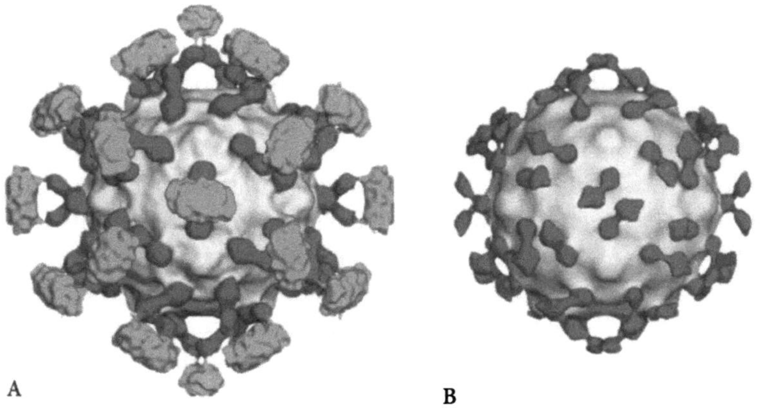

Fig. 6 A, B. Structure of full-length CAR and CAR D1D2 bound to coxsackievirus B3. **A, B** CryoEM reconstructions of coxsackievirus B3 (CVB3) bound to full-length CAR and the complete CAR ectodomain (CAR D1D2), respectively. [Reproduced (permission pending) from He et al. (2001). Interaction of coxsackievirus B3 with the full length coxsackievirus-adenovirus receptor. Nature Structural Biology 8(10) (in press)]

ectodomain relative to the CVB3 capsid is essentially superimposable for both structures. Therefore, it seems unlikely that dimerization of CAR could result from non-specific binding between cytoplasmic domains, as this would most probably generate a disordered electron density and also might force the CAR ectodomain to dissociate from the CVB3 capsid or alter the bending angle between the CAR D1 and D2 domains.

The formation of CAR dimers would have potentially important implications for both CVB and adenovirus infection. Receptor dimerization would probably increase the avidity of virus attachment to the host cell membrane, thereby increasing the efficiency of infection. At present, it is unknown whether CAR dimers pre-exist on the cell membrane or whether dimerization is induced by virus binding. If the latter scenario proves to be correct, then it is possible that dimerization might generate signals that contribute to post-binding events in infection. It should be noted again here that GPI-linked forms of CAR can serve as functional receptors for both CVB and adenovirus, indicating that dimerization through the cytoplasmic domain is not required for infection (Wang and Bergelson 1999). However, this does not exclude the possibility that dimeric receptors play a more important role in infections in vivo or in the normal cellular function of CAR.

5
Characterization of the Human CAR Gene, *CXADR*

5.1
Chromosomal Location of *CXADR*

The gene for human CAR, *CXADR*, is located on the long arm of chromosome 21 at position q21.1, about 15.96 M base pairs (bp) from the telomere and 4.55 Mbp from the centromere (Hattori et al. 2000; Antonarakis 2001). Transcription is in the direction of the telomere, and the transcribed region extends over about 57 kbp. Upstream of *CXADR* (toward the centromere) are several predicted genes, corresponding to expressed sequence tags (ESTs) of unknown function, and several pseudogenes. The first gene of known function, *USB25*, a ubiquitin-specific proteinase, is more that 16 Mbp away. On the downstream side, ANA (*BTG3/TOFA*) is transcribed toward the *CXADR* gene from the opposite strand, and the 3′ ends of the *CXADR* and ANA transcripts are separated by about 23.5 kbp (Fig. 7). Chromosome 21, which is perhaps best known for its participation in Down's syndrome, is relatively sparsely populated with genes, especially the proximal half of the long arm on which *CXADR* is located. The entire chromosome is thought to contain only about 230 known or predicted genes, whose coding sequences collectively account for about 3% of the total DNA sequence of chromosome 21 (Antonarakis 2001).

The genomic location of human *CXADR* was first inferred from an analysis of a panel of mouse–human somatic cell hybrids, each of which contained only a single human chromosome in the background of mouse A9 cells (Mayr and Freimuth 1997). Mouse A9 cells, a subline derived

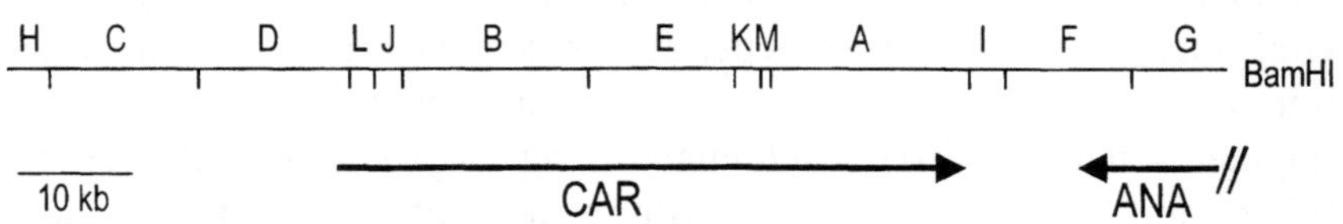

Fig. 7. Schematic diagram of the DNA insert contained in BAC *CXADR1*. The *Bam*HI restriction map for the 122-kb insert of BAC *CXADR1* is shown on the *top line*, oriented with the left and right ends pointing towards the centromere and telomere, respectively, of human chromosome 21. The size of the *CXADR* primary transcript and the direction of transcription are shown below the restriction map. Also indicated is the 3′ end of the ANA gene. The nucleotide sequence of the BAC *CXADR1* insert was determined (GenBank accession number AF200465)

from L929 cells, like most cultured rodent cell lines, express no or few adenovirus receptors. Among the cell panel, only cells containing human chromosome 21 were capable of binding human group A or C adenoviruses, suggesting that chromosome 21 encoded at least one essential protein component of the adenovirus receptor. This conclusion was further supported by analysis of several human–hamster hybrid cell lines containing human chromosome 21 fragments. Identification of the CAR cDNA (Bergelson et al. 1997; Tomko et al. 1997), followed shortly by reports of the complete sequence of human chromosome 21 (Hattori et al. 2000), allowed a precise determination of the location of the CAR gene, as well as several CAR pseudogenes, two of which also are located on chromosome 21 (Bowles et al. 1999). It should be noted that most if not all vertebrates, including mice, have a gene homologous to *CXADR*; however, in many established mouse and hamster cell lines this gene is not expressed.

5.2 Sequence Analysis of *CXADR*

As noted above, we had observed that mouse A9 and Chinese hamster ovary cell lines failed to bind radiolabeled adenovirus particles but that Ad2 virus did bind to derivatives of these cells which contained human chromosome 21. To isolate the human gene(s) which encoded receptor activity, high molecular weight genomic DNA from the chromosome 21-containing hybrid cells (A9–21 cells) was transfected into normal A9 cells along with a selectable neomycin phosphotransferase gene, which confers resistance to the cytotoxic drug G418. G418-resistant colonies that expressed adenovirus receptors were identified by their ability to bind modified sheep erythrocytes whose membranes were coated with purified viral fiber protein. Several independent, receptor-bearing A9 cell clones were isolated and shown to contain a similar pattern of DNA restriction fragments that hybridized to a human Alu repeat probe on Southern blots (Mayr and Freimuth 1997). One Alu repeat-containing fragment diagnostic of this locus was cloned and then used to probe a phage library of human (HeLa cell) genomic DNA. This approach ultimately led to the isolation of several phage clones whose HeLa DNA inserts overlapped to form a contig spanning about 30 kb. Hybridization probes derived from the ends of this contig were then used to screen a BAC library of large human genomic DNA fragments. One clone was isolated that contained a

~120-kb insert and both ends of the 30-kbp HeLa contig. When transfected into mouse A9 cells along with a neomycin phosphotransferase gene, this BAC clone, designated BAC-CXADR1, transformed a large fraction of G418-resistant colonies to express adenovirus receptor activity (see Fig. 5). Restriction fragments of this BAC clone hybridized to CAR cDNA on Southern blots.

To determine the sequence of the BAC-CXADR1 insert, *Bam*HI and *Bgl*II fragment subclones were prepared, and both strands of selected subclone inserts were sequenced using the dideoxy chain termination method and a combination of techniques including a nested deletion approach and primer walking as described elsewhere (Connelly et al. 1998; Anderson et al. 2001). Where overlapping fragments were not available, fragment order was determined by sequencing junction fragments produced by PCR. A 121,793-nucleotide (nt) insert sequence was derived that was consistent with restriction analyses of the original BAC clone and with restriction analyses of individual subclones (Fig. 7).

The ~57-kbp CAR gene lies near the middle of the 122-kbp *CXARD1* genomic fragment, suggesting that most of the signals required for regulating CAR expression lie within this segment. The overall G+C content for the *CXADR1* insert is 40.1%, which is almost identical to the average of 40.89% for the entire sequenced portion of chromosome 21 (Hattori et al. 2000). Thirty-eight percent of the fragment consists of interspersed repeat sequences most of which are SINE (23%) or MIR (21.4%) sequences, followed by LINE (9.8%) elements. This compares to 10.8% SINE and 15.5% LINE, and 38% total repetitive elements for chromosome 21 as a whole (Antonarakis 2001). The *CXADR1* segment contains 26 simple repeat sequences and 29 regions of low complexity. Inspection of the *CXADR1* sequence immediately upstream of the homology to the 5′ end of the CAR cDNA does not reveal a likely TATA- or CAT-box sequence. Indeed, this region is quite GC rich, and inspection of the CpG frequency revealed a region of high CpG content that overlaps *CXADR* exon 1 (Fig. 8). The promoters of housekeeping genes are universally associated with CpG islands (Antequera and Bird 1999). Therefore, while experimental evidence for the location of the *CXADR* promoter is lacking, the available evidence is consistent with the existence of a housekeeping-type promoter located at or immediately upstream of the region that encodes the 5′ end of the CAR cDNA.

Comparison of the current 2,434-nt CAR cDNA sequence (GenBank NM_001338) to the BAC-*CXADR1* insert sequence (GenBank AF200465)

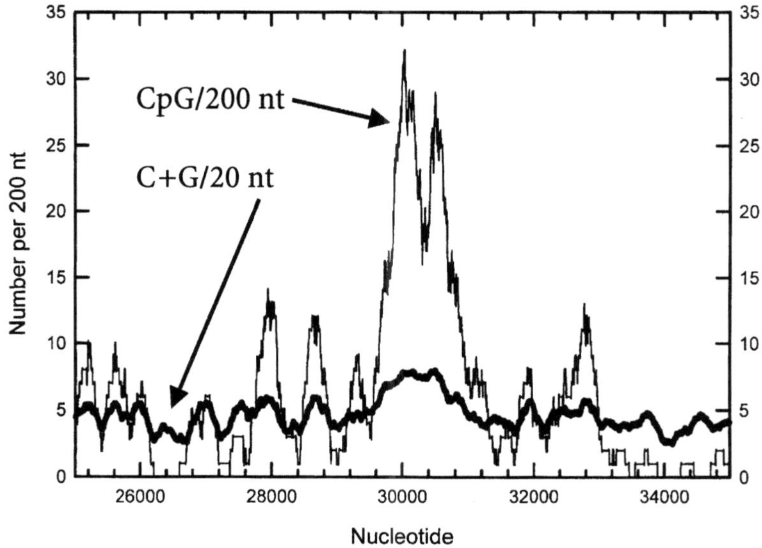

Fig. 8. CpG and C+G Content of the *CXADR* gene promoter region. The number of C and G residues and CpG dinucleotides in a 200-nt window was determined for the *CXADR1* insert. Shown is the region around the *CXADR* promoter, which represents the only major CpG island found in the insert. Peak CpG values occur at nucleotides ~30,030 and ~30,510 with respect to the beginning of the *CXADR1* insert. The 5′ end of the CAR cDNA is at nucleotide 30,239

revealed an exact match to seven regions spanning a ~53 kbp segment of the BAC insert with two exceptions, an *Eco*RI restriction site (GAATTC) at the beginning of the CAR cDNA sequence that is almost certainly derived from the cloning vector, and a few 'A-track' nucleotides at the 3′ end. This result indicated that the *CXADR1* insert contains the entire CAR gene, consistent with the fact that transfection experiments conferred Ad2 binding activity to mouse A9 cells and that the CAR gene is composed of seven exons (BOWLES et al. 1999). This structure is not atypical of human housekeeping genes. The first ATG in the cDNA sequence is 54 nt from the 5′ end of the cDNA (excluding the *Eco*RI site) and initiates the 1,098-nt CAR open reading frame.

The internal exon–intron boundaries all obey the AG/GT splice site rule (JACKSON 1991); exons 2 through 6 have an average length of 158 bp; translation is predicted to begin in exon 1 and it terminates at a TAG codon 263 nt into exon 7. However, four of the six introns are larger than average, especially intron 1, which is 33,886 nt, and intron 3, which is

11,780 nt. The average intron length for vertebrates is about 1,200 nt (HAWKINS 1988). The 3′ ends of the published cDNA sequences end at or in a sequence of 13 A residues at the beginning of a 30 residue A-rich segment in the *CXADR1* BAC that includes only three non-A nucleotides. This A-rich region represents one end of an *Alu* element. No polyA addition signal (AAUAAA) is found upstream of this polyA tract, suggesting that priming of the cDNA may have initiated within the *Alu* element rather than at the true 3′ end of the CAR mRNA.

CAR cDNA expression studies by TOMKO et al. revealed a range of CAR mRNA species in HeLa cells with sizes from 1.3 to 6 kb (TOMKO et al. 1997). Our own Northern blot studies revealed a major CAR species of about 6 kb (data not shown). This length is surprising because it would make the CAR mRNA six times the length required to encoded the open reading frame and one of the longest 3′ untranslated region (UTR) sequences known (PESOLE et al. 1997). The average size for mammalian 3′ UTR elements is about 740 nt. Comparison of a 4,000-nt segment of the BAC–*CXADR1* sequence distal to the beginning of exon 7 with the NCBI EST database revealed the presence of a number of cDNAs in the database that terminated at or just before nucleotide 87,294 in the BAC *CXADR1* sequence. Eleven nucleotides 5′ of this site is an AAUAAA sequence associated with the addition of polyA tails. These observations suggest that the CAR mRNA terminates at or near nt 87,294 in the *CXADR1* sequence, thus making the CAR cDNA ~5,610 nt in length and the CAR 3′ UTR 4,409 nt. Consistent with this observation, primers just within the predicted boundaries of CAR exon 7 amplified a ~4 kbp fragment from HeLa cell cDNA that was indistinguishable in size from the BAC amplified product, whereas PCR with a 3′ primer just distal to the predicted 3′ end of the CAR cDNA failed to yield a product (P. Freimuth and C.W. Anderson, unpublished results). There are no open reading frames in the 3′ UTR longer than 100 codons, although two ATG-initiated open reading frames longer than 200 codons are found in the *CXADR1* sequence that are not associated with known features (see Fig. 9). The 3′ UTR contains an AluY element and part of a FLAM C SINE element near the middle of the exon (at 85,332–85,624 bp and 85,756–84,030 bp in *CXARD1*, respectively). The orientation of the *Alu* element is such that its A track would serve as a template for RT–PCR amplification with an oligo-T primer, and this is probably the mechanism by which the partial CAR cDNA was cloned (BERGELSON et al. 1997; TOMKO et al. 1997). Approximately 5% of human mRNAs contain *Alu* sequences (YULUG et al. 1995). It is not known if the unusually long CAR 3′ UTR is physiologically important. The levels of CAR mRNA in many cells is quite low, however, and one function of the 3′ UTR may be to regulate mRNA stability. Several sequence elements in the 3′ UTRs of mRNAs, including a A+U-rich motif AUUUA, bind proteins that are thought to mediate mRNA degradation (Guhaniyogi and Brewer 2001). The 4,674-nt exon 7 of CAR has 18 of these elements, 11 of which are distal to the *Alu* A-track. This number of AUUUA elements is unusual in mRNA sequences. For example, the entire 12,500-nt-long mRNA for DNA-PKcs, the catalytic subunit of a protein kinase involved in DNA repair, which is quite stable, contains only 12 AUUUA elements. Nevertheless, 18 is about the number expected for a 4,600-nt segment of the nucleotide composition of exon 7, which like many mRNA 3′ UTR segments, is A+U rich (Guhaniyogi and Brewer 2001).

Comparison of the BAC insert sequence with the GenBank database also revealed exact matches to presumptive exons 4 and 5 of the human *ANA* (TOFA/BTG3) cDNA (Guéhenneux et al. 1997; Yoshida et al. 1998). The *CXADR* and *ANA* genes therefore are transcribed towards each other (Fig. 7). *ANA* was mapped to chromosome 21 q11.2 to q12.1, consistent with the location of the CAR gene. An EST database analysis of the segment in the CXADR1 sequence distal to the reported 3′ end of the 1,210-nt *ANA* cDNA (GenBank D64110) also revealed a set of cDNAs that terminated at or near nucleotide 110,834, that is 83 nt distal to the last reported nucleotide of the *ANA* cDNA. Fourteen nucleotides proximal to this site is an AATAAA sequence, consistent with position 110,834 being the approximate 3′ end of the *ANA* mRNA. Between this site and the predicted end of the CAR mRNA are 23,539 nt of intergenic sequence with an average G+C content of 38%.

Comparison of the remainder of the *CXADR1* sequence with the NCBI EST database revealed one additional area between nt 17,634 and 18,009 that is similar to a number (>100) of EST clones homologous to ribosomal protein L39. Because none of the EST sequences provided an exact match to the *CXADR1* sequence, this segment may represent a fragment of a pseudogene rather than an expressed sequence related to the L39 protein.

5.3 *CXADR* Polymorphism Analysis

Although no human disease has been associated with the CAR locus, it is possible that individuals might differ in their sensitivity to adenovirus or coxsackievirus infections because of variations in the CAR sequence. As noted above, the 345-aa CAR protein is composed of a signal sequence (aa 1–19), two extracellular immunoglobulin (Ig)-like domains (aa 34–127, 155–219), a transmembrane region (aa 238–258), and a cytoplasmic domain (aa 259–365). Inspection of the *CXADR* gene structure revealed that the signal sequence is encoded primarily in exon 1 (aa 1–14), the first Ig-like domain is encoded by exons 2 and 3 (aa 15–69 and aa 70–138), the second Ig-like domain is encoded by exons 4 and 5 (aa 139–190 and aa 191–231), and the transmembrane region and first part of the cytoplasmic domain by exon 6 (aa 232–277). The CAR protein domain organization and splicing pattern also are seen in a small subset of other Ig superfamily members termed the CTX family (Chretien et al. 1998). CTX family proteins also are found in lower vertebrates, and some family members appear to function in cell–cell adhesion during early stages of development. The human genome encodes a second CTX family protein, the A33 antigen, which is the closest known sequence homologue to CAR (Heath et al. 1997; Chretien et al. 1998). A33 is expressed on many colon and intestinal tumor cells (Barendswaard et al. 1998). Differential regulation of CAR in tumor vs normal cells, however, has not been reported.

About a dozen polymorphic nucleotides (SNPs) in the gene have been identified by comparing sequences from different individuals and EST sequences in the public databases, and a chromosome 21-specific SNP web database was established as a joint effort between the Division of Medical Genetics of the University of Geneva Medical School and the Swiss Institute of Bioinformatics (http://csnp.unige.ch/) (Deutsch et al. 2001). Unfortunately, at the current time, all of the CAR-related SNPs in that database are in the 3′ UTR except one, which is a silent change in the third position of a proline codon (CAR amino acid 40) in exon 2.

To examine more systematically potential genetic diversity in CAR, we examined the sequences of exons 2–6 encoding all of the mature extracellular receptor protein from a number of different individuals by re-sequencing PCR amplified genomic segments containing these exons (Fig. 9). DNAs from 90 individuals representative of the normal US population were obtained from the Coriell Institute Polymorphism Discovery Resource (Collins et al. 1998). Both strands of each amplified fragment were then sequenced by the dideoxy method using an automated sequencer, and the nucleotide traces were manually inspected for potential polymorphisms. For exons 2 and 3, which encode CAR D1 (the region of CAR that interacts with both CVB and adenovirus), DNAs from 90

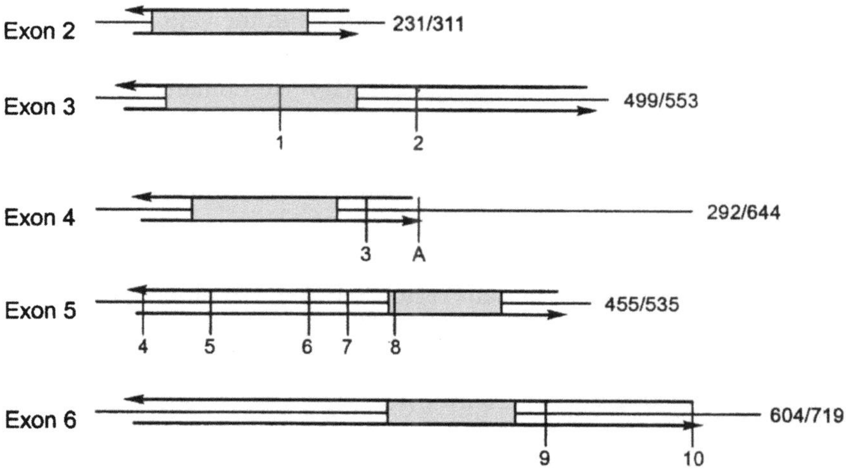

Fig. 9. Polymorphism analysis of *CXADR1* exons 2 through 6. Segments containing exons 2 through 6 from DNA from the Polymorphism Discovery Resource (COLLINS et al. 1998) were amplified by PCR and sequenced on both strands. Each exon is represented by a *box* that is positioned with respect to a *central line* representing the PCR amplified segment. The average region sequenced on both strands is represented by *double arrows*; the nucleotide length of this region and the length of the amplified region are given at the *right*. The positions of polymorphic nucleotides are indicated by *vertical lines*; numbers correspond to SNPs as described in Table 1; *A* refers to a sequence length/composition polymorphism (see text). A total of 2,081 nucleotides of the *CXADR*1 gene were analyzed of which 790 corresponded to exon sequence. Eleven potential polymorphisms were identified of which two were in exon sequence. Neither of these would affect the amino acid sequence of CAR

individuals were examined (Table 1); however, only two potential polymorphisms were identified. One of these (SNP 1) occurs in the third base of the codon for amino acid Gln111 (CA<u>A</u>). In the 180 chromosomes analyzed, A appeared at this position 170 times while G appeared 10 times for a minor allele frequency of 5.5%. Four individuals were heterozygotes while three where homozygous for the minor allele G. The second potential SNP is located 67 nt distal to the 3′ exon 3 boundary and was observed as a C/T heterozygote in only one individual. We did not detect polymorphisms at Pro40 (see above), indicating that this potential SNP is either very rare or represents a sequencing error.

Eight additional SNPs were identified in amplicon segments 4, 5 and 6, but only one of these was in exon sequence (SNP 8 in exon 5, see Table 1). The variant sequence was observed in only one of 24 individuals analyzed

Table 1. Single nucleotide polymorphisms in *CXADR* exons 2 through 6

No.	Location[a]	AF200465 (bp)	Sequence[b]	Allele distribution[c]	Sample size[d]
1	Exon 3	69,067	TTACA**A**CTGTC	AA 83 AG 4 GG 3	90
2	Intron 3	69,215	TGTAG**T**AGCAG	TT 86 TC 4 CC 0	90
3	Intron 4	76,354	CATGC**C**ATTGA	CC 23 CT 1 TT 0	24
4	Intron 4	77,623	AGCAG**C**TACAC	CC 21 AG 0 GG 3	24
5	Intron 4	77,696	CCTTG**A**TACAT	AA 23 AT 0 TT 1	24
6	Intron 4	77,802	TTGTT**A**AATTT	AA 14 AT 8 TT 2	24
7	Intron 4	77,844	CAATA**C**ATACT	CC 22 CT 1 TT 1	24
8	Exon 5	77,895	ATGAC**T**TCATC	TT 23 AT 1 AA 0	24
9	Intron 6	78,698	TGTAT**A**CCAAA	AA17 AG 6 GG 1	24
10	Intron 6	78,856	GTTGT**G**CCCCG	GG 23 GT 0 TT 1	24

[a] Position of the SNP with respect to *CXARD1* gene features is given (see also Fig. 9).
[b] Nucleotide sequence surrounding the SNP is given; central polymorphic nucleotide is in **bold** and is *underlined.*
[c] The distribution of alleles among the individuals analyzed is given; e.g., for SNP1 in exon 3, 83 individuals were homozygous for the major allele and had A at the polymorphic position in both chromosomes; four individuals were heterozygous with A in one chromosome and G in the other; three individuals were homozygous for the minor allele and had G at the polymorphic position on both chromosomes.
[d] Number of individuals analyzed.

as a A/T heterozygote in the third nucleotide of the codon for Thr193 (AC**T**), which appears in the second Ig-type domain (CAR D2). Three other SNPs (SNPs 4, 5, and 10, Table 1), all in intron sequence, were observed in only one individual, while the minor T allele of SNP 7, located proximal to the 5′ boundary of exon 5, was observed in two individuals, one of whom was homozygous, giving an allele frequency of 6%. Of the remaining three SNPs, SNP 4 is in intron 3 near the 5′ end of the exon 4 segment, and the minor allele was found in 3 of 24 individuals. SNP 6 is just proximal to exon 5, and the minor allele appeared in 10 of 24 individuals giving an allele frequency of 25%. SNP 9 occurs just distal to exon 6 and appeared with a minor allele frequency of 16%.

One sequence length and composition polymorphism was observed (designated A in Fig. 9) just distal to exon 4. Our reference sequence (GenBank AF200465) had the simple sequence $A_{10}T_5$ at this position, and individuals differed in the number of As and Ts in this segment. Because both strands cannot be sequenced through the PCR amplified product from heterozygotes, we did not attempt to quantitate the composition and frequency of this polymorphism.

Although our analysis did not include the CAR leader sequence nor the second half of the cytoplasmic domain, we conclude from our analysis that coding change SNPs in CAR that affect adenovirus binding are likely to be very rare. This may suggest an important or essential role for CAR in cell metabolism; however, because of the small size of the coding region of the gene, at most only about one cSNPs would be expected for the sample size analyzed.

6
Concluding Remarks

From these analyses it appears that CAR has several favorable characteristics that may account for why both CVB and adenoviruses have selected it as their major attachment receptor. First, CAR is well conserved in vertebrate animals, so it has been a constant target during the evolution of virus–host systems. This high level of conservation parallels the low frequency of genetic diversity of the CAR gene in humans. Second, CAR has a fairly wide tissue distribution; thus receptor availability is not a major factor in determining virus tropism. Third, in comparison to antibodies, CAR is a fairly small and compact molecule which may allow for interaction with binding sites on viral proteins that are relatively inaccessible to antibodies. Fourth, the possible dimerization of CAR either before or after interaction with virus could result in increased avidity of virus binding to the host cell membrane.

In the near future we can expect that additional cellular receptors other than CAR will be identified that interact with non-CAR binding serotypes such as those in subgroup B. Structural studies of these knob–receptor complexes should provide further insights into the constraints on altering receptor specificity and the range of possible receptor types with which knob might interact after alteration by in vitro evolution methods (PEREBOEV et al. 2001). Further study of these systems will provide a wealth of information on factors that govern molecular recognition, and they also may provide a means to alter the receptor specificity of adenoviral vectors for use in gene therapy and other applications.

References

Anderson CW et al. (2001) Frameshift mutation in *PRKDC*, the gene for DNA-PKcs, in the DNA repair-defective, human, glioma-derived cell line M059J. Radiat Res 156:2–9

Antequera F and Bird A (1999) CpG islands as genomic footprints of promoters that are associated with replication origins. Curr Biol 9:R661–7

Antonarakis SE (2001) Chromosome 21: from sequence to applications. Curr Opin Genet Dev 11:241–6

Bai M et al. (1993) Mutations that alter an Arg-Gly-Asp (RGD) sequence in the adenovirus type 2 penton base protein abolish its cell-rounding activity and delay virus reproduction in flat cells. J Virol 67:5198–205

Barendswaard EC et al. (1998) Rapid and specific targeting of monoclonal antibody A33 to a colon cancer xenograft in nude mice. Int J Oncol 12:45–53

Bell AC et al. (2001) Insulators and boundaries: versatile regulatory elements in the eukaryotic genome. Science 291:447–50

Benihoud K et al. (1999) Adenovirus vectors for gene delivery. Curr Opin Biotechnol 10:440–7

Bergelson JM et al. (1997) Isolation of a common receptor for Coxsackie B viruses and adenoviruses 2 and 5. Science 275:1320–3

Bewley MC et al. (1999) Structural analysis of the mechanism of adenovirus binding to its human cellular receptor, CAR. Science 286:1579–83

Boulanger PA and Puvion F (1973) Large-scale preparation of soluble adenovirus hexon, penton and fiber antigens in highly purified form. Eur J Biochem 39:37–42

Bowles KR et al. (1999) Genomic organization and chromosomal localization of the human Coxsackievirus B-adenovirus receptor gene. Hum Genet 105:354–9

Chretien I et al. (1998) CTX, a Xenopus thymocyte receptor, defines a molecular family conserved throughout vertebrates. Eur J Immunol 28:4094–104

Collins FS et al. (1998) A DNA polymorphism discovery resource for research on human genetic variation. Genome Res 8:1229–31

Connelly MA et al. (1998) The promoters for human DNA-PK_{cs} (*PRKDC*) and *MCM4*: divergently transcribed genes located at chromosome 8 band q11. Genomics 47: 71–83

Defer C et al. (1990) Human adenovirus-host cell interactions: comparative study with members of subgroups B and C. J Virol 64:3661–73

Deutsch S et al. (2001) A cSNP map and database for human chromosome 21. Genome Res 11: 300–7

Dmitriev I et al. (2000) Ectodomain of coxsackievirus and adenovirus receptor genetically fused to epidermal growth factor mediates adenovirus targeting to epidermal growth factor receptor-positive cells. J Virol 74:6875–84

Dmitriev I et al. (1998) An adenovirus vector with genetically modified fibers demonstrates expanded tropism via utilization of a coxsackievirus and adenovirus receptor-independent cell entry mechanism. J Virol 72:9706–13

Douglas JT et al. (1996) Targeted gene delivery by tropism-modified adenoviral vectors. Nat Biotechnol 14:1574–8

Durmort C et al. (2001) Structure of the fiber head of Ad3, a non-CAR-binding serotype of adenovirus. Virology 285:302–12

Freimuth P (1996) A human cell line selected for resistance to adenovirus infection has reduced levels of the virus receptor. J Virol 70:4081–5
Freimuth P et al. (1999) Coxsackievirus and adenovirus receptor amino-terminal immunoglobulin V-related domain binds adenovirus type 2 and fiber knob from adenovirus type 12. J Virol 73:1392–8
Guéhenneux F et al. (1997) Cloning of the mouse *BTG3* gene and definition of a new gene family (the *BTG* family) involved in the negative control of the cell cycle. Leukemia 11:370–5
Guhaniyogi J and Brewer G (2001) Regulation of mRNA stability in mammalian cells. Gene 265:11–23
Haisma HJ et al. (2000) Targeting of adenoviral vectors through a bispecific single-chain antibody. Cancer Gene Ther 7:901–4
Hattori M et al. (2000) The DNA sequence of human chromosome 21. The chromosome 21 mapping and sequencing consortium. Nature 405:311–9
Hautala T et al. (1998) An interaction between penton base and alpha v integrins plays a minimal role in adenovirus-mediated gene transfer to hepatocytes in vitro and in vivo. Gene Ther 5:1259–64
Hawkins JD (1988) A survey on intron and exon lengths. Nucleic Acids Research 16: 9893–9908
Heath JK et al. (1997) The human A33 antigen is a transmembrane glycoprotein and a novel member of the immunoglobulin superfamily. Proc Natl Acad Sci USA 94: 469–74
He Y et al. (2001) Interaction of coxsackievirus B3 with the full length coxsackievirus – adenovirus receptor. Nat Struct Biol 8:874–8
Henry LJ et al. (1994) Characterization of the knob domain of the adenovirus type 5 fiber protein expressed in *Escherichia coli*. J Virol 68:5239–46
Honda T et al. (2000) The coxsackievirus-adenovirus receptor protein as a cell adhesion molecule in the developing mouse brain. Brain Res Mol Brain Res 77:19–28
Hong JS and Engler JA (1996) Domains required for assembly of adenovirus type 2 fiber trimers. J Virol 70:7071–8
Hubbard SJ and Argos P (1994) Cavities and packing at protein interfaces. Protein Sci 3:2194–206
Jackson IJ (1991) A reappraisal of non-consensus mRNA splice sites. Nucleic Acids Res 19:3795–8
Kirby I et al. (2000) Identification of contact residues and definition of the CAR-binding site of adenovirus type 5 fiber protein. J Virol 74:2804–13
Kirby I et al. (2001) Adenovirus type 9 fiber knob binds to the coxsackie B virus-adenovirus receptor (CAR) with lower affinity than fiber knobs of other CAR-binding adenovirus serotypes. J Virol 75:7210–4
Kwong PD et al. (1998) Structure of an HIV gp120 envelope glycoprotein in complex with the CD4 receptor and a neutralizing human antibody. Nature 393: 648–59
Lortat-Jacob H et al. (2001) Kinetic analysis of adenovirus fiber binding to its receptor reveals an avidity mechanism for trimeric receptor-ligand interactions. J Biol Chem 276:9009–15
Louis N et al. (1994) Cell-binding domain of adenovirus serotype 2 fiber. J Virol 68: 4104–6

MAYR GA and FREIMUTH P (1997) A single locus on human chromosome 21 directs the expression of a receptor for adenovirus type 2 in mouse A9 cells. J Virol 71: 412–8

MYSZKA DG et al. (2000) Energetics of the HIV gp120-CD4 binding reaction. Proc Natl Acad Sci USA 97:9026–31

NEMEROW GR (2000) Cell receptors involved in adenovirus entry. Virology 274:1–4

PEREBOEV A et al. (2001) Phage display of adenovirus type 5 fiber knob as a tool for specific ligand selection and validation. J Virol 75:7107–13

PERSSON R et al. (1985) Virus-receptor interaction in the adenovirus system: characterization of the positive cooperative binding of virions on HeLa cells. J Virol 54:92–7

PESOLE G et al. (1997) Structural and compositional features of untranslated regions of eukaryotic mRNAs. Gene 205:95–102

PHILIPSON L et al. (1968) Virus-receptor interaction in an adenovirus system. J Virol 2:1064–75

ROBBINS PD et al. (1998) Viral vectors for gene therapy. Trends Biotechnol 16:35–40

ROBISON CS and WHITT MA (2000) The membrane-proximal stem region of vesicular stomatitis virus G protein confers efficient virus assembly. J Virol 74:2239–46

ROELVINK PW et al. (1996) Comparative analysis of adenovirus fiber-cell interaction: adenovirus type 2 (Ad2) and Ad9 utilize the same cellular fiber receptor but use different binding strategies for attachment. J Virol 70:7614–21

ROELVINK PW et al. (1998) The coxsackievirus-adenovirus receptor protein can function as a cellular attachment protein for adenovirus serotypes from subgroups A, C, D, E and F. J Virol 72:7909–15

ROELVINK PW et al. (1999) Identification of a conserved receptor-binding site on the fiber proteins of CAR-recognizing adenoviridae. Science 286:1568–71

SANTIS G et al. (1999) Molecular determinants of adenovirus serotype 5 fibre binding to its cellular receptor CAR. J Gen Virol 80:1519–27

SCHNELL MJ et al. (1996) Foreign glycoproteins expressed from recombinant vesicular stomatitis viruses are incorporated efficiently into virus particles. Proc Natl Acad Sci USA 93:11359–65

SETH P et al. (1994) Mechanism of enhancement of DNA expression consequent to cointernalization of a replication-deficient adenovirus and unmodified plasmid DNA. J Virol 68: 933–40

STITES SE (1997) Protein-protein interactions: interface structure, binding thermodynamics, and mutational analysis. Chem Rev 97:1233–1250

STOUTEN PF et al. (1992) New triple-helical model for the shaft of the adenovirus fibre. J Mol Biol 226:1073–84

TAINER JA et al. (1982) Determination and analysis of the 2 A-structure of copper, zinc superoxide dismutase. J Mol Biol 160:181–217

TOMKO RP et al. (1997) HCAR and MCAR: the human and mouse cellular receptors for subgroup C adenoviruses and group B coxsackieviruses. Proc Natl Acad Sci USA 94:3352–6

TREPEL M et al. (2000) Molecular adaptors for vascular-targeted adenoviral gene delivery. Hum Gene Ther 11:1971–81

VAN OOSTRUM J and BURNETT RM (1985) Molecular composition of the adenovirus type 2 virion. J Virol 56:439–48

van Raaij MJ et al. (2000) Dimeric structure of the coxsackievirus and adenovirus receptor D1 domain at 1.7 A resolution. Structure Fold Des 8:1147–55
van Raaij MJ et al. (1999) Structure of the human adenovirus serotype 2 fiber head domain at 1.5 A resolution. Virology 262:333–43
van Raaij MJ et al. (1999) A triple β-spiral in the adenovirus fibre shaft reveals a new structural motif for a fibrous protein. Nature 401:935–8
Wang X and Bergelson JM (1999) Coxsackievirus and adenovirus receptor cytoplasmic and transmembrane domains are not essential for coxsackievirus and adenovirus infection. J Virol 73:2559–62
Wickham TJ et al. (1993) Integrins α v β 3 and alpha v beta 5 promote adenovirus internalization but not virus attachment. Cell 73:309–19
Wilson JM (1996) Adenoviruses as gene-delivery vehicles. N Engl J Med 334:1185–1187
Xia D et al. (1994) Crystal structure of the receptor-binding domain of adenovirus type 5 fiber protein at 1.7 A resolution. Structure 2:1259–70
Yoshida Y et al. (1998) *ANA*, a novel member of *Tob/BTG1* family, is expressed in the ventricular zone of the developing central nervous system. Oncogene 16:2687–93
Yulug IG et al. (1995) The frequency and position of *Alu* repeats in cDNAs, as determined by database searching. Genomics 27:544–8

CHAPTER 12

Transactivation of Human *cdc2* Promoter by Adenovirus E1A

R. PADMANABHAN, A. TANIMOTO, Y. SASAGURI

R. PADMANABHAN
Department of Microbiology and Immunology, Georgetown University Medical Center, 3900 Reservoir Road, Washington DC, WA 20057, USA
e-mail: rp55@georgetown.edu

A. TANIMOTO
Department of Pathology, Toranomon Hospital
2-2-2 Toranomon, Minato-ku 105-8470, Tokyo, Japan

Y. SASAGURI
Department of Pathology, University of Occupational and Environmental Health, Kitakyushu, Japan

Abstract. Expression of the adenovirus oncoprotein E1A 12S induces the heterotrimeric transcription factor, NF-Y. NF-Y binds to the two CCAAT motifs upstream of the transcriptional start site of the human *cdc2* promoter and is required for activation of the promoter by E1A 12S in cycling cells. The observations that a number of eukaryotic cell cycle regulatory genes also contain the CCAAT motifs and NF-Y binds to them support the notion that E1A 12S could play an important role in deregulated expression of these genes through activation of NF-Y gene in cycling cells.

1
Introduction

Our understanding of functions of many eukaryotic genes in translation, transcription, and replication stems from studies of small DNA tumor viruses: adenovirus, simian virus 40, and polyomavirus and human papilloma virus. Because these viral genomes have limited coding capacities, they rely largely on the functions of cellular proteins for these essential processes in the viral life cycle. These viruses replicate in the permissive hosts through a lytic cycle pathway which systematically usurps the normal function of the host cell. They cause an abnormal cellular proliferation in normally quiescent or differentiated epithelial cells, shut off the host translational, transcriptional and replication machineries and redirect to viral genome-directed pathways that ultimately lead to release of virions and host cell lysis. In the lytic cycle of the virus, the sequential events starting from the expression of immediate early to early and late gene expression is exquisitely linked and regulated in the permissive cells. To understand the molecular events in the virus life cycle, a number of methods have been used in previous studies to analyze early events in viral gene expression uncoupled from late gene expression, DNA replication, and virion assembly by using inhibitors such as hydroxyurea which inhibit host cell and viral DNA replication. By using DNA transfection methods, it was possible to dissect the function of the early gene products of DNA tumor viruses expressed in normal cells (GRAHAM et al. 1974; TEMPLETON and ECKHART 1982; TREISMAN et al. 1981). These studies have provided valuable insight into the function of cellular genes involved in cell cycle control (CRESS and NEVINS 1996), DNA replication (STILLMAN 1996), and programmed cell death pathways (BARBER 2001; PIPAS and LEVINE 2001; SUBRAMANIAN and CHINNADURAI 2001; WHITE 1993). Early studies on cell transformation by DNA tumor viruses gave valuable insight into non-homologous recombination pathways that are common occurrences in mammalian cells (for a review see FECHTELER et al. 1995). For example, it was established early on that adenovirus (Ad) type 12-induced tumors in new born hamsters (TRENTIN et al. 1968) were associated with integration of Ad12 DNA into the hamster genome by a mechanism involving non-homologous recombination (DOERFLER 1968, 1975). The transformed rodent cells all have in common an integrated portion of the adenoviral DNA coding for early region 1 (DOERFLER 1968; GALLIMORE 1974; SHARP et al. 1974). Using a pioneering DNA transfection technique for introducing foreign DNA into mammalian cells (GRAHAM et al. 1974), a transformed human embryonic kidney (human 293) cell line in which the E1 region of Ad5 genome is integrated into the human genome was successfully established (GRAHAM et al. 1977). The fully transformed human 293 cells express the E1A-encoded, predominantly E1A 12S and 13Sgene products, and the E1B encoding the 55-kDa and 19-kDa proteins (for a review see SHENK and FLINT 1991). In the transformed monkey kidney (COS) cell line, the early region of SV40 genome coding for the large T and small t antigens is integrated into the host genome (for a review see COLE 1996).

2
Early Regions of DNA Tumor Viruses Encode Proteins That Deregulate the Eukaryotic Cell Cycle and Induce Cellular DNA Synthesis and Proliferation

The cyclin-dependent kinases and the heterodimeric protein kinases are some of the key cell cycle regulatory proteins from yeast to humans that are induced and activated by DNA tumor virus oncoproteins (BEACH et al. 1982; DRAETTA et al. 1987; FANG and NEWPORT 1991; LEE and NURSE 1987; MURRAY 1992; NURSE and BISSETT 1981; REED et al. 1985; RIABOWOL et al. 1989; SIMANIS and NURSE 1986) (for reviews, see DRAETTA 1990; MURRAY 1995; NURSE 1990; PINES 1995). In yeast, the catalytic subunit, $p34^{cdc2}$, is encoded by the *cdc2* gene in *Schizosaccharomyces pombe* and the *cdc28* gene in *Saccharomyces cerevisiae* (henceforth referred as Cdc2 for protein and *cdc2* for mRNA and gene). Cdc2, in association with a cyclin component, regulates both $G_1 \rightarrow S$ and $G_2 \rightarrow M$ phase progression. However, in higher eukaryotes, separate Cdks in association with their cognate cyclin components are involved in these steps: Cdc2 is primarily required for the G_2 to M phase transition (FANG and NEWPORT 1991; MURRAY 1995; RIABOWOL et al. 1989) and other Cdks (e.g. Cdk2) seem to be primarily involved in the $G_1 \rightarrow S$ transition leading to DNA replication of cellular chromosomes (FANG and NEWPORT 1991; PAGANO et al. 1993; TSAI et al. 1991). There is some evidence to suggest that the Cdc2 kinase may also be involved in phosphorylation of key cellular proteins that are required for $G_1 \rightarrow S$ transition. For example, an antisense oligonucleotide to the *cdc2* gene inhibited DNA synthesis in T lymphocytes suggesting this role for Cdc2 (FURUKAWA et al. 1990). A role for Cdc2 kinase in viral DNA replication has also been reported: Cdc2/cyclinA alone is sufficient for initiation of SV40 DNA synthesis in vitro (D'URSO et al. 1990). Phosphorylation of adenovirus DNA polymerase by Cdc2 kinase is required for initiation of adenovirus replication (RAMACHANDRA et al. 1993; RAMACHANDRA and PADMANABHAN 1993; for a review see RAMACHANDRA and PADMANABHAN 1995).

The kinase activity of Cdc2 is regulated by phosphorylation and dephosphorylation events which are coupled to the growth state of mammalian cells (for reviews see GRANA and REDDY 1995; MORGAN 1995; PINES 1995). The levels of *cdc2* mRNA and Cdc2 protein are very low when cells are growth arrested or undergo differentiation, development or senescence. These levels become elevated as cells enter the cell cycle by addition of serum growth factors (DALTON 1992; DRAETTA and BEACH 1988; D'URSO et al. 1990; KREK and NIGG 1989; LEE et al. 1988b) which suggests that Cdc2 is required for cell proliferation.

DNA tumor viruses normally infect cells that are in quiescence or in a terminally differentiated state. They activate the $G_1 \rightarrow S$ phase transition, concomitantly inducing many of the activities associated with host DNA synthesis and cell cycle progression. Very low levels of *cdc2* mRNA and protein in quiescent cells are dramatically increased by the expression of adenovirus E1A (DALTON 1992; DRAETTA et al. 1988; HOWE et al. 1990; WANG et al. 1991) and SV40-LT. OSHIMA et al. (1993) showed that primary neonatal rat lung epithelial cells express low levels of Cdc2 and cyclin A. However, cells immortalized by wild-type SV40 LT or mutant SV40-LT that was defective in binding to pRB induce expression of high levels of these cell cycle regulated proteins. Cells expressing a temperature-sensitive mutant SV40-LT still maintained high levels of *cdc2* mRNA and protein even after shift to restrictive temperature that arrests cell growth. These results suggest that even low levels of SV40-LT are sufficient to stimulate and maintain high levels of Cdc2 and cyclin A (OSHIMA et al. 1993).

The viral oncoproteins promote cell proliferation by their abilities to physically and/or functionally interact and interfere with negative regulators of the cell cycle. Early studies using immunoprecipitation methods using monoclonal antibodies against E1A 12S and 13S proteins revealed that these oncoproteins associate with a number of cellular proteins with sizes of p300, 130, 107, and 105 kDa (Harlow et al. 1986; Lillie et al. 1986). A common motif of five amino acids in E1A, LXCXE, is required for binding to the p105, p107, and p130 proteins. SV40-LT can also bind to these cellular proteins through the same motif which is interchangeable between E1A and SV40-LT proteins without affecting the binding properties (Egan et al. 1988; Moran and Zerler 1988; Whyte et al. 1989). The finding that the E1A-interacting p105 is a tumor suppressor protein is a landmark discovery in cancer research. p105 acquired its name of retinoblastoma susceptibility gene product, or pRB, because of the finding that the loss of function of p105 correlated with the development of retinoblastomas and osteosarcomas (Whyte et al. 1988; reviewed in Knudson 1996; Weinberg 1985). The cloning and characterization of these cellular proteins that interact with E1A proteins led to valuable insight into their role in eukaryotic cell cycle regulation (Ewen et al. 1991; Hannon et al. 1993; Li et al. 1993; Mayol et al. 1993; for reviews see Beck et al. 1998; Dyson and Harlow 1992).

Immunoprecipitation experiments using specific antibodies against SV40-LT also led to the discovery of an important cellular tumor suppressor protein, p53 (Lane and Crawford 1979; Linzer and Levine 1979), which is an important cellular target of SV40-LT. p53 protein is also targeted by adenovirus E1B 55-kDa protein (Sarnow et al. 1982), an early protein of adenovirus and by E6 protein of human papillomavirus (Crook et al. 1991; Werness et al. 1990). Thus, in contrast to adenovirus E1A and E1B 55-kDa proteins, each targeting a separate tumor suppressor protein, SV40-LT can bind to both tumor suppressor proteins through distinct domains (Ludlow 1993). Although E1A does not bind directly to p53, it also compromises the activity of p53 by binding to its coactivator, p300/CBP that is required for p53-dependent transcriptionl activation of $p21^{Waf1}$, resulting in the release of cells from G_1 arrest and causing stimulation of DNA synthesis (Avantaggiati et al. 1997; Gu et al. 1997; Lill et al. 1997; Scolnick et al. 1997).

The results of these studies revealed that DNA tumor virus oncoproteins evolved a common mechanism to target the tumor suppressor proteins p53 and pRB (Buchkovich et al. 1989; DeCaprio et al. 1988; Horowitz et al. 1990; Lee et al. 1988a; for reviews see Beck et al. 1998; Cress and Nevins 1996; Moran 1993; Nevins 1992).

Expression of the adenovirus early region 1-encoded proteins individually in quiescent cells revealed that the expression of E1A 12S alone can cause cell cycle progression from G_0/G_1 phase into S phase but it is not sufficient for full transformation unless E1B proteins are also expressed (Houweling et al. 1980; Quinlan 1994; Rao et al. 1992; Ruley 1983). The E1B 55-kDa and 19-kDa proteins can independently block the apoptotic function of p53, and the anti-apoptotic function of the E1B 19-kDa protein is also found in a cellular homologue, Bcl2 (Gottlieb and Oren 1996; Ko and Prives 1996; Levine 1997; Subramanian et al. 1995; White 1996).

Sequence comparisons of E1A 12S and 13S from several serotypes revealed that E1A has essentially three conserved regions (CR): CR1, CR2, and CR3. CR1 and CR2 are common among both E1A proteins whereas the CR3 consisting of 46 amino acid residues, is unique to the E1A 13S protein. The 46-amino acid region in E1A 13S is required and sufficient for transactivation of other adenoviral early genes encoded by the E1B, E2, E3, and E4 regions as well as a number of cellular genes (reviewed in Flint and Shenk 1989). The E1A 12S is a weak transcriptional activator compared to the 13S gene product (Montell 1982; Winberg and Shenk 1984). However, this 46 amino acid CR3 region of E1A 13S is not essential for activation of the cell cycle and proliferation in quiescent cells (Zerler et al. 1986).

E1A proteins do not have any DNA-binding properties (Ferguson et al. 1985). Transactivation by E1A is achieved through either direct protein–protein interactions with other DNA-binding proteins (Lillie and Green 1989; Liu and Green 1990) or through several indirect mechanisms such as induction of transcriptional activities of specific factors at the transcriptional level or at the level of post-translational modifications. Studies on activation of the adenovirus E2 promoter that regulates the expression of DNA replication proteins by E1A led to the discovery of cellular transcription factor E2F which plays a key role in cell cycle progression and proliferation (for reviews see Cress and Nevins 1996; Girling et al. 1993; Nevins 1992, 1993). Several proteins that are related to E2F and which have distinct functions have been identified (E2F-1 to -5) and have been classified within the E2F family (Dagnino et al. 1997; Good et al. 1995; Lukas et al. 1996; Muller et al. 1997; Tao et al. 1997).

There is a significant amount of experimental evidence for interaction of pRB or pRB-related proteins with distinct E2F family members in vivo and in vitro (for reviews see Cobrinik 1996; Cress and Nevins 1996; La Thangue 1996; Slansky and Farnham 1996). E2Fs exist in transcriptionally inactive complexes with pRB family proteins (WEINTRAUB et al. 1992, 1995). During normal cell cycle progression from G_0/G_1 phase to S phase the inactive E2F/pRB complexes are dissociated by sequential phosphorylation of pRB by cyclin D/Cdk4,6 in mid-G_1 and cyclin E/Cdk2 in late G_1 (BUCHKOVICH et al. 1989; CHEN et al. 1989; DECAPRIO et al. 1989; LUDLOW et al. 1990; MIHARA et al. 1989). After a round of cell division, pRB is dephosphorylated and in this form pRB associates with E2F in early G_1 phase. While the pRB–p105 plays a central role in cell cycle progression from G_1 to S phase, the E2F-4/p130 plays a role in the transition from G_0 to G_1 phase. The transcriptionally inactive E2F–pRB complexes are disrupted by several viral oncoproteins releasing E2Fs. The same motif in the CR2 region of E1A that is required for binding to pRB is also required for dissociation of the E2F–pRB complex (BAGCHI et al. 1990; BANDARA and LA THANGUE 1991; RAYCHAUDHURI et al. 1991; WHYTE et al. 1988) The ability to disrupt the inactive E2F–pRB complex is also conserved in other viral oncoproteins: large T antigens of mouse and monkey, and human polyomaviruses (DECAPRIO et al. 1988; DYSON et al. 1990) and E7 protein of human papillomaviruses (CHELLAPPAN et al. 1992). E2F-1 is also phosphorylated in vivo and in vitro phosphorylation by Cdc2 kinase has been shown to disrupt its association with pRB and promote its association with E4–19-kDa protein and activation of E2 promoter/reporter gene expression (FAGAN et al. 1994). It has been shown previously that transactivation of an E2F-dependent adenovirus E2 promoter/CAT reporter is pRB-binding domain dependent in quiescent primary baby rat kidney (BRK) cells (RAYCHAUDHURI et al. 1991; WANG et al. 1995).

3
Regulation of the Human *cdc2* Promoter by Adenovirus E1A 12S in Primary Quiescent or Serum-Starved Cells

The human *cdc2* promoter contains no classical TATA-like elements but has consensus binding sites for several cellular transcription factors including ATF/CREB-like (GACGACA), c-Myb (CAGTTG), four inverted Sp1 (CCGCCC), E2F-like (TTTCGCGC), and CCAAT box binding factor (also known as NF-Y) (DALTON 1992; KU et al. 1993) (Fig. 1). The 2.5 kb of the 5′-flanking region of the human *cdc2* gene was analyzed to determine its cell cycle-dependent expression using several promoter-reporter plasmids in cycling human embryonic lung fibroblasts (ICRF23) cells and non-cycling (G_0) BRK cells (DALTON 1992). The results indicated that E1A

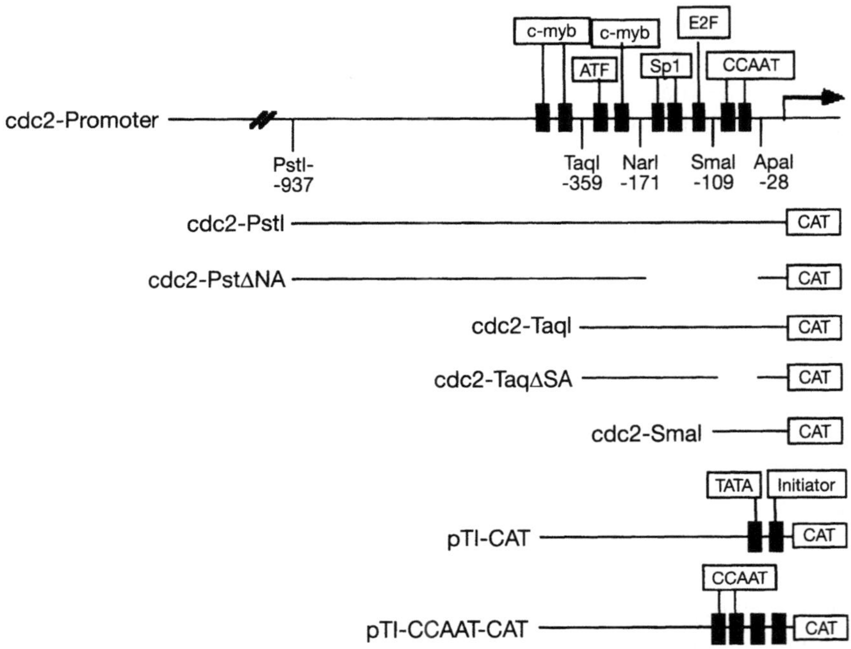

Fig. 1. Schematic diagram of the human *cdc2* promoter/CAT reporter plasmids. The relative positions of the cis-acting regulatory elements are fused to the CAT reporter gene. The *rightward arrow* indicates the major transcription start site. (Adapted from CHEN et al. 1996)

12S or -13S protein activated the *cdc2* promoter (2.5 kb) chloramphenicol acetyl transferase reporter (CAT) 10- to 20-fold in growth arrested BRK cells (DALTON 1992). Cotransfection of pRB and the *cdc2*/CAT reporter plasmid in cycling ICRF23 cells repressed the reporter expression 10- to 20-fold. While deletion of 2.5 kb up to the –245 region, still responded to pRB repression, deletion up to the *Sma*I site (–249 to –109 region) abrogated the ability of pRB to repress *cdc2* promoter/CAT activity. Within the region between –245 and –107, there is a high affinity E2F binding site at –125 (5′-TTTCGCGC-3′) and a low affinity E2F site at –107. When these mutant E2F sites were tested for pRB response in cycling cells, pRB repression of the mutant promoter was reduced by 80%. From these results, it was concluded that pRB repressed the human *cdc2* promoter activity. For optimum repression, the high affinity E2F binding site(s) was important. In addition, it was suggested that other sites may also be involved in negative regulation of the cdc2 promoter as 20% of the wild-type level of repression was still retained with the E2F binding site mutant (DALTON 1992).

However, in an in vivo genomic footprinting study, it was observed that the high affinity E2F site of the *cdc2* promoter was unoccupied in serum-starved human fibroblasts and after stimulation by the addition of serum (TOMMASI and PFEIFER 1995). On the other hand, the authors showed that the –20 element that differs from the consensus E2F binding site sequence by one nucleotide was involved in specific interaction with the E2F-4–p130 complex. This complex was present in G_0 phase but disappeared when cells entered S phase due to the addition of serum (TOMMASI and PFEIFER 1995). Therefore, it was proposed that the function of the –20 element is to recruit the E2F-4–p130 complex to negatively regulate the *cdc2* promoter in G_0 and G_1 cells. Mutation within the –20 element caused an increase in basal activity of the *cdc2* promoter suggesting that the –20 element negatively regulates the *cdc2* gene expression in the G_0/G_1 phase (TOMMASI and PFEIFER 1995).

4
Regulation of the Human *cdc2* Promoter by SV40-LT and E1A Oncoproteins in Cycling Cells

We sought to determine the potential of SV40-LT and E1A proteins, and their mutants defective in binding to tumor suppressor proteins, in transactivation of the human *cdc2* promoter in cycling cells. Because these cells are grown in the presence of serum, E2F is expected to be transcriptionally active. We first used several deletion constructs of the *cdc2* promoter/CAT reporter and determined their basal as well as SV40-LT stimulated activities for expression in monkey kidney (CV1) cells. Our results showed that deletion of the region between –937 and –359 increased the basal activity suggesting that this region contains a negative element for *cdc2* expression; this is consistent with an earlier observation (DALTON 1992). In addition, the region containing the Sp1, E2F, and the two CCAAT box motifs (located between –171 and –28) (see Fig. 1) is essential for basal activity (CHEN et al. 1996). The basal activity is stimulated several fold by SV40-LT and the two CCAAT elements are required for optimal transactivation. Moreover, the SV40-LT mutant defective in binding pRB or p53 still retained the ability to transactivate the human *cdc2* promoter near wild-type levels in CV-1 cells. This result is consistent with the observation that the SV40 mutant that is defective in binding pRB was still able to induce high levels of the *cdc2* mRNA and protein in immortalized primary neonatal rat lung epithelial cells (OSHIMA et al. 1993). Moreover, a new DNA-binding activity that could specifically recognize the two CCAAT motifs of the *cdc2* promoter was induced in COS cells transformed by SV40 or when SV40-LT was transiently expressed in CV-1 cells (CHEN et al. 1996). The two CCAAT motifs could activate expression of the CAT reporter 14-fold from a heterologous basic promoter that contained only the TATA box. Moreover, the –109 region (*cdc2-Sma*I in Fig. 1) of the *cdc2* promoter containing essentially the two CCAAT motifs could confer about 73% of transactivation achieved with the –359 region of the promoter containing ATF, E2F, Sp1 and two CCAAT motifs suggesting that the two CCAAT motifs play a major role in SV40-LT-mediated activation of the human *cdc2* promoter in cycling cells (CHEN et al. 1996). Mutation of the two CCAAT motifs within the –359 region of the promoter (*cdc2–Taq*I plasmid in Fig. 1) increased the basal activity of the promoter suggesting that the two CCAAT motifs are subject to repression in the absence of SV40-LT in cycling cells. However, the transactivation of this mutant promoter by SV40-LT was reduced from 13-fold to 2.6-fold again supporting the important role of the CCAAT motifs in the activation of the *cdc2* promoter in cycling cells (CHEN et al. 1996). Consistent with the transactivation of the *cdc2* promoter by SV40-LT transiently expressed in CV1 cells and earlier reports that the *cdc2* mRNA levels remained high in SV40-immortalized human fibroblasts (OSHIMA et al. 1993), we found that SV40-LT also induced the expression of *cdc2* mRNA in cells transfected with the SV40-LT expression plasmid in a dose-dependent manner (CHEN et al. 1996).

Next, we examined the pathway by which E1A activated *cdc2* gene expression. Previous studies showed that adenovirus E1A mutants which were not able to bind either the cellular p300/CBP or the pRB family of proteins still retained the ability to induce cellular DNA synthesis and Cdc2 protein as well as the kinase activity (HOWE et al. 1990; WANG et al. 1991). However, there is considerable evidence that both functional domains of E1A are required for cell proliferation or immortalization (GOPALAKRISHNAN et al. 1997; QUINLAN et al. 1988; SUBRAMANIAN et al. 1988; WANG et al. 1993, 1995) suggesting that there is crosstalk between the two domains. The simultaneous interaction between p300 and the pRB family of proteins with E1A in the same complexes is important for proliferation in quiescent primary cells (WANG et al. 1995; reviewed in BECK et al. 1998). Furthermore, in cycling cells, transcriptionally active E2F is

Table 1. The relative basal promoter/CAT reporter activities and fold activation values were obtained by coexpression of either E1A 13S or 12S followed by CAT assays and quantitation by liquid scintillation counting as described (TANIMOTO et al. 1998a). The promoter activity obtained with *cdc2*-TaqI(–359) was arbitrarily taken as 100% (adapted from TANIMOTO et al. 1998a)

cdc2 Promoter constructs	Relative basal activity	Fold activation	
		+13S	+12S
cdc2-PstI(–937)	39%	6.6	5.2
cdc2-TaqI(–359)	100%	4.1	3.1
cdc2-TaqIMut	175%	1.3	0.9
cdc2-SmaI(–109)	67%	5.4	3.2
cdc2-ΔSA359)	33%	1.2	1.4

already formed; under these conditions, we sought to determine whether there are other elements of the *cdc2* promoter involved in maintaining high levels of Cdc2 protein. To address this question, adenovirus E1A 13S and 12S expression plasmids and the various *cdc2* promoter/CAT reporter constructs were transfected into CV-1 cells and CAT assays were performed. The results showed that both E1A proteins were able to activate the *cdc2* promoter containing the –937 region upstream of the transcription start site five- to sevenfold (Table 1) (TANIMOTO et al. 1998a). The basal activity of the *cdc2*(–937) promoter was only 40% of the *cdc2*(–359) promoter confirming that the region between –937 and –359 negatively regulates the *cdc2* promoter as previously observed (CHEN et al. 1996). The *cdc2*(–359) and *cdc2*(–109) promoter/CAT plasmids were also activated four- to fivefold by 13S and threefold by 12S protein. Point mutations or deletion of both CCAAT motifs inhibited the transactivation by E1A proteins (Table 1) suggesting that the two CCAAT motifs are important for E1A-mediated activation of the *cdc2* promoter. The *cdc2*–TaqMutCCAAT promoter which contains point mutations in both CCAAT motifs still has consensus sites for ATF, c-Myb, E2F, and Sp1 suggesting that these sites do not play a significant role in activation by E1A in cycling cells. Our previous observation that the basal activity of the *cdc2* promoter was increased when the two CCAAT motifs were mutated was confirmed (Table 1) suggesting that wild-type motifs are involved in transcriptional repression of the *cdc2* promoter in cycling cells in the absence of SV40-LT or adenovirus

E1A (Chen et al. 1996; Tanimoto et al. 1998a). However, this increase in basal activity was not observed when both motifs were deleted (cdc2 TaqIΔSA; Table 1). Since both E1A proteins efficiently activated the *cdc2* promoter, the CR3 region of the E1A 13S protein did not contribute significantly to the activation of this cellular promoter. On the other hand, the transcriptional activation of the cellular *hsp70* gene by E1A proteins required the CCAAT motif around –70 with respect to the transcriptional start site and the CR3 region of E1A; the E1A 12S was found to be much weaker in activating the *hsp70* promoter compared with the 13S (Wu et al. 1986). In contrast to this report, another study found that the efficiencies of activation of the *hsp70* promoter by either of the two E1A proteins were not significantly different (less than twofold) (Simon et al. 1987). The discrepancies between these two studies of *hsp70* gene expression were attributed to differences in the experimental conditions (Simon et al. 1987).

Since there are two CCAAT motifs in the *cdc2* promoter at –76 and –44 (Dalton 1992), we analyzed the contribution of each of the motifs for basal activity and for the E1A-mediated activation by using single site mutants. The distal site (–76) and proximal site mutants had 14% and 2%, respectively, of the basal activity of the wild-type suggesting that both motifs are important for optimum basal activity of the promoter. However, each motif contributed almost equally to E1A-mediated transactivation (Tanimoto et al. 1998a).

Human 293 cells express the proteins encoded by the E1 region of Ad5 which includes not only the two E1A proteins under study but also the E1B 55-kDa and E1B 19-kDa proteins. Therefore, we sought to determine whether the nuclear proteins from human 293 cells have specific DNA binding to the two CCAAT motifs by electrophoretic mobility shift assays (EMSA). Specific binding of nuclear proteins to a DNA fragment containing both CCAAT motifs, either as wild-type or single site mutants, was observed, but when both motifs were mutated, the DNA binding was lost. The DNA binding could be competed out by a heterologous CCAAT motif from the thymidine kinase promoter but not by an oligonucleotide containing the E2F site suggesting that DNA binding was specific to DNA containing the CCAAT motif (Tanimoto et al. 1998a).

Next, we analyzed the effect of wild-type and mutant E1As in inducing DNA binding to the distal CCAAT motif of the *cdc2* promoter using EMSA in cycling rat liver epithelial cells (RLE). Nuclear extracts from wild-type RLE with low passage history (passage 13) had barely detectable DNA-binding activity. E1A 12S significantly induced a new DNA-binding activity having the same mobility as the nuclear proteins from human 293 cells. Mutants of E1A were used for infecting RLE cells and the nuclear extracts were analyzed by EMSA. These mutants were: two single site mutants, 12S.RG2 (an Arg → Gly substitution at position 2 of E1A that abolished the ability of E1A to bind p300) and 12S.YH47/928 (a Tyr → His substitution at residue 47 and a Cys → Gly substitution at residue 124 that abolished E1A's ability to interact with the pRB family of proteins but left p300 binding unperturbed) (Wang et al. 1995), a deletion mutant, E1A 12S.NCdl, in which the region between amino acid residues 86 and 120 was deleted (Zerler et al. 1987), and a double site mutant, dl01/08/520. The 12S.NCdl mutant bound to proteins of the pRB family and p300 individually but differed from the wild-type E1A 12S in that both sets of cellular proteins are not found in the same immune complex (Wang et al. 1995). The double mutant had both the N-terminal p300-binding domain and the pRB binding domain inactivated (Mymryk and Bayley 1994). The single site CCAAT motif mutants also induced DNA-binding activity in RLE cells that recognized the distal CCAAT motif which was indistinguishable from the wild-type; but the nuclear extract from double mutant-infected cells lost significant DNA-binding activity. The induction of this DNA-binding activity correlated well with the abilities of the single site and double site mutants, defective with regard to pRB and p300 binding, to induce Cdc2 kinase activity and DNA synthesis in quiescent primary BRK cells (Howe et al. 1990; Wang et al. 1991, 1993, 1995).

Next, we analyzed the activities of the basic promoter/CAT reporter containing only the TATA element and the two CCAAT motifs of the *cdc2* promoter by coexpression of the E1A wild-type and single site and double site mutants. Our results showed that the two CCAAT motifs increased the basal activity of the pTI-CAT basic promoter 18-fold when cloned upstream of the TATA element consistent with our previous report (Chen et al. 1996). E1A 12S transactivated the pTI-CCAAT/CAT reporter expression 40-fold over the level achieved in the absence of wild-type E1A 12S. The promoter activity was reduced by about 50% with single site mutants which were defective in binding either pRB or p300. However, the activation of the promoter by the double site E1A mutant was reduced to the same level of activation of E1A as that achieved with the basic promoter in the absence of CCAAT motifs (Tanimoto et al. 1998a).

5 Role of NF-Y in the Transactivation of the Human *cdc2* Promoter by E1A 12S Protein

The two CCAAT motifs of the human *cdc2* promoter conform to the consensus binding site, TGATTGG(T/C)(T/C)(A/G), derived from a random PCR-based selection procedure (BI et al. 1997), that is specifically recognized by the transcription factor NF-Y. The CCAAT pentanucleotide motifs are ubiquitously present in a wide variety of eukaryotic promoters, generally located at –60 to –100 upstream of the transcription start sites of yeast, plant, and vertebrate genes (for a review see MANTOVANI 1998). NF-Y, also known as CBF, α-CP1, and CP-1 (CHODOSH et al. 1988; DORN et al. 1987; HATAMOCHI et al. 1988; HOOFT van HUIJSDUIJNEN et al. 1987; MAITY et al. 1988), is a heterotrimeric protein consisting of three subunits, NF-YA, NF-YB, and NF-YC, all of which are required for DNA binding (SINHA et al. 1995). We sought to determine whether NF-Y is included in the nuclear proteins from human 293 cell extracts that specifically recognized the CCAAT motifs of the *cdc2* promoter in EMSA. Supershift EMSA assays using rabbit polyclonal antibodies raised against NF-YA and NF-YB subunits and the distal CCAAT motif or proximal motif as the labeled probe indicated that NF-Y was part of the DNA–protein complex that had the slowest electrophoretic mobility (complex I) (KAO et al. 1999). Nuclear proteins from the 293 cells binding to the distal motif gave rise to three DNA-protein complexes, I, II, and III (in the increasing order of their mobilities). With the proximal motif, there was only one DNA–protein complex with the same mobility as complex I similar to that observed with the cyclin A probe (KAO et al. 1999; TANIMOTO et al. 1998a). Complexes II and III which formed with the distal CCAAT motif of the *cdc2* promoter as the probe were competed out by the unlabeled distal motif oligonucleotide, suggesting that these are specific complexes although they did not include NF-Y as shown by supershift EMSA (KAO et al. 1999). Further work is required to identify the protein(s) in the nuclear extracts of 293 cells interacting with the distal CCAAT motif of the *cdc2* promoter.

Since an NF-Y containing binding activity specifically recognizing the CCAAT motif was induced by E1A 12S, we wanted to verify whether E1A 12S could activate the gene expression of all three subunits of the transcription factor NF-Y. Total RNAs isolated from RLE cells infected with adenovirus E1A 12S were subjected to Northern blot analysis using cDNA probes encoding A, B, and C subunits as well as cDNA encoding *cdc2* mRNA. The results of Northern blot analysis indicated that E1A 12S induced *cdc2* mRNA expression as previously shown (Dalton 1992; Draetta et al. 1988; Tanimoto et al. 1998a). In addition, we found that the steady state levels of mRNAs encoding all three subunits of NF-Y were elevated in E1A 12S expressing RLE cells (Kao et al. 1999).

Our results showed that adenovirus E1A 12S activated gene expression of all three subunits of NF-Y and induced a new DNA-binding activity that specifically recognized the CCAAT motif of the *cdc2* promoter. Next, we determined whether NF-Y has any functional role in the activation of the *cdc2* gene expression. We made use of the previous observation that a dominant negative mutant NF-YA subunit (NF-YA29) interfered with DNA binding of the wild-type NF-Y (Mantovani et al. 1994). As the expression of E1A 12S resulted in activation of the human *cdc2* promoter/CAT reporter in RLE cells (Tanimoto et al. 1998a), RLE cells were used for infection with adenovirus E1A 12S and transfection with the *cdc2*(–359) promoter/CAT reporter plasmid and the dominant negative mutant NF-YA29 expression plasmid in increasing dosage. CAT assays showed that with increasing dosage of NF-YA29 mutant expression, transactivation of the *cdc2* promoter was reduced to near the basal level of the promoter obtained in the absence of E1A. As expression of all three subunits of NF-Y were induced by E1A12S, the assembly of newly synthesized subunits of NF-Y would be disrupted by incorporation of the DNA-binding defective mutant, NF-YA29 subunit resulting in suppression of *cdc2* promoter activity. These results indicate that one mechanism by which E1A 12S activates the *cdc2* promoter in cycling cells is by induction and assembly of the heterotrimeric transcription factor NF-Y (Kao et al. 1999).

Our finding that NF-Y plays an important role in E1A 12S-mediated transactivation of *cdc2* gene expression in cycling cells suggests that other cell cycle-regulated genes that have NF-Y binding sites in their promoters (such as *thymidine kinase, cyclin A, cdc25c, hsp70 and human mdr1*) are also activated by E1A in a similar manner. E1A is known to activate the gene expression of *cyclin A, thymidine kinase,* and *hsp70* (Braithwaite et al. 1983; Simon et al. 1987; Wu et al. 1986; Zerfass et al. 1996). The *hsp70* gene is a cell cycle regulated gene and its expression is induced by serum (Wu and Morimoto 1985) in late G_1 and early S phase of the cell cycle (Milarski and Morimoto 1986). The activation of *hsp70* gene expression by E1A proteins requires the CCAAT motif located at –70 upstream of the transcription initiation site (Williams et al. 1989). Furthermore, RLE cells infected with E1A 12S virus induced NF-Y that specifically recog-

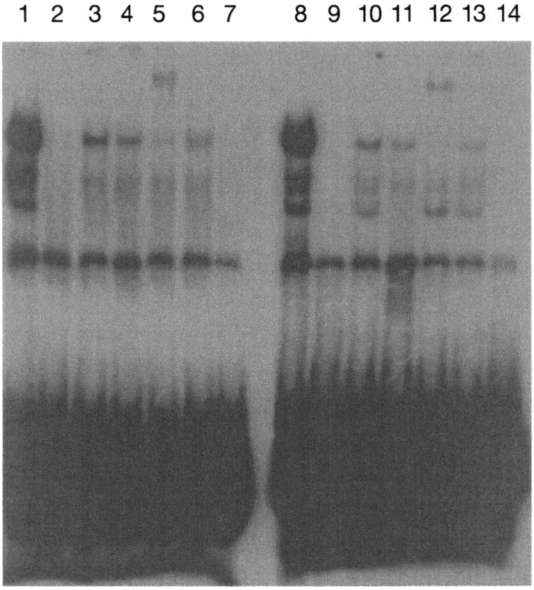

Fig. 2. Electrophoretic mobility shift assay. Rat liver epithelial (*RLE*) cells were infected with adenovirus encoding E1A 12S only, nuclear extracts were prepared, and analyzed by EMSA using the CCAAT motifs of the human *cdc2* (distal motif) and the human *hsp70* promoter as described previously (TANIMOTO et al. 1998a). *Lanes 1* and *8*: 293 cell extract as positive control; *lanes 2* and *9*: RLE extract (from uninfected cells); *lanes 3–7* and *10–14* represent infected cell lysates using the *cdc2* and *hsp70* probes, respectively. *Lanes 3* and *10*: no antibody; *lanes 4* and *11*: normal rabbit serum; *lanes 5* and *12*: anti-NF-YA antibody; *lanes 6* and *13*: anti-NF-YB antibody. *Lanes 7* and *14*: cold *cdc2* CCAAT motif oligonucleotide as homologous (*lane 7*) and heterologous (*lane 14*) competitor, respectively. The immunoreactivity of the commercial anti-NF-YB antibody from Rockland Co. was much weaker than the anti-NF-YA antibody (see KAO et al. 1999)

nized the CCAAT motifs of the *hsp70* promoter (Fig. 2). Our analysis of the DNA binding of 293 cell nuclear proteins to the CCAAT motifs of the *cyclin A* and *cdc25c* promoters by EMSA supported the possibility that NF-Y is involved in the transactivation of these genes as well (TANIMOTO et al. 1998b). Moreover, the DNA–protein complex that includes NF-Y formed on the two CCAAT motifs of the *cdc2* promoter could be efficiently competed with unlabeled CCAAT motif from the *thymidine kinase* promoter (TANIMOTO et al. 1998a) suggesting that common factors are involved in specific recognition of this motif. There is direct evidence that NF-Y binds to the CCAAT motifs of the *thymidine kinase* promoter (CHANG and LIU 1994), and the *cyclin A* promoter. In the latter case, the interaction is necessary for adhesion-dependent transcription of the *cyclin A* gene (KRAMER et al. 1997).

6
Does the NF-Y Activation of the Human *cdc2* Promoter Require a Coactivator?

There is increasing evidence that NF-Y performs its transcriptional activation function in cooperation with other accessory proteins to achieve optimal gene expression. These accessory proteins function by DNA–protein and/or protein–protein interactions. Some accessory coactivators recognize specific DNA elements that are located proximal to the NF-Y binding site and any mutations that alter the distance between these DNA elements affect their coactivator function. Other coactivators function through interactions with NF-Y without themselves having any DNA-binding activity.

For example, in the first category of coactivators of NF-Y, the promoter regulating the expression of the human multiple drug resistance (*hmdr1*) gene encoding the 170-kDa P-glycoprotein contains binding sites for NF-Y and Sp1, and these transcription factors have been shown to interact with the promoter. Mutagenesis of the NF-Y or the Sp1 binding site reduced the promoter activity suggesting that these factors cooperatively regulate transcription initiation from the *hmdr1* promoter (SUNDSETH et al. 1997). In the fibronectin gene promoter, an ATF/cAMP response element (CRE) site at −170 and a CCAAT box motif at −150 are conserved in rats, mice and humans. The CRE element is stimulated by cAMP (DEAN et al. 1988, 1989; HABENER 1990; YAMAMOTO et al. 1990) and adenovirus E1A 13S (WEINTRAUB and DEAN 1992) and by serum (DEAN et al. 1990). Mutational analysis of the CRE and NF-Y/NFI binding sites either by increasing the distance between the sites or by substitutions revealed that the CCAAT motif is essential for cAMP activation by ATF-2, suggesting that the factors synergistically activate fibronectin gene transcription in a cell type specific (HepG2 but not in NIH3T3) manner (PESCE et al. 1999). In hepatic cells, ATF-2 was shown to cooperate with the two CCAAT box binding proteins: NFI, the factor that was shown to stimulate adenovirus DNA replication (JONES et al. 1987) and NF-Y (ALONSO et al. 1996; PESCE et al. 1999). In the murine *mdr1b* promoter, however, besides NF-Y that binds to the Y box CCAAT motif, the expression of another class of CCAAT box binding protein, C/EBP-1β enhanced the *mdr1b* promoter activity suggesting that there is a cooperation between NF-Y and C/EBP-1β in transcriptional activation of murine *mdr1b* gene expression (YU et al. 1995). The C/EBP interaction with NF-Y also regulates serum albumin gene expression. The C/EBP and NF-Y binding sites are juxtaposed in the albumin promoter in such a way that binding of C/EBP reduces the binding of NF-Y. However, the interaction between these two factors synergistically activates the albumin promoter (MILOS and ZARET 1992). The regulation of the 3-hydroxy-3-methylglutaryl-coenzyme A (HMG-CoA) synthase and farnasyl diphosphate synthase gene promoters is achieved by an interaction between the sterol regulatory element binding proteins (SREBPs) and NF-Y. The spatial arrangement of the SREBP and NF-Y binding sites that are separated by 17 bp is critical, and SREBP and the heterotrimeric form of NF-Y physically interact to activate the HMG-CoA synthase gene promoter (DOOLEY et al. 1998).

In the second category of NF-Y-mediated transcriptional activation of a variety of genes, the coactivator itself does not bind to DNA but physically interacts with NF-Y. For example, in human T-cell lymphotropic virus type I (HTLV-1)-encoded Tax1 protein and NF-Y interact physically as well as functionally in activating transcription of the MHC class II DQ promoter (PISE-MASISON et al. 1997) although Tax1 protein itself is not a DNA-binding protein. In the regulation of *hsp70* promoter, it was shown in a recent study that NF-Y physically interacts with the 114-kDa protein CBF-HSP70 (IMBRIANO et al. 2001). Although attempts to demonstrate DNA binding of the CBF-HSP70 protein failed (IMBRIANO et al. 2001), the cDNA clone encoding this protein was originally isolated from an expression cDNA library using a multimeric CCAAT binding motif of the *hsp70* promoter as a probe (LUM et al. 1990, 1992). The CBF-HSP70 protein is required for the activation of the *hsp70* promoter by E1A proteins and the *hsp70* promoter-CAT reporter expression by E1A 13S activation was about fivefold greater than that achieved with E1A 12S (AGOFF and WU 1994).

We investigated whether this 114-kDa CBF-HSP70 could be involved in activation of the human *cdc2* promoter. Our results showed that: (1) E1A 12S induced the transcription of the endogenous *cbf/hsp70* gene in RLE cells expressing E1A 12S; (2) The 293-cell nuclear protein complexes containing NF-Y bind equally well to the CCAAT motif of the human *cdc2* and the *hsp70* promoters and this DNA binding could be efficiently competed out in heterologous competition experiments (Tanimoto et al. 1998a); and (3) CBF-HSP70 activated the human *cdc2*-(109)*Sma*I-CAT reporter expression in transient expression assays in HeLa cells; this activation was dependent on the CCAAT motif (Tanimoto et al. 1998a), suggesting that the 114-kDa CBF-HSP70 protein may also be involved as a coactivator for NF-Y in the transcriptional activation of *cdc2* gene expression in cycling cells.

Our previous studies suggested that an additional protein of approximately 110 kDa could potentially be involved as a transcriptional modulator of NF-Y. The experimental evidence for this notion came from our analysis of the DNA-binding activities of nuclear extracts from different mammalian cells transiently or stably expressing viral oncoproteins, SV40-LT or adenovirus E1A by EMSA and South-Western analysis. For these analyses, the DNA probe used was either the *Sma*I–*Apa*I fragment of the *cdc2* promoter which included both CCAAT motifs (Fig. 1) (Chen et al. 1996) or each of the two motifs as double-stranded oligonucleotides (Kao et al. 1999; Tanimoto et al. 1998a, 1998b). The DNA binding of a 110-kDa protein to the CCAAT motif was commonly observed by South-Western analysis when nuclear extracts from mammalian cells were analyzed when expressing SV40-LT, transiently in CV-1 cells, or stably in COS cells and in human fibroblasts reversibly immortalized by SV40-LT (IDH-4 cells) (Chen et al. 1996). The 110-kDa protein was also detected by the same method in CV-1 cells stably expressing E1A 12S or 13S protein, RLE cells expressing either E1A 12S transiently, or stably transfected with various other oncogenes as well as in spontaneously transformed RLE (C4T) cells (Kao et al. 1999; Tanimoto et al. 1998a, b). There is a good correlation between the induction of this 110-kDa protein and activation of the *cdc2* promoter and upregulation of the *cdc2* mRNA and protein. There is evidence that this 110-kDa protein associates with NF-Y by coimmunoprecipitation experiments using anti-NF-Y antibody followed by South-Western analysis. Independently, the UV-crosslinking experiment also revealed that a 110-kDa protein and a 45-kDa protein (consistent with the size for the NF-YA subunit) crosslinked to the CCAAT motif of the *cdc2* promoter (Kao et al. 1999). This 110-kDa protein is unlikely to be the 114-kDa CBF-HSP70 because the latter does not have any DNA-binding activity (Imbriano et al. 2001; C.Y-Kao, A. Tanimoto, Y. Sasaguri, R. Padmanabhan, unpublished results). The identity of this 110-kDa protein or its function in activation of the human *cdc2* promoter by SV40-LT or E1A is unknown at present.

7 Negative Regulation of the Human *cdc2* Gene Expression

7.1 Role of R-box in Transcriptional Repression of Cell Cycle Genes

An insight into the mechanisms involved in negative regulation of the human *cdc2* promoter, the DNA elements and the protein factors that interact with these elements have started to emerge during the last few years. For example, during macrophage differentiation and G_2 arrest that follows treatment of the human myeloid leukemic cell line, THP1, with the phorbol ester 12-O-tetradecanoylphorbol 13-acetate (TPA), expression of the *cdc2* gene is negatively regulated and this is mediated through a region between –2 and –22 base pairs (bp) (named the 'R-box') upstream of the major transcriptional start site. This downregulation of the *cdc2* promoter is independent of the E2F site at –129 but dependent on the distal CCAAT motif between –92 and –80 bp in the region upstream of the transcription start site (SUGARMAN et al. 1995). However, this TPA-mediated negative regulation was not specific to the NF-Y binding site because the B cell and macrophage-specific transcription factor binding site could function just as well without affecting the down-regulation of the *cdc2* promoter by TPA. These results suggested that the factors binding to the upstream activator sequence (UAS) such as CCAAT motif per se are not specific targets of negative regulation by TPA. It was proposed that TPA mediates negative regulation by uncoupling the upstream activator element and the core transcriptional machinery in a promoter-specific and cell-type specific manner (as CV-1 cells and the Jurkat cells were not active) (SUGARMAN et al. 1995). Although a 55-kDa protein induced by TPA seemed to be involved in this negative regulation by binding to the R-box (–2 to –22 bp region) of the *cdc2* promoter, the identity of this protein is unknown (SUGARMAN et al. 1995).

Recent studies are beginning to elucidate the identity of additional DNA elements and the role of proteins which interact with these elements within the region between –2 and –22 nucleotides upstream of the transcriptional start site of the *cdc2* and other cell cycle-regulated gene promoters. For example, analysis by in vivo footprinting of the *cdc25c* promoter, which is activated in the S-G_2 phase, revealed a novel sequence element known as 'cell cycle-dependent element' (CDE) 20 bp upstream of the transcription start site (LUCIBELLO et al. 1995). Genomic footprinting

R-Box

cdc2 -25 T TTAGCGCGGTGAGTTTGAAACTGCT -2

Cdc25c -20 GGCTGGCGGAAGGTTTGAATGG +1

cyclin A -39 TAGTCGCGGGATACTTGAACTG -18

CDE CHR

Consensus sequences:

E2F: TTTGCGCGC

4 5 6 7 8

CDE: N N T/G G/C G C G G N A/G

CHR: N G/A T/C T T G A A N N

Fig. 3. The elements in the vicinity of transcription start sites of cell cycle-regulated genes. The CDE and CHR elements of *cdc25c*, *cyclin A* and the *cdc2* promoters are underlined. The R-box identified as TPA-mediated negative regulatory element of *cdc2* promoter by SUGARMAN et al. (1995) is indicated. The consensus sequences for CDE and CHR elements were derived by ZWICKER et al. 1995b. (Modified from ZWICKER et al. 1995b)

results indicate that the CDE is protected by protein binding in G_0 but is free in G_2 when the *cdc25c* is expressed. In addition to the CDE, three Yc box motifs (of which the Yc box 1 and 2 are sufficient), which are bound by NF-Y, a constitutive binding site 7 (CBS 7) and a Sp1 site upstream of CDE were also identified as important for synergistic activation and cell cycle regulated repression of the *cdc25c* promoter (ZWICKER et al. 1995a).

A second element immediately downstream from the CDE, known as the 'cell cycle genes homology region' (CHR), is conserved in *cdc25c*, *cyclin A* and *cdc2* promoters (Fig. 3). Both elements are required for negative regulation as shown by mutational analysis and both are occupied in the G_0 and G_1 phase as shown by in vivo footprinting experiments (ZWICKER et al. 1995b). Expression patterns of *cyclin A* and *cdc2* suggest that these genes are coordinately activated preceding *cdc25c* expression. Mutagenesis of the CDE and CHR elements of all three promoters resulted in 10–22-fold loss of repression in G_0 cells but less than 2-fold in cycling cells (ZWICKER et al. 1995b). The identity of proteins that bind to the CDE and CHR elements are unknown.

In another study, nine sites were identified as in vivo protein binding sites for transcription factors throughout the cell cycle by genomic footprinting of the 800 bp *cdc2* promoter in quiescent fibroblasts and at different time points after addition of serum to allow cells to enter the cell cycle (TOMMASI and PFEIFER 1995). However, the protein-binding patterns changed only at the –60 and –20 elements as cells passed from quiescence to S phase. The identity of the transcription factor that binds to the –60 element is unknown. However, the –20 element includes an E2F-like site which differs from the E2F consensus site by only one nucleotide (Fig. 3) just upstream of the CDE and CHR elements of the *cdc2* promoter. An equivalent site is absent in *cyclin A* and *cdc25c* promoters. By genomic footprinting, the –20 element identified as being occupied in quiescent cells (TOMMASI and PFEIFER 1995) overlaps with the 'R-Box' element that is responsive to TPA-mediated repression (SUGARMAN et al. 1995). The –20 element is recognized by the E2F-4/p130 complex (TOMMASI and PFEIFER 1995) suggesting that binding of this complex to the E2F-like site negatively regulates *cdc2* gene expression by uncoupling interaction of the basal transcription machinery with the upstream transactivating sequences in quiescent and G_1 cells (TOMMASI and PFEIFER 1995). Therefore, one possible mechanism of E1A-mediated activation of these cell cycle genes would be derepression of promoters through disruption of the E2F-4/p130 complex in late $G_1 \rightarrow S$ phase at least partially (CORBEIL and BRANTON 1997) although other mechanisms might be concurrently involved.

7.2 Repression of *cdc2* Gene Expression by p53

Ectopic expression of p53 in a cell line that lacks endogenous p53 caused growth arrest at both G_1 and G_2/M phases of the cell cycle, induction of $p21^{waf1}$, a reduced expression of mitotic *cyclins A* and *cyclin B1* and reduced level of *cdc2* mRNA (INNOCENTE et al. 1999; SUGRUE et al. 1997). Transient expression of wild-type and mutant p53 indicated that the wild-type p53 repressed while mutant forms of p53 activated *cdc2* gene expression (YUN et al. 1999). Moreover, expression of adenovirus E1B 55K inhibited the wild-type p53-mediated repression of the *cdc2* promoter but E1B 19K expression had no effect. The p53-mediated repression required the UAS containing the CCAAT motif at position–76. Furthermore, NF-Y was shown to interact with this element and a dominant negative mutant

NF-Y reduced the *cdc2* promoter activity to a level not further decreased by p53 expression (YUN et al. 1999).

More insight into the mechanism of p53-mediated repression of the *cdc2* promoter is provided by a recent study. The p53-mediated activation of $p21^{Waf1}$ expression is required for repression of the *cdc2* promoter (TAYLOR et al. 2001). Furthermore, there is clear evidence that the previously identified E2F-4/p130 complex that binds the E2F-like site overlapping with the CDE (TOMMASI and PFEIFER 1995; ZWICKER and MULLER 1997) is induced by p53 due to inhibition of Cdks and reduced phosphorylation of p130 (TAYLOR et al. 2001). These results and others elaborated above suggest that p53-mediated repression of the *cdc2* promoter uncouples the interaction of UASs from the basal transcription machinery (SUGARMAN et al. 1995; TAYLOR et al. 2001; TOMMASI and PFEIFER 1995; YUN et al. 1999; ZWICKER and MULLER 1997).

8 Conclusions

The results from our laboratory and others, taken together, support the conclusion that adenovirus E1A functions as a positive regulator of *cdc2* gene expression by inducing the expression of NF-Y and a coactivator such as CBF-HSP70 (KAO et al. 1999; TANIMOTO et al. 1998a) and by at least partially disrupting the negatively acting E2F-4/p130 complex (see CORBEIL and BRANTON 1997). Because the establishment of a continuously proliferative state requires cooperation between E1A and the E1B 55-kDa proteins, the latter oncoprotein is likely to function in the release of cells from p53-mediated repression by inhibiting the formation and binding of the E2F-4/p130 complex or other negative regulatory proteins that mediate their function through the CDE–CHR elements and UASs and their binding factors (SUGARMAN et al. 1995; TAYLOR and STARK 2001; TOMMASI and PFEIFER 1995; YUN et al. 1999; ZWICKER and MULLER 1997). Further work is necessary to identify the cellular factors involved in positive and negative regulation of human *cdc2* gene expression.

Acknowledgements. Work in the authors' laboratory is supported by a grant from the NSF (MCB-9507034 to RP) and funds from the Department of Pathology, UOEH, Kitakyushu, Japan (YS). We thank M. Wetzel for carefully editing the manuscript.

References

Agoff SN and Wu B (1994) CBF mediates adenovirus E1a trans-activation by interaction at the C-terminal promoter targeting domain of conserved region 3. Oncogene 9:3707–11

Alonso CR, Pesce CG and Kornblihtt AR (1996) The CCAAT-binding proteins CP1 and NF-I cooperate with ATF-2 in the transcription of the fibronectin gene. J Biol Chem 271:22271–9

Avantaggiati ML, Ogryzko V, Gardner K, Giordano A, Levine AS and Kelly K (1997) Recruitment of p300/CBP in p53-dependent signal pathways. Cell 89:1175–84

Bagchi S, Raychaudhuri P and Nevins JR (1990) Adenovirus E1A proteins can dissociate heteromeric complexes involving the E2F transcription factor: a novel mechanism for E1A trans-activation. Cell 62:659–69

Bandara LR and La Thangue NB (1991) Adenovirus E1A prevents the retinoblastoma gene product from complexing with a cellular transcription factor. Nature 351:494–7

Barber GN (2001) Host defense, viruses and apoptosis. Cell Death Differ 8:113–26

Beach D, Durkacz B and Nurse P (1982) Functionally homologous cell cycle control genes in budding and fission yeast. Nature 300:706–9

Beck GRJ, Zerler B and Moran E (1998) Introduction to DNA tumor viruses: Adenovirus, Simian Virus 40, Polyomavirus. In: Human Tumor Viruses (DJ McCance, ed), pp 51–86. American Society for Microbiology Press, Washington DC

Bi W, Wu L, Coustry F, de Crombrugghe B and Maity SN (1997) DNA binding specificity of the CCAAT-binding factor CBF/NF-Y. J Biol Chem 272:26562–72

Braithwaite AW, Cheetham BF, Li P, Parish CR, Waldron-Stevens LK and Bellett AJ (1983) Adenovirus-induced alterations of the cell growth cycle: a requirement for expression of E1A but not of E1B. J Virol 45:192–9

Buchkovich K, Duffy LA and Harlow E (1989) The retinoblastoma protein is phosphorylated during specific phases of the cell cycle. Cell 58:1097–105

Chang ZF and Liu CJ (1994) Human thymidine kinase CCAAT-binding protein is NF-Y, whose A subunit expression is serum-dependent in human IMR-90 diploid fibroblasts. J Biol Chem 269:17893–8

Chellappan S, Kraus VB, Kroger B, Munger K, Howley PM, Phelps WC and Nevins JR (1992) Adenovirus E1A, simian virus 40 tumor antigen and human papillomavirus E7 protein share the capacity to disrupt the interaction between transcription factor E2F and the retinoblastoma gene product. Proc Natl Acad Sci USA 89:4549–53

Chen H, Campisi J and Padmanabhan R (1996) SV40 large T antigen transactivates the human cdc2 promoter by inducing a CCAAT box binding factor. J Biol Chem 271:13959–67

Chen PL, Scully P, Shew JY, Wang JY and Lee WH (1989) Phosphorylation of the retinoblastoma gene product is modulated during the cell cycle and cellular differentiation. Cell 58:1193–8

Chodosh LA, Olesen J, Hahn S, Baldwin AS, Guarente L and Sharp PA (1988). A yeast and a human CCAAT-binding protein have heterologous subunits that are functionally interchangeable. Cell 53:25–35

Cobrinik D (1996) Regulatory interactions among E2Fs and cell cycle control proteins. Curr Top Microbiol Immunol 208:31–61

COLE C (1996) Polyomavirinae. In: 'Fundamental Virology' (Fields B, Knipe D and Howley P, eds), pp 917–946. Lippencott-Raven, Philadelphia

CORBEIL HB and BRANTON PE (1997) Characterization of an E2F-p130 complex formed during growth arrest. Oncogene 15:657–68

CRESS WD and NEVINS JR (1996) Use of the E2F transcription factor by DNA tumor virus regulatory proteins. Curr Top Microbiol Immunol 208:63–78

CROOK T, TIDY JA and VOUSDEN KH (1991) Degradation of p53 can be targeted by HPV E6 sequences distinct from those required for p53 binding and trans-activation. Cell 67:547–56

DAGNINO L, FRY CJ, BARTLEY SM, FARNHAM P, GALLIE BL and PHILLIPS RA (1997) Expression patterns of the E2F family of transcription factors during murine epithelial development. Cell Growth Differ 8:553–63

DALTON S (1992) Cell cycle regulation of the human cdc2 gene. EMBO J 11:1797–1804

DEAN DC, BLAKELEY MS, NEWBY RF, GHAZAL P, HENNIGHAUSEN L and BOURGEOIS S (1989) Forskolin inducibility and tissue-specific expression of the fibronectin promoter. Mol Cell Biol 9:1498–506

DEAN DC, MCQUILLAN JJ and WEINTRAUB S (1990) Serum stimulation of fibronectin gene expression appears to result from rapid serum-induced binding of nuclear proteins to a cAMP response element. J Biol Chem 265:3522–7

DEAN DC, NEWBY RF and BOURGEOIS S (1988) Regulation of fibronectin biosynthesis by dexamethasone, transforming growth factor beta and cAMP in human cell lines. J Cell Biol 106:2159–70

DECAPRIO JA, LUDLOW JW, FIGGE J, SHEW JY, HUANG CM, LEE WH, MARSILIO E, PAUCHA E and LIVINGSTON DM (1988) SV40 large tumor antigen forms a specific complex with the product of the retinoblastoma susceptibility gene. Cell 54:275–83

DECAPRIO JA, LUDLOW JW, LYNCH D, FURUKAWA Y, GRIFFIN J, PIWNICA-WORMS H, HUANG CM and LIVINGSTON DM (1989) The product of the retinoblastoma susceptibility gene has properties of a cell cycle regulatory element. Cell 58:1085–95

DOERFLER W (1968) The fate of the DNA of adenovirus type 12 in baby hamster kidney cells. Proc Natl Acad Sci USA 60:636–43

DOERFLER W (1975) Integration of viral DNA into the host genome. Curr Top Microbiol Immunol 71:1–78

DOOLEY KA, MILLINDER S and OSBORNE TF (1998) Sterol regulation of 3-hydroxy-3-methylglutaryl-coenzyme A synthase gene through a direct interaction between sterol regulatory element binding protein and the trimeric CCAAT-binding factor/nuclear factor Y. J Biol Chem 273:1349–56

DORN A, BOLLEKENS J, STAUB A, BENOIST C and MATHIS D (1987) A multiplicity of CCAAT box-binding proteins. Cell 50:863–72

DRAETTA G (1990) Cell cycle control in eukaryotes: molecular mechanisms of cdc2 activation. Trends in Biochemical Sciences 15:378–383

DRAETTA G and BEACH D (1988) Activation of cdc2 protein kinase during mitosis in human cells: cell cycle-dependent phosphorylation and subunit rearrangement. Cell 54:17–26

DRAETTA G, BEACH D and MORAN E (1988) Synthesis of p34, the mammalian homolog of the yeast cdc2+/CDC28 protein kinase, is stimulated during adenovirus-induced proliferation of primary baby rat kidney cells. Oncogene 2:553–7

Draetta G, Brizuela L, Potashkin J and Beach D (1987) Identification of p34 and p13, human homologs of the cell cycle regulators of fission yeast encoded by cdc2+ and suc1+. Cell 50:319–25

D'Urso G, Marraccino RL, Marshak DR and Roberts JM (1990) Cell cycle control of DNA replication by a homologue from human cells of the p34cdc2 protein kinase. Science 250:786–91

Dyson N, Bernards R, Friend SH, Gooding LR, Hassell JA, Major EO, Pipas JM, Vandyke T and Harlow E (1990) Large T antigens of many polyomaviruses are able to form complexes with the retinoblastoma protein. J Virol 64:1353–6

Dyson N and Harlow E (1992) Adenovirus E1A targets key regulators of cell proliferation. Cancer Surv 12:161–95

Egan C, Jelsma TN, Howe JA, Bayley ST, Ferguson B and Branton PE (1988) Mapping of cellular protein-binding sites on the products of early-region 1A of human adenovirus type 5. Mol Cell Biol 8:3955–9

Ewen ME, Xing YG, Lawrence JB and Livingston DM (1991) Molecular cloning, chromosomal mapping and expression of the cDNA for p107, a retinoblastoma gene product-related protein. Cell 66:1155–64

Fagan R, Flint KJ and Jones N (1994) Phosphorylation of E2F-1 modulates its interaction with the retinoblastoma gene product and the adenoviral E4 19 kDa protein. Cell 78:799–811

Fang F and Newport JW (1991) Evidence that the G1-S and G2-M transitions are controlled by different cdc2 proteins in higher eukaryotes. Cell 66:731–42

Fechteler K, Tatzelt J, Huppertz S, Wilgenbus P and Doerfler W (1995) The mechanism of adenovirus DNA integration: studies in a cell-free system. Curr Top Microbiol Immunol 199:109–37

Ferguson B, Krippl B, Andrisani O, Jones N, Westphal H and Rosenberg M (1985) E1A 13 S and 12 S mRNA products made in Escherichia coli both function as nucleus-localized transcription activators but do not directly bind DNA. Mol Cell Biol 5:2653–61

Flint J and Shenk T (1989) Adenovirus E1A protein paradigm viral transactivator. Annu Rev Genet 23:141–61

Furukawa Y, Piwnica Worms H, Ernst TJ, Kanakura Y and Griffin JD (1990) cdc2 gene expression at the G1 to S transition in human T lymphocytes. Science 250: 805–8

Gallimore PH (1974) Viral DNA in transformed cells. II. A study of the sequences of adenovirus 2 dna in nine lines of transformed rat cells using specific fragments of the viral genome. J Mol Biol 89:49–72

Girling R, Bandara LR, Zamanian M, Sorensen TS, Xu FH and La Thangue NB (1993) DRTFI/E2F transcription factor: an integrator of cell-cycle events with the transcriptional apparatus. Biochem Soc Trans 21:939–42

Good L, Chen J and Chen KY (1995) Analysis of sequence-specific binding activity of cis-elements in human thymidine kinase gene promoter during G1/S phase transition. J Cell Physiol 163:636–44

Gopalakrishnan S, Douglas JL and Quinlan MP (1997) Immortalization of primary epithelial cells by E1A 12 S requires late, second exon-encoded functions in addition to complex formation with pRB and p300. Cell Growth Differ 8: 541–51

GOTTLIEB TM and OREN M (1996) p53 in growth control and neoplasia. Biochim Biophys Acta 1287:77–102

GRAHAM FL, SMILEY J, RUSSELL WC and NAIRN R (1977) Characteristics of a human cell line transformed by DNA from human adenovirus type 5. J Gen Virol 36:59–74

GRAHAM FL, VAN DER EB AJ and HEIJNEKER HL (1974). Size and location of the transforming region in human adenovirus type 5 DNA. Nature 251:687–91

GRANA X and REDDY EP (1995) Cell cycle control in mammalian cells: role of cyclins, cyclin dependent kinases (CDKs), growth suppressor genes and cyclin-dependent kinase inhibitors (CKIs). Oncogene 11:211–9

GU W, SHI X-L and ROEDER RG (1997) Synergistic activation of transcription by CBP and p53. Nature (Lond) 387:823–827

HABENER JF (1990) Cyclic AMP response element binding proteins: a cornucopia of transcription factors. Mol Endocrinol 4:1087–94

HANNON GJ, DEMETRICK D and BEACH D (1993) Isolation of the Rb-related p130 through its interaction with CDK2 and cyclins. Genes Dev 7:2378–91

HARLOW E, WHYTE P, FRANZA BR, Jr and SCHLEY C (1986) Association of adenovirus early-region 1 A proteins with cellular polypeptides. Mol Cell Biol 6:1579–89

HATAMOCHI A, GOLUMBEK PT, VAN SCHAFTINGEN E and DE CROMBRUGGHE B (1988) A CCAAT DNA binding factor consisting of two different components that are both required for DNA binding. J Biol Chem 263:5940–7

HOOFT VAN HUIJSDUIJNEN RA, BOLLEKENS J, DORN A, BENOIST C and MATHIS D (1987) Properties of a CCAAT box-binding protein. Nucleic Acids Res 15:7265–82

HOROWITZ JM, PARK SH, BOGENMANN E, CHENG JC, YANDELL DW, KAYE FJ, MINNA JD, DRYJA TP and WEINBERG RA (1990) Frequent inactivation of the retinoblastoma anti-oncogene is restricted to a subset of human tumor cells. Proc Natl Acad Sci USA 87:2775–9

HOUWELING A, VAN DEN ELSEN PJ and VAN DER EB AJ (1980) Partial transformation of primary rat cells by the leftmost 4.5% fragment of adenovirus 5 DNA. Virology 105:537–50

HOWE JA, MYMRYK JS, EGAN C, BRANTON PE and BAYLEY ST (1990) Retinoblastoma growth suppressor and a 300 kDa protein appear to regulate cellular DNA synthesis. Proc Natl Acad Sci USA 87:5883–7

IMBRIANO C, BOLOGNESE F, GURTNER A, PIAGGIO G and MANTOVANI R (2001) HSP-CBF is an NF-Y-dependent coactivator of the heat shock promoters CCAAT boxes. J Biol Chem 276:26332–9

INNOCENTE SA, ABRAHAMSON JL, COGSWELL JP and LEE JM (1999) p53 regulates a G2 checkpoint through cyclin B1. Proc Natl Acad Sci USA 96:2147–52

JONES KA, KADONAGA JT, ROSENFELD PJ, KELLY TJ and TJIAN R (1987) A cellular DNA-binding protein that activates eukaryotic transcription and DNA replication. Cell 48:79–89

KAO CY, TANIMOTO A, ARIMA N, SASAGURI Y and PADMANABHAN R (1999) Transactivation of the human cdc2 promoter by adenovirus E1A: E1A induces the expression and assembly of a heteromeric complex consisting of the CCAAT box binding factor, CBF/NF-Y and a 110 kDa DNA binding protein'. J Biol Chem 274:23043–23051

KNUDSON AG (1996) Hereditary cancer: two hits revisited. J Cancer Res Clin Oncol 122:135–40

KO LJ and PRIVES C (1996) p53: puzzle and paradigm. Genes Dev 10:1054–72

KRAMER A, CARSTENS CP, WASSERMAN WW and FAHL WE (1997) CBP/cycA, a CCAAT-binding Protein Necessary for Adhesion-dependent Cyclin A Transcription, Consists of NF-Y and a Novel Mr 115,000 Subunit. Cancer Res 57:5117–5121

KREK W and NIGG EA (1989) Structure and developmental expression of the chicken CDC2 kinase. EMBO J 8:3071–3078

KU DH, WEN SC, ENGELHARD A, NICOLAIDES NC, LIPSON KE, MARINO TA and CALABRETTA B (1993) c-myb transactivates cdc2 expression via Myb binding sites in the 5′-flanking region of the human cdc2 gene [published erratum appears in J Biol Chem 1993 Jun 15; 268(17):13010]. J Biol Chem 268:2255–9

LA THANGUE NB (1996) E2F and the molecular mechanisms of early cell-cycle control. Biochem Soc Trans 24:54–9

LANE DP and CRAWFORD LV (1979) T antigen is bound to a host protein in SV40-transformed cells. Nature 278:261–3

LEE EY, TO H, SHEW JY, BOOKSTEIN R, SCULLY P and LEE WH (1988a) Inactivation of the retinoblastoma susceptibility gene in human breast cancers. Science 241: 218–21

LEE MG, NORBURY CJ, SPURR NK and NURSE P (1988b) Regulated expression and phosphorylation of a possible mammalian cell-cycle control protein. Nature 333: 676–9

LEE MG and NURSE P (1987) Complementation used to clone a human homologue of the fission yeast cell cycle control gene cdc2. Nature 327:31–5

LEVINE AJ (1997) p53, the cellular gatekeeper for growth and division. Cell 88:323–31

LI Y, GRAHAM C, LACY S, DUNCAN AM and WHYTE P (1993) The adenovirus E1A-associated 130 kD protein is encoded by a member of the retinoblastoma gene family and physically interacts with cyclins A and E. Genes Dev 7:2366–77

LILL NL, GROSSMAN SR, GINSBERG D, DECAPRIO J and LIVINGSTON DM (1997) Binding and modulation of p53 by p300/CBP coactivators. Nature 387:823–7

LILLIE JW, GREEN M and GREEN MR (1986) An adenovirus E1A protein region required for transformation and transcriptional repression. Cell 46:1043–51

LILLIE JW and GREEN MR (1989) Transcription activation by the adenovirus E1A protein. Nature 338:39–44

LINZER DI and LEVINE AJ (1979) Characterization of a 54 K dalton cellular SV40 tumor antigen present in SV40-transformed cells and uninfected embryonal carcinoma cells. Cell 17:43–52

LIU F and GREEN MR (1990) A specific member of the ATF transcription factor family can mediate transcription activation by the adenovirus E1A protein. Cell 61:1217–24

LUCIBELLO FC, TRUSS M, ZWICKER J, EHLERT F, BEATO M and MULLER R (1995) Periodic cdc25 C transcription is mediated by a novel cell cycle-regulated repressor element (CDE). EMBO J 14:132–42

LUDLOW JW (1993) Interactions between SV40 large-tumor antigen and the growth suppressor proteins pRB and p53. Faseb J 7:866–71

LUDLOW JW, SHON J, PIPAS JM, LIVINGSTON DM and DECAPRIO JA (1990). The retinoblastoma susceptibility gene product undergoes cell cycle-dependent dephosphorylation and binding to and release from SV40 large T. Cell 60:387–96

Lukas J, Petersen BO, Holm K, Bartek J and Helin K (1996) Deregulated expression of E2F family members induces S-phase entry and overcomes p16INK4A-mediated growth suppression. Mol Cell Biol 16:1047–57

Lum LS, Hsu S, Vaewhongs M and Wu B (1992) The hsp70 gene CCAAT-binding factor mediates transcriptional activation by the adenovirus E1A protein. Mol Cell Biol 12:2599–605

Lum LS, Sultzman LA, Kaufman RJ, Linzer DI and Wu BJ (1990) A cloned human CCAAT-box-binding factor stimulates transcription from the human hsp70 promoter. Mol Cell Biol 10:6709–17

Maity SN, Golumbek PT, Karsenty G and de Crombrugghe B (1988). Selective activation of transcription by a novel CCAAT binding factor. Science 241:582–5

Mantovani R (1998) A survey of 178 NF-Y binding CCAAT boxes. Nucleic Acids Res 26:1135–43

Mantovani R, Li XY, Pessara U, Hooft van Huisjduijnen R, Benoist C and Mathis D (1994) Dominant negative analogs of NF-YA. J Biol Chem 269:20340–6

Mayol X, Grana X, Baldi A, Sang N, Hu Q and Giordano A (1993) Cloning of a new member of the retinoblastoma gene family (pRb2) which binds to the E1A transforming domain. Oncogene 8:2561–6

Mihara K, Cao XR, Yen A, Chandler S, Driscoll B, Murphree AL, T'Ang A and Fung YK (1989) Cell cycle-dependent regulation of phosphorylation of the human retinoblastoma gene product. Science 246:1300–3

Milarski KL and Morimoto RI (1986) Expression of human HSP70 during the synthetic phase of the cell cycle. Proc Natl Acad Sci USA 83:9517–21

Milos PM and Zaret KS (1992) A ubiquitous factor is required for C/EBP-related proteins to form stable transcription complexes on an albumin promoter segment in vitro. Genes Dev 6:991–1004

Montell C, Fisher EF, Caruthers MH and Berk AJ (1982) Resolving the functions of overlapping viral genes by site-specific mutagenesis at a mRNA splice site. Nature 295:380–4

Moran B and Zerler B (1988). Interactions between cell growth-regulating domains in the products of the adenovirus E1A oncogene. Mol Cell Biol 8:1756–64

Moran E (1993) Interaction of adenoviral proteins with pRB and p53. Faseb J 7:880–5

Morgan DO (1995) Principles of CDK regulation. Nature 374:131–4

Muller H, Moroni MC, Vigo E, Petersen BO, Bartek J and Helin K (1997) Induction of S-phase entry by E2F transcription factors depends on their nuclear localization. Mol Cell Biol 17:5508–20

Murray AW (1992) Creative blocks: cell-cycle checkpoints and feedback controls. Nature 359:599–604

Murray AW (1995) The genetics of cell cycle checkpoints. Curr Opin Genet Dev 5:5–11

Mymryk JS and Bayley ST (1994) Multiple pathways for activation of E2 A expression in human KB cells by the 243R E1A protein of adenovirus type 5. Virus Res 33:89–97

Nevins JR (1992) E2F: a link between the Rb tumor suppressor protein and viral oncoproteins. Science 258:424–9

Nevins JR (1993) Disruption of cell-cycle control by viral oncoproteins. Biochem Soc Trans 21:935–8

Nurse P (1990) Universal control mechanism regulating onset of M-phase. Nature 344:503–8

NURSE P and BISSETT Y (1981) Gene required in G1 for commitment to cell cycle and in G2 for control of mitosis in fission yeast. Nature 292:558–60

OSHIMA J, STEINMANN KE, CAMPISI J and SCHLEGEL R (1993) Modulation of cell growth, p34cdc2 and cyclin A levels by SV-40 large T antigen. Oncogene 8:2987–93

PAGANO M, PEPPERKOK R, LUKAS J, BALDIN V, ANSORGE W, BARTEK J and DRAETTA G (1993) Regulation of the cell cycle by the cdk2 protein kinase in cultured human fibroblasts. J Cell Biol 121:101–11

PESCE CG, NOGUES G, ALONSO CR, BARALLE FE and KORNBLIHTT AR (1999). Interaction between the (-170) CRE and the (–150) CCAAT box is necessaryfor efficient activation of the fibronectin gene promoter by cAMP and ATF-2. FEBS Lett 457: 445–51

PINES J (1995) Cyclins, CDKs and cancer. Semin Cancer Biol 6:63–72

PIPAS JM and LEVINE AJ (2001) Role of T antigen interactions with p53 in tumorigenesis. Semin Cancer Biol 11:23–30

PISE-MASISON CA, DITTMER J, CLEMENS KE and BRADY JN (1997) Physical and functional interaction between the human T-cell lymphotropic virus type 1 Tax1 protein and the CCAAT binding protein NF-Y. Mol Cell Biol 17:1236–43

QUINLAN MP (1994) Enhanced proliferation, growth factor induction and immortalization by adenovirus E1A 12 S in the absence of E1B. Oncogene 9:2639–47

QUINLAN MP, WHYTE P and GRODZICKER T (1988) Growth factor induction by the adenovirus type 5 E1A 12 S protein is required for immortalization of primary epithelial cells. Mol Cell Biol 8:3191–203

RAMACHANDRA M, NAKANO R, MOHAN PM, RAWITCH AB and PADMANABHAN R (1993) Adenovirus DNA polymerase is a phosphoprotein. J Biol Chem 268:442–8

RAMACHANDRA M and PADMANABHAN R (1993) Adenovirus DNA polymerase is phosphorylated by a stably associated histone H1 kinase. J Biol Chem 268:17448–56

RAMACHANDRA M and PADMANABHAN R (1995) Expression, nuclear transport and phosphorylation of adenovirus DNA replication proteins. Curr Top Microbiol Immunol 199:50–88

RAO L, DEBBAS M, SABBATINI P, HOCKENBERY D, KORSMEYER S and WHITE E (1992) The adenovirus E1A proteins induce apoptosis, which is inhibited by the E1B 19 kDa and Bcl-2 proteins. Proc Natl Acad Sci USA 89:7742–6

RAYCHAUDHURI P, BAGCHI S, DEVOTO SH, KRAUS VB, MORAN E and NEVINS JR (1991) Domains of the adenovirus E1A protein required for oncogenic activity are also required for dissociation of E2F transcription factor complexes. Genes Dev 5: 1200–11

REED SI, HADWIGER JA and LORINCZ AT (1985) Protein kinase activity associated with the product of the yeast cell division cycle gene CDC28. Proc Natl Acad Sci USA 82:4055–9

RIABOWOL K, DRAETTA G, BRIZUELA L, VANDRE D and BEACH D (1989) The cdc2 kinase is a nuclear protein that is essential for mitosis in mammalian cells. Cell 57: 393–401

RULEY HE (1983) Adenovirus early region 1 A enables viral and cellular transforming genes to transform primary cells in culture. Nature 304:602–6

SARNOW P, HO YS, WILLIAMS J and LEVINE AJ (1982) Adenovirus E1b-58kd tumor antigen and SV40 large tumor antigen are physically associated with the same 54 kd cellular protein in transformed cells. Cell 28:387–94

SCOLNICK DM, CHEHAB NH, STAVRIDI ES, LIEN MC, CARUSO L, MORAN E, BERGER SL and Halazonetis TD (1997) CREB-binding protein and p300/CBP-associated factor are transcriptional coactivators of the p53 tumor suppressor protein. Cancer Res 57:3693–6

SHARP PA, PETTERSSON U and SAMBROOK J (1974) Viral DNA in transformed cells. I. A study of the sequences of adenovirus 2 DNA in a line of transformed rat cells using specific fragments of the viral genome. J Mol Biol 86:709–26

SHENK T and FLINT J (1991) Transcriptional and transforming activities of the adenovirus E1A proteins. Adv Cancer Res 57:47–85

SIMANIS V and NURSE P (1986) The cell cycle control gene cdc2+ of fission yeast encodes a protein kinase potentially regulated by phosphorylation. Cell 45:261–8

SIMON MC, KITCHENER K, KAO HT, HICKEY E, WEBER L, VOELLMY R, HEINTZ N and NEVINS JR (1987) Selective induction of human heat shock gene transcription by the adenovirus E1A gene products, including the 12 S E1A product. Mol Cell Biol 7:2884–90

SINHA S, MAITY SN, LU J and DE CROMBRUGGHE B (1995) Recombinant rat CBF-C, the third subunit of CBF/NFY, allows formation of a protein-DNA complex with CBF-A and CBF-B and with yeast HAP2 and HAP3. Proc Natl Acad Sci USA 92:1624–8

SLANSKY JE and FARNHAM PJ (1996) Transcriptional regulation of the dihydrofolate reductase gene. Bioessays 18:55–62

STILLMAN B (1996) Cell cycle control of DNA replication. Science 274:1659–64

SUBRAMANIAN T, BOYD JM and CHINNADURAI G (1995) Functional substitution identifies a cell survival promoting domain common to adenovirus E1B 19 kDa and Bcl-2 proteins. Oncogene 11:2403–9

SUBRAMANIAN T and CHINNADURAI G (2001) Interaction of cellular apoptosis regulating proteins with adenovirus anti-apoptosis protein E1B-19 K. Methods Mol Biol 177:211–8

SUBRAMANIAN T, KUPPUSWAMY M, NASR RJ and CHINNADURAI G (1988) An N-terminal region of adenovirus E1A essential for cell transformation and induction of an epithelial cell growth factor. Oncogene 2:105–12

SUGARMAN JL, SCHONTHAL AH and GLASS CK (1995) Identification of a cell-type-specific and E2F-independent mechanism for repression of cdc2 transcription. Mol Cell Biol 15:3282–90

SUGRUE MM, SHIN DY, LEE SW and AARONSON SA (1997) Wild-type p53 triggers a rapid senescence program in human tumor cells lacking functional p53. Proc Natl Acad Sci USA 94:9648–53

SUNDSETH R, MACDONALD G, TING J and KING AC (1997) DNA elements recognizing NF-Y and Sp1 regulate the human multidrug-resistance gene promoter. Mol Pharmacol 51:963–71

TANIMOTO A, CHEN H, KAO CY, MORAN E, SASAGURI Y and PADMANBHAN R (1998a) Transactivation of the human cdc2 promoter by adenovirus E1A in cycling cells is mediated by induction of a 110 kDa CCAAT-box-binding factor. Oncogene 17: 3103–3114

TANIMOTO A, KAO CY, CHANG CC, SASAGURI Y and PADMANABHAN R (1998b) Deregulation of cdc2 gene expression correlates with overexpression of a 110 kDa CCAAT box binding factor in transformed cells. Carcinogenesis 19:1735–41

Tao Y, Kassatly RF, Cress WD and Horowitz JM (1997) Subunit composition determines E2F DNA-binding site specificity. Mol Cell Biol 17:6994–7007

Taylor WR, Schonthal AH, Galante J and Stark GR (2001) p130/E2F4 binds to and represses the cdc2 promoter in response to p53. J Biol Chem 276:1998–2006

Taylor WR and Stark GR (2001) Regulation of the G2/M transition by p53. Oncogene 20:1803–15

Templeton D and Eckhart W (1982) Mutation causing premature termination of the polyoma virus medium T antigen blocks cell transformation. J Virol 41:1014–24

Tommasi S and Pfeifer GP (1995) In vivo structure of the human cdc2 promoter: release of a p130-E2F-4 complex from sequences immediately upstream of the transcription initiation site coincides with induction of cdc2 expression. Mol Cell Biol 15:6901–13

Treisman R, Novak U, Favaloro J and Kamen R (1981) Transformation of rat cells by an altered polyoma virus genome expressing only the middle-T protein. Nature 292:595–600

Trentin JJ, Van Hoosier GL, Jr and Samper L (1968) The oncogenicity of human adenoviruses in hamsters. Proc Soc Exp Biol Med 127:683–9

Tsai L.-H, Harlow E and Meyerson M (1991) Isolation of the human cdk2 gene that encodes the cyclin A- and adenovirus E1A-associated p33 kinase. Nature 353: 174–177

Wang HG, Draetta G and Moran E (1991) E1A induces phosphorylation of the retinoblastoma protein independently of direct physical association between the E1A and retinoblastoma products. Mol Cell Biol 11:4253–65

Wang HG, Moran E and Yaciuk P (1995) E1A promotes association between p300 and pRB in multimeric complexes required for normal biological activity. J Virol 69:7917–24

Wang HG, Rikitake Y, Carter MC, Yaciuk P, Abraham SE, Zerler B and Moran E (1993) Identification of specific adenovirus E1A N-terminal residues critical to the binding of cellular proteins and to the control of cell growth. J Virol 67:476–88

Weinberg RA (1985) The action of oncogenes in the cytoplasm and nucleus. Science 230:770–6

Weintraub SJ, Chow KN, Luo RX, Zhang SH, He S and Dean DC (1995). Mechanism of active transcriptional repression by the retinoblastoma protein. Nature 375:812–5

Weintraub SJ and Dean DC (1992) Interaction of a common factor with ATF, Sp1, or TATAA promoter elements is required for these sequences to mediate transactivation by the adenoviral oncogene E1A. Mol Cell Biol 12:512–7

Weintraub SJ, Prater CA and Dean DC (1992) Retinoblastoma protein switches the E2F site from positive to negative element. Nature 358:259–61

Werness BA, Levine AJ and Howley PM (1990) Association of human papillomavirus types 16 and 18 E6 proteins with p53. Science 248:76–9

White E (1993) Regulation of apoptosis by the transforming genes of the DNA tumor virus adenovirus. Proc Soc Exp Biol Med 204:30–9

White E (1996) Life, death and the pursuit of apoptosis. Genes Dev 10:1–15

Whyte P, Buchkovich KJ, Horowitz JM, Friend SH, Raybuck M, Weinberg RA and Harlow E (1988) Association between an oncogene and an anti-oncogene: the adenovirus E1A proteins bind to the retinoblastoma gene product. Nature 334: 124–9

Whyte P, Williamson NM and Harlow E (1989) Cellular targets for transformation by the adenovirus E1A proteins. Cell 56:67–75

Williams GT, McClanahan TK and Morimoto RI (1989) E1A transactivation of the human HSP70 promoter is mediated through the basal transcriptional complex. Mol Cell Biol 9:2574–87

Winberg G and Shenk T (1984) Dissection of overlapping functions within the adenovirus type 5 E1A gene. EMBO J 3:1907–12

Wu BJ, Kingston RE and Morimoto RI (1986) Human HSP70 promoter contains at least two distinct regulatory domains. Proc Natl Acad Sci USA 83:629–33

Wu BJ and Morimoto RI (1985) Transcription of the human hsp70 gene is induced by serum stimulation. Proc Natl Acad Sci USA 82:6070–4

Yamamoto KK, Gonzalez GA, Menzel P, Rivier J and Montminy MR (1990) Characterization of a bipartite activator domain in transcription factor CREB. Cell 60:611–7

Yu L, Wu Q, Yang CP and Horwitz SB (1995) Coordination of transcription factors, NF-Y and C/EBP beta, in the regulation of the mdr1b promoter. Cell Growth Differ 6:1505–12

Yun J, Chae HD, Choy HE, Chung, J, Yoo HS, Han MH and Shin DY (1999) p53 negatively regulates cdc2 transcription via the CCAAT-binding NF-Y transcription factor. J Biol Chem 274:29677–82

Zerfass K, Spitkovsky D, Schulze A, Joswig S, Henglein B and Jansen-Durr P (1996) Adenovirus E1A activates cyclin A gene transcription in the absence of growth factors through interaction with p107. J Virol 70:2637–42

Zerler B, Moran B, Maruyama K, Moomaw J, Grodzicker T and Ruley HE (1986) Adenovirus E1A coding sequences that enable ras and pmt oncogenes to transform cultured primary cells. Mol Cell Biol 6:887–99

Zerler B, Roberts RJ, Mathews MB and Moran E (1987) Different functional domains of the adenovirus E1A gene are involved in regulation of host cell cycle products. Mol Cell Biol 7:821–9

Zwicker J, Gross C, Lucibello FC, Truss M, Ehlert F, Engeland K and Muller R (1995a) Cell cycle regulation of cdc25 C transcription is mediated by the periodic repression of the glutamine-rich activators NF-Y and Sp1. Nucleic Acids Res 23: 3822–30

Zwicker J, Lucibello FC, Wolfraim LA, Gross C, Truss M, Engeland K and Muller R (1995b) Cell cycle regulation of the cyclin A, cdc25 C and cdc2 genes is based on a common mechanism of transcriptional repression. EMBO J 14:4514–22

Zwicker J and Muller R (1997) Cell-cycle regulation of gene expression by transcriptional repression. Trends Genet 13:3–6

Chapter 13

Nuclear Perturbations Following Adenovirus Infection

W. C. Russell, D. A. Matthews

Abstract. Adenoviruses are processed and assembled in the nuclei of infected cells and thereby produce significant perturbations to their structure and function. As the complex interactions that occur in the nuclei of uninfected cells are not yet fully understood many of the changes seen on infection have been described mainly in morphological terms. This chapter attempts to place more recent findings into this context and

W. C. Russell
BMS Building, University of St Andrews, Fife KY16 9ST, Fife, Scotland, UK
e-mail: wcr@st-andrews.ac.uk

D. A. Matthews
Department of Pathology and Microbiology, School of Medical Sciences, University Walk, Bristol University, Bristol, BS8 ITD, UK

demonstrates that adenoviruses are able to hijack many cellular processes and enzymes to their advantage. In particular, modifications to nuclear PODs and nucleoli have more recently been explored in greater detail.

1
Introduction

Within a few years of the initial description of adenoviruses in 1953, a series of observations described the characteristic changes seen in the nuclei of infected cells. These studies were carried out using the techniques of cytochemistry (Boyer et al. 1957), thin section electron microscopy (Kjellen et al. 1955; Morgan et al. 1956) and fluorescent antibody microscopy (Pereira et al 1959).

Huebner and coworkers (1954) followed up the first description of the viruses (or agents as they were called) by demonstrating that they could be classified into serotypes on the basis of neutralization and complement fixation and then into subgroups on the basis of other criteria. Interestingly, these subgroups also showed clear differences in the nature of the changes seen in the nuclei of infected cells (Ginsberg and Dingle 1965).

The alterations in the nuclear structure involved the appearance of nuclear inclusions and 'margination of the chromatin'. To the skilled observer these changes provided a very useful diagnostic marker. However, it was not until the development of immunological and better biochemical techniques that some understanding of these changes could be obtained. Thus, by utilizing specific antisera against 'early' antigens and the major capsid proteins it was clear that some of the changes could be ascribed to the production of non-structural virus proteins (Hayashi and Russell 1968) both early and late in infection. Of particular note was the appearance in the nucleus of large rings which were later shown to be related to the accumulation of the virus non-structural DNA binding protein. The introduction of in situ hybridization methods, in turn, allowed virus transcription and replication events to be tied into some of the morphological and antigenic structures (Bridge and Pettersson 1995).

While there has been some headway over the last 25 years in understanding the nature of these nuclear perturbations, progress has been slow mainly because of the difficulties in trying to unravel the complexities of nuclear structure and function in the uninfected cell. This contribution will attempt to bring the reader up to date in our understanding of the events in the nuclei of infected cells.

2
Targeting the Virus to the Nucleus

Interaction of the virus with the plasma membrane of a susceptible cell involves high affinity binding to cell receptors via the knob portion of the virus fibre. Entry to the cell is via clathrin-mediated endocytosis following on from the interaction of the virus penton base with cellular integrins (for a review see RUSSELL 2000). These early events trigger a number of cellular pathways which facilitate passage of the virus genome out of the endosomes and into the nuclear pores (GREBER et al. 1997). However, the precise host biochemical transport system involved is unclear with some reports suggesting the involvement of microtubules (SUOMALAINEN et al. 2001) while another more recent one discounts the participation of the microtubular network (GLOTZER et al. 2001). Also, in an earlier investigation MATTHEWS and RUSSELL (1998a) had suggested that another cellular protein, p32, which appears to shuttle between the nucleus and mitochondria (BROKSTAD et al. 2001) could also be implicated. In any event, virus-like particles could be observed at the nuclear pores by electron microscopy (DALES and CHARDONNET 1973) within 1 h of infection and virus DNA and the core proteins V and VII could be detected within the nucleus at the same time post-infection (GREBER et al. 1997; MATTHEWS and RUSSELL 1998b). Virus capsid proteins could be recognized at the nuclear pore but it is not clear if they are implicated in the nuclear import or are merely bystanders (GREBER et al. 1993; SAPHIRE et al. 2000). A role for nuclear location sequences (NLSs) in the process has been suggested by the observation that adenovirus DNA nuclear import studied in vitro can be blocked by peptides with NLS motifs. It was also suggested by the same study that a heat shock protein Hsc70 was required for import (SAPHIRE et al. 2000). A number of adenovirus structural proteins have NLS motifs (RUSSELL and KEMP 1995) but whether they are involved in these early events is not known.

3
Initial Nuclear Events

The anchoring of the virus genome via its covalently linked terminal protein to discrete sites on the nuclear matrix has been described by FREDMAN and ENGLER (1993) using a biochemical approach. A multifunctional pyrimidine biosynthesis enzyme complex in association with the terminal protein has also been shown (ANGELETTI and ENGLER 1998) and moreover appears to be linked to sites of replication as determined by incorporation of bromodeoxyuridine. This nuclear location presumably facilitates the primary stages of virus transcription and subsequent replication perhaps by the proximity of the appropriate cellular enzymes.

It is pertinent to these observations that other (morphological) studies have suggested that the so-called PML oncogenic domains (PODs) are the sites of genomic attachment (see below).

4
Transcription, Splicing and Transport of Virus mRNAs

A number of studies have shown that virus transcription and DNA replication occur in multiple separate, but adjacent foci, in the nucleoplasm (BRIDGE et al. 1993). Splicing small nuclear ribonucleoproteins (SnRNPs) co-localize with the transcription sites early in infection (POMBO et al. 1994).These sites have been termed 'intrachromatin granules' (IGs) and seem to be involved in the sorting, regulation and export of different RNA species (PUVION and PUVION-DUTILLEUL 1996). By studying the distribution of splicing factors and virus RNA throughout infection it was evident that the IGs were dynamic structures. Early in infection they were present as 'speckles' which later formed ring structures and finally clusters (ASPEGREN et al. 1998). By analysis of virus protein production and early and late virus RNA it was evident that the appearance of ring structures corresponded to the switch from early to late transcription while the development of clusters was predominately associated with the synthesis of late virus proteins and the presence of late virus RNA and splicing factors. However, the IG clusters did not correspond to sites of late transcription per se (PUVION-DUTILLEUL et al. 1994) and probably reflected post-transcriptional events such as processing of spliced and unspliced mRNA, degradation of RNA and export of mRNA to the cytoplasm.

5
Sites of Replication

A number of studies have examined the location of virus DNA synthesis (MURTI et al. 1990; POMBO et al. 1994) and the results have been described mainly in morphological terms which have not always been easy to relate to other nuclear events. However, the consensus seems to be that virus

DNA synthesis occurs at highly organized multiple foci in the nuclei in association with the virus encoded DNA binding protein (DBP) and the cellular protein NF1 (Bosher et al. 1992). The progeny double-stranded DNA molecules are then displaced to adjacent sites where late transcription occurs. These replication and transcription sites have been amalgamated in a rather non-specific fashion and termed 'virus inclusion bodies' and presumably provide the foci for eventual virus maturation. In conjunction with the concomitant breakdown of cellular chromatin, these events seem to account for the major morphological changes seen on infection.

6
Effect of Adenovirus Infection on the Distribution of Nuclear PODs

Dot-like nuclear structures termed PODs (promyelocytic leukaemia – PML – oncogenic domains), nuclear domains or ND10 (nuclear domain 10) with about 20 per nucleus have been the subject of intense study (for a review see Maul et al. 2000). They appear to be involved in regulation of transcription, DNA repair, apoptosis, cell proliferation and oncogenesis as well as being associated with virus DNA synthesis (Sternsdorf et al. 1997). Numerous cellular proteins accumulate in these structures including PML, sp100 and CREB(CBP) – a transcriptional activator. PML seems to be necessary for the proper assembly of the PODs.

As noted above the incoming adenovirus genome appears to be associated with PODs (Ishov and Maul 1996) and early in infection disruption of the PODs occurs. The POD disruption involves the formation of elongated structures variously termed fibres, bars, tracks or filaments (Carvalho et al. 1995; Konig et al. 1999; Muller and Dejean 1999; Yu et al. 1999). This latter event appears to be related to the production of two early virus proteins E1B55K and E4orf3. An important role for E4orf3 in relocating E1B 55K to the nuclear structures has been demonstrated and it also seems to be required for the reorganization of the PODs and for the formation of different isoforms of PML (Konig et al. 1999; Leppard and Everett 1999). Sumolylation (sentrization) and phosphorylation of PML have also been implicated in these events (Everett et al. 1999) and similar modifications of Sp100 have been described by Muller and Dejean (1999). It may be significant that E1A as well as E1B has been described in PODs, and along with the transcription activator CBP point to an important role in regulation of virus transcription and transforma-

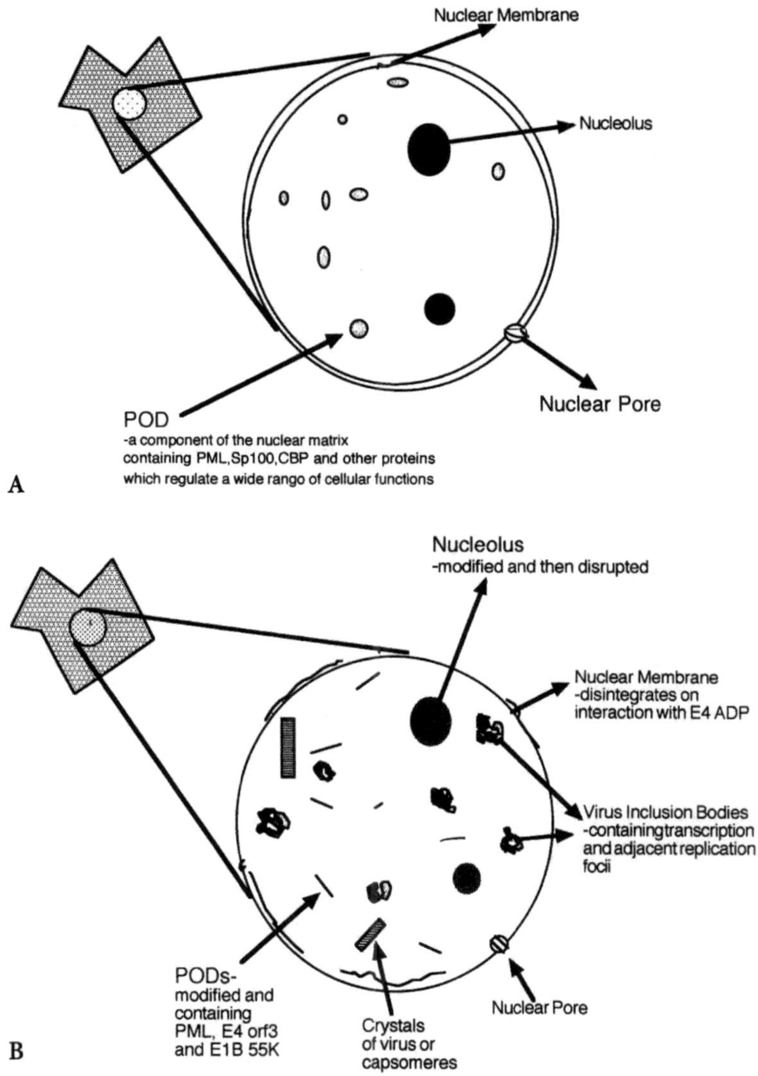

Fig. 1 A, B. The gross morphology of the cell nucleus. **A** A representation of a cell nucleus indicating some of its main features. PODs can be considered to be a component of the nuclear matrix which can be defined as a network or scaffold closely associated with the chromatin. The nuclear matrix is implicated in the precise regulation of many nuclear events. The relative location of chromatin is not depicted here. **B** A representation of a nucleus in an adenovirus infected cell. The term 'virus inclusion body' is utilized very loosely here to include a great variety of structures in the nucleoplasm which have been described in various studies at different stages of infection, e.g. interchromatin granules, transcription and replication foci, rings, clusters etc. (see text)

tion (Carvalho et al. 1995). It is interesting that during infection while PML is retained in the elongated POD structures, sp100 and other POD proteins are relocated to the virus inclusion bodies and seem to be important in supporting virus transcription and replication (Doucas et al. 1996) (see Fig. 1 for a representation of these changes). Disruption of the PODs is a characteristic of other DNA viruses, e.g. herpesviruses, but in these cases degradation rather than reorganization is the consequence of infection.

7
Translation and Virus Maturation

Following the activation of transcription and splicing, virus mRNA is synthesized and exported to the cytoplasm via the nuclear pores (Bridge et al. 1996) and virus proteins are translated and imported to the nuclear inclusion bodies where the virus is then assembled. This procedure requires that the integrity of the cellular machinery for processing RNA and protein is maintained and this is achieved by a variety of early virus gene products which facilitate virus metabolism at the expense of the cell (see chapter by Akusjärvi and Stévenin, this volume).

8
The Effects of Infection on the Nucleolus

The nucleolus is the site of rRNA synthesis, processing and incorporation into ribosomes prior to transport to the cytoplasm (reviewed by Scheer and Hock 1999). In recent years, however, it has been suggested that the nucleolus may play a role in a wider range of cellular functions such as mRNA transport (Pederson 1998). The nucleolus can be simply divided into three regions called the fibrillar centre (FC), the dense fibrillar component (DFC) and the granular component (GC). In the centre of most cell nucleoli is the FC where inactive ribosomal DNA (rDNA) is stored and rRNA synthesis is initiated near the boundary with the DFC. The FC is rich in RNA polymerase I, TATA binding protein and in the rRNA transcription factor known as upstream binding factor. As the rRNA is synthesized it progresses from the DFC to the GC being processed from the 48S precursor to the 28S, 26S and 18S components. Most rRNA processing is thought to occur in the DFC. Key proteins involved in processing of rRNA include B23, fibrillarin and nucleolin, but there are numerous other components. The GC is thought to be where the ribosomal components and the rRNA are brought together and so consists mainly of ribosomes maturing prior to export to the cytoplasm.

That adenovirus affects the nucleolus has been known for some time. Castiglia and Flint (1983) showed that rRNA processing and transport to the cytoplasm were progressively blocked during adenovirus infection. In addition, they noted that rRNA synthesis was inhibited during the late phase of virus infection. Later, Walton et al. (1989) observed that nucleolar antigen B23 became redistributed during the late phase of infection. Puvion-Dutilleul and Christensen (1993) used electron microscopy to demonstrate that the structure of the nucleolus was dramatically altered during adenovirus infection. Specifically they noted that the DFC was compacted and that fibrillarin could be detected outside the nucleolus in virus inclusion bodies in some cells. Finally, Rodrigues et al. (1996) noted that a variety of nucleolar antigens became redistributed during adenovirus infection although they did not relate the redistributed antigens to virus inclusion bodies. However, these studies did not address what specific adenovirus proteins might be involved in the disruption of the nucleolus or even what benefits might be gained from shutting off host cell rRNA production. One study by White et al. (1988) indicated that E1A protein was associated with the nucleolus in a subpopulation of virus-infected cells. In addition, they showed that during the early phase of adenovirus infection, the cellular Hsp70 co-localized with E1A in the nucleolus. To date, however, only adenovirus proteins IVa2 (Lutz et al. 1996) and V (Matthews and Russell 1998b) have been shown to associate with the nucleolus routinely in infected cells. Protein IVa2 has been implicated in packaging of adenovirus DNA (Zhang and Imperiale 2000) and in controlling the transcription from the major late promoter (Tribouley et al. 1994). On the other hand, protein V is a basic protein and a structural component of the adenovirus DNA-protein core complex. Protein V has been speculated to form a bridge between the viral DNA-core protein complex (consisting primarily of proteins V, Mu and VII) and the inside of the virus capsid (Matthews and Russell 1995). Interestingly, this protein contains multiple, distinct, nuclear and nucleolar targeting signals and is mainly concentrated in the DFC (Matthews 2001). The role of protein V during virus infection has not yet been elucidated, perhaps because there are no temperature sensitive mutants available. None of these observations, however, lead to obvious conclusions as to why these proteins might associate with the nucleolus.

Two recent publications have served to stimulate further interest in the role of the nucleolus in adenovirus infections. Firstly, MATTHEWS (2001) demonstrated that adenovirus protein V triggered nucleolar proteins B23 and nucleolin to redistribute from the nucleolus into the cytoplasm. This property of protein V appears to map to a region near the C terminus that is conserved in all mastadenoviruses sequenced so far. In the case of nucleolin and poliovirus infection, there is a precedent for this redistribution in that it has been implicated as playing a role in the replication of poliovirus RNA (WAGGONER and SARNOW 1998). In this report it was noted that poliovirus RNA induced redistribution of nucleolin to the cytoplasm where it may enhance replication of viral RNA as determined by in vitro poliovirus replication assays. However, it is clear that multiple adenovirus proteins (potentially along with virus DNA or RNA) play a role in the disruption of the nucleolus. For example, MATTHEWS (2001) demonstrated that whilst B23 is redistributed to the cytoplasm by expression of protein V in vitro, this is clearly not the case during adenovirus infection where B23 is redistributed but remains within the nucleus. In addition, in adenovirus-infected cells protein V and B23 do not co-localize indicating that protein V has indirect effects on the distribution of B23.

In addition, OKUWAKI et al. (2001) recently demonstrated that addition of B23 to in vitro adenovirus replication assays stimulated replication of viral DNA. OKUWAKI et al. proposed that B23 might be involved in remodelling of cellular chromatin normally (in particular rDNA) and that this ability to remodel chromatin might be utilized by adenovirus to remodel the adenovirus DNA-protein complex during replication and transcription. In support of this, WALTON et al. (1989) claimed to observe B23 in adenovirus replication centres and OKUWAKI et al. cited their own unpublished observations that at least a proportion of B23 co-localizes with the adenovirus single-stranded DNA binding protein (DBP – a major component of the adenovirus replication centres). This is consistent with data presented by MATTHEWS noting that B23 does not co-localize with protein V which, in turn, is excluded from DBP-rich centres in the infected cell (MATTHEWS and RUSSELL 1998b). Thus, like poliovirus, adenovirus appears to utilize nucleolar proteins to enhance viral genome replication. These nucleolar changes are depicted in Fig. 2.

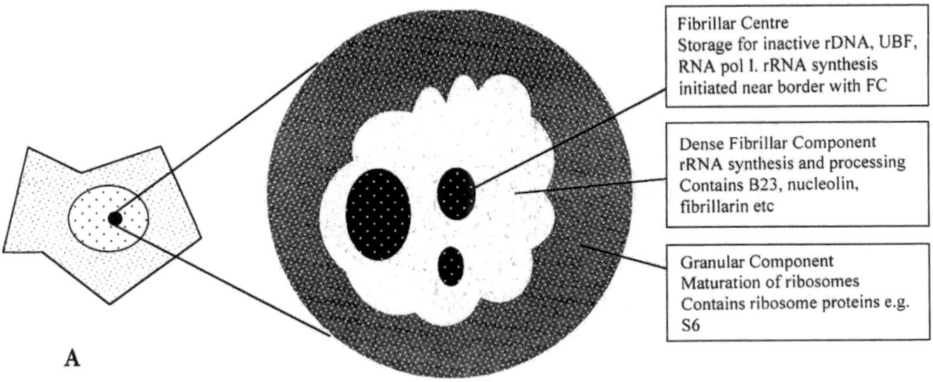

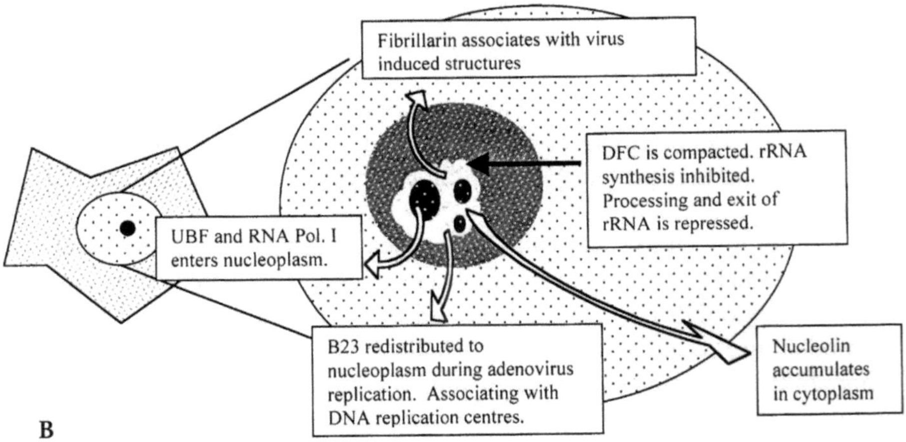

Fig. 2 A, B. The gross morphology of the nucleolus. **A** A representation of a nucleolus showing the three principal regions, i.e. the dense fibrillar component, the fibrillar centre and the granular component. The basic function of each region is outlined together with an indication of the distribution of some nucleolar proteins. It should be noted that the nucleolus and its subcompartments are not physically separated from each other or the nucleus and arise merely through the concentration of the relevant proteins and RNA. **B** A representation of the effects of adenovirus infection on nucleolar components. For details see text

9
The Coiled Body

The nucleolus is reported to be associated with another nuclear structure known as the coiled body or Cajal body (CB) (reviewed by GALL 2000).The CB contains snRNPs, transcription factors, kinases, the survival motor neuron protein and others. However, there does not appear to be any mRNA processing taking place in the CBs and so the precise function of the CB in cells is not yet understood but several suggestions have been made (GALL 2000). Interestingly, CBs are enriched in nucleolar proteins such as fibrillarin and the CB marker protein, p80 coilin, and can be detected in the nucleolus under a variety of conditions (LYON et al. 1997). It has been shown that adenovirus affects the distribution of p80 coilin (LYON et al. 1997; RODRIGUES et al. 1996). However, REBELO et al. (1996) noted that a similar effect could be produced in uninfected cells by cyclohexamide and concluded that adenovirus effects on the CB could be an indirect result of the shut off of host protein synthesis by adenovirus.

10
Late Nuclear Events

As infection proceeds there is a significant build up of virus particles which can lead to massive crystals of virus in the nucleus (KJELLEN et al. 1955). Since the assembly process can be relatively inefficient in some cell types, there is also an accumulation of the virus capsomeres which can themselves form characteristic crystals (WILLS et al. 1973). That such a build up of virus products can occur is an indication that the overall integrity of the nuclear machinery is retained even late in infection. However, at the very late stages of infection, apoptosis mechanisms (see RUSSELL 2000) overwhelm the nucleus leading to the degradation of the cellular DNA and the final disruption of the nuclear membrane (PUVION-DUTILLEUL et al. 1998). The release of virus from the nucleus at this stage also appears to be mediated by the action of another virus protein – ADP or adeno death protein (TOLLEFSON et al. 1996) – at the nuclear membrane.

11
Conclusions

Since adenovirus depends on the nucleus for its life cycle it is not surprising that there are considerable effects on the morphology of the nucleus and on the function of the normal machinery of cell metabolism. As well as hijacking some of the cellular regulatory centres apparently residing in the PODS, virus transcription and replication sequester cellular enzymes giving rise to separate transcription and replication locations leading to early and late translation and maturation of virus. Infection also triggers significant changes in nucleoli which presumably facilitate deregulation of cellular RNA metabolism in favour of virus-directed events.

Almost all of these nuclear events have been observed in studies using human adenoviruses types 2 and 5 and it is interesting that there has as yet been no explanation for the significantly different nuclear changes which have been observed during infection by different adenovirus serotypes (Ginsberg and Dingle 1965) and referred to in the Introduction. However, it is not unreasonable to propose that the different adenovirus serotypes will have different levels of expression of critical virus genes, such as those yielding the E1 and E4 proteins, which could influence the interaction with cellular proteins in the nuclear matrix. Similarly it can be expected that different host cells will provide a variation in the cellular POD proteins thus leading to different outcomes following infection.

References

Angeletti PC, Engler JA (1998) Adenovirus preterminal protein binds to the CAD enzyme at active sites of viral DNA replication on the nuclear matrix. J Virol 72:2896–2904

Aspegren A, Rabino C, Bridge E (1998) Organization of splicing factors in adenovirus-infected cells reflects changes in gene expression during the early to late phase transition. Exp Cell Res 245:203–213

Bosher J, Dawson A, Hay RT (1992) Nuclear factor I is specifically targeted to discrete subnuclear sites in adenovirus type 2-infected cells. J Virol 66:3140–3150

Boyer GS, Leuchtenberger C, Ginsberg HS (1957) Cytological and cytochemical studies of HeLa cells infected with adenoviruses. Journal of Experimental Medicine 105:195–216

Bridge E, Carmo-Fonseca M, Lamond A, Pettersson U (1993) Nuclear organization of splicing small nuclear ribonucleoproteins in adenovirus-infected cells. J Virol 67:5792–5802

Bridge E, Pettersson U (1995) Nuclear organization of replication and gene expression in adenovirus- infected cells. Curr Top Microbiol Immunol 199:99–117

Bridge E, Riedel KU, Johansson BM, Pettersson U (1996) Spliced exons of adenovirus late RNAs colocalize with snRNP in a specific nuclear domain. J Cell Biol 135: 303–314

Brokstad KA, Kalland KH, Russell WC, Matthews DA (2001) Mitochondrial Protein p32 Can Accumulate in the Nucleus. Biochem Biophys Res Commun 281: 1161–1169

Carvalho T et al. (1995) Targeting of adenovirus E1 A and E4-ORF3 proteins to nuclear matrix-associated PML bodies. J Cell Biol 131:45–56

Castiglia CL, Flint SJ (1983) Effects of adenovirus infection on rRNA synthesis and maturation in HeLa cells. Mol Cell Biol 3:662–671

Dales S, Chardonnet Y (1973) Early events in the interaction of adenoviruses with HeLa cells. IV. Association with microtubules and the nuclear pore complex during vectorial movement of the inoculum. Virology 56:465–483

Doucas V et al. (1996) Adenovirus replication is coupled with the dynamic properties of the PML nuclear structure. Genes Dev 10:196–207

Everett RD, Lomonte P, Sternsdorf T, van Driel R, Orr A (1999) Cell cycle regulation of PML modification and ND10 composition. J Cell Sci 112:4581–4588

Fredman JN, Engler JA (1993) Adenovirus precursor to terminal protein interacts with the nuclear matrix in vivo and in vitro. J Virol 67:3384–3395

Gall JG (2000) Cajal bodies: the first 100 years. Annu Rev Cell Dev Biol 16:273–300

Ginsberg HS, Dingle JH (1965) The Adenovirus Group. In: Horsfall FL, Tamm I (eds) Viral ad Rickettsial Infections of Man, 4th edn. Pitman Medical Publishing Co., London, pp 860–891

Glotzer JB, Michou AI, Baker A, Saltik M, Cotten M (2001) Microtubule-independent motility and nuclear targeting of adenoviruses with fluorescently labeled genomes. J Virol 75:2421–2434

Greber UF, Suomalainen M, Stidwill RP, Boucke K, Ebersold MW, Helenius A (1997) The role of the nuclear pore complex in adenovirus DNA entry. EMBO J 16:5998–6007

Greber UF, Willetts M, Webster P, Helenius A (1993) Stepwise dismantling of adenovirus 2 during entry into cells. Cell 75:477–486

Hayashi K, Russell WC (1968) A study of the development of adenovirus antigens by the immunofluorescent technique. Virology 34:470–480

Huebner RJ, Rowe WP, Ward TG, Parrott RH, Bell JA (1954) Adenoidal-pharyngeal-conjuctival agents. A newly recognised group of common viruses of the respiratory system. New England Journal of Medicine 251

Ishov AM, Maul GG (1996) The periphery of nuclear domain 10 (ND10) as site of DNA virus deposition. J Cell Biol 134:815–826

Kjellen L, Lagermalm G, Svedmyr A, Thorsson KG (1955) Crystalline-like patterns in the nuclei of cells infected with an animal virus. Nature 175:505–506

Konig C, Roth J, Dobbelstein M (1999) Adenovirus type 5 E4orf3 protein relieves p53 inhibition by E1B-55-kilodalton protein. J Virol 73:2253–2262

Leppard KN, Everett RD (1999) The adenovirus type 5 E1b 55 K and E4 Orf3 proteins associate in infected cells and affect ND10 components. J Gen Virol 80:997–1008

Lutz P, Puvion-Dutilleul F, Lutz Y, Kedinger C (1996) Nucleoplasmic and nucleolar distribution of the adenovirus IVa2 gene product. J Virol 70:3449–3460
Lyon CE, Bohmann K, Sleeman J, Lamond AI (1997) Inhibition of protein dephosphorylation results in the accumulation of splicing snRNPs and coiled bodies within the nucleolus. Exp Cell Res 230:84–93
Matthews DA (2001) Adenovirus protein V induces redistribution of nucleolin and B23 from nucleolus to cytoplasm. J Virol 75:1031–1038
Matthews DA, Russell WC (1995) Adenovirus protein-protein interactions: molecular parameters governing the binding of protein VI to hexon and the activation of the adenovirus 23K protease. J Gen Virol 76:1959–1969
Matthews DA, Russell WC (1998a) Adenovirus core protein V interacts with p32-a protein which is associated with both the mitochondria and the nucleus. J Gen Virol 79:1677–1685
Matthews DA, Russell WC (1998b) Adenovirus core protein V is delivered by the invading virus to the nucleus of the infected cell and later in infection is associated with nucleoli. J Gen Virol 79:1671–1675
Maul GG, Negorev D, Bell P, Ishov AM (2000) Review: properties and assembly mechanisms of ND10, PML bodies, or PODs. J Struct Biol 129:278–287
Morgan C, Howe C, Rose HM, Moore DH (1956) Structure and development of viruses observed in the electron microscope. IV Viruses of the RI-APC group. Journal of Biophysical and Biochemical Cytology 2:351–360
Muller S, Dejean A (1999) Viral immediate-early proteins abrogate the modification by SUMO-1 of PML and Sp100 proteins, correlating with nuclear body disruption. J Virol 73:5137–5143
Murti KG, Davis DS, Kitchingman GR (1990) Localization of adenovirus-encoded DNA replication proteins in the nucleus by immunogold electron microscopy. J Gen Virol 71:2847–2857
Okuwaki M, Iwamatsu A, Tsujimoto M, Nagata K (2001) Identification of nucleophosmin/B23, an acidic nucleolar protein, as a stimulatory factor for in vitro replication of adenovirus DNA complexed with viral basic core proteins. J Mol Biol 311:41–55
Pederson T (1998) The plurifunctional nucleolus. Nucleic Acids Res 26:3871–3876
Pereira HG, Allison AC, Balfour B (1959) Multiplication of adenovirus type 5 studied by infectivity titration and by the fluorescent antibody technique. Virology 7:300–314
Pombo A, Ferreira J, Bridge E, Carmo-Fonseca M (1994) Adenovirus replication and transcription sites are spatially separated in the nucleus of infected cells. EMBO J 13:5075–5085
Puvion E, Puvion-Dutilleul F (1996) Ultrastructure of the nucleus in relation to transcription and splicing: roles of perichromatin fibrils and interchromatin granules. Exp Cell Res 229:217–225
Puvion-Dutilleul F, Bachellerie JP, Visa N, Puvion E (1994) Rearrangements of intranuclear structures involved in RNA processing in response to adenovirus infection. J Cell Sci 107:1457–1468
Puvion-Dutilleul F, Besse S, Pichard E, Cajean-Feroldi C (1998) Release of viruses and viral DNA from nucleus to cytoplasm of HeLa cells at late stages of

productive adenovirus infection as revealed by electron microscope in situ hybridization. Biol Cell 90:5–38
Puvion-Dutilleul F, Christensen ME (1993) Alterations of fibrillarin distribution and nucleolar ultrastructure induced by adenovirus infection. Eur J Cell Biol 61: 168–176
Rebelo L, Almeida F, Ramos C, Bohmann K, Lamond AI, Carmo-Fonseca M (1996) The dynamics of coiled bodies in the nucleus of adenovirus-infected cells. Mol Biol Cell 7:1137–1151
Rodrigues SH, Silva NP, Delicio LR, Granato C, Andrade LE (1996) The behavior of the coiled body in cells infected with adenovirus in vitro. Mol Biol Rep 23: 183–189
Russell WC (2000) Update on adenovirus and its vectors. J Gen Virol 81:2573–2604
Russell WC, Kemp GD (1995) The role of adenovirus structural proteins in the regulation of adenovirus infection. In: Doerfler W, Bohm P (eds) Molecular Repertoire of Adenoviruses, vol 199 (part1). Springer Verlag, Berlin, pp 81–98
Saphire AC, Guan T, Schirmer EC, Nemerow GR, Gerace L (2000) Nuclear import of adenovirus DNA in vitro involves the nuclear protein import pathway and hsc70. J Biol Chem 275:4298–4304
Scheer U, Hock R (1999) Structure and function of the nucleolus. Curr Opin Cell Biol 11:385–390
Sternsdorf T, Grotzinger T, Jensen K, Will H (1997) Nuclear dots: actors on many stages. Immunobiology 198:307–331
Suomalainen M, Nakano MY, Boucke K, Keller S, Greber UF (2001) Adenovirus-activated PKA and p38/MAPK pathways boost microtubule-mediated nuclear targeting of virus. Embo J 20:1310–1319
Tollefson AE, Ryerse JS, Scaria A, Hermiston TW, Wold WS (1996) The E3-11.6-kDa adenovirus death protein (ADP) is required for efficient cell death: characterization of cells infected with adp mutants. Virology 220:152–162
Tribouley C, Lutz P, Staub A, Kedinger C (1994) The product of the adenovirus intermediate gene IVa2 is a transcriptional activator of the major late promoter. J Virol 68:4450–4457
Waggoner S, Sarnow P (1998) Viral ribonucleoprotein complex formation and nucleolar-cytoplasmic relocalization of nucleolin in poliovirus-infected cells. J Virol 72:6699–6709
Walton TH, Moen PT Jr, Fox E, Bodnar JW (1989) Interactions of minute virus of mice and adenovirus with host nucleoli. J Virol 63:3651–3660
White E, Spector D, Welch W (1988) Differential distribution of the adenovirus E1A proteins and colocalization of E1A with the 70-kilodalton cellular heat shock protein in infected cells. J Virol 62:4153–4166
Wills EJ, Russell WC, Williams JF (1973) Adenovirus-induced crystals: studies with temperature-sensitive mutants. J Gen Virol 20:407–412
Yu E, Joo YK, Lee I (1999) Dynamic redistribution of nuclear matrix proteins by adenovirus infection. Int J Mol Med 3:591–596
Zhang W, Imperiale MJ (2000) Interaction of the adenovirus IVa2 protein with viral packaging sequences. J Virol 74:2687–2693

Chapter 14

The Abortive Infection of Syrian Hamster Cells with Human Adenovirus Type 12

M. Hösel, D. Webb, J. Schröer, W. Doerfler

Abstract. Human adenovirus type 12 (Ad12) induces undifferentiated tumors in newborn Syrian hamsters, and this tumor model has been investigated in detail in our laboratory. One of the characteristics of the Ad12-hamster cell system is a strictly abortive infection cycle. In this chapter, we summarize previous and more recent results of studies on the interaction of Ad12 with the nonpermissive BHK21 hamster cell line. The block of Ad12 replication lies before viral DNA replication and late gene transcription which cannot be detected with the most sensitive techni-

M. Hösel, D. Webb, J. Schröer, W. Doerfler
Institut für Genetik, Universität zu Köln, 50931 Köln, Germany
e-mail: sokolora@scan.genetik.uni-koeln.de
e-mail: doerfler@scan.genetik.uni-koeln.de

M. Hösel, W. Doerfler
Institut für Klinische und Molekulare Virologie, Universität Erlangen-Nürnberg, 91054 Erlangen, Germany

Present address:
J. Schröer, Department of Molecular Biology, Princeton University, Princeton, NJ 08544, USA

ques. Ad12 adsorption, cellular uptake and transport of the viral DNA to the nucleus are less efficient in the nonpermissive hamster cells than in permissive human cells. However, most of the early functions of the Ad12 genome are expressed in BHK21 cells, though at a low level. In the downstream region, the first exon, of the major late promoter (MLP) of Ad12 DNA, a mitigator element of 33 nucleotide pairs in length has been identified which contributes to the inactivity of the MLP in hamster cells and its markedly decreased activity in human cells. The E1 functions of Ad2 or Ad5 are capable of partly complementing the Ad12 deficiencies in hamster cells in that Ad12 viral DNA replication and late gene transcription can proceed, e.g. in a BHK hamster cell line, BHK297-C131, which carries in an integrated form and constitutively expresses the E1 region of Ad5 DNA. Nevertheless, the late Ad12 mRNAs, which are synthesized in this system with the authentic nucleotide sequence, fail to be translated to structural viral proteins. Hence, infectious virions are not produced in the partly complementing system. Probably there is also a translational block for late Ad12 mRNAs in hamster cells. We have recently shown that the overexpression of the Ad12 preterminal protein (pTP) gene or of the E1A gene facilitates the synthesis of full-length, authentic Ad12 DNA in BHK21 cells infected with Ad12. Apparently the pTP has a hitherto unknown function in eliciting full cycles of Ad12 DNA replication even in nonpermissive BHK21 cells when sufficient levels of Ad12 pTP are produced. We pursue the possibility that the completely abortive infection cycle of Ad12 in hamster cells ensures the survival of Ad12-induced hamster tumor cells which all carry, integrated in their genomes, multiple copies of Ad12 DNA. In this way, the viral genomes are immortalized and expanded in a huge number of tumor cells.

1 Virus–Host Interactions

The outcome of a viral infection depends on a number of factors including the molecular and structural characteristics of the virus and the susceptibility of potential host cells. Initially, the host range of a virus is dependent on the appropriate interactions between the virus attachment proteins and the host receptors. Cells that do not allow virions to bind to or lack specific cell surface receptors are resistant to virus infection. In a productive adenovirus infection, the virions adsorb to the surface, enter the host cells, and the viral DNA is transported to the nucleus. Early genes are expressed and viral DNA replication is initiated. After DNA replication begins, the late genes encoding mainly the structural proteins of the virion are expressed, and the late phase of infection is characterized by continued DNA replication, synthesis of late proteins and virion assembly.

In general, cultured human epithelial cell lines can support the complete adenovirus infection cycle. The deficiencies observed in infections of monkey or rodent cells with human adenoviruses are only in some cases due to the lack of cell surface receptors but are rather caused by blocks in various steps essential for the virus–cell interactions. Some of the abortive infections with adenoviruses are due to an early block, when viral DNA replication is affected. There are a number of well-studied examples of abortive infections.

The enteric adenovirus type 40 (Ad40) replicates very poorly in human HeLa cells because of an early defect in Ad40 infection involving the E1B region (Mautner et al. 1989). Ad40-specific E1B mRNA is not detectable by Northern blotting in cytoplasmic RNA which has been isolated at early times after the infection of HeLa cells with Ad40. Moreover, in HeLa cells, the expression of the Ad40 E1B promoter was reduced 80-fold compared with the Ad5 promoter (Bailey et al. 1994). Deletion analyses revealed that sequences upstream of the Sp1 binding site down-modulated the Ad40 E1B promoter responsiveness. Two protein binding sites may be implicated in this effect (Mautner et al. 1999).

The infection of WI-38 cells, human diploid embryonic lung fibroblasts, by Ad41 is conditionally abortive in that Ad41 DNA replication cannot proceed at high serum concentrations in the culture medium. At low serum concentrations, however, the block in viral DNA replication can be overcome, and Ad41 will complete its multiplication cycle (Pieniazek et al. 1990b).

The infection of human continuous lymphoid cell lines with Ad2 represents another example of a conditionally abortive system (Lavery et al. 1987; Silver and Anderson 1988). Only mitotic lymphoid cells appear to be susceptible to Ad2 infection. In nonmitotic cells, infection is blocked very early, because most of the adsorbed Ad2 remains at the cell surface. This passive viral reservoir may contribute to the establishment of persistent adenovirus infections in lymphoid cells.

Abortive adenovirus infections with a late block in virus multiplication also exist. In such instances, viral DNA replicates, but late gene expression and/or virus assembly are affected. Thus, human Ad2 cannot grow in monkey cells. The adsorption, early gene expression, and viral DNA replication proceed in monkey cells; however, the expression of some of the late Ad2 proteins is markedly reduced due to a low concentration of some of the late viral mRNAs and decreased protein synthesis (Silverman et al. 1989). Moreover, defective splicing and/or transport of Ad2 late mRNAs may contribute to the abortive infection (Ross and Ziff 1994).

The infection of the Ad5-transformed human embryonic kidney cell line 293 with Ad41 leads to high yields of incomplete particles. The abortive character of this system appears to be caused by an assembly defect (Pieniazek et al. 1990a).

Since human Ad12 is highly oncogenic in newborn Syrian hamsters (Huebner et al. 1962; Trentin et al. 1962; Kuhlmann et al. 1982; Hilger-Eversheim and Doerfler 1997; Hohlweg et al. 2003), we have studied the interaction of Ad12 with Syrian hamster cells. Ad12 infects human cells productively, and new virions are generated. Syrian hamster (*Mesocricetus auratus*) cells are completely nonpermissive for the replication of Ad12 (Strohl 1969; Doerfler 1969). Ad12 virions can enter hamster cells, and viral DNA reaches the nucleus. At least some of the early viral mRNAs are synthesized albeit at reduced levels, but Ad12 DNA replication is totally blocked. The late structural genes are not detectably transcribed, and Ad12 virions are not produced. In contrast, human Ad2 or Ad5 can productively infect both human and hamster cells (Strohl et al. 1966; Rouse et al. 1966; Doerfler 1969). Obviously, the inability to multiply in hamster cells is not a deficiency common to all human adenoviruses but is specific for Ad12. In addition, the propagation of Ad12 in human cells is less efficient than the virus production in Ad2-infected human cells. Hence, it is sensible to reason that Ad12-specific defects or lack of functions are responsible for the inability to replicate in hamster cells and cause the delay in virus production in human cells. On the other hand, this replication defect can be viewed, at least to some extent, as a consequence of the species-specific insufficiency of Ad12 viral promoters in hamster cells. The deficiency of Ad12 replication in hamster cells is complex and probably involves several blocks both in early and late stages of the viral genome. The discovery that Ad12 can induce tumors in newborn Syrian hamsters has made this abortive system an interesting model for studies on the mechanism of DNA viral oncology (see chapter by Hohlweg and Doerfler, volume 273 this series).

2
Block in Ad12 DNA Replication. Partial Complementation of Ad12 Defects in Hamster Cells by Early Ad2 or Ad5 Functions

2.1
Early Events in the Ad12 Infection of BHK21 Cells

Productive infection by adenoviruses proceeds in a temporally precisely regulated manner. The earliest events include the binding to cellular receptors and entry into the cell, the uncoating of viral DNA, its delivery to the nucleus, and the expression of early adenovirus genes. The first transcription units, regions E1A and E1B, are activated shortly after infection and encode proteins involved in cellular transformation and in the transactivation of viral and cellular transcription units (for reviews see FLINT and SHENK 1989; BERK et al. 1979; NEVINS 1981). Regions E2A and E2B encode proteins required for adenovirus DNA replication, including the viral DNA binding protein (DBP), DNA polymerase (AdPol) and preterminal protein (pTP) (for a review see VAN DER VLIET 1995). Finally, regions E3 and E4 are involved in the suppression of host defense mechanisms (WOLD and GOODING 1991), in the transcriptional activation of other promoters (HARDY et al. 1989) or in the preferential transport of viral mRNAs from the nucleus to the cytoplasm (HALBERT et al. 1985). The late phase of adenovirus infection is induced by the onset of viral DNA replication and leads to the activation of the major late promoter (MLP) followed by the expression of late viral genes, synthesis of late proteins and virion assembly (ZIFF 1980; for a review see PERSSON and PHILIPSON 1982).

Abortive infection of baby hamster kidney (BHK21) cells by Ad12 cannot primarily be accounted for by defects in binding, entry and transport of Ad12 DNA to the nucleus. However, it is likely that the adsorption of Ad12 virions to the cytoplasmic membrane of BHK21 cells is reduced as compared to that of HeLa cells (D. Webb, M. Hösel, and W. Doerfler, unpublished results). Ad12 virions enter the cells and viral DNA reaches the nucleus where it can be integrated into the cellular genome (DOERFLER 1968; DOERFLER 1970; STABEL et al. 1980). Within 24 h of infection, up to 50% of the cell-bound Ad12 DNA has become associated with the chromatin of BHK21 cells (DOERFLER 1970; SCHRÖER et al. 1997). This finding may be significant for the ability of Ad12 to transform hamster cells. The mechanisms of Ad12 DNA integration during the abortive infection of BHK21 cells are still unknown.

Some of the early Ad12 mRNAs are synthesized in abortively infected cells. In vitro translational studies have shown that many of the early Ad12 mRNAs isolated from BHK21 cells can be translated into viral proteins which are similar to those translated in vitro from Ad12 mRNAs isolated from Ad12-productively infected human cells (Raška and Strohl 1972; Ortin and Doerfler 1975; Ortin et al. 1976; Esche et al. 1979). However, there are defects in early gene expression during the abortive infection. The in vitro translation of Ad12 mRNAs from Ad12-infected hamster cells has revealed the lack of a 34-kDa protein (Esche et al. 1979). Moreover, in immunoprecipitation analyses of cellular extracts, Ad12 E1A-specific proteins are detectable at 6 h post infection (p.i.) in KB cell extracts but not before 12 h p.i. in BHK21 cell extracts (Lucher 1990). The concentrations of Ad12 E1A and E1B proteins in abortively infected BHK21 cells are only approximately 10% of those in productively infected human KB or HeLa cells (Lucher 1990; Hösel et al. 2001b). Although some of the early genes of Ad12 are expressed in Ad12-infected BHK21 cells, the replication of Ad12 DNA is completely blocked.

As Ad2 or Ad5, which are both nononcogenic in newborn hamsters, can replicate efficiently in hamster cells (Strohl et al. 1966; Rouse et al. 1966; Doerfler 1969), the question arises of whether Ad2 or Ad5 functions can complement some of the Ad12-specific gene products which are unable to support Ad12 replication in hamster cells. To identify factors necessary for Ad12 DNA replication in hamster cells, BHK21 cells have been doubly infected with Ad2 and Ad12. Subsequently, viral DNA replication and gene expression have been examined. In double-infected BHK21 cells, both Ad2 and Ad12 DNAs replicate, and late Ad12 genes are transcribed. However, only Ad2 virions are assembled, the synthesis of Ad12 virions cannot be detected (Klimkait and Doerfler 1985, 1987).

The analysis has been further refined by superinfecting the Ad5-transformed cell line BHK297-C131 with Ad12. This cell line carries in an integrated form and constitutively expresses the E1 region of Ad5 (Visser et al. 1979). Similar to BHK21 cells doubly infected with Ad2 and Ad12, Ad12 DNA replication and late viral gene transcription can proceed in this cell line (Klimkait and Doerfler 1985). However, the production of newly synthesized Ad12 virions cannot be detected.

In another series of experiments, the E1A (*dl312*) and E1B (*dl313*) deletion mutants of Ad5 (Jones and Shenk 1979; Shenk et al. 1979) have been used for the double infection of BHK21 cells with Ad12. The results suggest that the Ad5-complementing functions seem to be predominantly, but not exclusively, located in the adenoviral E1B region. However, a supporting role of the E1A region is also possible (Klimkait and Doerfler 1987). Interestingly, the partial complementation of functions between Ad5 and Ad2 appear to be mutual. Ad5 functions are able to rescue the Ad12 defects in DNA replication and late gene transcription in hamster cells. Conversely, coinfection of hamster cells with Ad12 and the Ad5 deletion mutants *dl312* or *dl313* enables the Ad5 deletion mutants to replicate in spite of their defects. In the absence of Ad12 early functions, these deletion mutants cannot replicate in hamster cells. Somehow, the Ad12 early functions, although by themselves incapable of supporting the replication and late gene transcription of Ad12 DNA in hamster cells, can substitute for deletions in Ad5 DNA, which is then capable of Ad5 DNA replication and the execution of the complete Ad5-specific transcriptional and translational programs.

During complementation, the Ad2/5 E1 gene products may act directly on the Ad12 genome by substituting for the nonsufficient Ad12 E1 proteins. Ad12 recombinants in which the E1A and/or E1B regions are replaced with E1A and/or E1B of Ad5 are viable in human cells (Sawada et al. 1988). This observation confirms the notion that the E1A and/or E1B proteins of Ad5 can directly act on Ad12 DNA, at least in human cells. Alternatively, the Ad5 E1 functions may act indirectly via the stimulation of viral and/or cellular factors which are missing or deficient in BHK21 cells upon the abortive infection with Ad12.

2.2
Late Blocks in Ad12 Infection of Hamster Cells

The abortive infection of BHK21 cells with Ad12 is not only characterized by the absence of Ad12 DNA replication. The transcription of the late Ad12 genes or of the virus-associated (VA) RNA region encoding the VA RNA is also blocked in hamster cells (Ortin et al. 1976; Klimkait and Doerfler 1985, 1987; Jüttermann et al. 1989). The results of nuclear run-on and RNase protection experiments have provided no evidence for the presence of any transcripts initiated at the Ad12 MLP in Ad12-infected hamster cells (Zock and Doerfler 1993). The late transcriptional defect of Ad12 DNA in hamster cells has been, at least partly, localized to the viral MLP. A 33-nucleotide mitigator sequence in the downstream (first exon) region of the MLP is responsible for the inactivity of the Ad12 MLP in hamster cells. This negative regulatory element is located between nucleotides +320 and +352, relative to the site of initiation of transcription in the MLP of Ad12 DNA. By using the chloramphenicol acetyltransferase (CAT) gene as a reporter in transient expression experiments, it has been shown that the removal of the mitigator sequence from the Ad12 MLP-CAT construct activates the Ad12 MLP in hamster cells and increases its activity in permissive human cells (Zock and Doerfler 1990). Because a similar negative regulator has not been found in the MLP of Ad2 or Ad5, the mitigator seems to be specific for the Ad12 MLP and contributes essentially to the abortive infection of Ad12 in hamster cells. Because the mitigator is involved in the silencing of a major part of the viral genome, which carries the viral structural proteins necessary for the assembly of new virions, this shutdown of the major viral replicative and late transcription functions may be one of the important preconditions for the malignant transformation of hamster cells by Ad12.

The molecular mechanisms of action of the Ad12 mitigator remain unknown. Surprisingly, the mitigator element cannot elicit its functions in in vitro transcription experiments with nuclear extracts from hamster BHK21 or human HeLa cells. Apparently, intact nuclear topology and/or tightly bound nuclear elements, which cannot be eluted in nuclear extracts, are required for the recognition of the Ad12 mitigator. When the mitigator sequence in the MLP of Ad12 DNA was converted to the equivalent sequence in the MLP of Ad2 DNA, activity of the mutated Ad12 MLP was not attained in hamster cells. Conversely, when the Ad12 mitigator was inserted into the Ad2 MLP, the latter's functions in hamster cells were not compromised. Moreover, deletions in the 5′ upstream region of the Ad12 MLP provided evidence for the existence of additional sequences that codetermined the deficiency of the Ad12 MLP in hamster cells (Zock and Doerfler 1993). The amphifunctional YY1 protein from HeLa cells can bind specifically to the mitigator and to the upstream elements of the MLP of Ad12 DNA. Thus, the promoter defect is more complex than would be suggested by the presence of the downstream mitigator. Several promoter sites probably interact in the inactivation of the Ad12 MLP in hamster cells and thus contribute to the shutdown of major parts of this viral genome in hamster cells.

As mentioned before, the defect in the late gene transcription of Ad12 DNA in hamster cells can be complemented by E1 functions of Ad2 or of Ad5. Most of the late Ad12 specific RNA populations are detectable in BHK21 cells after coinfection with Ad12 and Ad2/5 or in Ad5-transformed BHK297-C131 cells which carry the integrated E1 region of Ad5 and express it constitutively (VISSER et al. 1979) after the superinfection with Ad12 (KLIMKAIT and DOERFLER 1985, 1987; SCHIEDNER et al. 1994). However, late Ad12 proteins cannot be detected, and Ad12 virions are not produced (KLIMKAIT and DOERFLER 1985, 1987; SCHIEDNER et al. 1994).

The late Ad12 mRNAs in the partly complementing BHK297-C131 cells are polyadenylated but are synthesized in reduced amounts when compared with those in Ad12-infected human cells. Moreover, the Ad12 mRNA derived from the late transcribed Ad12 fiber gene has been investigated for its structural properties by cloning of the cDNA transcript from this RNA and determining its nucleotide sequence. This RNA carries the authentic nucleotide sequence of the Ad12 fiber gene and the complete tripartite leader sequence. In addition, the Ad12 VA RNA, which is required for the efficient translation of viral mRNAs at late times after infection, as has been shown for Ad2 (THIMMAPPAYA et al. 1982; SCHNEIDER et al. 1984), is also synthesized in this complementing system with the correct nucleotide sequence. Nevertheless, the synthesis of detectable amounts of Ad12 virion-specific proteins, and in particular that of the main viral antigens, hexons and fibers, cannot be documented (SCHIEDNER et al. 1994). The failure of sequence-perfect late Ad12 RNAs to be translatable in BHK21 cells contributes significantly to the nonpermissivity of hamster cells for Ad12 even in the presence of the Ad2 or Ad5 complementing system. It is likely that the cellular functions, either structural or enzymatic, which are required for the efficient translation of late Ad12 mRNAs are missing in hamster cells. Alternatively, translation inhibitors for Ad12 late mRNAs may be present in BHK21 cells.

3
Cellular and Early Viral Factors in the Interaction of Ad12 with Hamster Cells: The Abortive Response

Studies on the interaction of Ad12 with Syrian hamster cells are of particular interest because of the high oncogenic potential of this virus in newborn hamsters (see chapter by HOHLWEG and DOERFLER, volume 273 this series; HILGER-EVERSHEIM and DOERFLER 1997; KUHLMANN et al. 1982; TRENTIN et al. 1962). Upon the injection of Ad12 into newborn hamsters, 70%–90% of the surviving animals develop tumors within 30–50 days of injection. Obviously, the absence of Ad12 replication in hamster cells is key to the transformation and viral oncogenesis in this system. Perhaps an even more detailed analyses of the abortive infection of hamster cells by Ad12 may help to elucidate the mechanisms of viral transformation and oncogenesis. Only scanty information is available on alterations in the expression of cellular functions in hamster cells upon the abortive infection with Ad12. Among the factors which might account for the inability of Ad12 to replicate in hamster cells are: the altered expression, nuclear localization or behavior of cellular proteins, the absence of essential cellular functions or conversely their overexpression in hamster cells.

During the productive infection with adenoviruses, adenoviral DNA replication is initiated within the inverted terminal repeats of the linear genome and proceeds by a strand displacement mechanism (for a review see van der Vliet 1995). Three viral gene products encoded by the E2 region of adenovirus DNA are required for adenoviral DNA replication: DBP, pTP, and the adenoviral DNA Pol (Adpol). DBP is encoded in the E2A region, pTP and Adpol are derived from the E2B region of Ad12 DNA. The initiation of adenoviral DNA replication is dependent on a protein-priming mechanism in that pTP becomes covalently bound to dCMP at the 5′ terminus and must bind to Adpol. After the initiation of replication at both termini, elongation proceeds by a displacement mechanism from one or both termini to the opposite end and by the addition of nucleotides to the free 3′-OH group of the covalently bound dCMP. Elongation requires Adpol and DBP (for a review see Challberg and Kelly 1989). In addition to viral proteins, at least three cellular proteins participate in adenoviral DNA replication. Nuclear factor II (NF II) is necessary at late stages of replication during chain elongation, whereas NFI and the nuclear octamer-binding protein NFIII/Oct-I are essential for the initiation of viral DNA replication (Nagata et al. 1982, 1983; Pruijn et al. 1986; Rosenfeld et al. 1987; O'Neil et al. 1988; Hatfield and Hearing 1993). NFI and NFIII bind specifically to the auxiliary origin of replication and seem to be necessary for the correct positioning of the pTP–Pol complex relative to the origin and for the stability of the pre-initiation complex. Together they can enhance viral DNA replication up to 200-fold (van der Vliet 1995).

The block of Ad12 DNA replication in hamster cells might be related to deficiencies in the accumulation or activity of the cellular proteins NFI and NFIII. Therefore, the expression levels of NFI and NFIII in abortively Ad12-infected BHK21 cells have been investigated and compared with those in productively Ad12-infected human HeLa cells or in the complementing Ad5-transformed hamster cell line BHK297-C131 infected with Ad12 (Hösel et al. 2001 a). NFI does not appear to be a limiting factor in the Ad12-BHK21 system, because there is no shortage of NFI in the nuclei of

Ad12-infected BHK21 cells when compared to Ad12-infected BHK297-C131 or HeLa cells or to BHK21 cells productively infected with Ad2. In contrast, the amounts of NFIII are markedly reduced in nuclear extracts prepared from mock-infected BHK21 cells as compared to hamster BHK297-C131 or human HeLa cells. Upon the abortive infection of BHK21 cells with Ad12, nuclear concentrations of NFIII remain low, whereas their productive infection with Ad2 rendered the amount of NFIII in the nuclei of BHK21 cells comparable to that in Ad12-infected HeLa or in BHK297-C131 cells. The intranuclear increase of NFIII protein in BHK21 cells infected with Ad2 correlates temporally with the initiation of viral DNA replication. Interestingly, the transcription of the NFIII gene in BHK21 cells does not change after the abortive infection with Ad12 or the productive infection with Ad2. Thus, it is likely, that the transport of NFIII from the cytoplasm into the nucleus rather than NFIII gene expression is deficient in Ad12-infected BHK21 cells. As the replication defect of Ad12 in hamster cells has many facets, it is not surprising that the overexpression of NFIII in BHK21 cells upon transfection with an NFIII construct and sufficient availability of NFIII by themselves cannot overcome the replication defect of Ad12 DNA (Hösel et al. 2001 a).

During the productive infection with adenoviruses, viral DNA replication and virion assembly take place in close association with the nuclear matrix of the cell (Younghusband and Maundrell 1982; Khittoo et al. 1986; Zhai et al. 1987; Schaack et al. 1990; Bridge et al. 1995). Adenoviruses replicate and accumulate DNA within distinct nuclear compartments, and host factors have also been shown to localize to these compartments (for a review see Bridge and Pettersson 1995). However, the role of the architecture and of structural components in the nuclei in adenovirus replication and viral gene expression are unknown.

One of the most extensively studied nuclear compartments – nuclear domain 10, also referred to as POD or promyelocytic leukemia protein (PML) bodies – is a multiprotein complex with an average frequency of ten domains per nucleus and a diameter of 0.3 – 0.5 μm (Ascoli and Maul 1991; Maul et al. 1996). These macromolecular complexes are present in all cultured cell lines and are presumed to be active in various aspects of cellular gene expression, cell cycle control, and the repair of DNA damage (Sternsdorf et al. 1997; Doucas et al. 1996). PML was first identified as part of a fusion product with the retinoic acid receptor, which arose in acute promyelocytic leukemia (APL) due to a t(15;17) chromosomal translocation. Disruption of the POD structure in lymphocytes from patients with APL is associated with cellular transformation and can be reversed by treatment with retinoic acid (Pandolfi et al. 1991; Kastner et al. 1992; Grignani, et al. 1994). The productive infection of human cells with adenoviruses leads to extensive morphological changes in the organization of the PML domains. Within 5–8 h of infection with Ad5, the PML domains begin to reorganize from dot-like to filamentous or track-like structures. These alterations are induced by the early viral gene product of open reading frame 3 from the E4 region (E4-ORF3) (Carvalho et al. 1995; Doucas et al. 1996). Similarly to the infection with adenoviruses, herpes simplex virus 1 or human cytomegalovirus (HCMV) infection leads to these modifications of the PML domains (Maul et al. 1993; Ishov et al. 1997). Virus replication commences in the immediate vicinity of this nuclear domain (Ishov and Maul 1996; Maul et al. 1996). As the proteins E1A, E1B-55 kDa of adenoviruses and the simian virus 40 large T antigen have been found in close association with these nuclear domains, the PML domains are suspected to be general targets for DNA tumor virus early proteins (Carvalho et al. 1995; Doucas et al. 1996). The exact function of the PML reorganization is, however, still unknown.

Are the PML domains also reorganized during the abortive infection of hamster cells with Ad12? By using specific antibodies raised against the PML protein (Stuurman et al. 1992), the arrangement of the PML domains has been investigated in the nuclei of mock-, Ad2- or Ad12-infected BHK21 cells as well as in Ad12-infected HeLa or in Ad12-infected complementing BHK297-C131 hamster cells (Hösel et al. 2001 a). Starting at 8 h p.i., the productive infection of HeLa cells leads to the structural reorganization of PML from spherical dots to filamentous structures. At late times after infection, all PML protein domains are reorganized. The same morphological changes are observed when BHK21 cells have been productively infected with Ad2. In contrast, upon the abortive infection of BHK21 cells with Ad12, almost no track-like structures of PML are detected. Most of the PML protein domains remain unaltered in the globular conformation as late as 40 h p.i. The absence of morphological alterations in the PML protein structures in Ad12-infected hamster cells is probably related to the inability of Ad12 DNA to replicate in hamster cells. In Ad12-infected BHK297-C131 cells, which allow the replication of Ad12 DNA, track-like PML protein structures are abundant, whereas in mock-infected BHK297-C131 cells, these structures remain globular.

The reorganization of the nuclear PML protein structures upon productive infection with adenoviruses is induced by the viral E4-ORF3

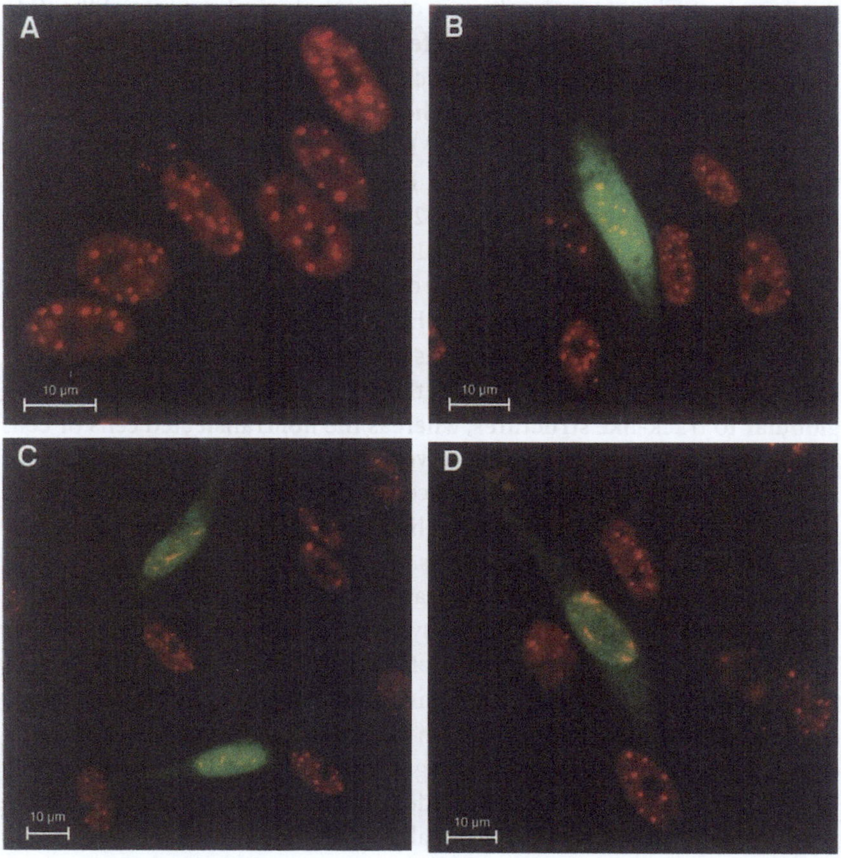

Fig. 1. Overexpression of the E4-ORF3 gene of Ad12 (**C**) or Ad2 (**D**) leads to the reorganization of the PML protein domains in BHK21 hamster cells. In untransfected cells (**A**) or in cells transfected with the pIRES2-EGFP vector (**B**), PML protein domains retain their spherical structures. All successfully transfected cells are *green* due to the expression of EGFP (enhanced expression of the green fluorescent protein). The PML protein is detectable by staining with primary anti-PML protein antibodies and secondary Cy3-anti mouse IgG. *Bars* designate magnifications. (Reproduced from Hösel et al. 2001 a)

gene product. Therefore, the absence of morphological changes of the PML domains in the Ad12-abortively infected BHK21 cells raised two questions: (a) is the E4-ORF3 gene product of Ad12 able to reorganize the PML structures in BHK21 cells?; and (b) how efficiently is this protein synthesized in Ad12-infected BHK21 cells? In fact, the E4-ORF3 of Ad12 is only weakly expressed in Ad12-infected BHK21 cells as compared to Ad12-infected human HeLa cells or Ad12-infected complementing BHK297-C131 hamster cells (Hösel et al. 2001a).

We have cloned the E4-ORF3 gene of Ad12 into the pIRES2-EGFP vector under the control of the strong early HCMV promoter and transfected this construct into BHK21 cells. In hamster cells transfected with either E4-ORF3 of Ad12 or E4-ORF3 of Ad2, the PML protein is rearranged from globular to track-like structures, whereas the nontransfected cells or cells transfected with the pIRES2-EGFP vector exhibit the unaltered globular PML protein structures (Fig. 1). Hence, the E4-ORF3 gene product of Ad12 is active in BHK21 cells and able to elicit the reorganization of the nuclear PML protein structures in these cells.

As the redistribution of the nuclear PML domains parallels adenoviral DNA replication in the productive system, we have investigated whether the overexpression of the E4-ORF3 of Ad12 in Ad12-infected BHK21 cells would permit Ad12 DNA replication in the abortive system. However, none of the applied transfection/infection protocols has led to the replication of Ad12 DNA. Apparently, the reorganization of the PML structures in BHK21 cells cannot by itself provide a permissive environment for the replication of Ad12 DNA (Hösel et al. 2001a).

4
Overexpression of the Ad12 pTP or E1A Gene Facilitates Ad12 DNA Replication in Nonpermissive BHK21 Hamster Cells

The adenovirus genome is characterized by two structural features that are functionally important for viral DNA replication. First, the two ends of the linear genome carry identical inverted repeats and contain the origin of viral DNA replication. Second, the 55-kDa terminal protein (TP) is covalently linked via a phosphodiester bond to the 5′ termini of mature adenovirus DNA through the β-hydroxyl group of a serine (Rekosh et al. 1977; Stillman 1982; Stow 1982; Hay et al. 1984). Early in vitro studies on DNA replication have revealed an 80-kDa precursor of the terminal protein (pTP) which is covalently attached to the 5′ ends of nascent DNA

(Challberg et al. 1980; Challberg and Kelly 1981) and subsequently processed by proteolytic cleavage to the mature TP late in infection (Stillman et al. 1981). During the initiation of adenovirus DNA replication, dCMP is transferred onto pTP to act as a primer for viral DNA synthesis. This protein priming reaction is catalyzed by the intimately associated adenovirus DNA polymerase (Adpol). In productively adenovirus-infected cells, pTP, but not TP, binds to Adpol and forms a stable complex which can be dissociated only under strong denaturing conditions (Lichy et al. 1982; Stillman et al. 1982; Webster et al. 1994). Both proteins can recognize the origin of replication separately, but the pTP–Adpol heterodimer does so with enhanced specificity (Temperley and Hay 1992). Moreover, the interaction with pTP appears to facilitate the nuclear transport of the adenovirus DNA polymerase (Zhao and Padmanabhan 1988).

The main role of adenovirus pTP is to act as the primer for viral DNA replication. pTP may also mediate the interaction of the origin of adenovirus DNA replication with the nuclear matrix in the course of infection. Moreover, pTP binds to specific sites on the nuclear matrix, and this matrix association presumably localizes the viral genome to particular nuclear compartments in which replication and transcription factors are concentrated (Fredman and Engler 1993).

As pTP plays a central role in the initiation of adenoviral DNA replication, the synthesis and activity of Ad12 pTP upon the abortive infection of hamster cells with Ad12 have been studied. Ad12 pTP and the Ad12 polymerase can associate in extracts from Ad12-infected BHK21 cells in vitro to form an initiation complex for replication although chain elongation never proceeds. Moreover, the initiation activity is markedly lower in BHK21 cell extracts than in extracts from Ad12-productively infected human KB cells (Lucher 1990; Chowrira and Lucher 1990).

The expression of Ad12-specific pTP in productively or abortively Ad12-infected cells as well as in BHK297-C131 cells, which can partly complement Ad12 DNA replication, has also been investigated. In abortively Ad12-infected BHK21 cells, pTP is produced in amounts barely detectable and markedly lower than in Ad12-infected HeLa or BHK297-C131 cells. The reduced levels of Ad12 pTP in abortively infected hamster cells are due to the insufficient transcription of the pTP gene of Ad12 (Hösel et al. 2001b).

The overexpression of pTP might facilitate the replication of Ad12 DNA in hamster cells. Hence, the pTP gene of Ad12 has been cloned and tether-

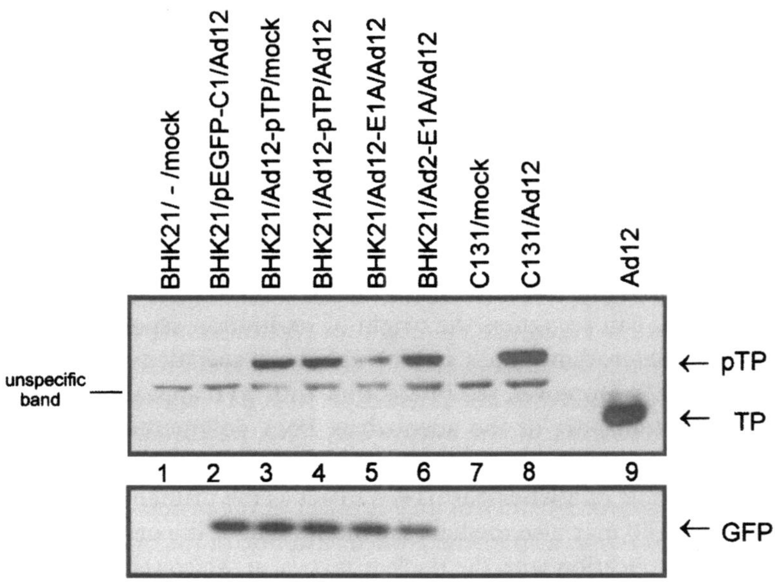

Fig. 2. Western blot analyses for the presence of the Ad12 pTP and TP in protein extracts from mock-transfected and mock-infected BHK21 cells (*lane 1*), from BHK21 cells transfected with the pEGFP-C1 vector and infected with Ad12 (*lane 2*), from BHK21 cells transfected with the Ad12-pTP and mock- (*lane 3*) or Ad12-infected (*lane 4*), from BHK21 cells transfected with the Ad12-E1A (*lane 5*) or Ad2-E1A (*lane 6*) and infected with Ad12, and from mock- (*lane 7*) or Ad12-infected BHK297-C131 cells (*lane 8*). Purified Ad12 virions were co-electrophoresed as control to identify the gel position of the Ad12 TP (*lane 9*). For reprobing, the membrane was stripped and subsequently incubated with antibodies against green fluorescent protein (*GFP*) to monitor the efficiency of transfection and GFP expression. (Reproduced from HÖSEL et al. 2001b)

Fig. 3A–I. Replication of full-length Ad12 DNA in BHK21 cells overexpressing the Ad12 pTP, the Ad12 E1A or the Ad2 E1A gene. BHK21 cells transfected with the pEGFP-C1 vector (**B**), the Ad12-pTP (**G**), the Ad12-E1A (**H**) or the Ad2-E1A construct (**I**) were infected with Ad12; the newly synthesized DNA was labeled by adding ^{3}H-thymidine to the medium. The total intracellular DNA was analyzed by zone velocity sedimentation in alkaline sucrose gradients (BURGER and DOERFLER 1974). DNA from mock-transfected and mock-infected BHK21 cells (**A**), from mock- (**D**) or Ad12-infected BHK297-C131 cells (**F**), from mock- (**C**) or Ad12-infected HeLa cells (**E**) was analyzed in control experiments. (Reproduced from HÖSEL et al. 2001b)

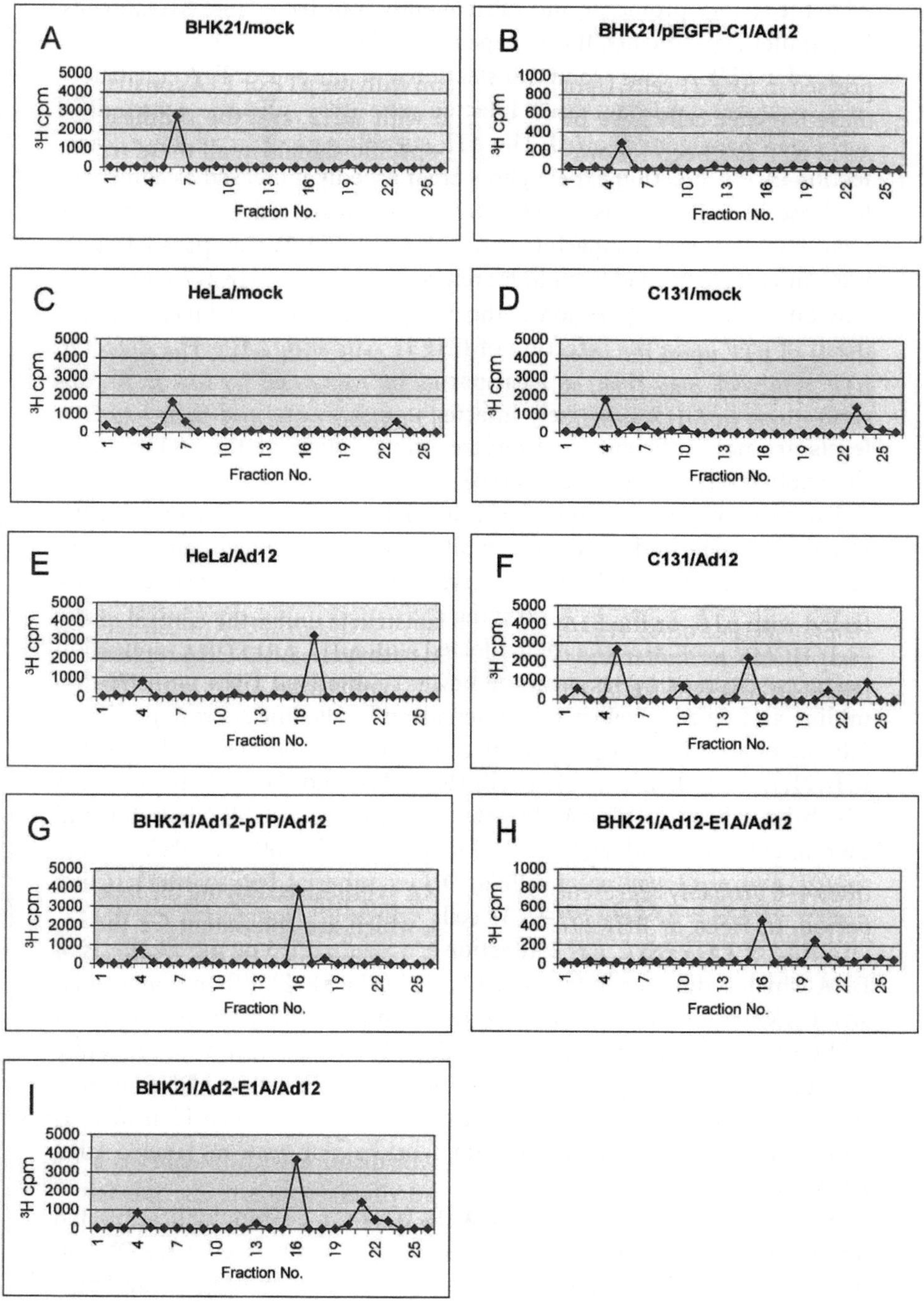
A
BHK21/mock
B
BHK21/pEGFP-C1/Ad12
C
HeLa/mock
D
C131/mock
E
HeLa/Ad12
F
C131/Ad12
G
BHK21/Ad12-pTP/Ad12
H
BHK21/Ad12-E1A/Ad12
I
BHK21/Ad2-E1A/Ad12
³H cpm
Fraction No.

ed to the HCMV promoter, and subsequently transfected into BHK21 cells. In parallel experiments, the E1A gene of Ad12 or Ad2 has been overexpressed in BHK21 cells. Upon transfection with the pTP or E1A constructs, these hamster cells have been infected with Ad12, and the synthesis of Ad12 pTP has been followed. The pTP-specific signals in all these transfection–Ad12 infection experiments are markedly increased as compared to those in BHK21 cells which have been transfected with the control pEGFP-C1 vector and superinfected with Ad12 (Fig. 2). Comparable signal intensities have been elicited in merely Ad12-infected BHK297-C131 cells. Obviously, the overexpression of the E1A gene triggers the increased synthesis of pTP upon the infection of BHK21 cells with Ad12. The defect in pTP synthesis may thus, at least partly, be explained by low E1A concentrations in Ad12-abortively infected hamster cells, and their low E1A levels are not sufficient to stimulate transcription of the pTP gene to threshold levels (Hösel et al. 2001 b).

Does the overproduction of pTP in BHK21 cells transfected with the pTP or E1A gene and subsequently infected with Ad12 facilitate the replication of Ad12 DNA in the abortive system? BHK21 cells have been transfected with pTP- or the E1A-carrying constructs under the control of the early HCMV promoter and then infected with Ad12. Ad12 DNA replication has been assessed by labeling the newly synthesized DNA with ^{3}H-thymidine and by zone velocity sedimentation in alkaline sucrose gradients (Hösel et al. 2001 b). Ad12 DNA is synthesized in neither mock-infected BHK21, HeLa or BHK297-C131 cells (Fig. 3 A, C and D) nor in BHK21 cells which have been previously transfected with the pEGFP-C1 vector and then infected with Ad12 (Fig. 3 B). The peaks of ^{3}H-labeled DNA in fractions 4–6 probably represent cellular DNA synthesized during the labeling period. In HeLa or BHK297-C131 cells, which are competent for the replication of Ad12 DNA, Ad12 infection is accompanied by the synthesis of DNA which sediments in fraction 17 or 15 (Fig. 3 E, F). When nonpermissive BHK21 cells were first transfected with the Ad12-pTP (Fig. 3 G), the Ad12-E1A (Fig. 3 H) or the Ad2-E1A (Fig. 3 I) construct, and subsequently infected with Ad12, a peak of newly synthesized, ^{3}H-labeled DNA in fraction 16 was detected. These viral peak fractions have been identified as authentic Ad12 DNA by Southern blot hybridization with ^{32}P-labeled Ad12 DNA (Fig. 4).

Moreover, the newly synthesized Ad12 DNA can be immuneprecipitated by an anti-Ad12 pTP serum. Hence, the de novo synthesized Ad12 DNA in Ad12-infected BHK21 cells, which overexpress pTP or E1A, is covalently

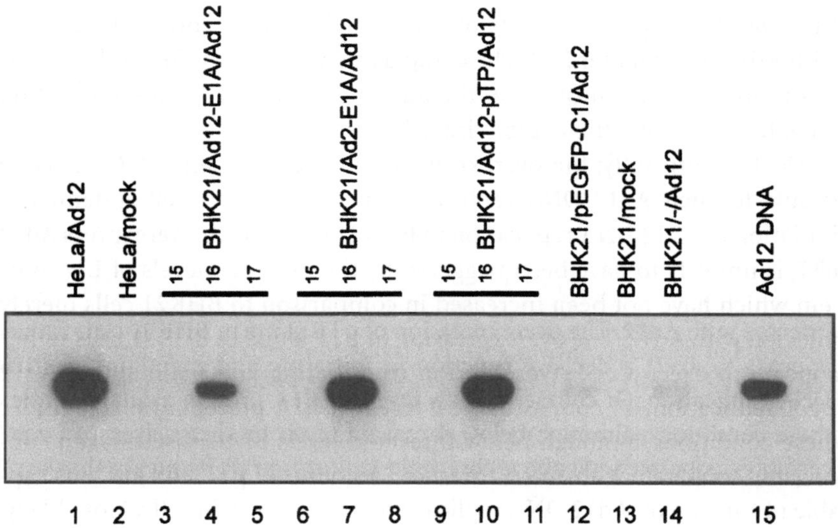

Fig. 4. Identification of Ad12 DNA in the peak gradient fractions of Fig. 3. The Ad12-peak ^{3}H fractions from Ad12-infected HeLa cells (*lane 1*), from Ad12-infected BHK21 cells which have previously been transfected with the Ad12-E1A construct (*lane 4*), with the Ad2-E1A construct (*lane 7*), or with the Ad12-pTP construct (*lane 10*) were analyzed by agarose gel electrophoresis followed by Southern blot hybridization using ^{32}P-labeled Ad12 DNA as the probe. These fractions were derived from alkaline sucrose density gradients. Corresponding fractions from mock-infected HeLa (*lane 2*), or BHK21 cells (*lane 13*), or from untransfected, merely Ad12-infected BHK21 cells (*lane 14*) or from pEGFP-C1-transfected and subsequently Ad12-infected BHK21 cells (*lane 12*) were co-electrophoresed. The very weak Ad12-specific signals apparent in *lanes 12 and 14* represent traces of parental Ad12 DNA. Purified Ad12 virion DNA was co-electrophoresed as control to identify authentic Ad12 DNA (*lane 15*). The fractions (Fig. 3) preceding and following each of the peak fractions 16 in BHK21/Ad12-E1A/Ad12 (*lanes 3, 5*), in BHK21/Ad2-E1A/Ad12 (*lanes 6, 8*), and in BHK21/Ad12-pTP/Ad12 (*lanes 9, 11*) were co-electrophoresed and shown to contain no Ad12 DNA at this film exposure time. (Reproduced from Hösel et al. 2001 b)

bound to pTP. This finding further documents the requirement for pTP as primer for Ad12 DNA replication in this system. We conclude that in BHK21 cells, which overexpress the pTP or the E1A gene of Ad12 or the E1A gene of Ad2, Ad12 DNA is capable of replicating to unit length viral DNA molecules which are covalently bound to pTP. Apparently, in an unaided Ad12 infection of BHK21 cells, the levels of the essential early viral proteins E1A and/or pTP do not suffice to support Ad12 DNA replication, although these levels are capable of complementing the replication of E1A *(dl312)* or

E1B *(del313)* deletion mutants of Ad5 in BHK21 cells abortively infected with Ad12 (see chapter 2.1). The compartmental distribution and the concentrations of these proteins at the nuclear sites of Ad12 DNA replication seem to remain insufficient for viral DNA replication to proceed.

Most significantly, the overexpression of the pTP gene of Ad12 alone is able to elicit Ad12 DNA replication in this abortive cell system. The initiation of Ad12 DNA replication in BHK21 cells, which overexpress Ad12 pTP, is unlikely to have been triggered by the marginal levels of E1A protein which have not been increased in comparison to BHK21 cells merely infected with Ad12. The overexpression of pTP alone in BHK21 cells rather appears to exert a decisive function in initiating and maintaining Ad12 DNA replication. Of course, the low levels of E1A protein available under these conditions, although below threshold levels by themselves, can conceivably cooperate with above threshold amounts of pTP and are thus capable of turning on Ad12 DNA replication directly or indirectly. More likely, Ad12 pTP may harbor new, previously unrecognized functions to promote Ad12 DNA replication even in the absence of functional E1A levels.

BHK21 cells, which overexpress the E1A genes of Ad12 or Ad2, produce sufficient amounts of Ad12 pTP upon Ad12 infection and allow the synthesis of full length Ad12 DNA in moderate amounts. Interestingly, the transcriptional levels of the Ad12 DBP gene or the Ad12 *Adpol* gene in these cells do not differ from those in merely Ad12-infected BHK21 cells. We therefore propose that the overexpression of the E1A gene of Ad12 or Ad2 facilitates Ad12 DNA replication in nonpermissive BHK21 cells due to the activation of the pTP gene. The E1A gene of Ad2 seems to be more effective in this respect than its Ad12 counterpart.

Although the overexpression of pTP or E1A overcomes the block in Ad12 DNA replication, only trace amounts of late Ad12 RNAs are detectable and late viral proteins are not synthesized. Perhaps, the transiently expressed pTP or E1A proteins are degraded and/or no longer functional late after infection and hence fail to stimulate late transcription of Ad12 DNA. Obviously, pTP or E1A can only partly complement the complex deficiencies of Ad12 in hamster cells (Hösel et al. 2001b).

The Ad12 · BHK21 abortive cell system demonstrates multiple blocks at different steps of viral replication and in this way facilitates viral DNA integration, oncogenic cell transformation and secures the long-term, transgenerational persistence of the integrated Ad12 genome in the Ad12-transformed hamster cells. We shall continue to analyze this complex abortive interaction of Ad12 with its nonpermissive, but tumor-responsive host.

Acknowledgments. This research was supported by the Deutsche Forschungsgemeinschaft through grant Do165/17-1, by SFB274-A1, by a collaborative grant, 03-12235-6, from the Bundesminsterium für Bildung und Wissenschaft with Coley Pharmaceutical GmbH, by the Wilhelm Sander Stiftung, München, and by the Ministerium für Wissenschaft und Forschung des Landes Nordrhein-Westfalen.

References

ASCOLI CA, MAUL GG (1991) Identification of a novel nuclear domain. J Cell Biol 112: 785–795

BAILEY A, ULLAH R, MAUTNER V (1994) Cell type specific regulation of expression from the Ad40 E1B promoter in recombinant Ad5/Ad40 viruses. Virology 202: 695–706

BERK AJ, LEE F, HARRISON T, WILLIAMS J, SHARP PA (1979) A pre-early adenovirus 5 gene product regulates synthesis of early viral messenger RNAs. Cell 17:935–944

BRIDGE E, XIA D-X, CARMO-FONSECA M, CARDINALI B, LAMOND AI, PETTERSSON U (1995) Dynamic organization of splicing factors in adenovirus-infected cells. J Virol 69:281–290

BRIDGE E, PETTERSSON U (1995) Nuclear organization of replication and gene expression in adenovirus-infected cells. Curr Top Microbiol Immunol 199/I:99–117

BURGER H, DOERFLER W (1974) Intracellular forms of adenovirus DNA. III. Integration of the DNA of adenovirus type 2 into host DNA in productively infected cells. J Virol 13:975–992

CARVALHO T, SEELER J-S, ÖHMAN K, JORDAN P, PETTERSSON U, AKUSJÄRVI G, CARMO-FONSECA M, DEJEAN A (1995) Targeting of adenovirus E1A and E4-ORF3 proteins to nuclear matrix-associated PML bodies. J Cell Biol 131:45–56

CHALLBERG MD, DESIDERIO SV, KELLY TJ (1980) Adenovirus DNA replication in vitro: characterization of a protein covalently linked to nascent DNA strands. Proc Natl Acad Sci USA 77:5105–5109

CHALLBERG MD, KELLY TJ (1981) Processing of the adenovirus terminal protein. J Virol 38:272–277

CHALLBERG MD, KELLY TJ (1989) Animal virus DNA replication. Annu Rev Biochem 58:671–717

CHOWRIRA BM, LUCHER LA (1990) Extracts of hamster cells abortively infected with human adenovirus type 12 are competent to support initiation of viral DNA replication. Virology 176:289–291

DOERFLER W (1968) The fate of the DNA of adenovirus type 12 in baby hamster kidney cells. Proc Natl Acad Sci USA 60:636–643

DOERFLER W (1969) Nonproductive infection of baby hamster kidney cells (BHK21) with adenovirus type 12. Virology 38:587–606

DOERFLER W (1970) Integration of the deoxyribonucleic acid of adenovirus type 12 into the deoxyribonucleic acid of baby hamster kidney cells. J Virol 6:652–666

DOUCAS V, ISHOV AM, ROMO A, JUGUILON H, WEITZMAN MD, EVANS RM, MAUL GG (1996) Adenovirus replication is coupled with the dynamic properties of the PML nuclear structure. Genes Dev 10:196–207

Esche H, Schilling R, Doerfler W (1979) *In vitro* translation of adenovirus type 12-specific mRNA isolated from infected and transformed cells. J Virol 30: 21–31
Flint SJ, Shenk T (1989) Adenovirus E1A protein paradigm viral transactivator. Annual Rev Genet 23:141–161
Fredman JN, Engler JA (1993) Adenovirus precursor to terminal protein interacts with the nuclear matrix in vivo and in vitro. J Virol 67:3374–3395
Grignani F, Fagioli M, Alcalay M, Longo L, Pandolfi PP, Donti E, Biondi A, LoCocco F, Grignani F, Pelicci PG (1994) Acute promyelocytic leukemia: from genetics to treatment. Blood 83:10–25
Halbert DN, Cutt JR, Shenk T (1985) Adenovirus early region 4 encodes functions required for efficient DNA replication, late gene expression and host cell shutoff. J Virol 56:250–257
Hardy S, Engel DA, Shenk T (1989) An adenovirus early region 4 gene product is required for induction of the infection-specific form of cellular E2F activity. Genes Dev 3:1062–1074
Hatfield L, Hearing P (1993) The NFIII/Oct-1 binding site stimulates adenovirus DNA replication in vivo and is functionally redundant with adjacent sequences. J Virol 67:3931–3939
Hay RT, Stow ND, McDougall JM (1984) Replication of adenovirus mini chromosomes. J Mol Biol 175:493–510
Hilger-Eversheim K, Doerfler W (1997) Clonal origin of adenovirus type 12-induced hamster tumors: nonspecific chromosomal integration sites of viral DNA. Cancer Res 57:3001–3009
Hösel M, Schröer J, Webb D, Jaroschevskaja E, Doerfler W (2001 a) Cellular and early viral factors in the interaction of adenovirus type 12 with hamster cells: the abortive response. Virus Res 81:1–16
Hösel M, Webb D, Schröer J, Schmitz B, Doerfler W (2001 b) The overexpression of the adenovirus type 12 pTP or E1A gene facilitates Ad12 DNA replication in non-permissive BHK21 hamster cells. J Virol 76:10041–10053
Hohlweg U, Hösel M, Dorn A, Webb D, Schramme A, Corzilius L, Niemann A, Hilger-Eversheim K, Schmitz B, Buettner R, Doerfler W (2002) Intraperitoneal dissemination of Ad12-induced primitive neuroectodermal hamster tumors: *de novo* methylation and transcription patterns of integrated viral and of cellular genes. Submitted
Hohlweg U, Doerfler W (2003) Tumorigenesis by adenovirus type 12 in hamsters. Curr Top Microbiol Immunol, volume 273, in press
Huebner RJ, Rowe WP, Lane WT (1962) Oncogenic effects in hamsters of human adenovirus types 12 and 18. Proc Natl Acad Sci USA 48:2051–2058
Ishov AM, Maul GG (1996) The periphery of nuclear domain 10 (ND10) as site of DNA virus deposition. J Cell Biol 134:815–826
Ishov AM, Stenberg RM, Maul GG (1997) Human cytomegalovirus immediate early interaction with host nuclear structures: definition of an immediate transcript enviroment. J Cell Biol 138:5–16
Jones N, Shenk T (1979) Isolation of adenovirus type 5 host range deletion mutants defective for transformation of rat embryo cells. Cell 17:683–689
Jüttermann R, Weyer U, Doerfler W (1989) Defect of adenovirus type 12 replication in hamster cells: absence of transcription of viral virus-associated and L1 RNAs. J Virol 63:3535–3540

KASTNER P, PEREZ A, LUTZ Y, ROCHETTE-EGLY C, GAUP MP, DURAND B, LANOTTE M, BERGER R, CHAMBON P (1992) Structure, localization and transcriptional properties of two classes of retinoic acid receptor alpha fusion proteins in acute promyelocytic leukemia (APL): structural similarities with a new family of oncoproteins. EMBO J 11:629–642

KHITOO G, DELORME L, DERY CV, TREMBLAY ML, WEBER JM, BIBOR-HARDY V, SIMARD R (1986) Role of the nuclear matrix in adenovirus maturation. Virus Res 4:391–403

KLIMKAIT T, DOERFLER W (1985) Adenovirus types 2 and 5 functions elicit replication and late expression of adenovirus type 12 DNA in hamster cells. J Virol 55: 466–474

KLIMKAIT T, DOERFLER W (1987) E1B functions of type C adenoviruses play a role in the complementation of blocked adenovirus type 12 DNA replication and late gene transcription in hamster cells. Virology 161:109–120

KUHLMANN I, ACHTEN S, RUDOLPH R, DOERFLER W (1982) Tumor induction by human adenovirus type 12 in hamsters: loss of the viral genome from adenovirus type 12-induced tumor cells is compatible with tumor formation. EMBO J 1:79–86

LAVERY D, FU SM, LUFKIN T, CHEN-KIANG S (1987) Productive infection of cultured human lymphoid cells by adenovirus. J Virol 61:1466–1472

LICHY JH, FIELD J, HORWITZ MS, HURWITZ J (1982) Separation of the adenovirus terminal protein precursor from its associated DNA polymerase: role of both proteins in the initiation of adenovirus DNA replication. Proc Natl Acad Sci USA 79:5225–5229

LUCHER LA (1990) Adenovirus type 12 tumor antigen synthesis differs during infection of permissive and non-permissive cells. J Gen Virol 71:579–583

MAUL GG, GULDNER HH, SPIVACK JG (1993) Modification of descrete nuclear domains induced by herpes simplex virus type 1 immediate early gene 1 product ICP0. J Gen Virol 74:2679–2690

MAUL GG, ISHOV AM, EVERETT RD (1996) Nuclear domain 10 as preexisting potential replication start sites of herpes simplex virus type-1. Virology 217:67–75

MAUTNER V, MACKAY N, STEINTHORSDOTTIR V (1989) Complementation of enteric adenovirus type 40 for lytic growth in tissue culture by E1B 55 K function of adenovirus types 5 and 12. Virology 171:619–622

MAUTNER V, BAILEY A, STEINTHORSDOTTIR V, ULLAH R, RINALDI A (1999) Properties of the adenovirus type 40 E1B promoter that contribute to its low transcriptional activity. Virology 265:10–19

NAGATA K, GUGGENHEIMER RA, ENOMOTO T, LICHY JH, HURWITZ J (1982) Adenovirus DNA replication in vitro: identification of a host factor that stimulates synthesis of the preterminal protein-dCMP complex. Proc Natl Acad Sci USA 79:6438–6442

NAGATA K, GUGGENHEIMER RA, HURWITZ J (1983) Adenovirus DNA replication in vitro: synthesis of full-length DNA with purified proteins. Proc Natl Acad Sci USA 80:4266–4270

NEVINS JR (1981) Mechanism of activation of early viral transcription by the adenovirus E1A gene product. Cell 26:213–220

O'NEILL EA, FLETCHER C, BURROW CR, HEINTZ N, ROEDER RG, KELLY TJ (1988) Transcriptional factor Oct-1 is functionally identical to the DNA replication factor NFIII. Science 241:1210–1213

Ortin J, Doerfler W (1975) Transcription of the genome of adenovirus type 12. I. Viral mRNA in abortively infected and transformed cells. J Virol 15:27–35

Ortin J, Scheidtmann K-H, Greenberg R, Westphal M, Doerfler W (1976) Transcription of the genome of adenovirus type 12. III. Maps of stable RNA from productively infected human cells and abortively infected and transformed hamster cells. J Virol 20:355–32

Pandolfi PP, Grignani F, Alcalay M, Mencarelli A, Biondi A, LoCoco F, Grignani F, Pelicci PG (1991) Structure and origin of the acute promyelocytic leukemia myl/RAR alpha cDNA and characterization of its retinoid-binding and transactivation properties. Oncogene 6:1285–1292

Persson H, Philipson L (1982) Regulation of adenovirus gene expression. Curr Top Microbiol Immunol 97:157–203

Pieniazek D, Pieniazek NJ, Macejak D, Coward J, Rayfield M, Luftig RB (1990a) Differential growth of human enteric adenovirus 41 (Tak) in continuous cell lines. Virology 174:293–249

Pieniazek D, Pieniazek NJ, Macejak D, Luftig RB (1990b) Enteric adenovirus 41 (Tak) requires low serum for growth in human primary cells. Virology 178:72–80

Prujin GJ, van Driel W, van der Vliet PC (1986) Nuclear factor III, a novel sequence-specific DNA-binding protein from HeLa cells stimulating adenovirus DNA replication. Nature 322:656–659

Raška K Jr, Strohl WA (1972) The response of BHK21 cells to infection with type 12 adenovirus. VI. Synthesis of virus-specific RNA. Virology 47:734–742

Rekosh DMK, Russell WC, Bellet AJD, Robinson AJ (1977) Identification of a protein linked to the ends of adenovirus DNA. Cell 11:283–295

Rosenfeld PJ, O'Neill E, Wides RJ, Kelly TJ (1987) Sequence-specific interactions between cellular DNA-binding proteins and the adenovirus origin of DNA replication. Mol Cell Biol 7:875–886

Ross D, Ziff E (1994) Defective processing of human adenovirus 2 late transcription unit mRNAs during abortive infections in monkey cells. Virology 202:107–115

Rouse HC, Strohl WA, Schlesinger RW (1966) Properties of cells derived from adenovirus-induced hamster tumors by long-term in vitro cultivation. I. Clonal stability of three biological characteristics. Virology 28:633–644.

Sawada Y, Raška K Jr, Shenk T (1988) Adenovirus type 5 and adenovirus type 12 recombinant viruses containing heterologous E1 genes are viable, transform rat cells, but are not tumorigenic in rats. Virology 166:281–284

Schaack J, Ho Y, Freimuth P, Shenk T (1990) Adenovirus terminal protein mediates both nuclear matrix association and efficient transcription of adenovirus DNA. Genes Devel 4:1197–1208

Schiedner G, Schmitz B, Doerfler W (1994) Late transcripts of adenovirus type 12 DNA are not translated in hamster cells expressing the E1 region of adenovirus type 5. J Virol 68:5476–5482

Schneider RJ, Weinberger C, Shenk T (1984) Adenovirus VAI RNA facilitates the initiation of translation in virus-infected cells. Cell 37:291–298

Schröer J, Hölker I, Doerfler W (1997) Adenovirus type 12 DNA firmly associates with mammalian chromosomes early after virus infection or after DNA transfer by the addition of DNA to the cell culture medium. J Virol 71:7923–7932

Shenk T, Jones N, Colby W, Fowlkes D (1979) Functional analysis of adenovirus 5 host-range deletion mutants defective for transformation of rat embryo cells. Cold Spring Harbor Symp Quant Biol 44:367–375

Silver L, Anderson CW (1988) Interaction of human adenovirus serotype 2 with human lymphoid cells. Virology 165:377–387

Silverman L, Cleghon V, Klessig DF (1998) Increased permissivity of monkey cells to human adenovirus multiplication is affected by culturing conditions and correlates with both synthesis of virion fiber protein and altered splicing of its mRNA. Virology 173:109–119

Stabel S, Doerfler W, Friis RR (1980) Integration sites of adenovirus type 12 DNA in transformed hamster cells and hamster tumor cells. J Virol 36:22–40

Sternsdorf T, Grotzinger T, Jensen K, Will H (1997) Nuclear dots: actors on many stages. Immunobiology 198:307–331

Stillman BW, Lewis JB, Chow LT, Mathews MB, Smart JE (1981) Identification of the gene and mRNA for the adenovirus terminal protein precursor. Cell 23:497–508

Stillman BW, Tamanoi F, Matthews MB (1982) Purification of an adenovirus-coded DNA polymerase that is required for initiation of DNA replication. Cell 31:613–623

Stow ND (1982) The infectivity of adenovirus genomes lacking DNA sequence from their left hand termini. Nucl Acids Res 10:5105–5119

Strohl WA, Rouse HC, Schlesinger RW (1966) Properties of cells derived from adenovirus-induced hamster tumors by long-term in vitro cultivation. II. Nature of the restricted response to type 2 adenovirus. Virology 28:645–658

Strohl WA (1969) The response of BHK21 cells to infection with type 12 adenovirus. Virology 39:642–665

Stuurman N, de Graaf A, Floore A, Josso A, Humbel B, de Jong L, van Driel RA (1992) A monoclonal antibody recognizing nuclear matrix-associated nuclear bodies. J Cell Sci 101:773–784

Tempereley SM, Hay RT (1992) Recognition of the adenovirus type 2 origin of replication by the virally encoded DNA polymerase and preterminal proteins. EMBO J 11:761–768

Thimmappaya B, Weinberger C, Schneider RJ, Shenk T (1982) Adenovirus VAI RNA is required for efficient translation of viral mRNA at late times after infection. Cell 31:543–551

Trentin JJ, Yabe Y, Taylor G (1962) The quest for human cancer viruses. A new approach to an old problem reveals cancer induction in hamsters by human adenoviruses. Science 137:835–841

Van der Vliet PC (1995) Adenovirus DNA replication. Curr Topics Microbiol Immunol 199/II:1–30

Visser L, van Maarschalkerweerd MW, Rozijn TH, Wassenaar ADC, Reemst AMCB, Sussenbach JS (1979) Viral DNA sequences in adenovirus-transformed cells. Cold Spring Harbor Symp Quant Biol 44:541–550

Webster A, Leith, IR, Hay RT (1994) Activation of adenovirus-coded protease and processing of preterminal protein. J Virol 68:7292–7300

Wold WSM, Gooding LR (1991) Region E3 of adenovirus: a cassette of genes involved in host immunosurveillance and virus-cell interactions. Virology 184:1–8

Younghusband HB, Maundrell K (1982) Adenovirus DNA is associated with the nuclear matrix of infected cells. J Virol 43:705–713

Zhai ZH, Nickerson JA, Krochmalnic G, Penman S (1987) Alterations in nuclear matrix structure after adenovirus infection. J Virol 61:1007–1018
Zhao L-J, Padmanabhan R (1988) Nuclear transport of adenovirus DNA polymerase is facilitated by interaction with preterminal protein. Cell 55:1005–1015
Ziff EB (1980) Transcription and RNA processing by the DNA tumor viruses. Nature 287:491–499
Zock C, Doerfler W (1990) A mitigator sequence in the downstream region of the major late promoter of adenovirus type 12 DNA. EMBO J 9:1615–1623
Zock C, Iselt A, Doerfler W (1993) A unique mitigator sequence determines the species specificity of the major late promoter in adenovirus type 12 DNA. J Virol 67:682–693

Subject Index

N

O

P

Current Topics in Microbiology and Immunology

Volumes published since 1998 (and still available)

Vol. 229: **Kelsoe, Garnett; Flajnik, Martin (Eds.):** Somatic Diversification of Immune Responses. 1998. 38 figs. IX, 221 pp. ISBN 3-540-63608-0

Vol. 230: **Kärre, Klas; Colonna, Marco (Eds.):** Specificity, Function, and Development of NK Cells. 1998. 22 figs. IX, 248 pp. ISBN 3-540-63941-1

Vol. 231: **Holzmann, Bernhard; Wagner, Hermann (Eds.):** Leukocyte Integrins in the Immune System and Malignant Disease. 1998. 40 figs. XIII, 189 pp. ISBN 3-540-63609-9

Vol. 232: **Whitton, J. Lindsay (Ed.):** Antigen Presentation. 1998. 11 figs. IX, 244 pp. ISBN 3-540-63813-X

Vol. 233/I: **Tyler, Kenneth L.; Oldstone, Michael B. A. (Eds.):** Reoviruses I. 1998. 29 figs. XVIII, 223 pp. ISBN 3-540-63946-2

Vol. 233/II: **Tyler, Kenneth L.; Oldstone, Michael B. A. (Eds.):** Reoviruses II. 1998. 45 figs. XVI, 187 pp. ISBN 3-540-63947-0

Vol. 234: **Frankel, Arthur E. (Ed.):** Clinical Applications of Immunotoxins. 1999. 16 figs. IX, 122 pp. ISBN 3-540-64097-5

Vol. 235: **Klenk, Hans-Dieter (Ed.):** Marburg and Ebola Viruses. 1999. 34 figs. XI, 225 pp. ISBN 3-540-64729-5

Vol. 236: **Kraehenbuhl, Jean-Pierre; Neutra, Marian R. (Eds.):** Defense of Mucosal Surfaces: Pathogenesis, Immunity and Vaccines. 1999. 30 figs. IX, 296 pp. ISBN 3-540-64730-9

Vol. 237: **Claesson-Welsh, Lena (Ed.):** Vascular Growth Factors and Angiogenesis. 1999. 36 figs. X, 189 pp. ISBN 3-540-64731-7

Vol. 238: **Coffman, Robert L.; Romagnani, Sergio (Eds.):** Redirection of Th1 and Th2 Responses. 1999. 6 figs. IX, 148 pp. ISBN 3-540-65048-2

Vol. 239: **Vogt, Peter K.; Jackson, Andrew O. (Eds.):** Satellites and Defective Viral RNAs. 1999. 39 figs. XVI, 179 pp. ISBN 3-540-65049-0

Vol. 240: **Hammond, John; McGarvey, Peter; Yusibov, Vidadi (Eds.):** Plant Biotechnology. 1999. 12 figs. XII, 196 pp. ISBN 3-540-65104-7

Vol. 241: **Westblom, Tore U.; Czinn, Steven J.; Nedrud, John G. (Eds.):** Gastroduodenal Disease and Helicobacter pylori. 1999. 35 figs. XI, 313 pp. ISBN 3-540-65084-9

Vol. 242: **Hagedorn, Curt H.; Rice, Charles M. (Eds.):** The Hepatitis C Viruses. 2000. 47 figs. IX, 379 pp. ISBN 3-540-65358-9

Vol. 243: **Famulok, Michael; Winnacker, Ernst-L.; Wong, Chi-Huey (Eds.):** Combinatorial Chemistry in Biology. 1999. 48 figs. IX, 189 pp. ISBN 3-540-65704-5

Vol. 244: **Daëron, Marc; Vivier, Eric (Eds.):** Immunoreceptor Tyrosine-Based Inhibition Motifs. 1999. 20 figs. VIII, 179 pp. ISBN 3-540-65789-4

Vol. 245/I: **Justement, Louis B.; Siminovitch, Katherine A. (Eds.):** Signal Transduction and the Coordination of B Lymphocyte Development and Function I. 2000. 22 figs. XVI, 274 pp. ISBN 3-540-66002-X

Vol. 245/II: **Justement, Louis B.; Siminovitch, Katherine A. (Eds.):** Signal Transduction on the Coordination of B Lymphocyte Development and Function II. 2000. 13 figs. XV, 172 pp. ISBN 3-540-66003-8

Vol. 246: **Melchers, Fritz; Potter, Michael (Eds.):** Mechanisms of B Cell Neoplasia 1998. 1999. 111 figs. XXIX, 415 pp. ISBN 3-540-65759-2

Vol. 247: **Wagner, Hermann (Ed.):** Immunobiology of Bacterial CpG-DNA. 2000. 34 figs. IX, 246 pp. ISBN 3-540-66400-9

Vol. 248: **du Pasquier, Louis; Litman, Gary W. (Eds.):** Origin and Evolution of the Vertebrate Immune System. 2000. 81 figs. IX, 324 pp. ISBN 3-540-66414-9

Vol. 249: **Jones, Peter A.; Vogt, Peter K. (Eds.):** DNA Methylation and Cancer. 2000. 16 figs. IX, 169 pp. ISBN 3-540-66608-7

Vol. 250: **Aktories, Klaus; Wilkins, Tracy, D. (Eds.):** Clostridium difficile. 2000. 20 figs. IX, 143 pp. ISBN 3-540-67291-5

Vol. 251: **Melchers, Fritz (Ed.):** Lymphoid Organogenesis. 2000. 62 figs. XII, 215 pp. ISBN 3-540-67569-8

Vol. 252: **Potter, Michael; Melchers, Fritz (Eds.):** B1 Lymphocytes in B Cell Neoplasia. 2000. XIII, 326 pp. ISBN 3-540-67567-1

Vol. 253: **Gosztonyi, Georg (Ed.):** The Mechanisms of Neuronal Damage in Virus Infections of the Nervous System. 2001. approx. XVI, 270 pp. ISBN 3-540-67617-1

Vol. 254: **Privalsky, Martin L. (Ed.):** Transcriptional Corepressors. 2001. 25 figs. XIV, 190 pp. ISBN 3-540-67569-8

Vol. 255: **Hirai, Kanji (Ed.):** Marek's Disease. 2001. 22 figs. XII, 294 pp. ISBN 3-540-67798-4

Vol. 256: **Schmaljohn, Connie S.; Nichol, Stuart T. (Eds.):** Hantaviruses. 2001, 24 figs. XI, 196 pp. ISBN 3-540-41045-7

Vol. 257: **van der Goot, Gisou (Ed.):** Pore-Forming Toxins, 2001. 19 figs. IX, 166 pp. ISBN 3-540-41386-3

Vol. 258: **Takada, Kenzo (Ed.):** Epstein-Barr Virus and Human Cancer. 2001. 38 figs. IX, 233 pp. ISBN 3-540-41506-8

Vol. 259: **Hauber, Joachim, Vogt, Peter K. (Eds.):** Nuclear Export of Viral RNAs. 2001. 19 figs. IX, 131 pp. ISBN 3-540-41278-6

Vol. 260: **Burton, Didier R. (Ed.):** Antibodies in Viral Infection. 2001. 51 figs. IX, 309 pp. ISBN 3-540-41611-0

Vol. 261: **Trono, Didier (Ed.):** Lentiviral Vectors. 2002. 32 figs. X, 258 pp. ISBN 3-540-42190-4

Vol. 262: **Oldstone, Michael B.A. (Ed.):** Arenaviruses I. 2002, 30 figs. XVIII, 197 pp. ISBN 3-540-42244-7

Vol. 263: **Oldstone, Michael B. A. (Ed.):** Arenaviruses II. 2002, 49 figs. XVIII, 268 pp. ISBN 3-540-42705-8

Vol. 264/I: **Hacker, Jörg; Kaper, James B. (Eds.):** Pathogenicity Islands and the Evolution of Microbes. 2002. 34 figs. XVIII, 232 pp. ISBN 3-540-42681-7

Vol. 264/II: **Hacker, Jörg; Kaper, James B. (Eds.):** Pathogenicity Islands and the Evolution of Microbes. 2002. 24 figs. XVIII, 228 pp. ISBN 3-540-42682-5

Vol. 265: **Dietzschold, Bernhard; Richt, Jürgen A. (Eds.):** Protective and Pathological Immune Responses in the CNS. 2002. 21 figs. X, 278 pp. ISBN 3-540-42668-X

Vol. 266: **Cooper, Koproski (Eds.):** The Interface Between Innate and Acquired Immunity, 2002, 15 figs. XIV, 116 pp. ISBN 3-540-42894-1

Vol. 267: **Mackenzie, John S.; Barrett, Alan D. T.; Deubel, Vincent (Eds.):** Japanese Encephalitis and West Nile Viruses. 2002. 66 figs. X, 418 pp. ISBN 3-540-42783-X

Vol. 268: **Zwickl, Peter; Baumeister, Wolfgang (Eds.):** The Proteasome-Ubiquitin Protein Degradation Pathway. 2002, 17 figs. X, 213 pp. ISBN 3-540-43096-2

Vol. 269: **Koszinowski, Ulrich H.; Hengel, Hartmut (Eds.):** Viral Proteins Counteracting Host Defenses. 2002, 47 figs. XII, 325 pp. ISBN 3-540-43261-2

Vol. 270: **Beutler, Bruce; Wagner, Hermann (Eds.):** Toll-Like Receptor Family Members and Their Ligands. 2002, 31 figs. X, 192 pp. ISBN 3-540-43560-3

Vol. 271: **Koehler (Ed.):** Anthrax. 2002, 14 figs. X, 169 pp. ISBN 3-540-43497-6

MIX
Papier aus verantwortungsvollen Quellen
Paper from responsible sources
FSC® C105338
FSC
www.fsc.org

If you have any concerns about our products,
you can contact us on
ProductSafety@springernature.com

In case Publisher is established outside the EU,
the EU authorized representative is:
Springer Nature Customer Service Center GmbH
Europaplatz 3, 69115 Heidelberg, Germany

Printed by Libri Plureos GmbH
in Hamburg, Germany